Cognitive Radio Mobile Ad Hoc Networks

2 Dynamic Spectrum Access

As the first chapter of this book, *Chapter 1*, authored by *F. R. Yu, H. Tang, M. Huang, and Z. Li*, introduces challenges related to spectrum sensing in CR-MANETs. Then, the authors propose a fully distributed and scalable cooperative spectrum sensing scheme based on recent advances in consensus algorithms. In the proposed scheme, the secondary users can maintain coordination based on only local information exchange without a centralized common receiver. The authors use the consensus of secondary users to make the final decision. The proposed scheme is essentially based on recent advances in consensus algorithms that have taken inspiration from complex natural phenomena including flocking of birds, schooling of fish, swarming of ants, and honeybees. Unlike the existing cooperative spectrum sensing schemes, there is no need for a centralized receiver in the proposed schemes, which make them suitable in distributed MANETs. Simulation results show that the proposed consensus schemes can have significant lower missing detection probabilities and false alarm probabilities in cognitive radio MANETs. It is also demonstrated that the proposed scheme not only has proven sensitivity in detecting the primary user's presence but also has robustness in choosing a desirable decision threshold.

Chapter 2, authored by *Y. Song and J. Xie*, studies the spectrum handoff problem in CR-MANETs. Since unlicensed users are considered as temporary visitors to the licensed spectrum, they are required to vacate the spectrum when a licensed user reclaims it. Due to the randomness of the appearance of licensed users, disruptions to both licensed and unlicensed communications are often difficult to prevent, which may lead to low throughput of both licensed and unlicensed communications. In this chapter, a proactive spectrum handoff framework for CR ad hoc networks is proposed to address these concerns. In the proposed framework, channel switching policies and a proactive spectrum handoff protocol are proposed to let unlicensed users vacate a channel *before* a licensed user utilizes it to avoid unwanted interference. Network coordination schemes for unlicensed users are also incorporated into the spectrum handoff protocol design to realize channel rendezvous. Moreover, a distributed channel selection scheme to eliminate collisions among unlicensed users in a multi-user spectrum handoff scenario is proposed. In the proposed framework, unlicensed users coordinate with each other without using a common control channel, which is highly adaptable in a spectrum-varying environment. The authors compare the proposed proactive spectrum handoff protocol with a reactive spectrum handoff protocol, under which unlicensed users switch channels *after* collisions with licensed transmissions occur under different channel coordination schemes. Simulation results show that the proactive spectrum handoff outperforms the reactive spectrum handoff approach in terms of higher throughput and fewer collisions to licensed users. Furthermore, the distributed channel selection can achieve substantially higher packet delivery rate in a multi-user spectrum handoff scenario, compared with existing channel selection schemes. In addition, the authors propose a novel three-dimensional discrete-time Markov chain to characterize the process of reactive spectrum handoffs and analyze the performance of unlicensed users. They validate the numerical results obtained from the proposed Markov model against simulation and investigate other parameters of interest in the spectrum handoff scenario. The proposed analytical model can be applied to various practical network scenarios.

Chapter 3, authored by *I. Macaluso, T. K. Forde, O. Holland, and K. E. Nolan*, introduces the concept of *environment–mobility* interaction map. CR-MANETs are likely to be complex radio systems. We already know that no single MANET solution can address all environments that may be encountered; such is the rationale of an ad hoc network that it must address the networking demands of unforeseen scenarios. Rather, a CR-MANET should be viewed as a feature-rich radio system, i.e, one which has access to a range of radio and network components, each suited to different demands. Such a reconfigurable system requires cognitive functionality to self-architect the radios when they are deployed in addition to the cognitive functionality required for the various layers to self-organise. However, any cognitive decision-making process requires awareness of the world for which it is trying to optimise the system. This chapter introduces the concept of an *environment–mobility interaction* map, a persistent internal representation of the network which captures the presence of areas in the network's environment in which particular, sustained, mobility dynamics are observed. Such a self-generated map enables the cognitive MANET to plan a response to challenges brought about by these network dynamics.

Chapter 4, authored by *K. Navaie, H. Yanikomeroglu, M. G. Khoshkholgh, A. R. Sharafat, and H. Nikoofar*, investigates the impact of the primary service communication activity as well as other system parameters on the interference level at the secondary service receiver. The achieved capacity of the secondary service is directly related to the interference level at the secondary service receiver as well as the secondary service adopted sub-channel selection policy. The achievable capacity of the secondary service in such systems are obtained under different sub-channel selection policies in fading environments. Two general sub-channel selection policies are studied in this chapter: *uniform sub-channel selection* and *non-uniform sub-channel selection*. Uniform sub-channel selection fits into the cases where a priori knowledge on sub-channels state information is not available at the secondary transmitter. For cases with available a priori knowledge on sub-channels state information, a variety of non-uniform sub-channel selection policies are studied. The authors then present results on the scaling law of the opportunistic spectrum sharing in DS-CDMA/OFDM systems with multiple users. Numerical results are presented to compare different sub-channel selection policies.

3 Medium Access Control

A major issue in CR-MANETs is the medium access control (MAC). There are many new challenges, such as the multi-channel hidden terminal problem and the fact that the time-varying channel availability differs for different secondary users, in the MAC layer. To overcome these challenges, *Chapter 5*, authored by *X. Zhang and H. Su*, presents an efficient Cognitive Radio-EnAbled Multi-channel MAC (CREAM-MAC) protocol, which integrates the cooperative sequential spectrum sensing at physical layer and the packet scheduling at MAC layer, over the wireless cognitive radio networks. Under the proposed CREAM-MAC protocol, each secondary user is equipped with a cognitive radio-enabled transceiver and multiple channel sensors. The cooperative sequential spectrum sensing scheme improves the accuracy of spectrum sensing and further protects the primary users. The proposed CREAM-MAC enables the secondary users to best utilize the unused frequency spectrum while avoiding the collisions among secondary users and between secondary users and primary users. The authors develop the Markov chain model and $M/G^Y/1$ queueing model to rigorously study the proposed CREAM-MAC protocol for both the saturation networks and the non-saturation networks. Extensive simulations are conducted to validate the developed protocol and analytical models.

Chapter 6, authored by *V. B. Mišić and J. Mišić*, investigates the performance of a cognitive personal area network (CPAN) in which spectrum sensing is linked to packet transmissions. Efficient CPAN operation may be achieved if each data transmission is taxed by requiring the transmitting node to participate in cooperative sensing for a prescribed time period. In this approach, each node is allowed to transmit a single packet in one transmission cycle, but must then 'pay' for it by spectrum sensing, which ensures fairness with respect to transmission but also distributes the sensing burden to all nodes. The authors describe a probabilistic model of the integrated system and evaluate its performance with respect to packet transmissions and spectrum sensing. The authors discuss two modifications that involve centralized and distributed selection of the channels to be sensed. They also propose an adaptive algorithm to determine the tax coefficient and show that it offers superior data transmission performance while not affecting the sensing accuracy.

Chapter 7, authored by *T. Chen, H. Zhang, and Z. Zhao*, introduces the concept of the control channel cloud to solve the control channel problem in dynamic spectrum access (DSA)-based ad hoc networks. A DSA-based ad hoc network is an infrastructure-less wireless network based on DSA and featured by self-organization, self-configuration, and self-healing. One of the challenges in such a network is the common control channel problem, which is caused by the opportunistic spectrum sharing nature of secondary users in the network. Without a common control channel, it is a challenge to coordinate the behaviors of SU nodes in a DSA-based ad hoc network. The control channel cloud approach, which relies only on the local information exchange to function, aligns the control channel of SU nodes to the same channel in a distributed way if a common control channel exists. It provides a simple but scalable way to synchronize the control channel in a dynamically changed radio environment. The convergence of the proposed approach is proved. The performance of proposed algorithms is studied by simulation.

4 Topology Control and Routing

Chapter 8, authored by *Q. Guan, F. R. Yu, and S. Jiang*, studies topology control and routing in CR-MANETs. CR technology will have significant impacts on upper layer performance such as topology control and routing in wireless networks, especially in MANETs. The dynamic spectrum availability issue imposes more challenges on routing in CR-MANETs. Since the spectrum availability is affected by primary user activities and the mobility of cognitive users, cognitive routing is required to be forward-looking rather than reactive. To this end, a topology control and routing framework is presented in this chapter, where cognitive routing is enabled by topology control. In the framework, topology control serves as a middleware and a cross-layer module residing between routing and CR module. Prediction techniques can be used to construct a smart network topology, which provisions cognition capability to routing. Particularly, the author present a distributed prediction-based Cognitive topology control (PCTC) scheme to demonstrate the framework and verify its feasibility.

Chapter 9, authored by *J. Li, Y. Zhou, and L. Lamont*, presents a survey of existing routing schemes for CR-MANETs. The authors describe a CR-MANET model and present a novel adaptive routing design for the CR-MANET, referred to as ARDC, algorithmically and through examples. ARDC is based on the graph modeling approach, and its most significant contribution is that ARDC adapts to dynamic changes in the network topology much more computationally efficient than other CR-MANET routing schemes. At last, some further research directions on CR-MANET routing are identified.

Chapter 10, authored by *Y. Yang, C. Han, and B. Gao*, presents analysis for delays for both multihop cognitive radio networks and single-hop cognitive radio networks. For multihop cognitive radio networks, we analyze the amount of time that a packet spends to travel over the intermittent relaying links over multiple relaying hops and characterize it with the metric called *information propagation speed*. Optimal relaying node placement strategies are derived to maximize information propagation speed. For single-hop cognitive radio networks, we will analyze how delay is affected by multiple cognitive radio design options, including the number of channels to be aggregated, the duration of transmission, the channel separation constraint on channel aggregation, and the time needed for spectrum sensing and protocol handshake. How these different options may affect the delay under different secondary and primary user traffic loads is revealed. Methods for computing optimal cognitive radio design and operation strategy are derived.

5 Multimedia Transmissions

Chapter 11, authored by *H. Luo, S. Ci, D. Wu, Z. Feng, and H. Tang*, studies the multimedia transmissions problem in CR-MANETs. Most current research only considers spectrum utilization and effectiveness at MAC and PHY layers, ignoring the system performance of upper layers. Therefore, in this chapter, the authors aim to improve the user experience of secondary users for wireless video services over cognitive radio networks. They propose a quality-driven cross-layer optimized system to maximize the expected user-perceived video quality at the receiver end, under the constraint of packet delay bound. By formulating network functions such as encoder behavior, cognitive MAC scheduling, transmission, as well as modulation and coding into a distortion-delay optimization framework, important system parameters residing in different network layers are jointly optimized in a systematic way to achieve the best user-perceived video quality for secondary users in cognitive radio networks. Furthermore, the proposed problem is formulated into a MIN-MAX problem and solved by using dynamic programming. The performance enhancement of the proposed system is evaluated through extensive experiments based on H.264/AVC.

6 Applications of Cognitive Radio Mobile Ad Hoc Networks

Chapter 12, authored by *D. Niyato, E. Hossain, and Teerawat Issariyakul*, considers an adaptive networking platform using WiFi/WiMAX technologies for cognitive vehicle-to-roadside communications, which can be used to transfer safety messages and provide Internet access for mobile users inside vehicles. The proposed platform is based on a heterogeneous multihop cluster-based vehicular network, where a vehicular node can choose to play the role of a gateway or a client. Gateway nodes communicate directly with a roadside base station through a WiMAX link. Client nodes connect to the gateways through WiFi links. Traffic from client nodes are relayed by the gateways to a roadside base station. The vehicular nodes are the self-interest (i.e., rational) and have capability to learn and adapt decision to achieve their objectives independently. A decision-making framework is proposed for this WiFi/WiMAX platform. This distributed decision-making framework, which enables the vehicular nodes with cognitive capability, is modeled and analyzed using game theory. Also, a Q-learning algorithm is used in vehicular nodes to provide the cognitive capability to learn and adapt their decision. Dynamics of Q-learning algorithm can be modeled as an evolutionary game.

Chapter 13, authored by *Z. Dong, S. Sengupta, S. Anand, K. Hong, R. Chandramouli, and K. P. Subbalakshmi*, studies to use CR-MANETs in health care. Low-cost automated health monitoring system sees a high demand with the President's proposal on health care reform. Legacy health care monitoring systems demand a great amount of resources such as health care personnel and medical equipments. This increases the cost of health care making it unaffordable to the majority of our society. This chapter introduces an architecture and design of a health care automation network. The health care automation network uses a cognitive radio-based infrastructure to monitor real time patients' vital signs, collect and document medical information. The health care automation network can be implemented in hospitals or in senior communities. This network can leverage the existing infrastructure and reduce the cost of implementation. Research challenges in development of cognitive radio health care automation network are also discussed.

Chapter 14, authored by *O. Cabral, J. M. Ferro, and F. J. Velez*, proposes an application of CR technology for interoperability between High-Speed Downlink Packet Access (HSDPA) and Wi-Fi. This scenario involves the end-user traveling in public transportation system and requesting multimedia services to the operator. The inter-operability between HSDPA and Wi-Fi (IEEE 802.11e standard) Radio Access Technologies (RAT) is first addressed, a topology in which the user has access to both RATs was considered, together with a Common Radio Resource Management (CRRM) to manage the connections. The authors reached the conclusion that the CRRM enables to increase the system throughput when the load thresholds are set to 0.6 for HSDPA and 0.53 for Wi-Fi. Then, spectrum aggregation is implemented in HSDPA. A Resource Allocation (RA) algorithm allocates user packets to the available radio resources (in this case Node Bs operating at 2 and a 5 GHz are available) in order to satisfy user requirements. Simulation results show that gains up to 22% may be achieved. The authors have also sought the most efficient way to manage routing packets inside the Wi-Fi network. The proposal which uses links with higher throughputs enables to reach the best results, with gains up to 300% in the packet delivery ratio. Finally, the authors discuss the challenges that need to be addressed in order to materialise the envisaged cognitive radio scenario in public transportation.

Chapter 15, authored by *B. R. Tamma, B. S. Manoj, and R. Rao*, presents an application of the Cognitive Networking paradigm to the design and development of autonomous Cognitive Access Point (CogAP) for Wi-Fi hotspots and home wireless networks. In these environments, the authors typically use only one AP per service provider/residence for providing wireless connectivity to the users. Here it can reduce the cost of autonomic network control by equipping the same AP with a cognitive functionality. The authors first present the architecture of autonomous CogAP which consists of two main modules: Traffic sensing module and Cognitive controller module. The traffic sensing module uses an efficient packet sampling scheme to characterize traffic from all Wi-Fi channels with single wireless interface. The cognitive controller module consists of two sub-modules: traffic predictor and cognitive decision engine. The Neural Network-based traffic predictor module makes use of the historical traffic traces for traffic prediction on all channels. The cognitive decision engine makes use of traffic forecasts to dynamically decide which channel is best for CogAP to operate on. They have built a prototype CogAP device using off-the-shelf hardware components and obtained better performance with respect to state-of-the-art channel selection strategies.

7 Game Theoretic Approach for Modeling and Optimization

Chapter 16, authored by *S. Maharjan, Y. Zhang, and S. Gjessing*, presents an extensive summary of the related work that uses economic approaches such as game theory and/or price theory/market theory in CR networks. Efficient resource allocation is one of the key concerns of implementing cognitive radio networks. Game theory has been extensively used to study the strategic interactions between primary and secondary users for effective resource allocation. The concept of spectrum trading has introduced a new direction for the coexistence of primary and secondary users through economic benefits to primary users. The use of price theory and market theory from economics has played a vital role to facilitate economic models for spectrum trading. So, it is important to understand the feasibility of using economic approaches as well as to realize the technical challenges associated with them for implementation of cognitive radio networks. With this motivation, The authors present an extensive summary of the related work that uses economic approaches to model the behavior of primary and secondary users for spectrum sharing and discuss the associated issues. The authors also propose some open directions for future research on economic aspects of spectrum sharing in cognitive radio networks.

Chapter 17, authored by *S. K. Das, V. Gardellin, and L. Lenzini*, studies the self-coexistence problem. One of the major challenges for implementing cognitive radio networks is to guarantee self-coexistence among devices, which means address interference issues among devices operating under the same set of rules and sharing the same resources. Among the several mathematical tools used to address the self-coexistence problem, the authors recognize the game theoretic approach as the most powerful. In this chapter, first the authors present an overview of cognitive radio technology focusing on the importance of guaranteed self-coexistence among cognitive devices. Then, they analyze the pros and cons of several game theoretic approaches proposed in the literature in order to model the self-coexistence problem. This chapter is concluded by describing non-cooperative and cooperative game paradigms to model the self-coexistence problem in cognitive radio networks.

8 Conclusion

A summary of the contributed chapters has been provided that will be helpful to follow the rest of this book. These chapters essentially feature some of the major advances in the research on cognitive radio mobile ad hoc networks for the next generation wireless communications systems. Therefore, the book will be useful to both researchers and practitioners in this area. The readers will find the rich set of references in each chapters particularly valuable.

Ottawa, ON, Canada

F. Richard Yu

Contents

Part I Dynamic Spectrum Access

Contributors

S. Anand Department of Electrical and Computer Engineering, Stevens Institute of Technology, Hoboken, NJ, USA, asanthan@stevens.edu

Orlando Cabral Instituto de Telecomunicações, Covilhã, Portugal, orlandoc@ubi.pt

Rajarathnam Chandramouli Department of Electrical and Computer Engineering, Stevens Institute of Technology, Hoboken, NJ, USA, mouli@stevens.edu

Tao Chen VTT, Oulu, Finland, tao.chen@vtt.fi

Song Ci College of Engineering, University of Nebraska-Lincoln, Omaha, NE, USA, sci@engr.unl.edu

Sajal K. Das University of Texas at Arlington, Arlington, TX, USA, das@uta.edu

Ziqian Dong Department of Electrical and Computer Engineering, New York Institute of Technology, New York, NY, USA, ziqian.dong@nyit.edu

Zhenghua Feng Virginia Polytechnic Institute and State University, Blacksburg, VA, USA, Zhenhua@vt.edu

Zhiyong Feng Beijing University of Posts and Telecommunications, Beijing, China, fengzy@bupt.edu.cn

João M. Ferro Instituto de Telecomunicações, Covilhã, Portugal, ferro@lx.it.pt

Timothy K. Forde CTVR, Trinity College, Dublin, Ireland, fordeti@tcd.ie

Bo Gao Virginia Polytechnic Institute and State University, Blacksburg, VA, USA, bgao@vt.edu

Vanessa Gardellin Italian National Research Council, Pisa, Italy, v.gardellin@iit.cnr.it

Stein Gjessing University of Oslo, Oslo, Norway, steing@ifi.uio.no

Quansheng Guan School of Electrical and Information Engineering, South China University of Technology, Guangzhou, China, guan.quansheng@mail.scut.edu.cn

Chuan Han Virginia Polytechnic Institute and State University, Blacksburg, VA, USA, hanc@vt.edu

Oliver Holland CTR, King's College, London, UK, oliver.holland@kcl.ac.uk

Kai Hong Department of Electrical and Computer Engineering, Stevens Institute of Technology, Hoboken, NJ, USA, khong@stevens.edu

Ekram Hossain Department of Electrical and Computer Engineering, University of Manitoba, Winnipeg, MB, Canada, ekram@ee.umanitoba.ca

Minyi Huang School of Mathematics and Statistics, Carleton University, Ottawa, ON, Canada, mhuang@math.carleton.ca

Teerawat Issariyakul TOT Public Company Limited, Bangkok, Thailand, iteerawat@hotmail.com

Shengming Jiang School of Electrical and Information Engineering, South China University of Technology, Guangzhou, China, shmjiang@scut.edu.cn

Mohammad G. Khoshkholgh Wireless Innovation Laboratory (WIL), Department of Electrical and Computer Engineering, Tarbiat Modares University, Tehran, Iran, khoshkholgh@modares.ac.ir

Louise Lamont Communications Research Centre Canada, Ottawa, ON, Canada, louise.lamont@crc.gc.ca

Luciano Lenzini University of Pisa, Pisa, Italy, l.lenzini@iet.unipi.it

Jun Li Communications Research Centre Canada, Ottawa, ON, Canada, jun.li@crc.gc.ca

Zhiqiang Li Department of Systems and Computer Engineering, Carleton University, Ottawa, ON, Canada, zlia@sce.carleton.ca

Haiyan Luo University of Nebraska-Lincoln, Omaha, NE, USA, haiyan.luo@huskers.unl.edu

Irene Macaluso CTVR, Trinity College, Dublin, Ireland, macalusi@cs.tcd.ie

Sabita Maharjan Simula Research Laboratory, Lysaker, Norway, sabita@simula.no

B.S. Manoj Indian Institute of Space Science and Technology (IIST), Trivandrum, India, bsmanoj@iist.ac.in

Peter Mason Defense R&D Canada, Ottawa, ON, Canada, peter.mason@drdc-rddc.gc.ca

Jelena Mišić Department of Computer Science, Ryerson University, Toronto, ON, Canada, jmisic@scs.ryerson.ca

Vojislav B. Mišić Department of Computer Science, Ryerson University, Toronto, ON, Canada, vmisic@scs.ryerson.ca

Keivan Navaie School of Electronic & Electrical Engineering, University of Leeds, Leeds, UK, keivan.navaie@ieee.org

Hamidreza Nikoofar Mobile Communication Company of Iran (MCI), Tehran, Iran, hr.nikoofar@mci.ir

Dusit Niyato School of Computer Engineering, Nanyang Technological University, Singapore, dniyato@ntu.edu.sg

Keith E. Nolan CTVR, Trinity College, Dublin, Ireland, keith.nolan@tcd.ie

Ramesh Rao University of California San Diego, La Jolla, CA, USA, rrao@ucsd.edu

Shamik Sengupta Department of Mathematics and Computer Science, John Jay College of Criminal Justice, City University of New York, New York NY, USA, ssengupta@jjay.cuny.edu

Ahmad R. Sharafat Wireless Innovation Laboratory (WIL), Department of Electrical and Computer Engineering, Tarbiat Modares University, Tehran, Iran, sharafat@modares.ac.ir

Yi Song The University of North Carolina at Charlotte, Charlotte, NC, USA, ysong13@uncc.edu

Hang Su Networking and Information Systems Laboratory, Department of Electrical and Computer Engineering, Texas A&M University, College Station, TX, USA, su@tamu.edu

K.P. Subbalakshmi Department of Electrical and Computer Engineering, Stevens Institute of Technology, Hoboken, NJ, USA, ksubbala@stevens.edu

Bheemarjuna Reddy Tamma Indian Institute of Technology Hyderabad, India, tbr@iith.ac.in

Helen Tang Defense R&D Canada, Ottawa, ON, Canada, helen.tang@drdc-rddc.gc.ca

Hui Tang Chinese Academy of Sciences, Beijing, China, tangh@hpnl.ac.cn

Fernando J. Velez Instituto de Telecomunicações, Covilhã, Portugal, fjv@ubi.pt

Dalei Wu University of Nebraska-Lincoln, Omaha, NE, USA, dwu@huskers.unl.edu

Jiang Xie The University of North Carolina at Charlotte, Charlotte, NC, USA, Linda.Xie@uncc.edu

Yaling Yang Virginia Polytechnic Institute and State University, Blacksburg, VA, USA, yyang8@vt.edu

Halim Yanikomeroglu Department of Systems and Computer Engineering, Broadband Communications and Wireless Systems (BCWS) Centre, Carleton University, Ottawa, ON, Canada, halim@sce.carleton.ca

F. Richard Yu School of Information Technology, Department of Systems and Computer Engineering, Carleton University, Ottawa, ON, Canada, richard_yu@carleton.ca

Honggang Zhang Zhejiang University, Hangzhou, China, honggangzhang@zju.edu.cn

Xi Zhang Networking and Information Systems Laboratory, Department of Electrical and Computer Engineering, Texas A&M University, College Station, TX, USA, xizhang@ece.tamu.edu

Yan Zhang Simula Research Laboratory, Lysaker, Norway, yanzhang@simula.no

Zhifeng Zhao Zhejiang University, Hangzhou, China, zhaozf@zju.edu.cn

Yifeng Zhou Communications Research Centre Canada, Ottawa, ON, Canada, yifeng.zhou@crc.gc.ca

About the Editor

F. Richard Yu is currently an Assistant Professor in the Department of Systems and Computer Engineering, School of Information Technology, at Carleton University, Ottawa, ON, Canada. He received the Ph.D. degree in electrical engineering from the University of British Columbia, Vancouver, BC, Canada, in 2003. From 2002 to 2004, he was with Ericsson, Lund, Sweden, where he worked on the research and development of third-generation cellular networks. From 2005 to 2006, he was with a startup company in California, where he worked on the research and development in the areas of advanced wireless communication technologies and new standards. He joined the School of Information Technology and the Department of Systems and Computer Engineering, Carleton University, Ottawa, ON, Canada, in 2007. His research interests include cross-layer design, security, and quality-of-service provisioning in wireless networks.

He serves on the editorial boards of several journals, including the IEEE Communications Surveys and Tutorials, Springer Wireless Networks, the EURASIP Journal on Wireless Communications Networking, Ad Hoc and Sensor Wireless Networks, the Wiley Journal on Security and Communication Networks, and the International Journal of Wireless Communications and Networking. Dr. Yu has served on the Technical Program Committee (TPC) of numerous conferences and as the Publication Chair of the Institute for Computer Sciences, Social-Informatics and Telecommunications Engineering QShine 2010, Co-chair of the 2009 International Conference on Ultra Modern Telecommunications Workshop on Cognitive Wireless Communications and Networking, and TPC Co-chair of the 2012 IEEE VTCs-WN, the 2011 IEEE Globecom-CRN, the 2011 IEEE INFOCOM-GCN, the 2010 IEEE INFOCOM-CWCN, the 2009 IEEE International Wireless Communications and Mobile Computing Conference, the 2008 IEEE VTCf Track 4, and the 2007 International Workshop on Wireless Networking for Intelligent Transportation Systems.

He received the Early Researcher Award in 2011, Leadership Opportunity Fund Award from the Canada Foundation of Innovation in 2009, Outstanding Contribution Award from the IEEE/IFIP TrustCom 2010, and best paper awards at IEEE/IFIP TrustCom 2009 and International Conference on Networking 2005.

Chapter 1
Distributed Consensus-Based Cooperative Spectrum Sensing in Cognitive Radio Mobile Ad Hoc Networks

F. Richard Yu, Helen Tang, Minyi Huang, Peter Mason, and Zhiqiang Li

Abstract In cognitive radio mobile ad hoc networks (CR-MANETs), secondary users can cooperatively sense the spectrum to detect the presence of primary users. In this chapter, we propose a fully distributed and scalable cooperative spectrum sensing scheme based on recent advances in consensus algorithms. In the proposed scheme, the secondary users can maintain coordination based on only local information exchange without a centralized common receiver. We use the consensus of secondary users to make the final decision. The proposed scheme is essentially based on recent advances in consensus algorithms that have taken inspiration from complex natural phenomena including flocking of birds, schooling of fish, swarming of ants, and honeybees. Unlike the existing cooperative spectrum sensing schemes, there is no need for a centralized receiver in the proposed schemes, which make them suitable in distributed CR-MANETs. Simulation results show that the proposed consensus schemes can have significant lower missing detection probabilities and false alarm probabilities in CR-MANETs. It is also demonstrated that the proposed scheme not only has proven sensitivity in detecting the primary user's presence but also has robustness in choosing a desirable decision threshold.

1.1 Introduction

Recently, there has been tremendous interest in the field of cognitive radio (CR), which has been introduced in [1]. CR is an enabling technology that allows unlicensed (secondary) users to operate in the licensed spectrum bands. This can help to overcome the lack of available spectrum in wireless communications and achieve significant improvements over services offered by current wireless networks. It is designed to sense the changes in its surroundings, thus learns from its environment and performs functions that best serve its users. This is a very crucial feature of CR networks which allow users to operate in licensed bands without a license [2]. To achieve this goal, spectrum sensing is an indispensable part in cognitive radio.

F.R. Yu (✉)
School of Information Technology, Department of Systems and Computer Engineering, Carleton University, Ottawa, ON, Canada
e-mail: richard_yu@carleton.ca

F.R. Yu (ed.), *Cognitive Radio Mobile Ad Hoc Networks*,
DOI 10.1007/978-1-4419-6172-3_1,

There are three fundamental requirements for spectrum sensing. In the first place, the unlicensed (secondary) users can use the licensed spectrum as long as the licensed (primary) user is absent at some particular time slot and some specific geographic location. However, when the primary user comes back into operation, the secondary users should vacate the spectrum instantly to avoid interference with the primary user. Hence, a first requirement of cognitive radio is that the continuous spectrum sensing is needed to monitor the existence of the primary user. Also, since cognitive radios are considered as lower priority and they are secondary users of the spectrum allocated to a primary user, the second fundamental requirement is to avoid the interference to potential primary users in their vicinity [3]. Furthermore, primary user networks have no requirement to change their infrastructure for spectrum sharing with cognitive radios. Therefore, the third requirement is for secondary users to be able to independently detect the presence of primary users.

Taking those three requirements into consideration, such spectrum sensing can be conducted non-cooperatively (individually), in which each secondary user conducts radio detection and makes decision by itself. However, the sensing performance for one cognitive user will be degraded when the sensing channel experiences fading and shadowing [4]. In order to improve spectrum sensing, several authors have recently proposed collaboration among secondary users [3, 5–7], which means a group of secondary users perform spectrum sensing by collaboration. As a result, it shows that collaboration may enhance secondary spectrum access significantly [5].

Our research is focused on the distributed cooperative spectrum sensing (DCSS) in cognitive radio and, more precisely, the distributed cooperative schemes of spectrum sensing in a Cognitive Radio Mobile Ad-hoc NETworks (CR-MANETs).

In the first place, at present, distributed cooperative detection problems are discussed in [6, 8–10]. In a typical wireless distributed detection problem, each sensor or secondary user individually forms its own discrete messages based on its local measurement and then reports to a fusion center via wireless reporting channels. In certain models [10], however, there is, in general, no direct communication among the sensors. A sensor may indirectly obtain information about other sensors, but this is achieved by feedback from a common fusion center. Nevertheless, a centralized fusion center may not be available in some CR-MANETs. Moreover, as indicated in [11], gathering the entire received data at one place may be very difficult under practical communication constraints. In addition, authors of [4] study the reporting channels between the cognitive users and the common receiver. The results show that there are limitations for the performance of cooperation when the reporting channels to the common receiver are under deep fading.

Based on recent advances in consensus algorithms [12], we propose a new scheme in distributed cooperative spectrum sensing called distributed consensus-based cooperative spectrum sensing (DCCSS).

The main contributions of this work include as follows:

- We propose a consensus-based spectrum sensing scheme, which is a fully distributed and scalable scheme. Unlike many existing schemes [21, 22, 46], there is no need for a common receiver to do data fusion and to reach the final decision.

Since it is rare to have a centralized node in MANETs, in the proposed scheme, a secondary user needs only to setup local interactions without centralized information exchange.
- Unlike most decision rules, such as OR-rule or n-out-of-N, adopted in existing spectrum sensing schemes, we use consensus from secondary users. The proposed scheme has self-configuration and self-maintenance capabilities.
- Since the CR paradigm imposes human-like characteristics (e.g., learning, adaptation, and cooperation) in wireless networks, the bio-inspired consensus algorithm used in this work can provide some insight into the design of future CR-MANETs.

Extensive simulation results illustrate the effectiveness of the proposed scheme. It is shown that the proposed scheme can have both lower missing detection probability and lower false alarm probability compared to the existing schemes. In addition, it is able to make better detection when secondary users undergo worse fading (lower average SNR). Last but not the least, with the help of this scheme, a fixed threshold is feasible, which can take active effect in different fading channels.

The rest of the chapter is organized as follows. Section 1.2 describes the research background of this research, which includes spectrum sensing in cognitive radios, cooperative spectrum sensing, and centralized/distributed cooperative spectrum sensing. Section 1.3 presents system models, spectrum sensing model, fixed/random graphs theories, and consensus notions. In Section 1.4, the distributed consensus-based cooperative spectrum sensing scheme is proposed based on fixed graphs, together with the network models. Going further, the distributed consensus-based cooperative spectrum sensing scheme based on random graphs is described in Section 1.5. In Section 1.6, the simulation results and discussions are presented. Finally, we conclude this chapter in Section 1.7.

1.2 Background

This section is intended to cover the topics regarding the research background. They include the introduction of cognitive radio, functionalities of cognitive radio, differences of individual spectrum sensing, and cooperative spectrum sensing, followed by the introduction of centralized distributed cooperative spectrum sensing and distributed consensus-based cooperative spectrum sensing.

1.2.1 Introduction of Spectrum Sensing in Cognitive Radio

The idea of cognitive radio is first presented officially in an article by Joseph Mitola and Gerald Q. Maguire, Jr. [13]. It is a novel approach in wireless communications that Mitola later describe in his PhD dissertation as:

> The point in which wireless Personal Digital Assistants (PDAs) and the related networks are sufficiently computationally intelligent about radio resources and related computer-to-computer communications to detect user communications needs as a function of use context, and to provide radio resources and wireless services most appropriate to those needs.

It is thought of as an ideal goal toward which a software-defined radio platform should evolve: a fully reconfigurable wireless blackbox that automatically changes its communication variables in response to network and user demands.

The above citation originates from the following fact. On one hand, the growing number of wireless standards is occupying more and more naturally limited frequency bandwidth for exclusive use as licensed bands. However, large part of licensed bands is unused for what concerns a large amount of both time and space: even if a particular range of frequencies is reserved for a standard, at a particular time and at a particular location it could be found free. The Federal Communication Commission (FCC) estimates that the variation of use of licensed spectrum ranges from 15 to 85%, whereas according to Defence Advance Research Projects Agency (DARPA) only the 2% of the spectrum is in use in United States at any given moment. It is then clear that the solution to these problems can be found dynamically looking at spectrum as a function of time and space.

With the high demand of bit transmission rate for 4G or IMT-advanced high-speed wireless applications, there are several approaches to increase the system capacity as stated in the following equation:

$$C = n \cdot B \cdot \log_2(1 + \mathrm{SNR}) \tag{1.1}$$

The first approach is using MIMO to increase n, so that capacity may have a gain proportionally. The second approach is trying to increase *SNR*. The third one is focusing on the bandwidth. Cognitive radio is among the third category and thrives to fully utilize the frequency.

1.2.1.1 Functionalities of Cognitive Radios

The main functionalities of cognitive radios are [14]

- *Spectrum Sensing (SS)*: Detecting the unused spectrum and sharing it without harmful interference with other users, it is an important requirement of the cognitive radio network to sense spectrum holes and detecting primary users is the most efficient way to detect spectrum holes. Spectrum sensing techniques can be classified into three categories:
  - Transmitter detection: cognitive radios must have the capability to determine if a signal from a primary transmitter is locally present in a certain spectrum, there are several approaches proposed:
    - Matched filter detection
    - Energy detection
    - Cyclostationary feature detection
  - Cooperative detection: refers to spectrum sensing methods where information from multiple cognitive radio users is incorporated for primary user detection.
  - Interference-based detection.

- *Spectrum Management (SMa)*: Capturing the best available spectrum to meet user communication requirements. Cognitive radios should decide on the best spectrum band to meet the quality of service requirements over all available spectrum bands, therefore spectrum management functions are required for cognitive radios, these management functions can be classified as spectrum analysis and spectrum decision.
- *Spectrum Mobility (SMo)*: Defined as the process when a cognitive radio user exchanges its frequency of operation. Cognitive radio networks target to use the spectrum in a dynamic manner by allowing the radio terminals to operate in the best available frequency band, maintaining seamless communication requirements during the transition to better spectrum.
- *Spectrum Sharing (SSh)*: Providing the fair spectrum scheduling method, which is one of the major challenges in open spectrum usage in the spectrum sharing. It can be regarded to be similar to generic media access control MAC problems in existing systems.

1.2.1.2 Individual and Cooperative Spectrum Sensing

Spectrum sensing can be conducted either non-cooperatively (individually), in which each secondary user conducts radio detection and makes decision by itself, or cooperatively, in which a group of secondary users perform spectrum sensing by collaboration. No matter in which way, the common topology of such a cognitive radio network can be depicted as in Fig. 1.1. Individual spectrum sensing is conducted by secondary users on its own, and each user has a local observation and a local decision accordingly. Thus, in Fig. 1.1, each secondary user performs the spectrum sensing locally and no communication is between one another nor is the common receiver (fusion center). In such a condition, cognitive radio sensitivity can only be improved [6] by enhancing radio RF front-end sensitivity, exploiting digital signal processing gain for specific primary user signal, and network cooperation where users share their spectrum sensing measurements. However, if the sensing channels are facing deep fading or shadowing, then affected individuals will not be able to detect the presence of the primary user, which leads to missing detection failure.

In order to improve the performance of spectrum sensing, several authors have recently proposed cooperation among secondary users [2, 4, 5, 15]. Cooperative spectrum sensing has been proposed to exploit multi-user diversity in sensing process. It is usually performed in three successive stages: sensing, reporting and broadcasting. In the sensing stage, every cognitive user performs spectrum sensing individually. This can be shown as in Fig. 1.1, where secondary users try to collect the signal of interest through sensing channels. In the reporting stage, all the local sensing observations are reported to a common receiver via reporting channels (see Fig. 1.1) and the latter will make a final decision on the absence or the presence of the primary user. Finally, the final decision is broadcasted via broadcast channels to all the secondary users concerned, which include not only the ones involved into

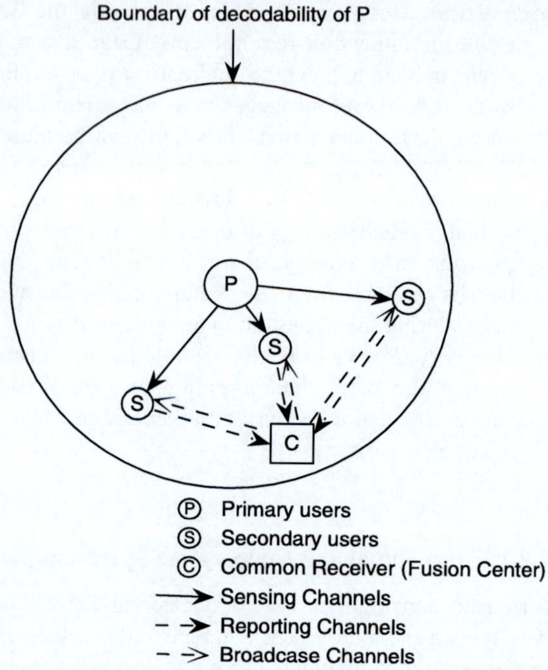

Fig. 1.1 A typical cognitive radio network

the sensing stage but also those that do not have sensing capabilities but want to participate into the spectrum sharing stage.

There are several advantages offered by cooperative spectrum sensing over the non-cooperative ones [5, 11, 16–22]. If a secondary user is in the condition of deep shadowing and fading, it is very difficult for a secondary user to distinguish a white space from a deep shadowing effect. Therefore, a non-cooperative spectrum sensing algorithm may not work well in this case, and a cooperative scheme can solve the problem by sharing the spectrum sensing information among secondary users. Moreover, because of the hidden terminal problem, it is very challenging for single cognitive radio sensitivity to outperform the primary user receiver by a large margin in order to detect the presence of primary users. For this reason, if secondary users spread out in the spatial distance, and any one of them detects the presence of primary users, then the whole group can gain benefit by collaboration.

Ghasemi and Sousa [5] quantify the performance of spectrum sensing in fading environments and study the effect of cooperation. The simulation results in [5] indicate that significant performance enhancements can be achieved through cooperation. Ganesan and Li [16] study the possibility to forward the signal with higher SNR to the one on the boundary of decidability region of the primary user. The performance is evaluated under correlated shadowing and user compromise in [11]. When the exchange of observations from all secondary users to the common receiver is not applicable, Peh and Liang [17] show that it is still worth doing by cooperating a certain number of users with relatively higher SNR. Moreover, in [18], a linear-quadratic (LQ) fusion strategy is designed with the consideration of the correlation between the nodes. In order to further reduce the computational complexity, Quan et al. [19] propose a heuristic approach so as to develop an optimal linear framework during cooperation. Sensing-throughput tradeoff is analyzed in [20] for both multiple mini-slots and multiple secondary users cooperative sensing.

1.2.1.3 Centralized Cooperative Spectrum Sensing

Although some research activities have been conducted in cooperative spectrum sensing, most of them use a common receiver (fusion center) to do data fusion for the final decision whether or not the primary user is present. However, a common receiver may not be available in some CR-MANETs. Moreover, as indicated in [11], gathering the entire received data at one place may be very difficult under practical communication constraints. In addition, Sun et al. [4] study the reporting channels between the cognitive users and the common receiver. The results show that there are limitations for the performance of cooperation when the reporting channels to the common receiver are under deep fading. In summary, the use of a centralized fusion center in CR-MANETs may have the following problems (see Fig. 1.1):

- Every secondary user needs to join/establish the connection with the common receiver, which requires a network protocol to implement.
- Some secondary users need a kind of relay routes to reach the common receiver if they are far away from the latter.
- Communication errors or packet drops can affect the performance of such a network if more users have worse reporting channels (e.g., Rayleigh Fading) to reach the common receiver.
- There should be a reliable wireless broadcast channel for the common receiver to inform each of every user once there is a decision made.
- The current centralized network does not fit for the average calculation of all the estimated sensing energy levels, because it requires the common receiver to correctly receive all the local estimated sensing results. Otherwise, the decision precision cannot be guaranteed.

1.2.2 Mobile Ad Hoc Networks

In recent years, MANETs have become a popular subject because of their self-configuration and self-organization capabilities. Each device in a MANET is free to move independently in any direction and will therefore change its links to other devices frequently. Wireless nodes can establish a dynamic network without the need of a fixed infrastructure. A node can function both as a network router for routing packets from the other nodes and as a network host for transmitting and receiving data. MANETs are particular useful when a reliable fixed or mobile infrastructure is not available. Instant conferences between notebook PC users, military applications, emergency operations, and other secure-sensitive operations are important applications of MANETs due to their quick and easy deployment.

1.2.2.1 Self-Organization of MANETs

Due to the lack of centralized control, MANETs nodes cooperate with each other to achieve a common goal [23]. The major activities involved in self-organization are neighbor discovery, topology organization, and topology reorganization. Through periodically transmitting beacon packets or promiscuous snooping on the channels, the activities of neighbors can be acquired. Each node in MANETs maintains the topology of the network by gathering the local or entire network information. MANETs need to update the topology information whenever the networks change such as participation of new node, failure of node, and links. Therefore, self-organization is a continuous process that has to adapt to a variety of changes or failures.

1.2.3 Distributed Consensus-Based Cooperative Spectrum Sensing Scheme

In this work, we will present a distributed consensus-based cooperative spectrum sensing scheme without using a common receiver. Our scheme is based on recent advances in consensus algorithms [12], or more precisely, bio-inspired mechanisms, which have become important approaches to handle complex communication networks [24–26]. An important motivational background of this area is initially related to the study of complex natural phenomena including flocking of birds, schooling of fish, and swarming of ants and honeybees (see the survey [27]). The investigation of such biological systems has generated fundamental insights into understanding the relation between group decision making at the higher level and the individual animals' communication at the lower level [28–32], and in fact consensus seeking in animal colonies is vital for group survival [32]. Such collective animal behavior has motivated many effective yet simple control algorithms for the coordination of multi-agent systems in engineering. Recently, consensus problems have played a crucial role in spacial distributed control models [12, 33], wireless sensor networks [34], and stochastic seeking with noise measurement [35]. Since these algorithms are usually constructed based on local communication of neighboring agents, they have low implementation complexity and good robustness, and the overall system may still function when local failure occurs.

The main highlights of this scheme are as follows:

- It is a fully distributed and scalable scheme. Unlike the existing schemes [21, 22, 46], there is no need for a common receiver to do the data fusion for the final decision. A secondary user only needs to set up neighborhood with those users having desired channel characteristics, such as Line of Sight ones, or even with probabilistic link failures.

- Unlike most decision rules, such as OR-rule or 1-out-of-N, adopted in the existing schemes, we use the consensus of secondary users to make the final decision. Therefore, the proposed scheme can leverage the detection results among users in a severe wireless fading networks.
- The proposed spectrum sensing scheme uses a consensus algorithm to cope with two underlying network models, one with *fixed* bidirectional graphs and one with *random* graphs.

Our consensus-based approach is different from those used in distributed/decentralized detection problems [8–10, 36]. In a typical distributed detection problem [8, 9, 36], each sensor individually forms its own discrete messages based on its local measurement and then reports to a fusion center, and there is in general no direct communication among the sensors. In certain models [10], a sensor may indirectly obtain information about other sensors, but this is achieved by feedback from a common fusion center.

1.3 Secondary Users Network Modeling

This section is organized in the following order. First, a network topology in distributed consensus-based cooperative spectrum sensing is presented. Then, the local spectrum sensing model is discussed in details. At last, the network model and consensus notions are presented.

1.3.1 Network Topology in Distributed Consensus-Based Cooperative Spectrum Sensing

As shown in Fig. 1.2, no common receiver is necessary compared with Fig. 1.1, and secondary users are communicating with each other via communication channels that are in good radio coverage of each of secondary users. Secondary users that are far away from each other do not have direct communication channels due to poor radio signal quality.

There are two stages in the proposed cognitive radio consensus schemes. In the first stage, secondary users use a spectrum sensing model to make measurements about primary users at the beginning of detection. This is done via sensing channels in Fig. 1.2. We denote the local measurement of user i as Y_i. In the second stage, secondary users establish communication links with their own neighbors to locally exchange information among them, and then calculate the obtained data so as to make a local decision whether primary users are around. The above process in the second stage is done iteratively. At the initial time instant $k = 0$, each user i sets $x_i(0) = Y_i$ as the initial value of the local state variable. Next, at time $k = 0, 1, 2, \ldots$, according to the real-time network topology (or local wireless neighborhood), users mutually transmit and receive their states and then use local

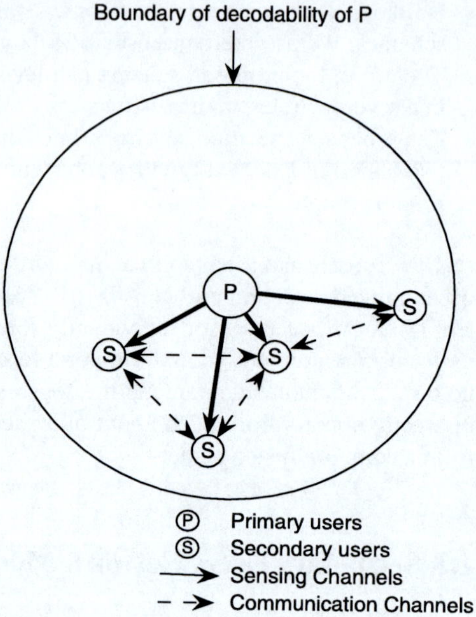

Fig. 1.2 A topology of distributed consensus-based cooperative spectrum sensing

computation rules to generate updated states $x_i(k+1)$. Those iterations are done repeatedly until all the individual states $x_i(k)$ converge toward a common value x^*.

Before we introduce the detailed algorithms used in our consensus scheme, the common spectrum sensing model used in the first stage and the network model used in the second stage are to be presented, followed by the formal definition of the spectrum sensing consensus scheme.

1.3.2 The Spectrum Sensing Model

In the first stage, secondary users make measurements about primary users at the beginning of each time slot. Three kinds of methods are widely used for spectrum sensing [6]: matched filter, energy detector, and cyclostationary feature detector.

- *Matched Filter*
  The optimal way for any signal detection is a matched filter [37], since it maximizes received signal-to-noise ratio. However, a matched filter effectively requires demodulation of a primary user signal. This means that cognitive radio has a priori knowledge of primary user signal at both PHY and MAC layers, e.g., modulation type and order, pulse shaping, and packet format. Such information might be pre-stored in CR memory, but the cumbersome part is that for demodulation it has to achieve coherency with primary user signal by performing timing and carrier synchronization, even channel equalization. This is

still possible since most primary users have pilots, preambles, synchronization words, or spreading codes that can be used for coherent detection. For examples: TV signal has narrowband pilot for audio and video carriers; CDMA systems have dedicated spreading codes for pilot and synchronization channels; OFDM packets have preambles for packet acquisition. The main advantage of matched filter is that due to coherency it requires less time to achieve high processing gain [38]. However, a significant drawback of a matched filter is that a cognitive radio would need a dedicated receiver for every primary user class.

- *Energy Detector*
  One approach to simplify matched filtering approach is to perform non-coherent detection through energy detection. This sub-optimal technique has been extensively used in radiometry. There are several drawbacks of energy detectors that might diminish their simplicity in implementation. First, a threshold used for primary user detection is highly susceptible to unknown or changing noise levels. Even if the threshold would be set adaptively, presence of any in-band interference would confuse the energy detector. Furthermore, in frequency selective fading it is not clear how to set the threshold with respect to channel notches. Second, energy detector does not differentiate between modulated signals, noise, and interference. Since, it cannot recognize the interference, it cannot benefit from adaptive signal processing for canceling the interferer. Furthermore, spectrum policy for using the band is constrained only to primary users, so a cognitive user should treat noise and other secondary users differently. Lastly, an energy detector does not work for spread spectrum signals: direct sequence and frequency hopping signals, for which more sophisticated signal processing algorithms need to be devised. In general, we could increase detector robustness by looking into a primary signal footprint such as modulation type, data rate, or other signal feature.
- *Cyclostationary Feature Detection*
  Modulated signals are in general coupled with sine wave carriers, pulse trains, repeating spreading, hoping sequences, or cyclic prefixes which result in built-in periodicity. Even though the data is a stationary random process, these modulated signals are characterized as cyclostationary, since their statistics, mean and autocorrelation, exhibit periodicity. This periodicity is typically introduced intentionally in the signal format so that a receiver can exploit it for: parameter estimation such as carrier phase, pulse timing, or direction of arrival. This can then be used for detection of a random signal with a particular modulation type in a background of noise and other modulated signals.

In summary, Matched filter is optimal theoretically, but it needs the prior knowledge of the primary system, which means higher complexity and cost to develop adaptive sensing circuits for different primary wireless systems. Energy detection is suboptimal, but it is simple to implement and does not have too much requirement on the position of primary users. Cyclostationary feature detection can detect the signals with very low SNR, but it still requires some prior knowledge of the primary user [4].

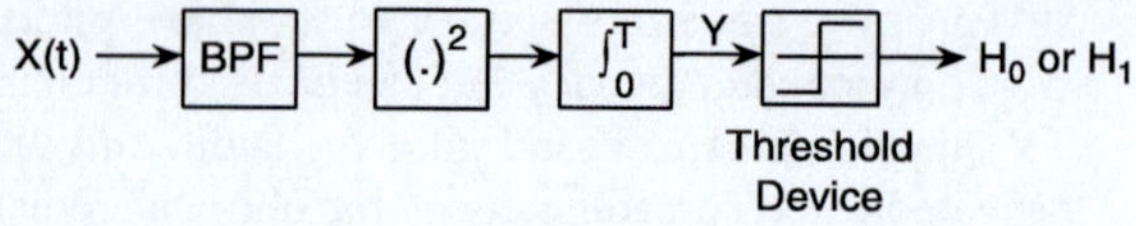

Fig. 1.3 Block diagram of an energy detector

In this chapter, we consider the modeling scenario where the prior knowledge of the primary user is unknown. For implementation simplicity, an energy detection spectrum sensing method [5] is used. Fig. 1.3 shows the block diagram of an energy detector. The input band pass filter (BPF) selects the center frequency f_s and the bandwidth of interest W. This filter is followed by a squaring device and subsequently an integrator over a period of T. The output Y of the integrator is the received energy at the secondary user and its distribution depends on whether the primary user signal is present or not. The goal of spectrum sensing is to decide between the following two hypotheses:

$$x(t) = \begin{cases} n(t), & H_0 \\ h \cdot s(t) + n(t), & H_1 \end{cases} \tag{1.2}$$

where $x(t)$ is the signal received by the secondary user, $s(t)$ is the primary user's transmitted signal, $n(t)$ is the additive white Gaussian noise (AWGN), and h is the amplitude gain of the channel. We also denote by γ the signal-to-noise ratio (SNR). The output of integrator in Fig. 1.3 is Y, which serves as the decision statistic. Following the work of [39], Y has the following form:

$$Y = \begin{cases} \chi^2_{2TW}, & H_0 \\ \chi^2_{2TW}(2\gamma), & H_1 \end{cases} \tag{1.3}$$

where χ^2_{2TW} and $\chi^2_{2TW}(2\gamma)$ denote random quantities with central and non-central chi-square distributions, respectively, each with $2TW$ degrees of freedom and a non-centrality parameter of 2γ for the latter distribution. For simplicity we assume that the time-bandwidth product, TW, is an integer number, which is denoted by m.

Under Rayleigh fading, the gain h is random, and the resulting SNR γ would have an exponential distribution, so in this case the distribution of the output energy depends on the average SNR ($\overline{\gamma}$). When the primary user is absent, Y is still distributed according to χ^2_{2TW}. When the primary user is present, Y may be denoted as the sum of two independent random variables [40, 41]:

$$Y = Y_\chi + Y_e, \qquad H_1, \tag{1.4}$$

where the distribution of Y_χ is χ^2_{2TW-2} and Y_e has an exponential distribution with parameter $2(\overline{\gamma} + 1)$.

As a summary, after T seconds, each secondary user i detects the energy and gets the measurement $Y_i \in \mathbb{R}^+$.

1.3.3 The Network Model and Consensus Notions

In the second stage, secondary users establish communication links with its neighbors to locally exchange information among them. In our scheme, the network formed by the secondary users can be described by a standard graph model. For simplicity, this can be represented by an undirected graph (to be simply called a graph) $\boldsymbol{G} = (\mathcal{N}, \mathcal{E})$ [42] consisting of a set of nodes $\{i = 1, 2, \ldots, n\}$ and a set of edges $\mathcal{E} \subset \mathcal{N} \times \mathcal{N}$. Denote each edge as an unordered pair (i, j). Thus, if two secondary users are connected by an edge, it means they can mutually exchange information. A path in $\boldsymbol{G}$ consists of a sequence of nodes $i_1, i_2, \ldots, i_l,\ l \geq 2$, such that $(i_m, i_{m+1}) \in \mathcal{E}$ for all $1 \leq m \leq l-1$. The graph $\boldsymbol{G}$ is connected if any two distinct nodes in $\boldsymbol{G}$ are connected by a path. For convenience of exposition, we often refer node i as secondary user i. The two names, secondary user and node, will be used interchangeably. The secondary user j (resp., node j) is a neighbor of user i (resp., node i) if $(j, i) \in \mathcal{E}$, where $j \neq i$. Denote the neighbors of node i by $\mathcal{N}_i = \{j | (j, i) \in \mathcal{E}\} \subset \mathcal{N}$. The number of elements in $\mathcal{N}_i$ is denoted by $|\mathcal{N}_i|$ and called the degree of node i.

Throughout this chapter, the analysis is for undirected graphs, because we only deal with good duplex wireless links by which two adjacent nodes can establish communication (being connected) with each other. That is, the graph $\boldsymbol{G}$ is connected, and the information exchange between two neighboring nodes is bidirectional.

The Laplacian of the graph $\boldsymbol{G}$ is defined as $\boldsymbol{L} = (l_{ij})_{n \times n}$, where

$$l_{ij} = \begin{cases} |\mathcal{N}_i|, \text{ if } j = i \\ -1, \quad \text{if } j \in \mathcal{N}_i \\ 0, \quad\ \ \text{otherwise} \end{cases} \tag{1.5}$$

The matrix $\boldsymbol{L}$ defined by (1.5) is positive semi-definite. Further, if $\boldsymbol{G}$ is a connected undirected graph, then $\text{rank}(\boldsymbol{G}) = n - 1$ (see, e.g., [27]).

Since the cooperative spectrum sensing problem is viewed as a consensus problem where the users locally exchange information regarding their individual detection outcomes before reaching an agreement, we give the formal mathematical definition of consensus as follows.

The underlying network turns out to consist of secondary users reaching a consensus via local communication with their neighbors on a graph $\boldsymbol{G} = (\mathcal{N}, \mathcal{E})$.

For the n secondary users distributed according to the graph model $\boldsymbol{G}$, we assign them a set of state variables x_i, $i \in \mathcal{N}$. Each x_i will be called a consensus variable, and in the cooperative spectrum sensing context, it is essentially used by node i for its estimate of the energy detection. By reaching consensus, we mean the individual states x_i asymptotically converge to a common value x^*, i.e.,

$$x_i(k) \to x^* \text{ as } k \to \infty \tag{1.6}$$

for each $i \in \mathcal{N}$, where k is the discrete time, $k = 0, 1, 2, \ldots$, and $x_i(k)$ is updated based on the previous states of node i and its neighbors.

The special cases with $x^* = \text{Ave}(x) = (1/n)\sum_{i=1}^{n} x_i(0)$, $x^* = \max_{i=1}^{n} x_i(0)$, and $x^* = \min_{i=1}^{n} x_i(0)$ are called average-consensus, max-consensus, and min-consensus, respectively. It is worth mentioning that the existing spectrum sensing algorithm with the OR-rule can be viewed as a form of max-consensus. This chapter is intended to propose a cooperative spectrum sensing scheme in the framework of average consensus.

1.4 Distributed Consensus-Based Cooperative Spectrum Sensing in Fixed Graphs

In this chapter, let us assume the secondary users have established duplex wireless connections with their desired neighbors, and the connections remain working until the consensus is reached. This kind of topology is called as a fixed graph. Based on this assumption, we are going to propose the spectrum sensing consensus algorithm as follows.

1.4.1 The Consensus Algorithm

We denote for user i, its measurement Y_i at time $k = 0$ by $x_i(0) = Y_i \in \mathbb{R}^+$. The state update of the consensus variable for each secondary user occurs at discrete time $k = 0, 1, 2, \ldots$, which is associated with a given sampling period. From $k = 0, 1, 2, \ldots$, the iterative form of the consensus algorithm can be stated as follows [27]:

$$x_i(k+1) = x_i(k) + \varepsilon \sum_{j \in \mathcal{N}_i} (x_j(k) - x_i(k)) \tag{1.7}$$

where

$$0 < \varepsilon < (\max_i |\mathcal{N}_i|)^{-1} \triangleq 1/\Delta \tag{1.8}$$

The number Δ is called the maximum degree of the network.

This algorithm can be written in the vector form:

$$\boldsymbol{x}(k+1) = \boldsymbol{P}\boldsymbol{x}(k) \tag{1.9}$$

where $\boldsymbol{P} = \boldsymbol{I} - \varepsilon \boldsymbol{L}$. Notice that the upper bound in (1.8) for ε ensures that $\boldsymbol{P}$ is a stochastic matrix, and in fact one can further show that $\boldsymbol{P}$ is ergodic when $\boldsymbol{G}$ is connected[1]. Since G is an undirected graph, all row sums and column sums of L are equal to zero. Hence P is a doubly stochastic matrix (i.e., P is a nonnegative matrix and all of its row sums and column sums are equal to one).

We also point out that (1.9) uses only a particular construction of the coefficient matrix for the consensus algorithm, which is based on the graph Laplacian L. As long as each node has the prior knowledge of an upper bound of the maximum degree Δ of the network, the iteration may be implemented and there is no necessity for neighboring nodes to exchange information regarding the network structure. Also, it is possible to construct P in other forms. An alternative choice of P may be based on the so-called Metropolis weights [34] by taking

$$\tilde{p}_{ij} = \begin{cases} \frac{1}{1+\max\{d_i,d_j\}} & \text{if } (j,i) \in \mathcal{E}, \\ 1 - \sum_{j \in \mathcal{N}_i} \tilde{p}_{ij} & \text{if } i = j, \\ 0 & \text{otherwise} \end{cases}$$

where $d_i = |\mathcal{N}_i|$ is the degree of node i. If G is a connected graph and we define $\widetilde{P} = (\tilde{p}_{ij})_{n \times n}$, then $\widetilde{P}$ is an ergodic doubly stochastic matrix. When $\widetilde{P}$ is used in (1.9) in place of P, the state average will still be preserved as an invariant during the iterations and our convergence analysis below is still valid. Notice that when $\widetilde{P}$ is used in the consensus algorithm, it is only required that any two neighboring nodes report to each other their degrees, and the knowledge of the maximum degree of the network is no longer needed.

We cite a theorem concerning the convergence property of the consensus algorithm.

Theorem 1 (see, e.g., [27]) *Consider a network of secondary users,*

$$x_i(k+1) = x_i(k) + u_i(k) \tag{1.10}$$

with topology G applying the distributed consensus algorithm (1.7), where $u_i(k) = \varepsilon \sum_{j \in \mathcal{N}_i}(x_j(k) - x_i(k))$, $0 < \varepsilon < 1/\Delta$, and Δ is the maximum degree of the network. Let G be a connected undirected graph. Then

1. *A consensus is asymptotically reached for all initial states;*
2. *P is doubly stochastic, and an average consensus is asymptotically reached with the limit $x^* = (1/n)\sum_{i=1}^{n} x_i(0)$ for the individual states.*

According to Theorem 1, if we choose ε such that $0 < \varepsilon < 1/\Delta$, then an average consensus is ensured and the final common value $x^* = (1/n)\sum_{i=1}^{n} x_i(0)$

[1] For some network topologies, it is possible to have an ergodic matrix $P = I - \varepsilon L$ when $\varepsilon = 1/\Delta$. For instance, if ε is taken as $1/\Delta$ and meanwhile it is ensured that P has at least one positive diagonal entry, then it can be shown that P is an ergodic stochastic matrix.

will be the average of the initial vector $\boldsymbol{x}(\mathbf{0})$, or equivalently, the average of $\boldsymbol{Y}^T = \{Y_1, Y_2, \ldots, Y_n\}$, which has been obtained during the energy detection stage.

Finally, by comparing the average-consensus result x^* with a pre-defined threshold λ based on Fig. 1.3, every secondary user i gets the final data fusion locally:

$$\text{Decision } \boldsymbol{H} = \begin{cases} 1, \; x^* > \lambda \\ 0, \text{ otherwise} \end{cases} \tag{1.11}$$

1.4.2 Performance of the Consensus Algorithm

It is quite apparent that the convergence rate is yet another interesting issue in evaluating the performance of the spectrum sensing consensus algorithm. This is due to the fact that secondary users must continuously detect the presence of primary users and back up as soon as possible on recognizing such incident. From this point of view, the speed of reaching a consensus is the key in the design of the network topology as well as the analysis of the performance of a consensus algorithm for a given spectrum sensing network. For the *connected* undirected graph $\boldsymbol{G}$, the above algorithm can ensure exponential convergence rate, where the error can be parameterized in the form $O(e^{-\delta t})$ with the exponent $\delta > 0$. To have some bound estimate for the parameter δ, we first recall that $\boldsymbol{P} = \boldsymbol{I} - \varepsilon \boldsymbol{L}$. Since L is a positive semi-definite matrix, denote its n eigenvalues by

$$0 = \lambda_1 < \lambda_2 \leq \ldots \leq \lambda_n. \tag{1.12}$$

Here $\lambda_2 > 0$ since the undirected graph $\mathbf{G}$ is *connected* which ensures that the rank of $\boldsymbol{L}$ is equal to $n-1$ ([43]). The second smallest eigenvalue λ_2 of $\boldsymbol{L}$ is usually called the algebraic connectivity of the undirected graph $\boldsymbol{G}$. Then the second largest absolute value of the eigenvalues of $\boldsymbol{P}$ is determined as $\alpha(\varepsilon) = \max\{|1-\varepsilon\lambda_2|, |1-\varepsilon\lambda_n|\}$, which can be verified to satisfy $\alpha(\varepsilon) < 1$. By using standard results in nonnegative matrix theory (see, e.g., [44]), we can obtain an upper bound for δ. In fact, we can take δ as any value in the interval $(0, -\ln\alpha(\varepsilon))$. We also remark that similar convergence rate estimates can be carried out when general weight matrices in averaging are used.

Since $\boldsymbol{P}$ has a unit eigenvalue, we see that the difference between the first two largest absolute values of the eigenvalues of $\boldsymbol{P}$ is given as $g(\varepsilon) = 1 - \alpha(\varepsilon)$, which is customarily called the spectral gap of $\boldsymbol{P}$. In general, the greater is $g(\varepsilon)$, the greater is the upper bound $-\ln\alpha(\varepsilon)$ for the exponent δ, and the faster is the convergence of the consensus algorithm. In practical implementations, it is desirable to choose a suitable value for ε to increase the spectral gap $g(\varepsilon)$ while $\boldsymbol{P}$ is ensured to be ergodic. We will discuss the convergence rate in the simulation part of this chapter.

1.5 Distributed Consensus-Based Cooperative Spectrum Sensing in Random Graphs

In the previous section, it has been assumed that any two neighboring nodes can reliably exchange data at all times. Hence the network topology remains unchanged during the overall time period of interest. This kind of network modeling may not be accurate in certain situations. For example, fading of wireless signals can cause packet errors, which will result in wireless link failures for that period. Furthermore, even under LOS channels, moving objects between neighboring nodes may temporarily affect signal reception. For the above reasons, in this chapter, we consider a more realistic inter-node communication model with random link failures. Unlike the previous model, which is based on fixed bidirectional graphs, the new model is based on random graphs. Nevertheless, similar to the previous fixed topology scenario, for the random graph-based modeling below, we still consider bidirectional links when two nodes can communicate.

1.5.1 Random Graph Modeling of the Network Topology

Before characterizing random connectivity of the network of all secondary users, let us first introduce a fixed undirected graph $\boldsymbol{G} = (\mathcal{N}, \mathcal{E})$ which describes the maximal set of communication links when there is no link failure. Due to the random link failures, at time k the inter-user communication is described by a subgraph of $\boldsymbol{G}$ denoted by $\boldsymbol{G}(k) = (\mathcal{N}, \mathcal{E}(k))$ where $\mathcal{E}(k) \subset \mathcal{E}$; the edge $(j,i) \in \mathcal{E}(k)$ if and only if nodes j and i can communicate at time k where $(j,i) \in \mathcal{E}$. Thus, the (undirected) graph $\boldsymbol{G}(k)$ is generated as the outcome of random link failures. Note that an edge (j,i) never appears in $\boldsymbol{G}(k)$ if it is not an edge of $\boldsymbol{G}$. The neighbor set of node i is $\mathcal{N}_i(k) = \{j|(j,i) \in \mathcal{E}(k)\}$ at time k. The number of elements in $\mathcal{N}_i(k)$ is denoted by $|\mathcal{N}_i(k)|$. At time $k \geq 0$, the adjacency matrix of $\boldsymbol{G}(k)$ is defined as $\boldsymbol{A}(k) = (\alpha_{ji}(k))_{1 \leq j,i \leq |\mathcal{N}|}$, where $\alpha_{ji}(k) = 1$ if $(j,i) \in \mathcal{E}(k)$, and $\alpha_{ji}(k) = 0$ otherwise. It is clear that the graph $\boldsymbol{G}(k)$ is completely characterized by the random matrix $\boldsymbol{A}(k)$.

Concerning the statistical properties of link failures, we assume that for all links (each associated with an edge in the graph $\boldsymbol{G}$) fail independently with the same probability $p \in (0,1)$. For notational simplicity we use the same parameter p to model the failure probability. The generalization of the modeling and analysis to link-dependent failure probabilities is straightforward.

1.5.2 The Algorithm with Random Graphs

For the random link failure-prone model, the two spectrum sensing stages introduced in the previous chapter are still applicable. In the first stage, each node performs the radio detection and computes the measurements according to (1.2).

During the second stage, at time k each node exchanges states information with its neighbors and performs the corresponding computation to generate its state update $x_i(k+1)$. Let Δ be the maximum degree of the graph $\boldsymbol{G}$ and take $\varepsilon \in (0, 1/\Delta)$.

The state of user $i \in \mathcal{N}$ is updated by the rule

$$x_i(k+1) = x_i(k) + \varepsilon \sum_{j \in \mathcal{N}_i(k)} [x_j(k) - x_i(k)] \tag{1.13}$$

where ε is a pre-determined constant step size. If $\mathcal{N}_i(k) = \emptyset$ (empty set), (1.13) reduces to $x_i(k+1) = x_i(k)$.

Theorem 2 *Under the independent link failure assumption, the algorithm (1.13) ensures average-consensus, i.e.,* $\lim_{k\to\infty} x_i(k) = (1/n)\sum_{j=1}^{n} x_j(0)$ *for all* $i \in \mathcal{N}$, *with probability 1. If, in addition,* $E|x(0)|^2 < \infty$ *and* $x(0)$ *is independent of the sequence of adjacency matrices* $\mathbf{A}(k)$, $k = 0, 1, \ldots,$ *then each* $x_i(k)$ *converges to* $(1/n)\sum_{j=1}^{n} x_j(0)$ *in mean square with an exponential convergence rate.*

Proof We can write the algorithm (1.13) in the vector form

$$\boldsymbol{x}(k+1) = [\boldsymbol{I} - \varepsilon \boldsymbol{L}(k)]\boldsymbol{x}(k)$$

where $\boldsymbol{L}(k)$ is the Laplacian of the graph $\boldsymbol{G}(k)$. For a vector $z = (z_1, \ldots, z_n)^T$, denote the Euclidean norm $|z| = \left(\sum_{i=1}^{n} z_i^2\right)^{1/2}$. For any given sample point, we can show that $\boldsymbol{M}(k) = \boldsymbol{I} - \varepsilon \boldsymbol{L}(k)$ is a symmetric aperiodic stochastic matrix so that it has all its eigenvalues within the interval $(-1, 1]$ (see, e.g., [44]), and therefore $\boldsymbol{M}(k)$ determines a paracontracting map [34, 45] in the sense $\boldsymbol{M}(k)z \neq z$ if and only if $|\boldsymbol{M}(k)z| < |z|$. For $\boldsymbol{M}(k)$, we denote its fixed point subspace $\mathcal{H}(\boldsymbol{M}(k)) = z \in \mathbb{R}^n | \boldsymbol{M}(k)z = z\}$.

By the assumption on the independent link failures, we see that with probability 1, $\boldsymbol{G}(k) = \boldsymbol{G}$ for an infinite number of times k. Let Ω denote the underlying probability sample space. Thus, after excluding a set A_0 of zero probability, for all $\omega \in \Omega \backslash A_0$, $\boldsymbol{G}(k) = \boldsymbol{G}$ infinitely often with the associated Laplacian being $\boldsymbol{L}(k) = \boldsymbol{L}$. Hence, for each $\omega \in \Omega \backslash A_0$, $x(k)$ converges to a point in the space $\mathcal{H}(\boldsymbol{I} - \varepsilon \boldsymbol{L}) = \{z \in \mathbb{R}^n | \boldsymbol{L}z = 0\}$ when $k \to \infty$. Furthermore, $z \in \mathbb{R}^n | \boldsymbol{L}z = 0\} = \text{span}\{1_n\}$ since $\boldsymbol{G}$ is a connected undirected graph.

On the other hand, it is straightforward to check that $(1/n)\sum_{j=1}^{n} x_j(k)$ remains as a constant since $\boldsymbol{M}(k)$ is a doubly stochastic matrix (i.e., nonnegative matric with all row sums and column sums equal to 1). Now it follows that each $x_i(k)$ converges to $(1/n)\sum_{j=1}^{n} x_j(0)$ with probability one, as $k \to \infty$.

We continue to analyze mean square convergence. Since $E|\boldsymbol{x}(0)|^2 < \infty$ and $\sup_{i \in \mathcal{N}, k \geq 0} |x_i(k)| \leq \max_{i \in \mathcal{N}} |x_i(0)| \leq |\boldsymbol{x}(0)|$, by the probability 1 convergence of $x_i(k)$, it follows from dominated convergence results in probability theory that $x_i(k)$ also converges to $(1/n)\sum_{j=1}^{n} x_j(0)$ in mean square.

Now, we proceed to give an estimation of the mean square convergence rate within the random network model. Denote $\text{Ave}(\boldsymbol{x}(0)) = (1/n)\sum_{j=1}^{n} x_j(0)$. It is straightforward to show that

$$x(k+1) - \text{Ave}(x(0))\mathbf{1}_n = \left[I - (1/n)\mathbf{1}_n\mathbf{1}_n^T\right][I - \varepsilon L(k)][x(k) - \text{Ave}(x(0))\mathbf{1}_n] \tag{1.14}$$

$$\equiv B(k)[x(k) - \text{Ave}(x(0))\mathbf{1}_n] \tag{1.15}$$

In fact, for each $\omega \in \Omega$, by the eigenvalue distribution of the matrices $(1/n)\mathbf{1}_n\mathbf{1}_n^T$ and $L(k)$, we can show that $B^T(k)B(k)$, and subsequently $E[B^T(k)B(k)]$, have n real eigenvalues on the interval [0, 1]. We use a contradiction argument to show that the largest eigenvalue ρ of $E[B^T(k)B(k)]$ is less than 1. Suppose $\rho = 1$ for $E[B^T(k)B(k)]$; then there exists a real-valued vector $x \neq 0$ such that

$$x^T E[B^T(k)B(k)]x = x^T x \tag{1.16}$$

By the fact $x^T[B^T(k)B(k)]x \leq x^T x$, the equality (1.16) leads to

$$x^T[B^T(k)B(k)]x = x^T x \tag{1.17}$$

with probability 1. On the other hand, by the link failure assumption, there exists a set $A_1 \subset \Omega$ such that $P(A_1) > 0$ and for each $\omega \in A_1$, the associated matrix value $B(k) = I - \varepsilon L$. Without the loss of generality, we can assume A_1 has been chosen in such a manner that for any $\omega \in A_1$ (1.17) also holds.

By noticing the fact that for any $z \in \mathbb{R}^n$,

$$z^T[B^T(k)B(k)]z \leq z^T(I - \varepsilon L)^2 z \leq z^T z \tag{1.18}$$

we obtain from (1.17) that

$$x^T(I - \varepsilon L)^2 x = x^T x \tag{1.19}$$

Hence, (1.19) implies that x is the eigenvector of $I - \varepsilon L$ associated with the eigenvalue 1, which further implies that $x \in \text{span}\{\mathbf{1}_n\}$. Denote $x = c\mathbf{1}_n$ where c is a constant. By substituting $x = c\mathbf{1}_n$ into the left hand side of (1.17), we obtain $x^T[B^T(k)B(k)]x = 0$ for each $\omega \in \Omega$, which contradicts with (1.17) and the fact $x \neq 0$. Hence, we conclude that the largest eigenvalue ρ of $E[B^T(k)B(k)]$ is in the interval [0, 1).

Finally, by elementary calculation we obtain the convergence rate estimate

$$E|x(k) - \text{Ave}(x(0))\mathbf{1}_n|^2 \leq \rho^k E|x(0) - \text{Ave}(x(0))\mathbf{1}_n|^2 \tag{1.20}$$

□

In fact, we have the simplified expression:

$$\begin{aligned} \boldsymbol{B}^T(k)\boldsymbol{B}(k) &= [\boldsymbol{I} - \varepsilon \boldsymbol{L}(k)] \left[\boldsymbol{I} - (1/n)\mathbf{1}_n\mathbf{1}_n^T\right]^2 [\boldsymbol{I} - \varepsilon \boldsymbol{L}(k)] \\ &= [\boldsymbol{I} - \varepsilon \boldsymbol{L}(k)] \left[\boldsymbol{I} - (1/n)\mathbf{1}_n\mathbf{1}_n^T\right] [\boldsymbol{I} - \varepsilon \boldsymbol{L}(k)] \\ &= [\boldsymbol{I} - \varepsilon \boldsymbol{L}(k)]^2 - (1/n)\mathbf{1}_n\mathbf{1}_n^T \end{aligned}$$

and therefore, ρ is also given as the largest eigenvalue of the positive semi-definite matrix $E[\boldsymbol{I} - \varepsilon \boldsymbol{L}(k)]^2 - (1/n)\mathbf{1}_n\mathbf{1}_n^T$.

1.6 Simulation Results and Discussions

In this section, we present and discuss the simulation results of the distributed consensus-based scheme.

1.6.1 Distributed Consensus-Based Cooperative Spectrum Sensing

1.6.1.1 Simulation Setup

In the simulations, we assume that all secondary users are experiencing i.i.d. Rayleigh fading without spatial correlation. Each secondary user uses an energy detector. We simulate the output Y of the energy detector directly in our simulations. When the primary user is absent, Y is a random quantity with chi-square distribution. When the primary user is present, Y may be denoted as the sum of two independent random variables [40, 41]. The parameters of Y depend on the average SNR in the Rayleigh fading (see (1.3) and (1.4)). The simulations are done in three test conditions. In the first condition, every user has the same average $SNR(\overline{\gamma})$, which is 10 dB. In the second condition, each user has different average $SNR(\overline{\gamma})$ varying from 5 dB to 9 dB. In the third condition, each user has different average $SNR(\overline{\gamma})$ varying from 5 dB to 15 dB. The relevant information of primary users, such as the position, the moving direction and the moving velocity, is unknown to the secondary users.

We compare the performance of the proposed scheme with that of an existing OR-rule cooperative sensing scheme [21, 22, 46], which is better than AND-rule and MAJORITY-rule in many cases of practical interest [22, 46]. In the OR-rule cooperative sensing scheme, each secondary user makes local spectrum sensing decision, which is a binary variable – a "one" denotes the presence of a primary user, and a "zero" denotes its absence. Then, all of the local decisions are sent to a data collector to sum up all local decision values. If the sum is greater than or equal to 1, a primary user is believed to be present.

In the first stage of spectrum sensing, after time synchronization, every secondary user performs energy detection with $TW = 5$ individually to get local measurement

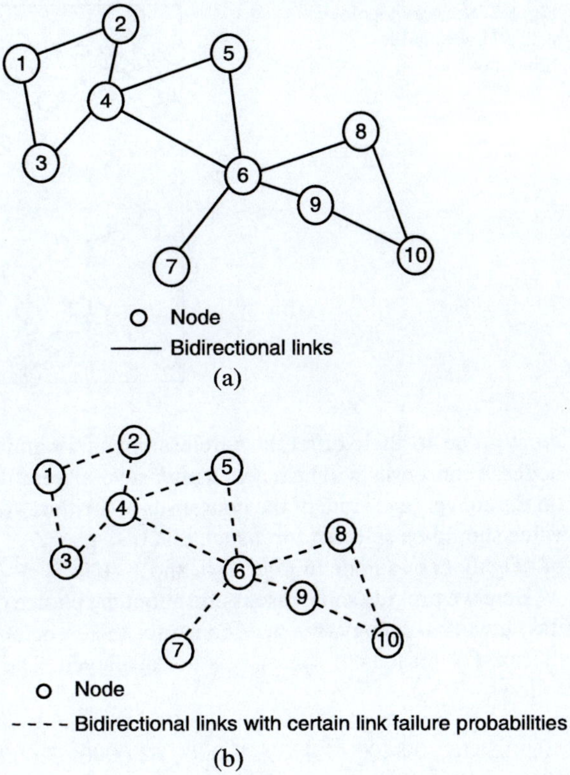

Fig. 1.4 Network topology with 10 nodes in the simulations. (**a**) A fixed graph and (**b**) A random graph

Y_i at the selected center frequency f_s and the bandwidth of interest W. To set up the initial energy vector $\boldsymbol{X}(0)$, we set $x_i(0) = Y_i$.

In the second stage, the existing method and the proposed consensus algorithm (1.7) are conducted based on fixed graph models, while the proposed consensus algorithm (1.13) is run based on random graph models. For fixed graphs, the basic requirement is to set up duplex wireless channels. In the simulations, we consider a network topology with 10 secondary users that establish a graph, $\boldsymbol{G} = \{\mathcal{N}, \mathcal{E}\}$, as shown in Fig. 1.4a. For random graphs, we use the same set of nodes as in Fig. 1.4b, but replace solid lines with dotted ones, which have probabilities of link failure of 40% (refer to Fig. 1.4b). The links in those figures stand for bidirectional wireless links. With regard to link failure probabilities, they mean both directions will fail to work in case of link failure. We also consider a network topology with 50 nodes in the simulations, which is shown in Fig. 1.5. All of the 50 nodes are located randomly. The links in the 50-node network have probabilities of failure of 40%.

1.6.1.2 Convergence of the Consensus Algorithm

Figs. 1.6a and 1.6b show the estimated primary user energy in the network with a 10-node fixed graph. We can observe that, although the initially sensed energy varies

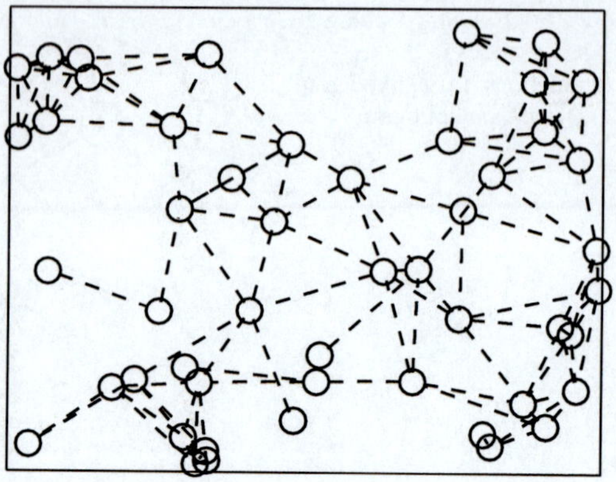

Fig. 1.5 Network topology with 50 nodes in the simulations

greatly due to their different wireless channel conditions for different secondary nodes, a consensus will be reached after several iterations. The step size ε has effects on the convergence rate of the consensus algorithm. According to (1.7) and (1.13), a value should be selected for ε such that $0 < \varepsilon < \Delta^{-1}$. Since the maximum number of neighbors of a node in Figs. 1.4a and 1.4b is 5, $\Delta = 5$. Then, $0 < \varepsilon < 0.2$.

Here we provide some discussion about the choice of the parameter ε. First, given the network topology, we may construct the associated Laplacian $\boldsymbol{L}$ as a 10×10 matrix. For reasons of space, $\boldsymbol{L}$ is not displayed. The eigenvalue of $\boldsymbol{L}$ are listed as follows:

$$0,\ 0.3416,\ 0.8400,\ 1.4239,\ 2.0000,\ 2.0000,\ 3.0000,\ 3.1373,\ 4.9411,\ 6.3161 \tag{1.21}$$

On the interval $(0,\ 0.2)$, the spectral gap $g(\varepsilon)$ may be shown to be

$$g(\varepsilon) = 1 - 0.3416\varepsilon \tag{1.22}$$

which monotonically decreases on $(0,\ 0.2)$. We note that for this specific network topology, when $\varepsilon = 0.2$, the resulting matrix $\boldsymbol{P} = \boldsymbol{I} - \varepsilon\boldsymbol{L}$ is ergodic. On the interval $(0, 0.2]$ the spectral gap is maximized at $\varepsilon = 0.2$.

In below we select two values for ε, 0.1 and 0.19, in Fig. 1.6a and Fig. 1.6b, respectively. We can see that the algorithm converges faster when $\varepsilon = 0.19$ than that when $\varepsilon = 0.1$, which is due to the fact that $\varepsilon = 0.19$ corresponds to a larger spectral gap $g(0.19)$.

After about five iterations in Fig. 1.6b, the difference between the nodes is less than 1 dB, which indicates that a consensus is achieved. Figure 1.7 shows the estimated primary user energy in the network with a random graph when $\varepsilon = 0.19$. Comparing Fig. 1.7 with Fig. 1.6b, we can see that the algorithm converges more slowly in the random graph case due to the random link failure in the CR network. In Fig. 1.7, after about 10 iterations, the difference between the nodes is less than 1 dB, which indicates that a consensus is achieved.

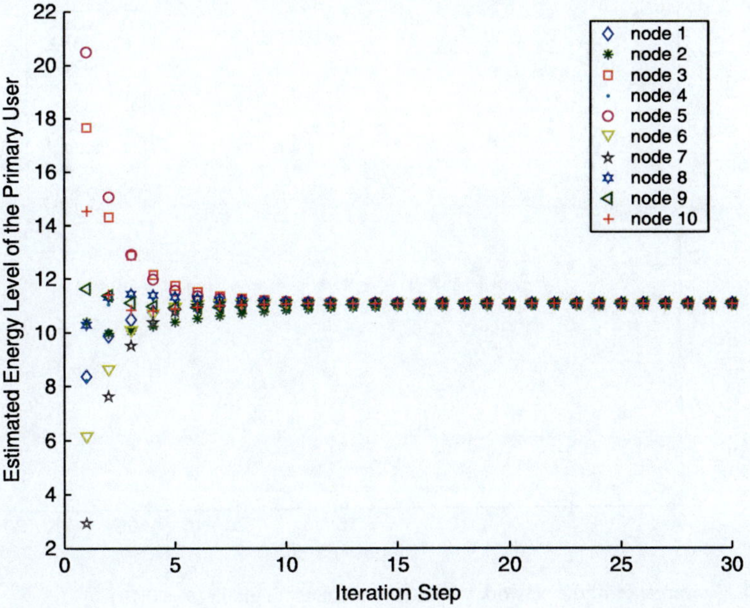

(a)

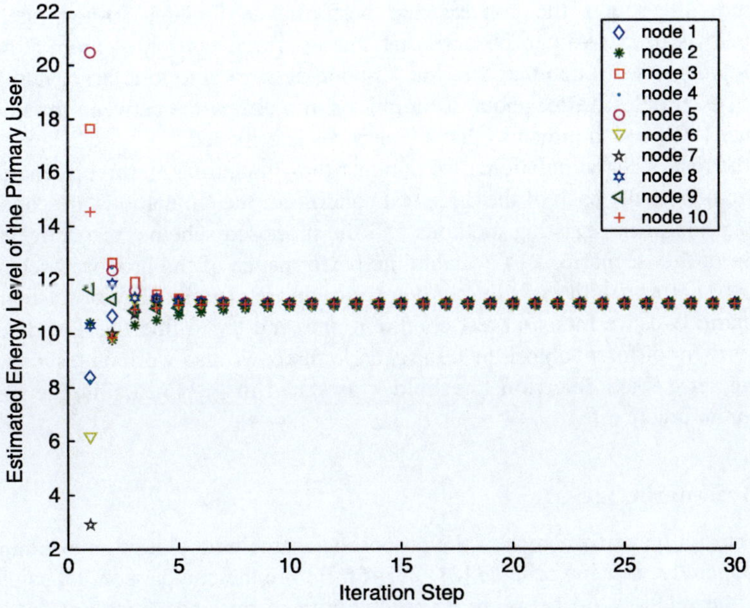

(b)

Fig. 1.6 Convergence of the network with a 10-node fixed graph. (**a**) Fixed graph ($\varepsilon = 0.1$) and (**b**) fixed graph ($\varepsilon = 0.19$)

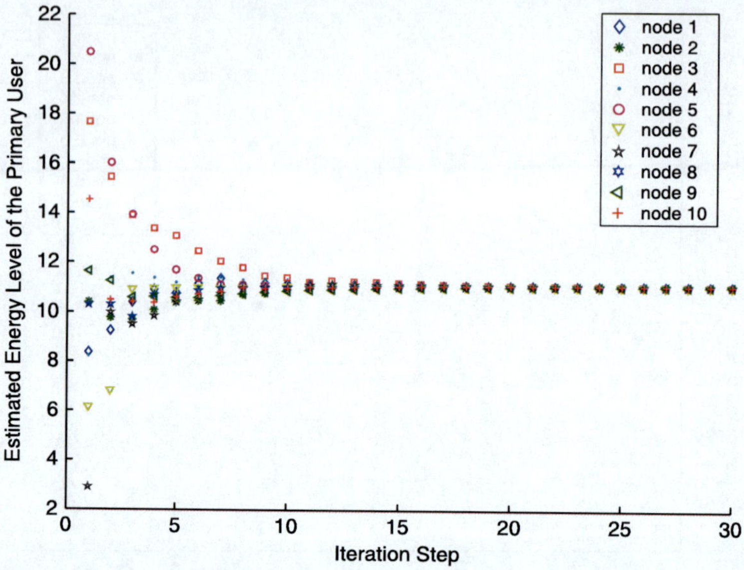

Fig. 1.7 Convergence of the network with a 10-node random graph ($\varepsilon = 0.19$)

Figure 1.8 shows the convergence performance for the 50-node network. $\varepsilon = 0.15$ is used. We can observe that the algorithm converges more slowly in the 50-node network compared to the 10-node network due to a larger number of nodes. Nevertheless, after about 30 iterations, the difference between the nodes is less than 1 dB, which indicates that a consensus is achieved.

In the rest of the simulations, we conduct the simulations in three scenarios. In scenario one, under each of the three test conditions, the simulations are conducted by using one of the existing methods and the proposed scheme, respectively. The purpose of this scenario is to evaluate the performance of the proposed scheme in terms of P_m (probability of missing detection) and P_f (probability of false alarm). In scenario two, we focus on test condition one, and try to find the best detection sensitivity for different algorithms. In scenario three, we also work on test condition one and set a fixed detection threshold λ as stated in (1.11) to simulate the real situation in practice.

1.6.1.3 Scenario One

We compare the performance of the proposed scheme with that of an existing OR-rule cooperative sensing scheme [21, 22, 46]. Before the comparison, let us discuss briefly the relationship between P_m (probability of missing detection)= $1 - P_d$ (probability of detection) and P_f (probability of false alarm). The fundamental tradeoff between P_m and P_f has different implications in the context of spectrum sensing [5]. A high P_m will result in the missing detection of primary users with high probability, which in turn increases the interference to primary users. On the other

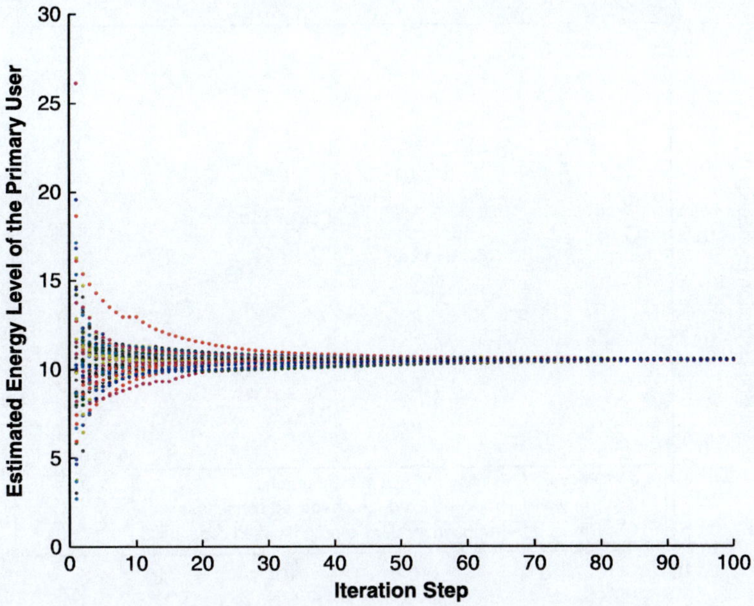

Fig. 1.8 Convergence of the network with a 50-node random graph ($\varepsilon = 0.15$)

hand, a high P_f will result in low spectrum utilization since false alarms increase the number of missed opportunities (white spaces). As expected, P_f is independent of γ since under H_0 there is no primary signal.

Figures 1.9 and 1.10 show P_f vs. P_m. We can see that the proposed algorithm has better performance than the existing OR-rule cooperative sensing scheme. The numbers beside the curves are the corresponding thresholds λ in dB. In Fig. 1.9, where each secondary user has the same average SNR 10 dB, if the threshold λ is in the range of 11.4 to 12 dB, both P_f and P_m can simultaneously drop below the probability of 10^{-2} for the proposed consensus algorithm in both fixed and random graphs. Also, the results are the same between the fixed and random models. In comparison, to reach the same goal, the existing OR-rule method must set λ to be around 14.8 dB, which has far worse P_m (10^{-2} vs. 10^{-3}) with regard to the same P_f level (10^{-2}).

In condition two, secondary users undergo different average SNR varying from 5 dB to 9 dB. In condition three, secondary users undergo different average SNR varying from 5 dB to 15 dB. The similar results are demonstrated in Figs. 1.10 and 1.11 for conditions two and three, respectively.

1.6.2 Scenario Two

Next, we examine the performance of detection probabilities P_d to find out the sensitivity in detecting the primary user's presence. Figure 1.12 shows P_d (detection

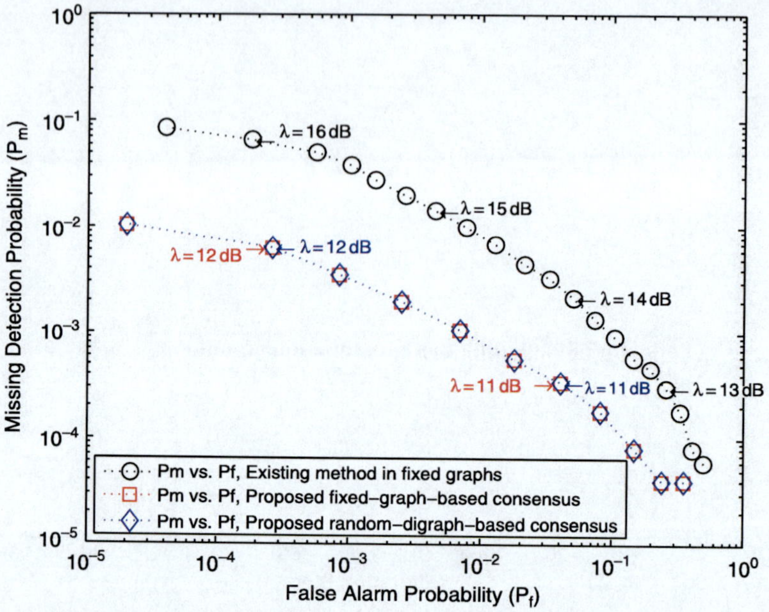

Fig. 1.9 Results in simulation scenario one under test condition one: Missing detection probability (P_m) vs. false alarm probability (P_f) (Each secondary user has the same average SNR, $\overline{\gamma} = 10\,\text{dB}$)

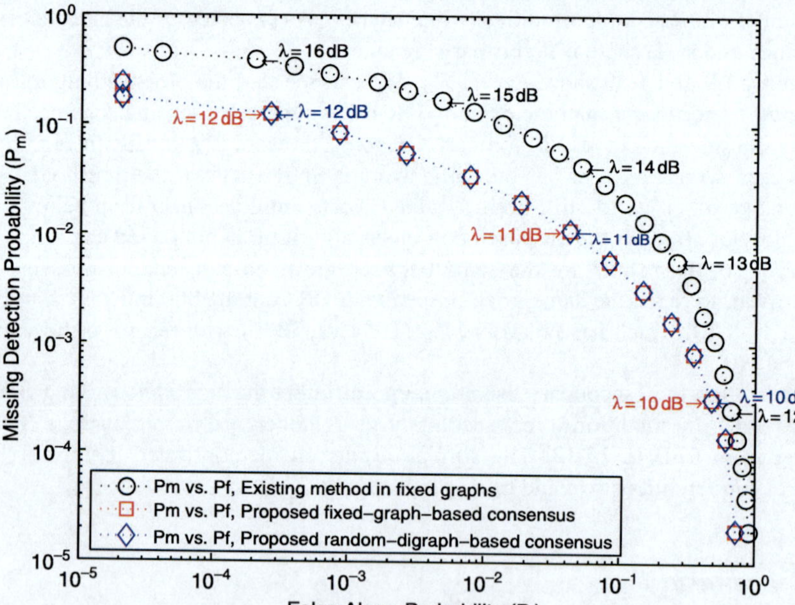

Fig. 1.10 Results in simulation scenario one under test condition two: Missing detection probability (P_m) vs. false alarm probability (P_f) (Each secondary user has different average SNR varying from 5 to 9 dB)

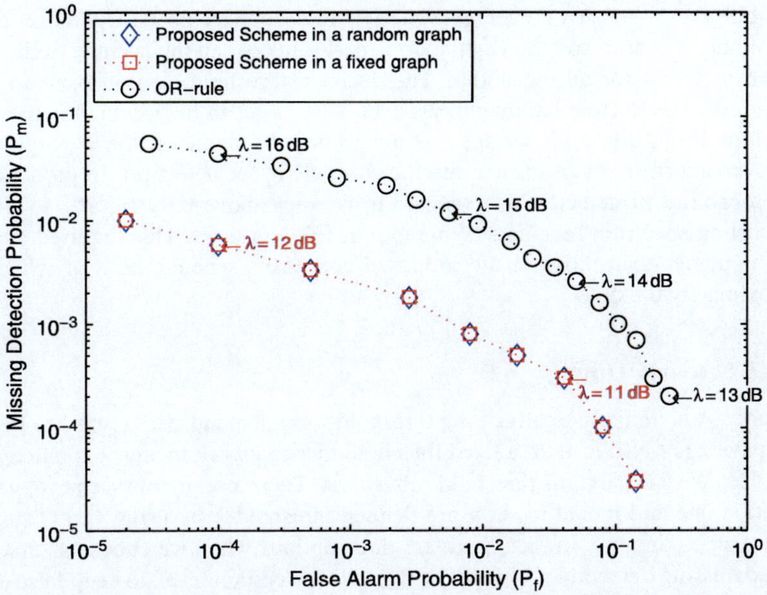

Fig. 1.11 Results in simulation scenario one under test condition three: Missing detection probability (P_m) vs. false alarm probability (P_f) (Each secondary user has different average SNR varying from 5 to 15 dB)

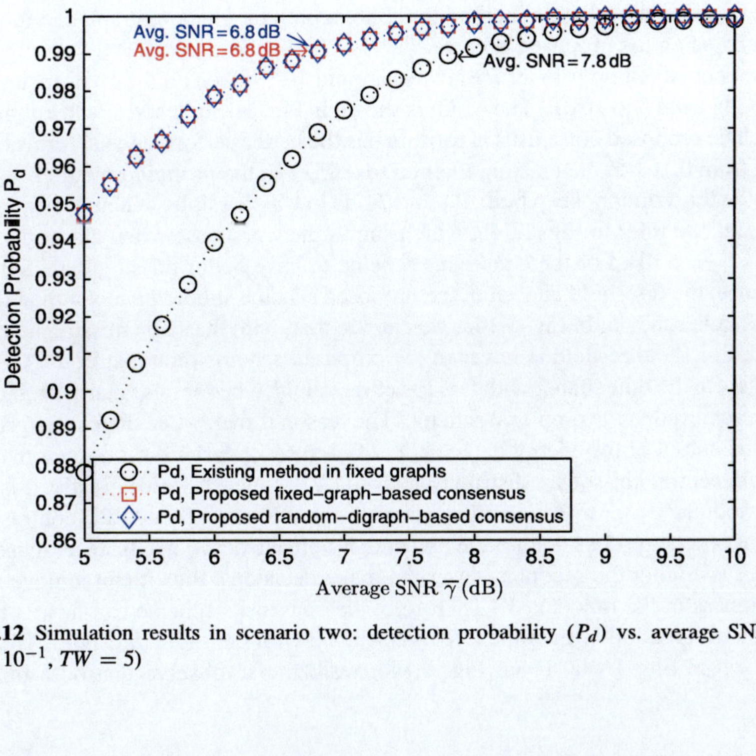

Fig. 1.12 Simulation results in scenario two: detection probability (P_d) vs. average SNR ($\bar{\gamma}$) ($P_f = 10^{-1}$, $TW = 5$)

probability = $1 - P_m$) vs. average SNR ($\bar{\gamma}$) of secondary users. Condition one is used in this scenario, and the simulation is performed when the average SNR varies from 5 to 10 dB for all the nodes. The decision threshold, λ, is chosen so as to keep $P_f = 10^{-1}$. Time-bandwidth product, TW, is set to be 5, which is the same as before. From Fig. 1.12, we see that the proposed scheme can have a significant improvement in terms of the required average SNR for detection. In particular, if the probability of detection is expected to be kept above 0.99 (or $P_m < 10^{-2}$), the existing spectrum sensing scheme requires $\bar{\gamma} = 7.8$ dB. This required average SNR is higher than those in the proposed consensus scheme, both of which are approximately 6.8 dB.

1.6.2.1 Scenario Three

In reality, it is unlikely to adjust the threshold λ on demand with regard to the different average SNR. Rather, a fixed threshold that can work in any $\bar{\gamma}$ is much more desirable. We can call it as threshold robustness. Therefore, in this scenario, we use condition one and intend to set a pre-defined threshold λ by using (1.11) so as to achieve a certain goal. In fact, there are three options when we choose such a goal to keep missing detection probability (P_m) below a certain level, to keep false alarm probability P_f around a certain level, or to keep both P_m and P_f as low as possible.

We first try to keep P_m below 10^{-2} when all the 10 users undergo the same $\overline{\gamma}$ varying from 5 to 10 dB. Fig. 1.13a shows a fixed λ that lets P_m below 10^{-2} for the average SNR ranging from 5 to 10 dB. As the result, the worst P_f decreases from 0.586 by using the existing method to 0.356 in both the random graph and the fixed graph by using the proposed scheme.

The second option is to let P_f always around 10^{-1} when all the 10 users undergo $\overline{\gamma}$ varying from 5 to 10 dB. The result is shown in Fig. 1.13b, where P_f keeps around 10^{-1}. The proposed consensus algorithm has the better performance in terms of P_m, down from 0.161 in the existing method to 0.0527 in the proposed method.

In the third option, keep both P_m and P_f as low as possible. When determining a threshold, we refer to Fig. 1.14a, which shows the worst case when all the 10 users suffers $\overline{\gamma} = 5$ dB. For the consensus scheme to have better missing detection performance, the threshold chosen in the proposed scheme should be lower than that in the OR-rule scheme. In Fig. 1.14a, we can see that, with the same missing detection probability, the threshold is lower in the proposed scheme than that in the OR-rule scheme. On the other hand, with this lower threshold, a better false alarm probability can be achieved in the proposed scheme. The reason is that, when there is no primary user, the output of the energy detector, Y, of each secondary user is a random quantity with central chi-square distribution (see (1.2)). Since Y varies greatly, it is easy for a secondary user to have a false alarm in the OR-rule scheme. By contrast, the consensus scheme does not use the raw data Y to make decisions. Instead, it uses the consensus among the secondary users to make decisions, thus it can remove some randomness in the raw data Y. Therefore, the consensus scheme can have a better false alarm probability than the OR-rule scheme with the same threshold. This can be shown in Fig. 1.14a. From Fig. 1.14a, we can also observe that both missing

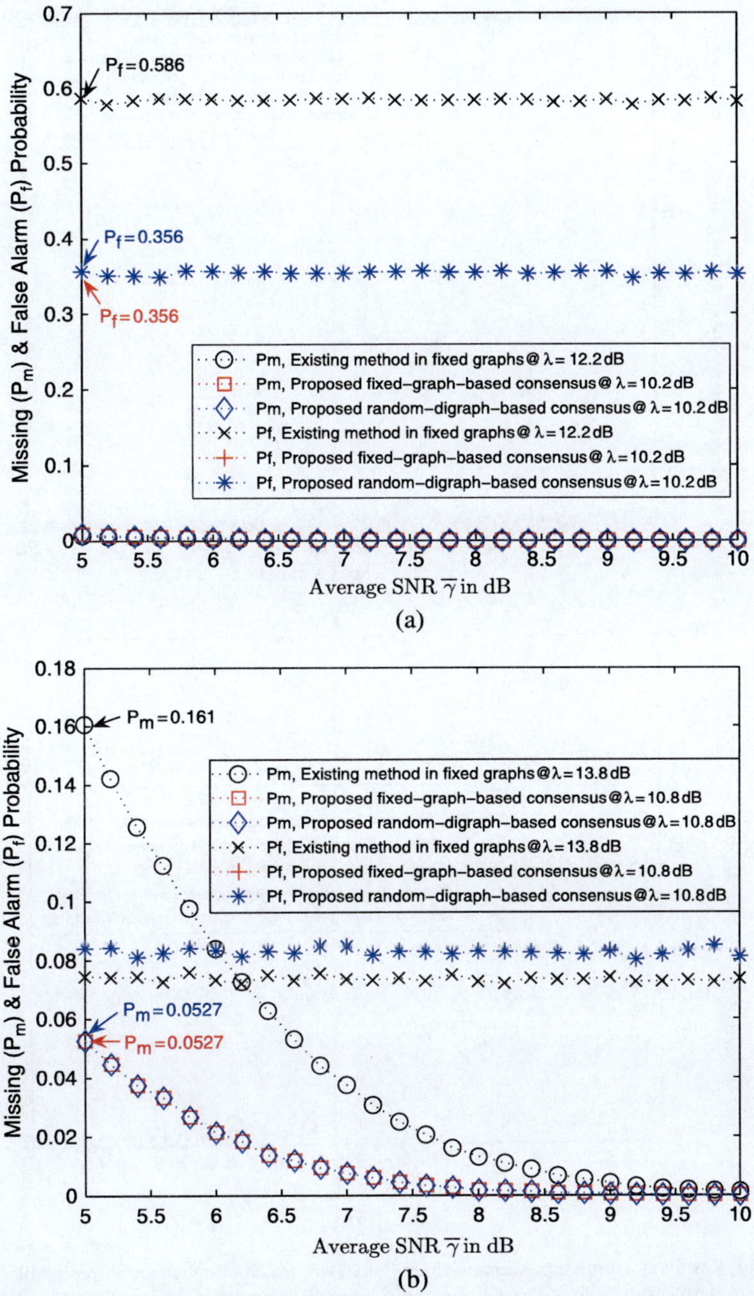

Fig. 1.13 Results in simulation scenario three: Part One. (**a**) Missing detection probability (P_m) and false alarm probability (P_f) vs. average SNR ($\overline{\gamma}$) with fixed threshold λ to keep P_m below 10^{-2}, when all the ten users undergo same $\overline{\gamma}$ varying from 5 to 10 dB and (**b**) Missing detection probability (P_m) and false alarm probability (P_f) vs. average SNR ($\overline{\gamma}$) with fixed threshold λ to keep P_f below 10^{-1}, when all the ten users undergo same $\overline{\gamma}$ varying from 5 to 10 dB

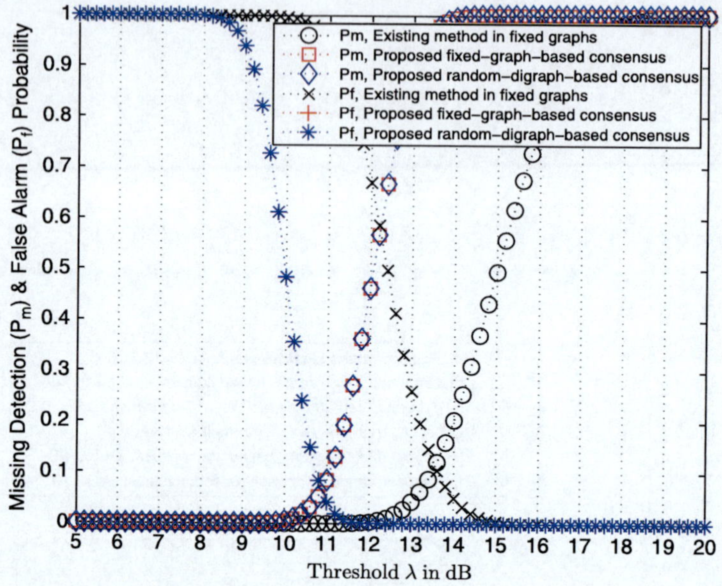

(a)

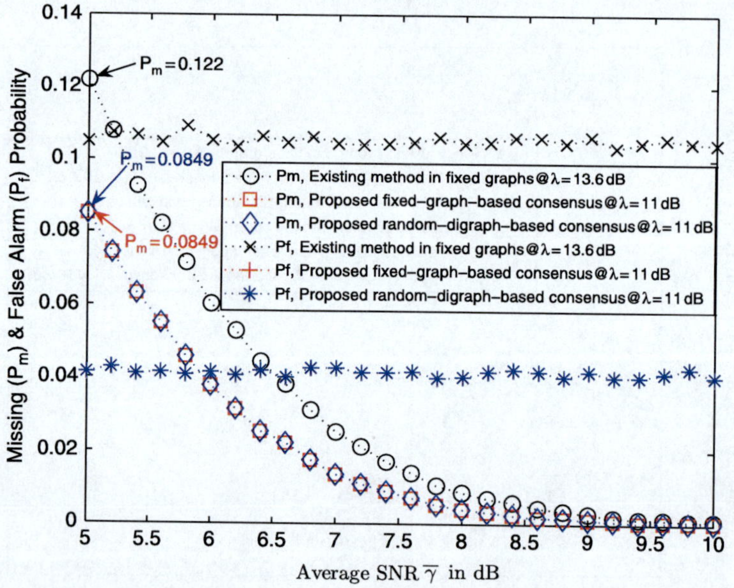

(b)

Fig. 1.14 Results in simulation scenario three: Part Two. (**a**) Missing detection probability (P_m) and false alarm probability (P_f) vs. threshold λ in dB when same $\overline{\gamma} = 5$dB for all users. (**b**) Missing detection probability (P_m) and false alarm probability (P_f) vs. average SNR ($\overline{\gamma}$) with fixed threshold λ to keep both P_m and P_f below a certain level, when all the 10 users undergo same $\overline{\gamma}$ varying from 5 to 10 dB detection and false alarm probabilities are low when the threshold is round 11 dB for the consensus scheme and when the threshold is around 13.6 dB for the OR-rule scheme. In Fig. 1.14a, if we compare the performance of the consensus scheme with a threshold 11 dB to that of the OR-rule scheme with a threshold 13.6 dB, we can see that both missing detection and false alarm probabilities are lower in the consensus scheme than those in the OR-rule scheme. We choose $\lambda = 11$ dB for the proposed consensus algorithm and $\lambda = 13.6$ dB for the existing method to conduct our numerical studies. Fig. 1.14b illustrates the result of such a fixed λ. It is seen that both P_m and P_f have better performance for the proposed algorithm than those of the existing method. P_m and P_f drops to a relatively low level. This highlights the overall advantage in so-called threshold robustness for the proposed consensus algorithm. That is, for a given λ, the proposed consensus algorithm can output less P_m and P_f than those of the existing method. The algorithm works well in both fixed graphs and random ones. Another observation in scenario three is, when the average SNR rises, P_m drops for a given threshold λ, but P_f remains more or less at the same level. This means, for a fixed λ, P_m is subject to the change of the average SNR. In contrast, P_f is stable, because this parameter deals with the condition of H_0, where only the collective noises exist.

1.7 Conclusion

In this chapter, we have presented a fully distributed and scalable scheme for spectrum sensing based on recent advances in consensus algorithms. Cooperative spectrum sensing is modeled as a multi-agent coordination problem. Secondary users can maintain coordination based on only local information exchange without a centralized receiver. Simulation results were presented to show the effectiveness of the proposed consensus-based scheme. It is shown that both missing detection probability and false alarm probability can be significantly reduced in the proposed scheme compared to those in the existing schemes.

Also, as the real network topologies undergo random changes and the primary user may randomly enter and leave the network, a protocol is necessary to quickly decide when the consensus is considered to be practical reached. If the secondary users cannot efficiently form a decision in finite steps, the energy measurements obtained at the beginning may become obsolete. To address this finite time detection issue, in implementations a certain toleration threshold may be used by the users. A secondary user may stop the iteration if it finds the difference between the states of each neighbor and itself has fallen below the threshold. The choice of threshold depends on empirical studies. Our simulation indicates that the threshold may be chosen to be around a fraction of 1 dB or close to 1 dB.

One limitation of the proposed scheme is that the choice of the step size ε depends on the maximum number of neighbors of a node in the network. In other words, each node needs to have the prior knowledge of an upper bound of the maximum degree of the network. To solve this problem, an alternative approach may be used, which is based on so-called Metropolis weights [34]. This approach does not need the knowledge of the maximum degree of the network. Future work is in progress in this direction. We also want to simplify the data format of detection statistics from each secondary user to save the wireless bandwidth. In addition, as energy detection does not work well for spread spectrum signals, other approaches will be studied to deal with such networks.

References

1. J. Mitola, *Cognitive radio: An integrated agent architecture for software defined radio*. Doctor of Technology Thesis, Royal Inst. Technol. (KTH), Stockholm, Sweden, 2000.
2. G. Ganesan and Y. Li, "Cooperative spectrum sensing in cognitive radio, part I: Two user networks," *IEEE Trans. Wireless Commun.*, vol. 6, pp. 2204–2213, 2007.
3. S. Haykin, "Cognitive radio: Brain-empowered wireless communications," *IEEE J. Sel. Areas Commun.*, vol. 23, pp. 201–220, 2005.
4. C. Sun, W. Zhang, and K. B. Letaief, "Cluster-based cooperative spectrum sensing in cognitive radio systems," in *Proc. IEEE ICC'07*, pp. 2511–2515, 2007.
5. A. Ghasemi and E. Sousa, "Collaborative spectrum sensing for opportunistic access in fading environments," in *Proc. IEEE DySPAN'05*, pp. 131–136, 2005.
6. D. Cabric, S. Mishra, and R. Brodersen, "Implementation issues in spectrum sensing for cognitive radios," in *Proc. Thirty-Eighth Asilomar Conference on Signals, Systems and Computers*, vol. 1, pp. 772–776, 2004.
7. J. Hillenbrand, T. Weiss, and F. Jondral, "Calculation of detection and false alarm probabilities in spectrum pooling systems," *IEEE Commun. Lett.*, vol. 9, no. 4, pp. 349–351, 2005.
8. J.-F. Chamberland and V. V. Veeravalli, "Wireless sensors in distributed detection applications," *IEEE Signal Proc. Mag.*, vol. 24, pp. 16–25, 2007.
9. R. Niu and P. Varshney, "Performance analysis of distributed detection in a random sensor field," *IEEE Trans. Signal Proc.*, vol. 56, no. 1, pp. 339–349, 2008.
10. V. Veeravalli, "Decentralized quickest change detection," *IEEE Trans. Inform. Theory*, vol. 47, no. 4, pp. 1657–1665, 2001.
11. S. Mishra, A. Sahai, and R. Brodersen, "Cooperative sensing among cognitive radios," in *Proc. IEEE ICC'06*, pp. 1658–1663, 2006.
12. W. Ren, R. Beard, and E. Atkins, "A survey of consensus problems in multi-agent coordination," in *Proc. American Control Conference'05*, pp. 1859–1864, 2005.
13. J. Mitola and G. Q. Maguire, "Cognitive radio: Making software radios more personal," *IEEE Pers. Commun.*, vol. 6, pp. 13–18, 1999.
14. I. Akyildiz, W. Lee, M. Vuran, and S. Mohanty, "Next generation/dynamic spectrum access/cognitive radio wireless networks: A survey," *Comput Networks*, vol. 50, no. 13, pp. 2127–2159, 2006.
15. G. Ganesan and Y. Li, "Cooperative spectrum sensing in cognitive radio - part II: Multiuser networks," *IEEE Trans. Wireless Commun.*, vol. 6, pp. 2214–2222, 2007.
16. G. Ganesan and Y. G. Li, "Agility improvement through cooperative diversity in cognitive radio," in *Proc. IEEE GLOBECOM'05*, pp. 2505–2509, 2005.
17. E. Peh and Y.-C. Liang, "Optimization for cooperative sensing in cognitive radio networks," in *Proc. IEEE WCNC'07*, pp. 27–32, 2007.
18. J. Unnikrishnan and V. V. Veeravalli, "Cooperative sensing for primary detection in cognitive radio," *IEEE J. Sel. Topics Signal Proc.*, vol. 2, no. 1, pp. 18–27, 2008.
19. Z. Quan, S. Cui, and A. H. Sayed, "Optimal linear cooperation for spectrum sensing in cognitive radio networks," *IEEE J. Sel. Topics Signal Proc.*, vol. 2, no. 1, pp. 28–40, 2008.
20. Y.-C. Liang, Y. Zeng, E. Peh, and A. T. Hoang, "Sensing-throughput tradeoff for cognitive radio networks," *IEEE Trans. Wireless Commun.*, vol. 7, no. 4, pp. 1326–1337, 2008.

21. R. Chen, J.-M. Park, and K. Bian, "Robust distributed spectrum sensing in cognitive radio networks," in *Proc. INFOCOM 2008. The 27th Conference on Computer Communications. IEEE*, pp. 1876–1884, 2008.
22. W. Zhang and K. Ben Letaief, "Cooperative communications for cognitive radio networks," *Proc. IEEE*, vol. 97, no. 5, pp. 878–893, 2009.
23. C. S. R. Murthy and B. S. Manoj, *Ad Hoc Wireless Networks: Architectures and Protocols*. Upper Saddle River, NJ: Prentice Hall, 2004.
24. T. Nakano and T. Suda, "Applying biological principles to designs of network services," *Appl. Soft Comput.*, vol. 7, no. 3, pp. 870–878, 2007.
25. I. Carreras, I. Chlamtac, F. D. Pellegrini, and D. Miorandi, "Bionets: Bio-inspired networking for pervasive communication environments," *IEEE Trans. Veh. Technol.*, vol. 56, pp. 218–229, 2007.
26. F. Dressler, Ö. B. Akan, and A. Ngom, "Guest Editorial - Special Issue on Biological and Biologically-inspired Communication," *Springer Trans. on Computational Systems Biology (TCSB)*, vol. LNBI 5410, 2008.
27. R. Olfati-Saber, J. Fax, and R. Murray, "Consensus and cooperation in networked multi-agent systems," *Proc. IEEE*, vol. 95, no. 1, pp. 215–233, 2007.
28. J.-M. Amé, J. Halloy, C. Rivault, C. Detrain, and J. L. Deneubourg, "Collegial decision making based on social amplification leads to optimal group formation," *Proc. Natl. Acad. Sci.*, vol. 103, no. 15, pp. 5835–5840, 2006.
29. L. Conradt and T. J. Roper, "Consensus decision making in animals," *Trends Ecol. Evol.*, vol. 20, pp. 449–456, 2005.
30. T. Vicsek, "A question of scale," *Nature*, vol. 441, p. 421, 2001.
31. I. D. Couzin, "Collective cognition in animal groups," *Trends Cogn. Sci.*, vol. 13, pp. 36–43, 2008.
32. P. K. Visscher, "How self-organization evolves?" *Nature*, vol. 421, pp. 799–800, 2003.
33. W. Ren and R. Beard, "Consensus seeking in multiagent systems under dynamically changing interaction topologies," *IEEE Trans. Auto. Control*, vol. 50, no. 5, pp. 655–661, 2005.
34. L. Xiao, S. Boyd, and S. Lall, "A scheme for robust distributed sensor fusion based on average consensus," in *Proc. Fourth International Symposium on Information Processing in Sensor Networks*, pp. 63–70, 2005.
35. M. Huang and J. H. Manton, "Stochastic consensus seeking with measurement noise: Convergence and asymptotic normality," in *Proc. American Control Conference'08*, pp. 1337–1342, 2008.
36. W. Irving and J. Tsitsiklis, "Some properties of optimal thresholds in decentralized detection," *IEEE Trans. Auto. Control*, vol. 39, no. 4, pp. 835–838, 1994.
37. J. Proakis and M. Salehi, *Digital Communications*. New York, NY: McGraw-hill 1995.
38. A. Sahai, N. Hoven, and R. Tandra, "Some fundamental limits on cognitive radio," in *Allerton Conference on Communication, Control, and Computing*, Citeseer, 2004.
39. H. Urkowitz, "Energy detection of unknown deterministic signals," *Proc. IEEE*, vol. 55, no. 4, pp. 523–531, 1967.
40. F. Digham, M.-S. Alouini, and M. Simon, "On the energy detection of unknown signals over fading channels," in *Proc. IEEE ICC'03*, vol. 5, pp. 3575–3579, 2003.
41. V. Kostylev, "Energy detection of a signal with random amplitude," in *IEEE Proc. ICC'02*, vol. 3, pp. 1606–1610, 2002.
42. M. Huang and J. H. Manton, "Coordination and consensus of networked agents with noisy measurements: Stochastic algorithms and asymptotic behavior," *SIAM J. Control and Optimization*, vol. 48, pp. 134–161, 2009.
43. C. Godsil and G. Royle, *Algebraic Graph Theory*. New York, NY: Springer, 2001.
44. E. Seneta, *Non-negative Matrices and Markov Chains*. New York, NY: Springer, 1981.
45. L. Elsner, I. Koltracht, and M. Neumann, "On the convergence of asynchronous paracontractions with applications to tomographic reconstruction from incomplete data," *Linear Algebra and its Applications*, vol. 130, pp. 65–82, 1990.
46. A. Ghasemi and E. Sousa, "Opportunistic spectrum access in fading channels through collaborative sensing," *J Commun*, vol. 2, no. 2, p. 71, 2007.

Chapter 2
On the Spectrum Handoff for Cognitive Radio Ad Hoc Networks Without Common Control Channel

Yi Song and Jiang Xie

Abstract Cognitive radio (CR) technology is a promising solution to enhance the spectrum utilization by enabling unlicensed users to exploit the spectrum in an opportunistic manner. Since unlicensed users are considered as temporary visitors to the licensed spectrum, they are required to vacate the spectrum when a licensed user reclaims it. Due to the randomness of the appearance of licensed users, disruptions to both licensed and unlicensed communications are often difficult to prevent, which may lead to low throughput of both licensed and unlicensed communications. In this chapter, a proactive spectrum handoff framework for CR ad hoc networks is proposed to address these concerns. In the proposed framework, channel switching policies and a proactive spectrum handoff protocol are proposed to let unlicensed users vacate a channel *before* a licensed user utilizes it to avoid unwanted interference. Network coordination schemes for unlicensed users are also incorporated into the spectrum handoff protocol design to realize channel rendezvous. Moreover, a distributed channel selection scheme to eliminate collisions among unlicensed users in a multi-user spectrum handoff scenario is proposed. In our proposed framework, unlicensed users coordinate with each other without using a common control channel, which is highly adaptable in a spectrum-varying environment. We compare our proposed proactive spectrum handoff protocol with a reactive spectrum handoff protocol, under which unlicensed users switch channels *after* collisions with licensed transmissions occur under different channel coordination schemes. Simulation results show that our proactive spectrum handoff outperforms the reactive spectrum handoff approach in terms of higher throughput and fewer collisions to licensed users. Furthermore, our distributed channel selection can achieve substantially higher packet delivery rate in a multi-user spectrum handoff scenario, compared with existing channel selection schemes. In addition, we propose a novel three-dimensional discrete-time Markov chain to characterize the process of reactive spectrum handoffs and analyze the performance of unlicensed users. We validate the numerical results obtained from our proposed Markov model against simulation

J. Xie (✉)
The University of North Carolina at Charlotte, Charlotte, NC, USA
e-mail: Linda.Xie@uncc.edu

F.R. Yu (ed.), *Cognitive Radio Mobile Ad Hoc Networks*,
DOI 10.1007/978-1-4419-6172-3_2,

and investigate other parameters of interest in the spectrum handoff scenario. Our proposed analytical model can be applied to various practical network scenarios.

2.1 Introduction

The rapid growth of wireless devices has led to a dramatic increase in the need of spectrum access from wireless services. However, according to Federal Communications Commission (FCC) [1], up to 85% of the assigned spectrum is underutilized due to the current fixed spectrum allocation policy. In order to overcome the imbalance between the increase in the spectrum access demand and the inefficiency in the spectrum usage, FCC has suggested a new paradigm for dynamically accessing the assigned spectrum where the spectrum is not used [2]. Cognitive radio (CR) is a key technology to realize dynamic spectrum access (DSA) that enables an unlicensed user (or, secondary user) to adaptively adjust its operating parameters and exploit the spectrum which is unused by licensed users (or, primary users) in an opportunistic manner [3].

The CR technology allows secondary users (SUs) to seek and utilize "spectrum holes" in a time and location-varying radio environment without causing harmful interference to primary users (PUs). This opportunistic use of the spectrum leads to new challenges to make the network protocols adaptive to the varying available spectrum [4]. Specifically, one of the most important functionalities of CR networks is *spectrum mobility*, which enables SUs to change the operating frequencies based on the availability of the spectrum. Spectrum mobility gives rise to a new type of handoff called *spectrum handoff*, which refers to the process that when the current channel used by a SU is no longer available, the SU needs to pause its ongoing transmission, vacate that channel, and determine a new available channel to continue the transmission. Compared with other functionalities (*spectrum sensing*, *spectrum management*, and *spectrum sharing*) [4] of CR networks, spectrum mobility is less explored in the research community. However, due to the randomness of the appearance of PUs, it is extremely difficult to achieve fast and smooth spectrum transition leading to minimum interference to legacy users and performance degradation of secondary users during a spectrum handoff. This problem becomes even more challenging in ad hoc networks where there is no centralized entity (e.g., a spectrum broker [4]) to control the spectrum mobility.

2.1.1 Spectrum Handoff in Cognitive Radio Networks

Related work on spectrum handoffs in CR networks falls into two categories based on the moment when SUs carry out spectrum handoffs. One approach is that SUs perform spectrum switching and radio frequency (RF) front-end reconfiguration *after* detecting a PU [5–9], namely the *reactive* approach. Although the concept of this approach is intuitive, there is a non-negligible sensing and reconfiguration delay which causes unavoidable disruptions to both the PU and the SU transmissions. Another approach is that SUs predict the future channel availability status and perform spectrum switching and RF reconfiguration *before* a PU occupies the channel based on observed channel usage statistics [10–13], namely the *proactive* approach. This approach can dramatically reduce the collisions between SUs and PUs by letting SUs vacate channels before a PU reclaims the channel. In the existing proposals of the proactive approach, a predictive model for dynamic spectrum access based on the past channel usage history is proposed in [10]. A cyclostationary detection and Hidden Markov Models for predicting the channel idle times are proposed in [11]. In [12], a binary time series for the spectrum occupancy characterization and prediction is proposed. In [13], a novel spectrum handoff scheme called voluntary spectrum handoff is proposed to minimize SU disruption periods during spectrum handoffs. In [14], the error of prediction of the channel usage is considered in designing an intelligent dynamic spectrum access mechanism. In [15], an experimental cognitive radio test bed is presented. It uses sensing and channel usage prediction to exploit temporal white space between primary WLAN transmissions.

2.1.2 Common Control Channel in Cognitive Radio Networks

A common control channel (CCC) is used for supporting the network coordination and channel-related information exchange among SUs. In the prior proposals of the above two spectrum handoff approaches, the network coordination and rendezvous issue (i.e., before transmitting a packet between two nodes, they first find a common channel and establish a link) is either not considered [7, 8, 11, 12, 14, 15] or simplified by using a global common control channel (CCC) [5, 6, 10, 13]. A SU utilizing a channel without coordinating with other SUs may lead to the failure of link establishment [16]. Therefore, network coordination has significant impact on the performance of SUs. Although a global CCC simplifies the network coordination among SUs [17], there are several limitations when using this approach in CR networks. First of all, it is difficult to identify a global CCC for all the secondary users throughout the network since the spectrum availability varies with time and location. Second, the CCC is influenced by the primary user traffic because a PU may suddenly appear on the current control channel. For these reasons, IEEE 802.22 [18], the first standard based on the use of cognitive radio technology on the TV band between 41 and 910 MHz, does not utilize a dedicated channel for control signaling, instead dynamically choosing a channel which is not used by legacy users [19].

In this chapter, we investigate the network scenario where no CCC exists and its impact on the spectrum handoff design in CR ad hoc networks. Since when no CCC exists in the network, message exchange among SUs is not always feasible. Thus, the spectrum handoff design becomes more challenging than the scenario with a CCC. Currently, several proposals have been proposed to accomplish network coordination without a CCC in ad hoc networks. Based on the number of users making link agreements simultaneously, the proposed network coordination schemes can be categorized into (1) single rendezvous coordination schemes [20–22] (i.e., only one pair of SUs in a network can exchange control information and establish a link at one time) and (2) multiple rendezvous coordination schemes [23–25] (i.e., multiple pairs of SUs in a network can use different channels to exchange control information and establish multiple links at the same time). Thus, we utilize these two types of network coordination schemes and incorporate them into the spectrum handoff design for CR ad hoc networks.

2.1.3 Channel Selection in Cognitive Radio Networks

Even though the channel allocation issue has been well studied in traditional wireless networks (e.g., cellular networks and wireless local area networks (WLANs)), channel allocation in CR networks, especially in a spectrum handoff scenario, still lacks sufficient research. When SUs perform spectrum handoffs, a well-designed channel selection method is required to provide fairness for all SUs as well as to avoid multiple SUs to select the same channel at the same time. Currently, the channel selection issue in a multi-user CR network is investigated mainly using game theoretic approaches [26–29], while properties of interest during spectrum handoffs, such as SU handoff delay and SU service time, are not studied. Furthermore, most of the prior work on channel allocation in spectrum handoffs [7, 10] only considers a two-secondary-user scenario, where a SU greedily selects the channel which either results in the minimum service time [7] or has the highest probability of being idle [10]. In [13], only one pair of SUs is considered and the channel selection issue is ignored. However, if multiple SUs perform spectrum handoffs at the same time, these channel selection methods will cause definite collisions among SUs. Hence, the channel selection method aiming to prevent collisions among SUs in a multi-secondary-user spectrum handoff scenario is ignored in the prior work.

2.1.4 Analytical Model for Spectrum Handoff in Cognitive Radio Networks

An analytical model is of great importance for performance analysis because it can provide useful insights into the operation of spectrum handoffs. However, there have been limited studies on the performance analysis of spectrum handoffs in CR networks using analytical models. The performance analysis of all prior works on spectrum handoffs is simulation based with the exception of [7] and [9]. In [7] and [9], a preemptive resume priority queueing model is proposed to analyze the total service time of SU communications for proactive and reactive-decision spectrum handoffs. However, in both [7] and [9], only one pair of SUs is considered in a network, while the interference and interactions among SUs are ignored, which may greatly affect the performance of the network. In all the above proposals, a common and severe limitation is that the authors assume that the detection of PUs is perfect (i.e., a SU transmitting pair can immediately perform channel switching if a PU is detected to appear on the current channel, thus the overlapping of SU and PU transmissions is negligible). However, since the power of a transmitted signal is much higher than the power of the received signal in wireless medium due to path loss, instantaneous collision detection is not possible for wireless communications. Thus, even if only a portion of a packet is collided with another transmission, the whole packet is wasted and needs to be retransmitted. Without considering the retransmission, the performance conclusion may be inaccurate, especially in wireless communications. Unfortunately, it is not easy to simply add retransmissions in the existing models. In this chapter, we model the retransmissions of the collided packets in our proposed Markov model.

2.1.5 Contributions

This chapter studies the spectrum handoff issues in cognitive radio networks without the existence of a CCC. The contributions of our work are as follows:

- Due to the spectrum-varying nature of CR networks, we consider more practical coordination schemes instead of using a CCC to realize channel rendezvous. We incorporate two types of channel rendezvous and coordination schemes into the spectrum handoff design and compare the performance of our proposed spectrum handoff protocol with the reactive spectrum handoff approach under different coordination schemes.
- Based on the observed channel usage statistics, we propose proactive spectrum handoff criteria and policies for SUs using a probability-based prediction method. SUs equipped with the prediction capability can proactively predict the idleness probability of the spectrum band in the near future. Thus, harmful interference between SUs and PUs can be diminished and SU throughput is increased. In addition, by considering channel rendezvous and coordination schemes, we propose a proactive spectrum handoff protocol for SUs based on our proposed handoff criteria and policies.
- With the aim of eliminating collisions among SUs and achieving short spectrum handoff delay, we propose a novel distributed channel selection scheme especially designed for multi-user spectrum handoff scenarios. Our proposed channel selection scheme does not involve centralized controller and only need SUs to broadcast their sensed channel availability information once, which drastically reduces the message exchange overhead.
- We propose a novel three-dimensional discrete-time Markov model to characterize the process of reactive spectrum handoffs and analyze the performance of SUs. We implement one of the considered network coordination schemes in our model. Since instantaneous collision detection is not feasible for wireless communications, we consider the retransmissions of the collided SU packets in spectrum handoff scenarios. We also consider the spectrum sensing delay and its impact on the network performance.

2.1.6 Organization

The rest of this chapter is organized as follows. In Section 2.2, network coordination schemes and assumptions considered in this chapter are introduced. In Section 2.3, the details of the proposed proactive spectrum handoff framework are given. In Section 2.4, the algorithm of the proposed distributed channel selection scheme is presented. Simulation results of our proposed spectrum handoff framework are presented in Section 2.5. In Section 2.6, a three-dimensional discrete-time Markov model is proposed, followed by the conclusions in Section 2.7.

2.2 Network Coordination and Assumptions

2.2.1 Single Rendezvous Coordination Scheme

We consider a network scenario where N SUs form a CR ad hoc network and opportunistically access M orthogonal licensed channels. For the single rendezvous coordination scheme[1], we use Common Hopping as the channel coordination scheme [20]. Figure 2.1 illustrates the operations of Common Hopping, under which the channels are time slotted and SUs communicate with each other in a synchronous manner. This is similar to the frequency hopping technique used in Bluetooth [30]. When no packet needs to be transmitted, all the SU devices hop through channels using the same hopping sequence (e.g., the hopping pattern cycles through channels $1, 2, \ldots, M$). The length of a time slot (i.e., the dwelling time on each channel during hopping) is denoted as β. If a pair of SUs wants to initiate a transmission, they first exchange request-to-send (RTS) and clear-to-send (CTS) packets during a time slot. Then, after the SU transmitter successfully receives the CTS packet, they pause the channel hopping and remain on the same channel for data transmissions, while other non-transmitting SUs continue hopping. After the data being successfully transmitted, the SU pair rejoins the channel hopping.

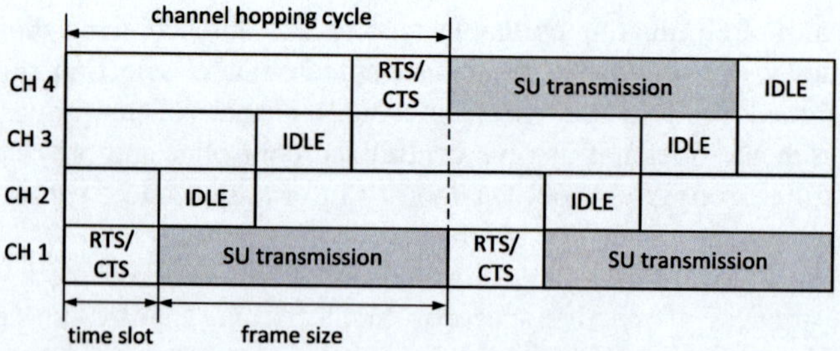

Fig. 2.1 An example of the single rendezvous coordination scheme

[1] Only one pair of SUs can exchange control information and establish a link at one time.

2.2.2 Multiple Rendezvous Coordination Scheme

Unlike in the single rendezvous coordination scheme that only one pair of SUs can make an agreement in one time slot, in the multiple rendezvous coordination scheme, multiple SU pairs can make agreements simultaneously on different channels[2]. A typical example of this type of coordination schemes is McMAC [23]. Figure 2.2 depicts the operations of McMAC. Instead of using the same channel hopping sequence for all the SUs, in McMAC, each SU generates a distinct pseudo-random hopping sequence (e.g., in Fig. 2.2, the channel hopping sequence for user *A* is 2-4-1-3 and for user *B* is 3-2-1-4). When a SU is idle, it follows its default hopping sequence to hop through the channels. If a SU intends to send data to a receiver, it temporarily tunes to the current channel of the receiver and sends a RTS during the time slot (i.e., in Fig. 2.2, SUs *AB* and *CD* are two transmitting pairs that intend to initiate new transmissions at the same time). Then, if the receiver replies with a CTS, both the transmitter and the receiver stop channel hopping and start a data transmission on the same channel. When they finish the data transmission, they resume to their default channel hopping sequences. In this chapter, we consider the scenario where SU nodes are aware of each other's channel hopping sequences [23].

In this chapter, we assume that stringent time synchronization among SUs for channel hopping can be achieved without the need to exchange control messages on a CCC in both cases. We consider a synchronization scheme similar to the one used in [23] that every SU includes a time stamp in every packet it sends. Then, a SU transmitter obtains the clock information of the intended SU receiver by listening to the corresponding channel and estimates the rate of clock drift to realize time synchronization. Various schemes have been proposed to calculate the rate of clock drift for synchronization [31].

In both types of coordination schemes, we assume that any SU data packet is transmitted at the beginning of a time slot and ends at the end of a time slot. This implies that the length of a SU data packet, δ, is a multiple of the time slot. This

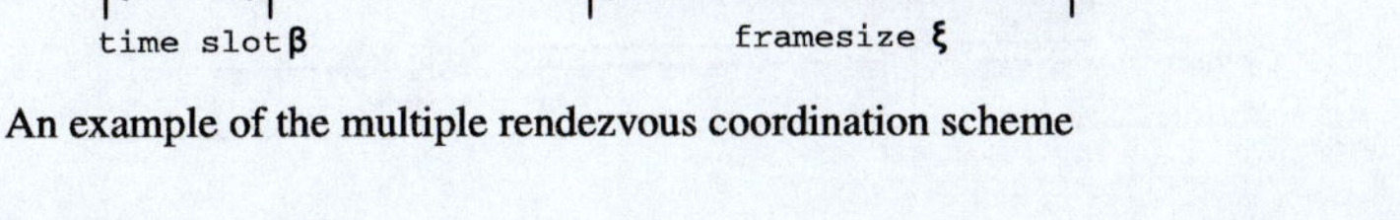

Fig. 2.2 An example of the multiple rendezvous coordination scheme

[2] Multiple pairs of SUs can use different channels to exchange control information and establish multiple links at the same time.

assumption is commonly used in time-slotted systems [32, 33]. We further define that a SU data packet is segmented into frames and each frame contains c time slots. The length of a frame is denoted as ξ, so $\xi = c\beta$. As shown in Fig. 2.1, at the end of a frame, the two SUs can either rejoin the channel hopping when a data transmission ends or start another data transmission by exchanging RTS/CTS packets.

2.2.3 Network Assumptions

In this chapter, we model each licensed channel as an ON-OFF process [12]. As shown in Fig. 2.3, each rectangle represents a PU data packet being transmitted on a channel (i.e., the ON period) and the other blank areas represent the idle periods (i.e., the OFF period). The length of a rectangle indicates the packet length of a PU data packet. Therefore, a SU can only utilize a channel when no PU transmits at the same time. In Fig. 2.3, t_0 represents the time a SU starts channel prediction. Thus, for the ith channel at any future time t $(t > t_0)$, the status of the channel is denoted as $N_i(t)$ which is a binary random variable with values 0 and 1 representing the idle and the busy state, respectively. We also assume that each PU is an $M/G/1$ system [7, 34], that is, the PU packet arrival process follows the Poisson process with the average arrival rate $\bar{\lambda}_i$ and the length of a data packet follows an arbitrary probability density function (pdf) $f_{L_i}(l)$.

Due to the fact that the power of a transmitted signal is much higher than the power of the received signal in wireless medium, instantaneous collision detection is not possible for wireless nodes. Thus, we assume that if a SU frame collides with a PU packet, the wasted frame can only be retransmitted at the end of the frame. In addition, in our proposed spectrum handoff protocol, we assume that each SU is equipped with two radios. One is used for data and control message transmission, namely the transmitting radio. The other is applied to scan all the channels in the band and to obtain the channel occupancy information, namely the scanning radio. The scanning radio has two major functions for the proposed protocol: (1) observe the channel usage and store the channel statistics in the memory for future channel availability prediction and (2) confirm that the newly selected channel is idle for SU transmissions.

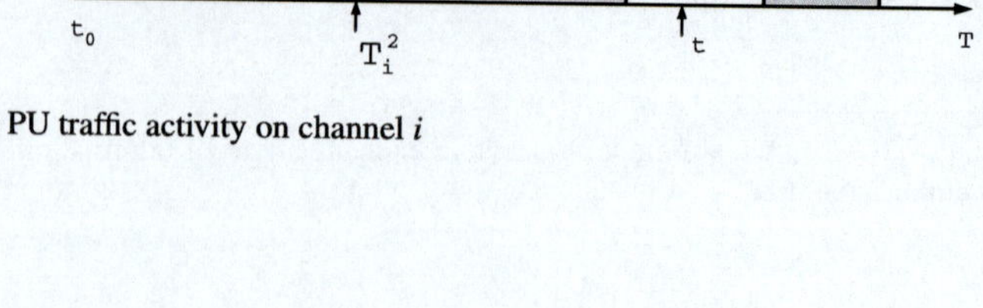

Fig. 2.3 The PU traffic activity on channel i

2.3 Proactive Spectrum Handoff Protocol

2.3.1 Proposed Spectrum Handoff Criteria and Policies

By utilizing the observed channel usage statistics, a SU can make predictions of the channel availability before the current transmission frame ends. Based on the prediction, the SU decides whether to stay in the present channel or switch to a new channel or stop the ongoing transmission. We propose two criteria for determining whether a spectrum handoff should occur: (1) the predicted probability that the current and a candidate channel (i.e., a channel that can be selected for continuing the current data transmission) is busy or idle and (2) the expected length of the channel idle period. Based on these criteria, we design spectrum handoff policies.

Figure 2.3 shows the PU user traffic activity on channel i, where X_i^k and T_i^k represent the inter-arrival time and arrival time of the kth packet, respectively. Consistent with the assumption that PU packets arrive in a Poisson stream fashion [34], X_i^k is exponentially distributed with the average arrival rate $\bar{\lambda}_i$ packets per second and the PU packet length follows the pdf $f_{L_i}(l)$. According to Fig. 2.3, for any future time t, the probability that the ith channel is busy or idle can be written as follows:

$$\begin{aligned}
&\Pr(N_i(t) = 1) \text{ if } && T_i^k < t \text{ and } T_i^k + L_i^k \geq t, && k \geq 1 \\
&\Pr(N_i(t) = 0) \text{ if } && T_i^k + L_i^k < t \text{ and } T_i^{k+1} \geq t, && k \geq 1 \\
& && T_i^{k+1} \geq t, && k = 0
\end{aligned} \tag{2.1}$$

where L_i^k denotes the length of the kth PU data packet on channel i. Therefore, the probability that channel i is idle at any future time t can be obtained by (2.2).

$$\begin{aligned}
\Pr(N_i(t) = 0) &= \int_0^\infty \left[\sum_{k=1}^{\infty} \Pr\left(T_i^k + L_i < t | k\right) \Pr\left(T_i^{k+1} \geq t | k\right) \Pr(k) + \Pr\left(T_i^1 \geq t\right) \Pr(k = 0) \right] f_{L_i}(l)\, \mathrm{d}l \\
&= \int_0^\infty \left\{ \sum_{k=1}^{\infty} \left[\frac{(\bar{\lambda}_i (t - L_i))^k}{k!} \mathrm{e}^{-\bar{\lambda}_i (t - L_i)} \right] \left(\frac{(\bar{\lambda}_i t)^k}{k!} \mathrm{e}^{-\bar{\lambda}_i t} \right) \frac{(\bar{\lambda}_i t)^k}{k!} \mathrm{e}^{-\bar{\lambda}_i t} + \mathrm{e}^{-2\bar{\lambda}_i t} \right\} f_{L_i}(l)\, \mathrm{d}l
\end{aligned} \tag{2.2}$$

Let t_{off} represent the duration of the OFF period. For the ith channel, the cumulative distribution function (CDF) of the duration of the OFF period is

$$\begin{aligned}
\Pr(t_{\text{off}} < x) &= \int_0^\infty \int_0^{l+x} \bar{\lambda}_i \mathrm{e}^{-\bar{\lambda}_i t} f_{L_i}(l)\, \mathrm{d}t \mathrm{d}l \\
&= \int_0^\infty \left(1 - \mathrm{e}^{-\bar{\lambda}_i (l+x)}\right) f_{L_i}(l)\, \mathrm{d}l
\end{aligned} \tag{2.3}$$

Hence, based on the above prediction, the policy that a SU should switch to a new channel is

$$\Pr(N_i(t) = 0) < \tau_L \tag{2.4}$$

where τ_L is the probability threshold below which a channel is considered to be busy and the SU needs to carry out a spectrum handoff, that is, the current channel is no longer considered to be idle at the end of the frame transmission. In addition, the policies that a channel j becomes a candidate channel at time t are

$$\begin{cases} \Pr(N_j(t) = 0) \geq \tau_H \\ \Pr(t_{j,\mathrm{off}} > \eta) \geq \theta \end{cases} \tag{2.5}$$

where τ_H is the probability threshold for a channel to be considered idle at the end of the current frame, η is the length of a frame plus a time slot (i.e., $\eta = \xi + \beta$), and θ is the probability threshold for a channel to be considered idle for the next frame transmission. The second criterion in (2.5) means that, in order to support at least one SU frame, the probability that the duration of the idleness of the jth channel to be longer than a frame size must be higher than or equal to θ.

2.3.2 *Proposed Spectrum Handoff Protocol Details*

The proposed spectrum handoff protocol is based on the above proposed spectrum handoff policies. It consists of two parts. The first part, namely *Protocol 1* (the pseudo-code of Protocol 1 is presented in Algorithm 1[3]), describes how a SU pair initiates a new transmission. Regardless of the coordination schemes used during channel hopping, if a data packet arrives at a SU, the SU predicts the availability of the next hopping channel (in the single rendezvous coordination scheme case) or the hopping channel of the receiver (in the multiple rendezvous coordination scheme case) at the beginning of the next slot. Based on the prediction results, if the channel satisfies the policies in (2.5) for data transmissions, the transmitter sends a RTS packet to the receiver on the same hopping channel as the receiver at the beginning of the next time slot. Upon receiving the RTS packet, the intended SU receiver replies a CTS packet in the same time slot. Then, if the CTS packet is successfully received by the SU transmitter, the two SUs pause the channel hopping and start the data transmission on the same channel. Note that if more than one pair of SUs contend the same hopping channel for data transmission, an algorithm that eliminates SU collisions is proposed in Section 2.4.

[3] DAT is the flag for data transmission requests, DSF is the data-sending flag, t is the beginning of the next slot, and k is the next hopping channel in the single rendezvous coordination scheme or the hopping channel for the receiver in the multiple rendezvous coordination scheme.

Algorithm 1 Protocol 1: starting a new transmission

```
Register initiation: DAT:=0, DSF:=0;
predicting Pr(N_k(t) = 0), Pr(t_{k,off} > η);
if Pr(N_k(t) = 0) ≥ τ_H AND Pr(t_{k,off} > η) ≥ θ
   DAT := 1;
end if
if DAT=1
   sending RTS;
end if
upon receiving CTS
   DSF := 1;
if DSF=1
   DSF := 0;
   transmitting a data frame;
   DAT := 0 when transmission ends;
end if
```

The second part, namely *Protocol 2* (the pseudo-code of Protocol 2 is presented in Algorithm 2[4]), is on the proactive spectrum handoff during a SU transmission. The goal of our proposed protocol is to determine whether the SU transmitting pair needs to carry out a spectrum handoff and then switch to a new channel by the time a frame transmission ends. Using the proposed protocol, the SU transmitting pair can avoid disruptions with PUs when PUs appear.

Based on the observed channel usage information, a SU transmitter checks the spectrum handoff policy in (2.4) for the current channel by predicting the channel availability at the end of the frame. If the policy is not satisfied, this means that the current channel is still available for the next frame transmission. Then, the SU transmitting pair does not perform a spectrum handoff and keeps staying on the same channel. However, if the policy is satisfied, the *channel-switching* (CSW) flag is set, that is, the current channel is considered to be busy during the next frame time and the SUs need to perform a spectrum handoff by the end of the frame to avoid harmful interference to a PU who may use the current channel. After the CSW is set, the two SUs rejoin the channel hopping in the next time slot after the previous frame. In the proposed distributed channel selection algorithm (which is explained in detail in Section 2.4), the SUs that need to perform spectrum handoffs at the same time are required to update the predicted channel availability information to other SUs. Hence, the SUs need to hop to the same channel to inform neighboring SUs. Note that in the single rendezvous coordination scheme, all SUs that do not transmit data follow the same hopping sequence. Therefore, when the CSW flag is set, all SUs that need to perform a spectrum handoff pause the current transmission and resume the channel hopping with the same sequence, so they will hop to the same channel. However, in the multiple rendezvous coordination scheme, each SU

[4] CSW is the channel switching flag, NUC and LSC are the number and the list of the candidate channels for data transmissions, respectively, and channel i is the current channel. As similar in Protocol 1, DAT is the flag for data transmission requests and DSF is the data-sending flag.

Algorithm 2 Protocol 2: spectrum handoff during a transmission

Register initiation: CSW:=0, DSF:=0, NUC:=0, LSC:=∅;
for $j := 0, j \leq M$ **do**
  predicting $\Pr(N_j(t) = 0)$, $\Pr(t_{j,off} > \eta)$;
end for
if $\Pr(N_i(t) = 0) < \tau_L$ AND DAT=1
  CSW := 1;
end if
if CSW=1
  for $k := 0, k \leq M$ **do**
    if $\Pr(N_k(t) = 0) \geq \tau_H$ AND $\Pr(t_{off} > \eta) \geq \theta$
      NUC := NUC+1;
      LSC(NUC) := k;
    end if
  end for
end if
if LSC=∅
  transmission stops and launch Protocol 2;
elseif LSC $\neq$ ∅
  start scanning radio;
  launch channel selection algorithm in LSC;
  sending CSR;
end if
upon receiving CSA **then**
  switch to the selected channel and start scanning radio;
if channel is busy
  transmission stops and launch Protocol 2;
else DSF := 1 CSW:=0;
end if
if DSF=1
  DSF := 0;
  transmitting a data frame;
  DAT := 0 when transmission ends;
end if

follows a default hopping sequence which may not be the same as other's hopping sequence. In order to be able to exchange channel availability information among SUs on the same channel, in our proposed protocol, SUs are required to follow the same hopping sequence only when performing spectrum handoffs.

On the other hand, the SU transmitter checks the criteria in (2.5) for available handoff candidate channels in the band. If no channel is available, then the ongoing transmission stops immediately at the end of the frame. The two SUs hop to the next channel for one more time slot and check the channel availability based on the criteria in (2.5) at the beginning of the next time slot for both the single rendezvous and the multiple rendezvous coordination schemes. However, if the set of the handoff candidate channels is not empty, the SU transmitter triggers a distributed channel selection algorithm (which is explained in detail in Section 2.4) and sends a *channel-switching-request* (CSR) packet containing the newly selected channel information in the next time slot. Upon receiving the CSR packet, the SU receiver replies with a *channel-switching-acknowledgement* (CSA) packet. If the CSA packet is successfully received by the SU transmitter, this means that the channel switching agreement between the two SU nodes has been established. Thus, both SU nodes switch to the selected channel and start the data transmission for the next frame. The handoff delay of a spectrum handoff is defined as the duration from the time a SU vacates the current channel to the time it resumes the transmission. Note that there is a possibility that the prediction is not correct and there is a PU on the channel which the SUs switch to. Hence, at the beginning of the frame, the SU transmitting pair restarts the scanning radio to confirm that the selected channel is idle. If the channel is sensed busy, the two SUs immediately resume the channel hopping and launch Protocol 2.

2.4 Distributed Channel Selection Algorithm

2.4.1 Procedure of the Proposed Channel Selection Algorithm

The channel selection issue should be handled with caution to avoid collisions among SUs. On one hand, preventing SU collisions is more important in the spectrum handoff scenario than in general channel allocation scenarios [29] due to the fact that collisions among SUs lead to data transmission failures, thus they may result in long spectrum handoff delay, which has deteriorating effect on delay-sensitive network applications. Additionally, the channel selection algorithm also should be executed fast in order to achieve short handoff delay. On the other hand, since no centralized network entity exists in CR ad hoc networks to manage the spectrum allocation, the channel selection algorithm should be applied in a distributed manner to prevent SU collisions.

Our goal is to design a channel selection scheme for the spectrum handoff scenario in CR ad hoc networks that can eliminate collisions among SUs in a distributed fashion. Based on the protocols described in Section 2.3.2, there are two cases in preventing collisions among SUs. The first case is that during the channel hopping phase, if more than one SU transmitters want to initiate new data transmissions, a collision occurs when they send RTS packets on the same channel at the same time, namely the type 1 collision. The second case is that when more than one SU pairs perform spectrum handoffs at the same time, a collision occurs when they select the same channel to switch to, namely the type 2 collision. Once a collision happens, all packets involved are wasted and need to be retransmitted. Since the spectrum handoff delay of an on-going transmission is more critical than the packet waiting time of a new transmission (i.e., the duration from the time a new packet arrives until it is successfully transmitted), the type 2 collision should be prevented with higher priority than the type 1 collision.

Figure 2.4 describes an example of the proposed channel selection scheme, where three SUs, *A*, *B*, and *C*, perform spectrum handoffs at the same time. In the parenthesis, the candidate channels are ordered based on the criterion for channel selection

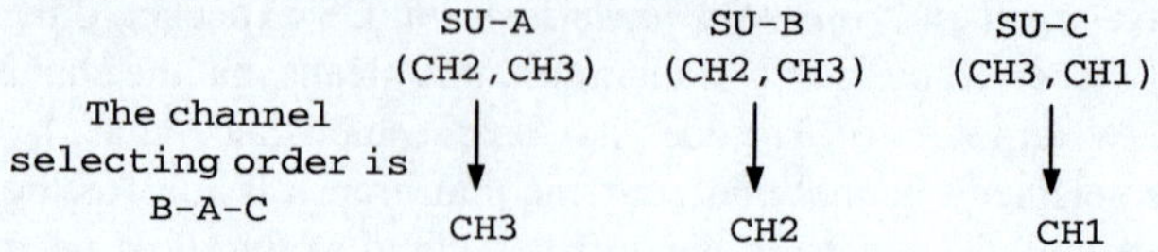

Fig. 2.4 An example of the proposed channel selection scheme

(e.g., the probability that a channel is idle). The proposed channel selection procedure is summarized as follows:

Step 1 Pseudo-random Sequence Generation: At each time slot, a pseudo-random channel selecting sequence is generated locally that all SU transmitters involved in spectrum handoffs should follow to choose channels. In Fig. 2.4, the channel selecting sequence for all SUs is *B-A-C*. Since the sequence is generated with the same seed (e.g., the time stamp), every SU generates the same channel selecting sequence at the same time slot. However, the selecting sequences are different at different time slots.

Step 2 Channel Information Update: For both the single rendezvous coordination scheme and the multiple rendezvous coordination scheme, all SUs follow the same sequence to hop through the channels during spectrum handoffs. Hence, when a SU needs to perform a spectrum handoff at the beginning of a time slot, it broadcasts the sensed channel availability information to neighboring SU nodes on the current hopping channel if it is idle. To avoid collisions of the broadcast messages, a time slot is further divided into W mini slots, W is an integer defined by the system. A SU broadcasts the channel availability information only in the corresponding mini slot based on the selecting sequence generated in Step 1. In the example shown in Fig. 2.4, SU-B broadcasts the channel availability information in the first mini slot, SU-A broadcasts in the second mini slot, and SU-C broadcasts in the third mini slot. If the broadcasting process cannot finish within one time slot due to many SUs performing spectrum handoffs at the same time, it should continue in the next time slot until all SUs broadcast the channel information messages. Hence, a SU can obtain the channel availability information predicted by all the neighboring SUs who need to perform spectrum handoffs.

Step 3 Channel Selection: Every SU who needs to perform a spectrum handoff computes the target handoff channel for its spectrum handoff based on the selecting sequence and the criterion for channel selection. The pseudo-code of the algorithm for computing the target channel is presented in Algorithm 3, where C_i denotes the target handoff channel for SU_i. In the example shown in Fig. 2.4, based on the selecting sequence, SU-B selects the first channel (i.e., channel 2) in its available channel list. Thus, the remaining SUs delete channel 2 in their available channel lists. Then, SU-A selects channel 3 so on and so forth. Therefore, for each SU, the proposed channel selection algorithm terminates until an available channel is selected or all available channels are depleted. If the target channel exists, then the SU selects it to resume its data transmission; otherwise, the SU waits for the next time slot to perform the spectrum handoff. Since the selecting sequence and the channel availability information are known to every SU who perform the spectrum handoff at the same time, the target channel for each SU (i.e., C_k, $k \in [1, N]$) is also known. Thus, the collision among SUs can be avoided.

Algorithm 3 Computing the Target Channel for SU k

```
Input: selecting sequence s, the list of candidate channels l_n, n ∈ [1, N]
Output: target channel C_k
for i := 1, i ≤ N do            // starting from the first SU in s
  if s(i) ≠ k
    if l_s(i) = ∅                 // if the list of candidate channels of SU s(i) is empty
      C_s(i) := NULL
    elseif l_s(i) ≠ ∅             // if the list of candidate channels of SU s(i) is not empty
      C_s(i) := arg max_{j∈l_s(i)} (Pr(N_j(t) = 0))
    end if
    for m := i + 1, m ≤ N do
      if C_s(i) ∈ l_s(m)   // if C_s(i) is in the list of candidate channels of SU s(m), i < m ≤ N
        l_s(m) := l_s(m) − C_s(i)   // remove the channel from the list
      end if
    end for
  elseif s(i) = k
    if l_k = ∅                    // if the list of candidate channels of SU k is empty
      return C_k := NULL break          // no available channel for SU k
    elseif l_k ≠ ∅                      // if the list of candidate channels of SU k is not empty
      return C_k := arg max_{j∈l_k} (Pr(N_j(t) = 0)) break
                        // SU k selects the channel that has the highest probability of being idle
    end if
  end if
end for
```

2.4.2 *Fairness and Scalability of the Proposed Channel Selection Scheme*

The above procedure shows that our proposed channel selection scheme can avoid collisions among SUs and it is a fully distributed algorithm. In addition, from the above discussion, we observe that an important feature of the proposed distributed channel selection scheme is fairness. Unlike the previous definition of fairness as equal channel capacity for every user [29], in this chapter, we define fairness as equal average handoff delay for every SU. This is because that, from the network performance point of view, handoff delay is the most significant metric to evaluate a spectrum handoff protocol. Thus, letting every SU have equal average handoff delay is fair. We define the spectrum handoff delay as the duration from the moment a SU starts to perform a spectrum handoff to the moment it resumes the data transmission. Figure 2.5 shows the simulation result of the average handoff delay of the SUs when they use the proposed channel selection scheme under the single rendezvous coordination scheme. We deploy 20 SU nodes in the network with different arrival rate

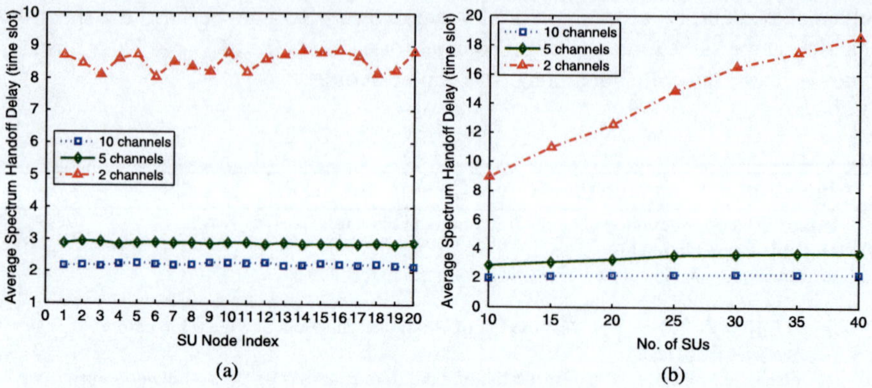

Fig. 2.5 Fairness and scalability of the proposed channel selection scheme. **(a)** Fairness of the proposed channel selection scheme; **(b)** Scalability of the proposed channel selection scheme

which is a uniform random variable in the range of [0, 500] (unit: packet/second). It is shown in the figure that SUs achieve approximately the same average spectrum handoff delay in the same scenario, which indicates that our proposed channel selection scheme is fair to all SUs.

On the other hand, for CR ad hoc networks where nodes membership may change over time, an important issue is the scalability of the proposed channel selection algorithm when the network size increases. Even though the number of SUs in a network may vary, as illustrated in Algorithm 3, only those SUs who are involved in the spectrum handoff process at the same time will activate the algorithm, which may not be a large number. In addition, from the number of broadcasted messages during the second step of the proposed channel selection scheme, our proposed channel selection algorithm will not result in excessive overhead when the network size increases. Because the number of channel information message updates affects the spectrum handoff delay (i.e., more channel information messages updated results in longer spectrum handoff delay), Fig. 2.5 shows the simulation result of the average spectrum handoff delay under different network sizes. It is shown that when the network size changes from 10 SUs to 40 SUs (i.e., the network size increases 300%), the spectrum handoff delay only increases 14.5%, 16%, and 105% for the cases when the number of channels is 10, 5, and 2, respectively.

2.5 Performance Evaluation of the Proposed Proactive Spectrum Handoff Framework

2.5.1 Simulation Setup

In this section, we adjust the spectrum handoff criteria and policies proposed in Section 2.3.1 to a time-slotted system and evaluate the performance of the proposed proactive spectrum handoff framework. In order for the system to be stable, we assume that the inter-arrival time of SU packets follow a biased geometric distribution, where the probability mass function (pmf) of the biased geometric distributed inter-arrival time is given by [35]:

$$p(N = n) = \begin{cases} 0 & n < a \\ x(1-x)^{(n-a)} & n \geq a \end{cases} \tag{2.6}$$

where n is the number of time slots between packet arrivals, $a \geq 0$ represents the minimum number of time slots between two adjacent packets, and x is the probability that a packet arrives during one time slot (i.e., x is the normalized arrival rate of data packets, that is, $x = \lambda\beta$, where λ is the arrival rate in terms of packet/second). Based on this model, if we set a as the packet length, then a new packet will not be generated until the previous packet finishes its transmission.

Accordingly, we modify the prediction criteria proposed in Section 2.3 based on the biased geometric distributed inter-arrival time model. Denote the starting slot of the prediction as slot 0 and the slot for prediction as slot n. As shown in Fig. 2.6a, the probability that no PU arrival occurs between slot 1 and n and channel k is idle at slot n ($n \geq 1$) is given by

$$P_0 = 1 - \sum_{i=1}^{n} x(1-x)^{(i-1)} \tag{2.7}$$

where x is the normalized arrival rate. As shown in Fig. 2.6b, the probability that only one PU packet arrives between slot 1 and n ($n > L$) and channel k is idle at slot n is

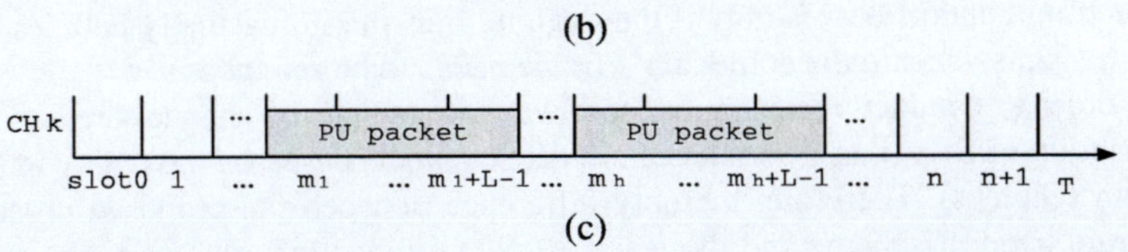

Fig. 2.6 The PU activity on channel k. (a) No PU packet arrives between slot 1 and n; (b) Only one PU packet arrives between slot 1 and n; (c) h PU packet arrives between slot 1 and n

$$P_1 = \sum_{m=1}^{n-L}\left[1 - \sum_{i=1}^{n-m-L+1} x(1-x)^{(i-1)}\right] x(1-x)^{(m-1)} \tag{2.8}$$

where m is the time slot at which a PU transmission starts and L is the length of a PU packet. Similarly, in Fig. 2.6c, m_i denotes the time slot at which the ith PU transmission starts. Thus, the probability that h PU packets arrives ($h \in [1, U]$), where U is the maximum number of PU packets that could arrives between slot 1 and n ($n > hL$) and channel k is idle at slot n is

$$P_h = \sum_{m_h=h}^{n-hL}\left[1 - \sum_{i=1}^{n-m_h-hL+1} x(1-x)^{(i-1)}\right] x^h(1-x)^{(m_h-h)} \tag{2.9}$$

Therefore, the total probability that channel k is idle at slot n is obtained as follows:

$$\Pr(N_k(n) = 0) = \sum_{i=0}^{U} P_i \tag{2.10}$$

Second, due to the memoryless property of geometric distribution, the probability that the duration of the idleness is longer than η slots on channel k is given by

$$P(t_{k,off} > \eta) = 1 - \sum_{i=1}^{\eta} x(1-x)^{(i-1)} \tag{2.11}$$

In this chapter, we exclude the effect of the channel switching delay (i.e., RF configuration delay), but it can be easily taken into account when necessary.

2.5.2 The Proposed Proactive Spectrum Handoff Scheme

We first compare the proposed proactive spectrum handoff scheme with the reactive spectrum handoff approach. In the reactive spectrum handoff approach, a SU transmits a packet without predicting the availability of the current channel at the moment when a frame ends (i.e., using the policy in (2.4)). That is, a SU does not change the current channel by the end of a frame if the previous frame is successfully received. A spectrum handoff occurs only if the ongoing transmission actually collides with a PU transmission and the collided SU frame needs to be retransmitted.

In order to conduct a fair comparison, we assume that the channel prediction is a capability of SUs (i.e., SUs select candidate channels based on the policy in (2.5) in both schemes). Therefore, the only difference between the proposed proactive spectrum handoff scheme and the reactive spectrum handoff scheme is the mechanism to trigger the spectrum handoffs. In addition, in order to solely investigate the performance of the two spectrum handoff schemes, we adopt a general random

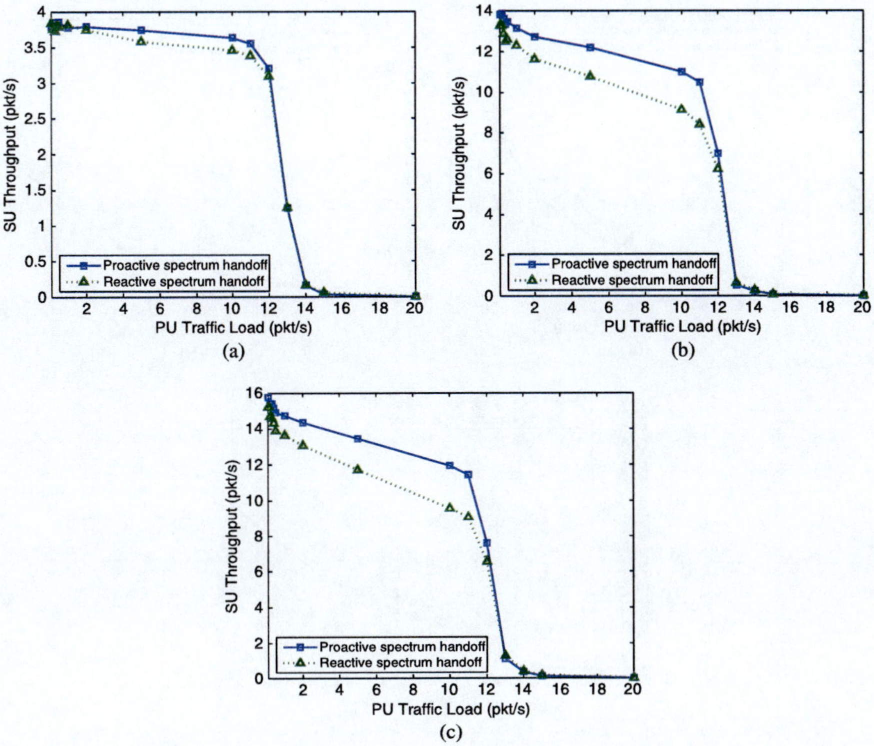

Fig. 2.7 Simulation results of SU throughput. (**a**) The SU packet arrival rate $\lambda_s = 5$ packets/s; (**b**) The SU packet arrival rate $\lambda_s = 100$ packets/s; (**c**) The SU packet arrival rate $\lambda_s = 500$ packets/s

channel selection scheme (i.e., a SU randomly selects a channel from its candidate channels) in both schemes.

Figures 2.7, 2.8 and 2.9 illustrate the performance results of the two spectrum handoff schemes under different SU and PU traffic loads, when the network coordination scheme is the single rendezvous coordination scheme, where there are 10 SUs and 10 channels in the network. A SU using our proposed proactive spectrum handoff scheme will stop the data transmission on a channel which is likely to have a PU and switch to a channel which has less probability a PU appears. We choose the throughput of SUs, collision rate (i.e., the number of collisions between SUs and PUs per SU packet transmitted), and the number of collisions between SUs and PUs per second as the performance metrics.

Figure 2.7 shows the SU throughput when SUs use different spectrum handoff schemes under varying SU and PU traffic load. It is shown that when both SU traffic and PU traffic are light (e.g., $\lambda_s = 5$ packets/s and $\lambda_p = 0.5$ packets/s), the SU throughput is similar in both schemes. This is because when the traffic is light, collisions between SUs and PUs are much fewer than the case when the traffic is heavy. SUs have less probability of retransmitting a packet for both cases, thus the

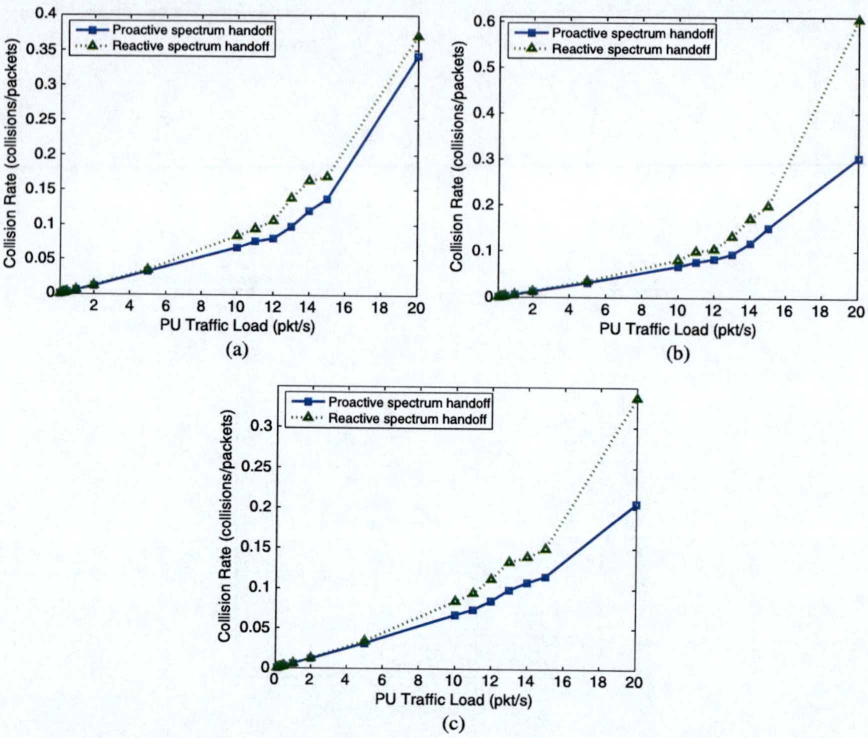

Fig. 2.8 Simulation results of collision rate. **(a)** The SU packet arrival rate $\lambda_s = 5$ packets/s; **(b)** The SU packet arrival rate $\lambda_s = 100$ packets/s; **(c)** The SU packet arrival rate $\lambda_s = 500$ packets/s

performance differences between the proactive spectrum handoff scheme and the reactive spectrum handoff scheme are not very obvious. However, when the SU traffic and PU traffic are heavy (e.g., $\lambda_s = 500$ packets/s and $\lambda_p = 10$ packets/s), the proactive spectrum handoff scheme outperforms the reactive scheme in terms of 30% higher throughput. Figures 2.8 and 2.9 show the collision rate and the number of collisions per second, respectively. From Fig. 2.8, it is shown that collision rate increases as PU traffic load increases. In addition, proactive spectrum handoff always outperforms reactive spectrum handoff in terms of lower collision rate and fewer number of collisions per second.

2.5.2.1 The Effect of the Number of SUs and PU channels

Figures 2.10 and 2.11 show the SU throughput and collision rate under varying number of SUs and PU channels, respectively. The results are generated in the scenario where the arrival rate of SU packets is saturated (i.e., $\lambda_s = 500$ packets/s) and the arrival rate of PU packet is equal to 10 packets/s. In both figures, our

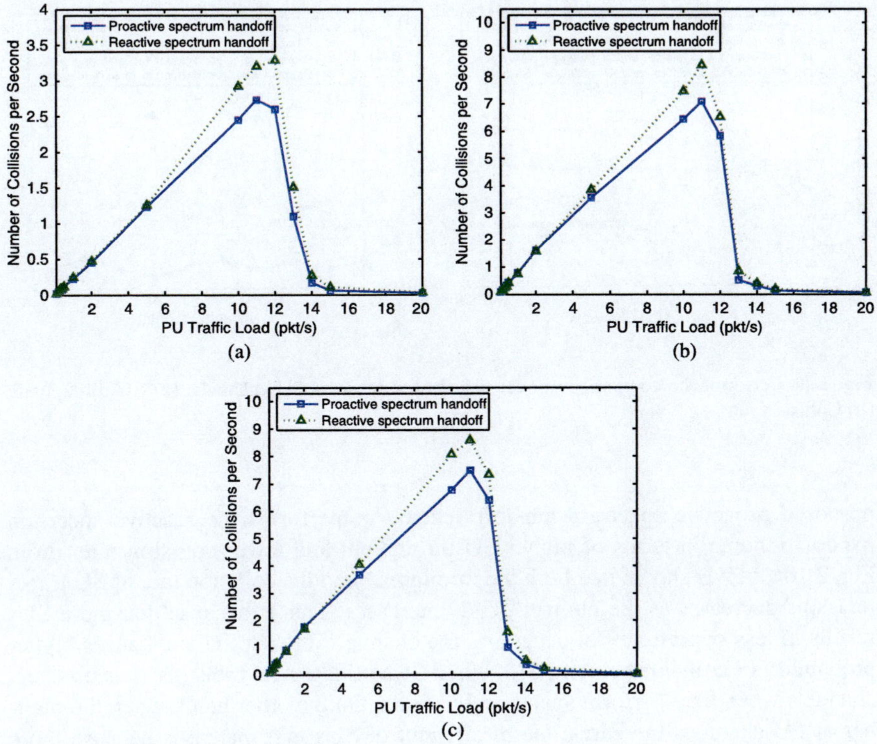

Fig. 2.9 Simulation results of number of collisions per second. (**a**) The SU packet arrival rate $\lambda_s = 5$ packets/s; (**b**) The SU packet arrival rate $\lambda_s = 100$ packets/s; (**c**) The SU packet arrival rate $\lambda_s = 500$ packets/s

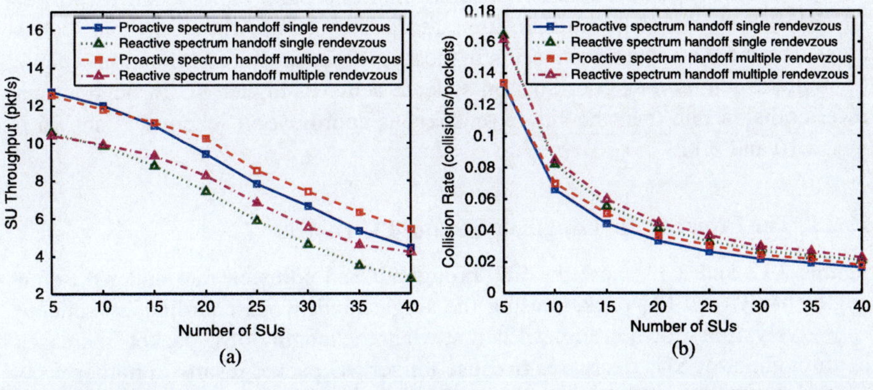

Fig. 2.10 Performance comparison under different number of SUs. (**a**) SU throughput; (**b**) Collision rate

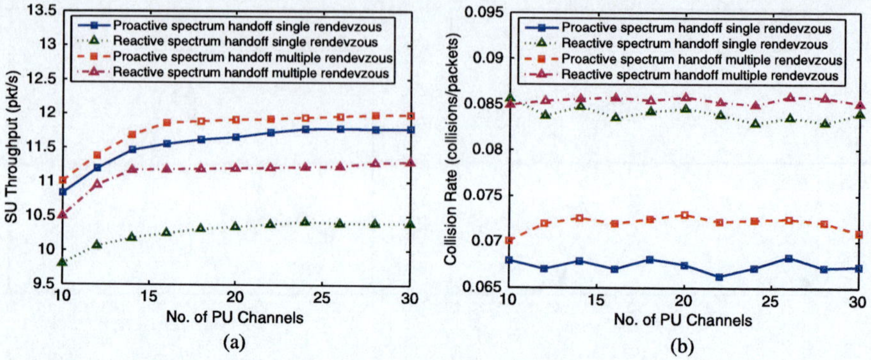

Fig. 2.11 Performance comparison under different number of PU channels. (**a**) SU throughput; (**b**) Collision rate

proposed proactive spectrum handoff scheme outperforms the reactive spectrum handoff scheme in terms of higher SU throughput and lower collision rate. From Fig. 2.10a, b, it is shown that both the throughput and the collision rate of SU transmissions decreases as the number of SU increases. This is because that more SUs results in less opportunity of accessing the channel for each SU and causes higher probability of collisions among SUs when SUs initiate new transmissions or select channels when they perform spectrum handoffs. On the other hand, when the number of PU channels increases, the throughput of SUs first increases because more channels can be used for data transmissions. Then, the SU throughput becomes stable because increasing the number of channels does not help increasing the chance of data transmissions of SU packets after a certain threshold. The collision rate (i.e., the number of collisions between SUs and PUs per SU packet transmitted) remains relative stable to the change of the number of PU channels. Since in the multiple rendezvous coordination scheme, multiple pairs of SUs can use different channels to establish multiple links at the same time while only one pair is allowed to initiate a data transmission in the single rendezvous coordination scheme, the multiple rendezvous coordination scheme achieves higher SU throughput and lower collision rate than the single rendezvous coordination scheme, as shown in Figs. 2.10 and 2.11.

2.5.2.2 The Effect of the Length of SU and PU Packets

Figures 2.12 and 2.13 show the SU throughput and collision rate under different lengths of SU and PU packets using the single rendezvous coordination scheme, respectively. It is shown in Fig. 2.12a that when the length of SU packets increases, the throughput of SUs decreases because longer SU packet results in higher probability of collisions with PUs and leads to fewer SU packets transmitted during a certain amount of time. Therefore, it is illustrated in Fig. 2.12b that the collision rate increases when the length of SU packets increases. On the other hand, the

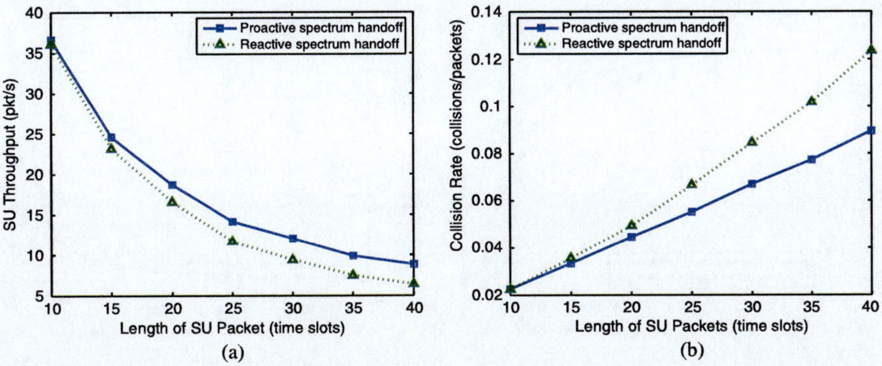

Fig. 2.12 Performance comparison under varying SU packet length. (**a**) SU throughput; (**b**) Collision rate

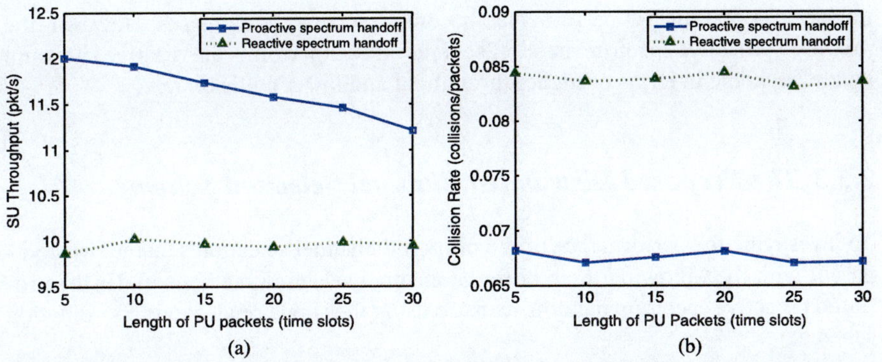

Fig. 2.13 Performance comparison under varying PU packet length. (**a**) SU throughput; (**b**) Collision rate

length of PU packets does not significantly affect the SU performance because we assume that once a SU frame collides with a PU packet, the whole frame needs to be retransmitted. Thus, the effect of the length of PU packets on SU performance is not significant.

2.5.2.3 The Effect of Spectrum Sensing Errors

Figure 2.14 shows the effect of spectrum sensing errors on the performance of different spectrum handoff schemes using the single rendezvous coordination scheme. We use a coefficient χ to indicate the level of imperfect spectrum sensing, where $\chi \in [0, 1]$ represents the probability that the result of spectrum sensing is wrong (the spectrum sensing errors include both miss detection and false alarm [36]). When $\chi = 0$, it means that the spectrum sensing is perfect and there is no error. Whereas when $\chi = 1$, it means that the spectrum sensing is completely incorrect. It is shown

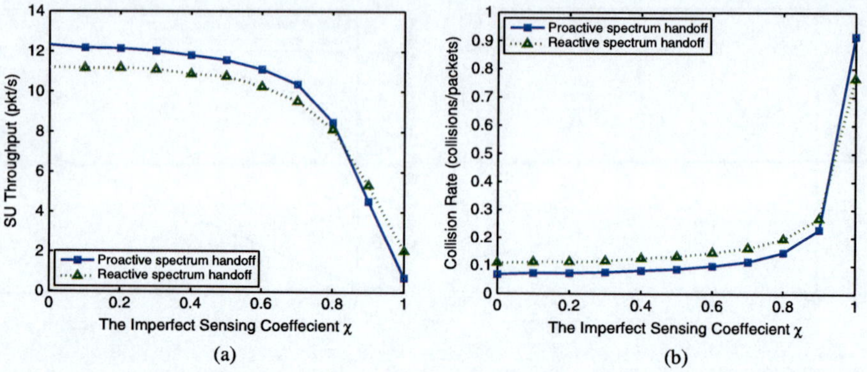

Fig. 2.14 Performance comparison under imperfect spectrum sensing. (**a**) SU throughput; (**b**) Collision rate

in Fig. 2.14 that the SU performance becomes worse as χ increases. However, the proposed proactive spectrum handoff scheme still outperforms the reactive spectrum handoff scheme in terms of higher throughput and lower collision rate.

2.5.3 The Proposed Distributed Channel Selection Scheme

To investigate the performance of the proposed channel selection scheme, we compare it with the following three different channel selection methods under the proposed proactive spectrum handoff scenario using the single rendezvous coordination scheme:

- *Random channel selection*: A SU randomly chooses a channel from its predicted available channels.
- *Greedy channel selection*: In this method, only one pair of SUs is considered in the network. The SUs can obtain all the channel usage information and predict the service time on each channel. Thus, when a spectrum handoff occurs, a SU selects a pre-determined channel that leads to the minimum service time [7].
- *Local bargaining*: In this method, SUs form a local group to achieve a collision-free channel assignment. To make an agreement among SUs, a four-way handshake is needed between neighbors (i.e., request, acknowledgment, action, acknowledgment). Since one of the SUs is the initiating node which serves as a group header, the total number of control messages exchanged is $2N_{\mathrm{LB}}$, where N_{LB} is the number of SUs need to perform spectrum handoffs [29].

Since for channel selection schemes, reducing the number of collisions among SUs is the primary goal, we consider the SU throughput, average SU service time, collisions among SUs, and average spectrum handoff delay as the performance metrics.

2.5.3.1 One-Pair-SU Scenario

Figure 2.15a, b shows the SU throughput and the average service time of different channel selection schemes in a one-pair-SU scenario, respectively. Because only one pair of SUs exists in the network, there is no collision among SUs. Thus, in this scenario, the greedy channel selection scheme performs the best among all the schemes. This is because that the handoff target channel a SU transmitter selects is pre-determined based on channel observation history. Hence, no signaling message is needed between the SU transmitting pair. While in other schemes, the SU transmitter needs to inform the receiver about the newly selected channel. Thus, the throughput is lower and the average service time is longer than the greedy scheme. However, among the three schemes other than the greedy scheme, our proposed channel selection scheme has the best performance in terms of higher throughput and shorter total service time.

2.5.3.2 Multiple-Pair-SU Scenario

Figure 2.16a, b, shows the SU throughput and the average service time of different channel selection schemes in a 10-pair-SU scenario, respectively. In the greedy channel selection method, all pairs of SUs always select the same pre-determined channel for spectrum handoffs. Therefore, the greedy method always leads to collisions among SUs. The throughput of SUs using the greedy method is almost zero. Because the proposed channel selection scheme can totally eliminate collisions among SUs, the throughput is higher and the average service time is shorter than the other channel selection schemes.

Figure 2.17a, b shows the performance under different number of SUs, when there are 10 channels and the SU and PU traffic load is 500 packet/s and 10 packet/s, respectively. In Fig. 2.17, we only show the local bargaining method, random channel selection, and the proposed channel selection. We exclude the greedy method because the greedy method constantly achieves zero throughput. Thus, its average

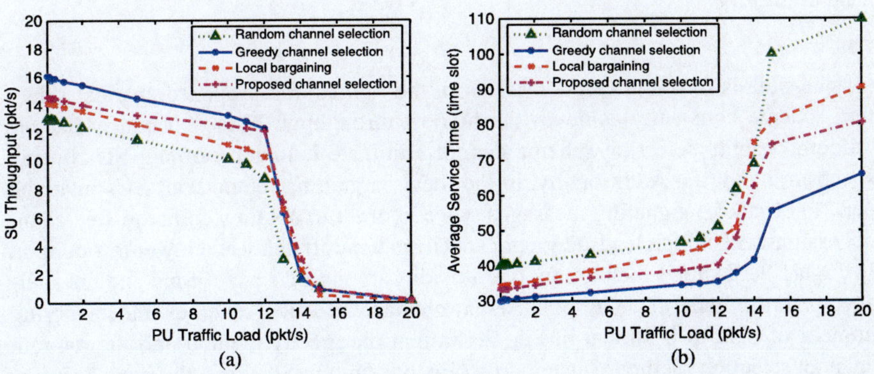

Fig. 2.15 Performance of the channel selection schemes in a one-pair-SU scenario. **(a)** SU throughput in a two-SU scenario; **(b)** SU average service time in a two-SU scenario

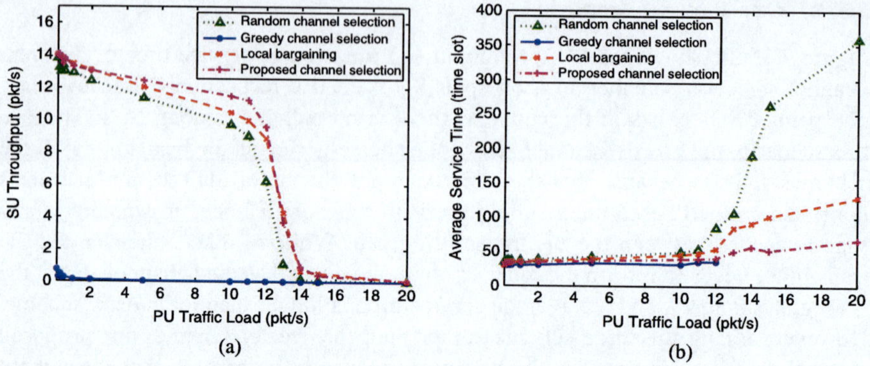

Fig. 2.16 Performance of the channel selection schemes in a 10-pair-SU scenario. **(a)** SU throughput in a multi-SU scenario; **(b)** SU average service time in a multi-SU scenario

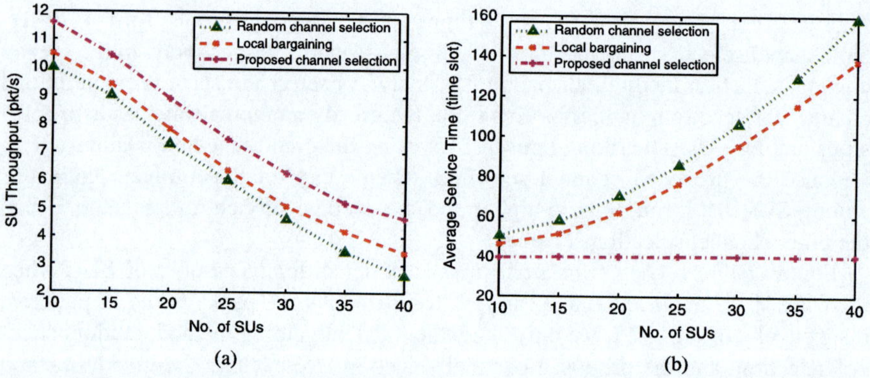

Fig. 2.17 Performance of the channel selection schemes in a multiple-pair-SU scenario under varying number of SUs. **(a)** SU throughput in a multi-SU scenario; **(b)** SU average service time in a multi-SU scenario

service time is meaningless. As shown in the figures, the proposed channel selection scheme constantly achieves the highest throughput. This is because that the random channel selection scheme cannot eliminate collisions among SUs during spectrum handoffs. Additionally, in the local bargaining method, all SUs involved need to broadcast signaling messages twice in order to obtain a collision-free channel assignment, which leads to longer spectrum handoff delay and lower throughput.

Figure 2.18 shows the number of collisions among SUs per second and the average spectrum handoff delay of different channel selection schemes under varying number of SUs. It is shown in Fig. 2.18a that the greedy method and the random channel selection method cause more collisions among SUs than the local bargaining and the proposed channel selection method. While on the other hand, the local bargaining method cause much longer average spectrum handoff delay than the

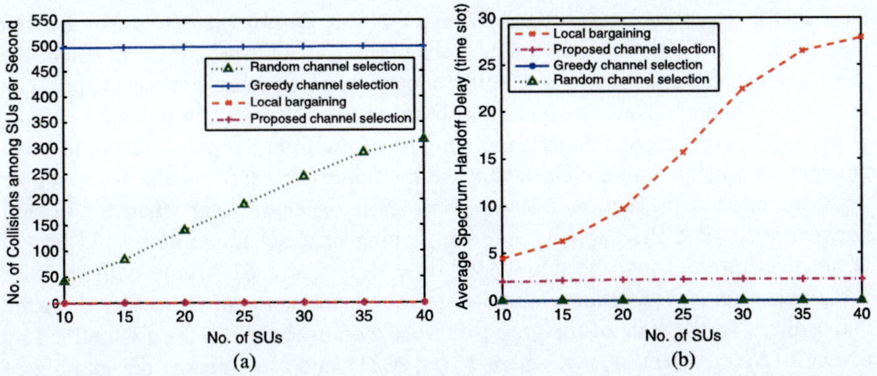

Fig. 2.18 Performance of the channel selection schemes in a multiple-pair-SU scenario under varying number of SUs. (**a**) Number of collisions among SUs per second; (**b**) Average spectrum handoff delay

proposed channel selection scheme as shown in Fig. 2.18b. Therefore, the proposed channel selection scheme is the most suitable one for spectrum handoff scenarios.

2.6 The Proposed Three Dimensional Discrete-time Markov Model

In this section, we develop a Markov model to analyze the performance of the reactive spectrum handoff process based on the single rendezvous coordination scheme. For simplicity, we assume that there are only two SUs in the network. We also ignore the propagation delay or any processing time in our analysis.

2.6.1 The Proposed Markov Model

Based on the time-slotted channels, any action of a SU can only be taken at the beginning of a time slot. In addition, the status of a SU in the current time slot only relies on its immediate past time slot. Such discrete-time characteristics allow us to model the status of a SU using Markov chain analysis. The status of a SU in a time slot can only be one of the following:

- *Idle*: no packet arrives at a SU.
- *Transmitting*: the transmission of a SU does not collide with PU packets in a time slot, i.e., successful transmission.
- *Collided*: the transmission of a SU collides with PU packets in a time slot, i.e., unsuccessful transmission.
- *Backlogged*: a SU has a packet to transmit in the buffer but fails to access a channel.

Note that there are two cases that a SU can be in the *Backlogged* status. In the first case, when a SU pair initiates a new transmission, if multiple SU pairs select the same channel for transmissions, a collision among SUs occurs, and no SU pair can access the channel. Thus, the packet is backlogged. Similarly, in the second case, when a SU pair performs a spectrum handoff, if multiple SU pairs select the same channel, a collision among SUs occurs and the frame in each SU is also backlogged.

As mentioned in Section 2.1, we consider the scenario that when a collision between a SU and PU happens, the overlapping of a SU frame and a PU packet is not negligible. Thus, the number of time slots that a SU frame collides with a PU packet is an important parameter to the performance of SUs. Based on the above analysis, the state of the proposed Markov model at time slot t is defined by a vector $(N_t(t), N_c(t), N_{f(t)})$, where $N_t(t)$, $N_c(t)$, and $N_{f(t)}$ denote the number of time slots including the current slot that are successfully transmitted in the current frame, the number of time slots including the current slot that are collided with a PU packet in the current frame, and the number of frames that have been successfully transmitted plus the current frame that is in the middle of a transmission at time slot t, respectively. Therefore, $N_t(t)+N_c(t) \leq c$. Figure 2.19 shows the state transition diagram of our proposed three-dimensional Markov chain. There are totally $(h+1)$ tiers in the state transition diagram. For each tier, it is a two-dimensional Markov chain with a fixed $N_f(t)$. Table 2.1 summarizes the notations used in our Markov model.

From Fig. 2.19, it is observed that the proposed Markov model accurately capture the status of a SU in a time slot. The state $(N_t(t) = 0, N_c(t) = 0, N_f(t) = 0)$ in Fig. 2.19 represents that a SU is in the *Idle* status. Similarly, the states $(N_t(t) \in [1, c], N_c(t) = 0, N_f(t) \in [1, h])$ represent the *Transmitting* status, i.e., no collision. The states $(N_t(t) \in [0, c-1], N_c(t) \in [1, c], N_f(t) \in [1, h])$ represent the *Collided* status. At last, the states $(N_t(t)=0, N_c(t)=0, N_f(t) \in [1, h])$ represent the *Backlogged* status, where $(N_t(t)=0, N_c(t)=0, N_f(t)=1)$ is the *Backlogged* status during a new transmission. As shown in Fig. 2.19, the feature of the common frequency-hopping sequence scheme is captured in our model that a SU can only start a new transmission when there is a channel available.[5]

2.6.2 Derivation of Steady-State Probabilities

To obtain the steady-state probabilities of the states in the three-dimensional Markov chain shown in Fig. 2.19, we first get the one-step state transition probability.[6] Thus, the non-zero one-step state transition probabilities for any $0 < i_0 < c$, $0 < j_0 < c$, and $0 < k_0 < h$ are given as follows:

---

[5] In the following discussion, we use the terms "states" in our proposed Markov model and the "status" of a SU in a time slot interchangeably. We also use the notations $(N_t(t+1)=i, N_c(t+1)=j, N_f(t+1)=k)$ and (i, j, k) to represent a state interchangeably.

[6] We denote the one-step state transition probability from time slot t to time slot $t+1$ as $P(i_1, j_1, k_1|i_0, j_0, k_0) = P(N_t(t+1) = i_1, N_c(t+1) = j_1, N_f(t+1) = k_1|N_t(t) = i_0, N_c(t) = j_0, N_f(t) = k_0)$.

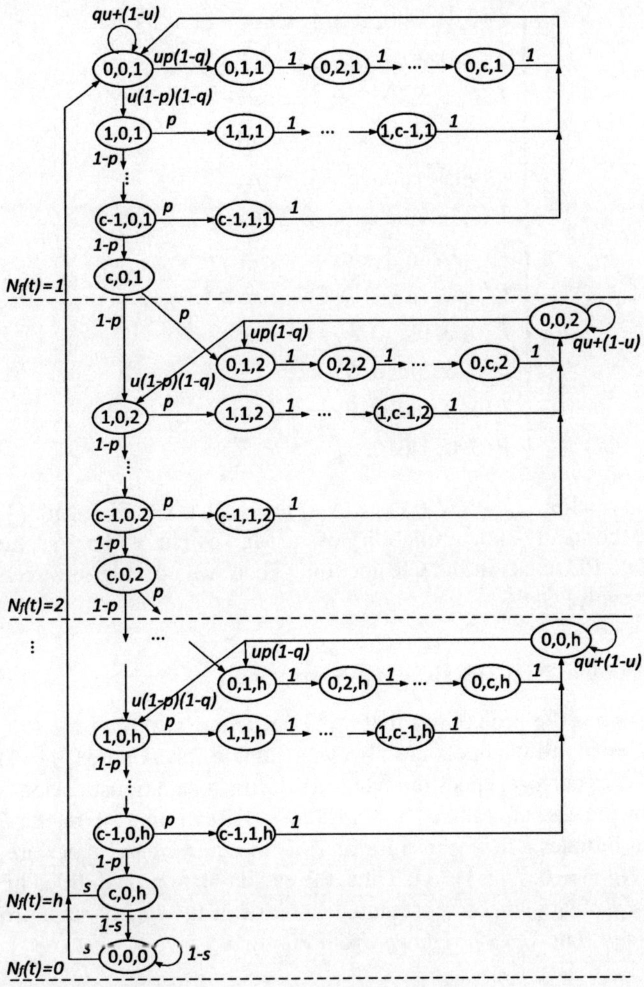

Fig. 2.19 The transition diagram of the proposed Markov model

Table 2.1 Notations used in the Markov analysis

Symbol	definition
p	Probability that a PU packet arrives in a time slot
s	Probability that a SU packet arrives in a time slot
h	Number of frames in a SU packet
c	Number of time slots in a frame
q	Probability of a collision among SUs
u	Probability that at least one channel is idle

$$\begin{cases} P(0,0,k_O|0,0,k_O) = qu + (1-u) \\ P(1,0,k_O|0,0,k_O) = u(1-p)(1-q) \\ P(0,1,k_O|0,0,k_O) = up(1-q) \\ P(i_O, j_O+1, k_O|i_O, j_O, k_O) = 1 \\ P(i_O, 1, k_O|i_O, 0, k_O) = p \\ P(i_O+1, 0, k_O|i_O, 0, k_O) = 1-p \\ P(1, 0, k_O+1|c, 0, k_O) = 1-p \\ P(0, 1, k_O+1|c, 0, k_O) = p \\ P(0,0,0|c,0,h) = 1-s \\ P(0,0,1|c,0,h) = s \\ P(0,0,0|0,0,0) = 1-s \\ P(0,0,1|0,0,0) = s \end{cases} \tag{2.12}$$

Let $P_{(i,j,k)} = \lim_{t\to\infty} P(N_t(t)=i, N_c(t)=j, N_f(t)=k), i \in [0,c], j \in [0,c]$, $k \in [0,h]$ be the steady-state probability of the Markov chain. We first study a simple case where no PU exists in the CR network. Then, we consider the scenario where SUs coexist with PUs.

2.6.2.1 Case One: No PU Exists in a Network

In this case, since the probability that a PU packet arrives in a time slot is equal to zero (i.e., $p = 0$), all channels are always available for SUs (i.e., $u = 1$) and a SU does not need to perform spectrum handoffs during a data transmission. Thus, a SU cannot be in the *Collided* state. In addition, a SU can only be in the *Backlogged* state when it initiates a new transmission (i.e., the *Backlogged* states are reduced to $(N_t(t)=0, N_c(t)=0, N_f(t)=1)$. Thus, the steady-state probabilities of the *Transmitting* and *Idle* state can be represented in terms of the steady-state probability of the *Backlogged* state $P_{(0,0,1)}$. Hence, from Fig. 2.19,

$$P_{(i,0,k)} = (1-q)P_{(0,0,1)}, \text{ for } 1 \le i \le c, 1 \le k \le h \tag{2.13}$$

$$P_{(0,0,0)} = \frac{(1-s)(1-q)}{s} P_{(0,0,1)} \tag{2.14}$$

Since $\sum_i \sum_j \sum_k P_{(i,j,k)}=1$, we can calculate the steady-state probability of every state in the Markov chain. Note that the probability of a collision among SUs, q, depends on the channel selection scheme. The derivation of q is given in Section 2.4.

2.6.2.2 Case Two: SUs Coexist with PUs in a Network

If the probability that a PU packet arrives in a time slot is not equal to zero (i.e., $p \neq 0$), collisions between SUs and PUs may occur when a SU transmits a frame. Thus, the steady-state probabilities of the *Collided* states are not zero. Similar to the no-PU case, we represent the steady-state probabilities in terms of $P_{(0,0,1)}$. First of all, for the first tier in Fig. 2.19, we can obtain the steady-state probabilities of all the *Transmitting* states in terms of $P_{(0,0,1)}$, that is,

$$P_{(i,0,1)} = u(1-q)(1-p)^i P_{(0,0,1)}, \text{ for } 1 \le i \le c \tag{2.15}$$

Then, for the *Collided* states with $i = 0$,

$$P_{(0,j,1)} = up(1-q)P_{(0,0,1)}, \text{ for } 1 \le j \le c \tag{2.16}$$

For the *Collided* states with $i > 0$,

$$P_{(i,j,1)} = u(1-q)p(1-p)^i P_{(0,0,1)}, \text{ for } 1 \le i \le c-1, 1 \le j \le c \tag{2.17}$$

For the kth ($k > 1$) tier, we first derive $P_{(1,0,k)}$ and $P_{(0,1,k)}$

$$P_{(1,0,k)} = (1-p)P_{(c,0,k-1)} + u(1-p)(1-q)P_{(0,0,k)} \tag{2.18}$$

$$P_{(0,1,k)} = pP_{(c,0,k-1)} + up(1-q)P_{(0,0,k)} \tag{2.19}$$

Then, the steady-state probabilities of the *Transmitting* states when $i > 1$ can be represented as

$$P_{(i,0,k)} = (1-p)^{i-1} P_{(1,0,k)}, \text{ for } 1 < i \le c \tag{2.20}$$

Similar to the derivation method for the first tier, for the *Collided* states with $i = 0$,

$$P_{(0,j,k)} = P_{(0,1,k)}, \text{ for } 1 \le j \le c \tag{2.21}$$

For the *Collided* states with $i > 0$,

$$P_{(i,j,k)} = p(1-p)^{i-1} P_{(1,0,k)}, \text{ for } 1 \le i \le c-1, 1 \le j \le c \tag{2.22}$$

Then, for the *Backlogged* state in the kth tier,

$$\sum_{i=0}^{c-1} P_{(i,c-i,k)} = u(1-q)P_{(0,0,k)} \tag{2.23}$$

Combining (2.18) through (2.23), we obtain the following equations using basic mathematical manipulations:

$$P_{(1,0,k)} = \frac{1}{(1-p)^{c-1}} P_{(c,0,k-1)} \tag{2.24}$$

$$P_{(0,1,k)} = \frac{p}{(1-p)^{c}} P_{(c,0,k-1)} \tag{2.25}$$

$$P_{(0,0,k)} = \frac{1-(1-p)^{c}}{u(1-q)(1-p)^{c}} P_{(c,0,k-1)} \tag{2.26}$$

Then, from (2.20),

$$P_{(c,0,k-1)} = (1-p)^{c-1} P_{(1,0,k-1)} \tag{2.27}$$

Combining (2.24) and (2.27), we find the following relationship:

$$P_{(c,0,k)} = P_{(c,0,k-1)} \tag{2.28}$$

Thus,

$$P_{(c,0,k)} = u(1-q)(1-p)^{c} P_{(0,0,1)} \tag{2.29}$$

Equation (2.29) indicates the steady-state probabilities of the states in the kth tier are independent of k. Now, we have all the steady-state probabilities of the states in all tiers except the state (0, 0, 0). At last, for the *Idle* state,

$$P_{(0,0,0)} = \frac{1-s}{s} u(1-q)(1-p)^{c} P_{(0,0,1)} \tag{2.30}$$

Similarly, since $\sum_i \sum_j \sum_k P_{(i,j,k)} = 1$, we can get the steady-state probability of every state in the Markov chain. If we denote Θ as the normalized throughput of SU transmissions, Θ is the summation of the steady-state probabilities of all the *Transmitting* states in our proposed Markov model. That is,

$$\Theta = \sum_{k=1}^{h} \sum_{i=1}^{c} P_{(i,0,k)} \tag{2.31}$$

2.6.3 The Probability That at Least One Channel Is Idle

In the above derivations, u and q are unknown. In this section, we calculate the probability that at least one channel is idle, u. Without loss of generality, we associate a PU with one channel and model the activity of a PU on a channel as an ON/OFF process [32]. SUs can only exploit the channels when the channels are idle (i.e., in the OFF period). We assume that the buffer in each PU can store at most one packet at a time. Once a packet is stored at a buffer, it remains there until it is

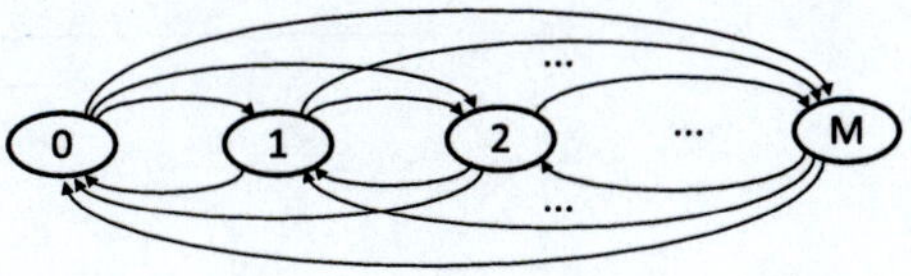

Fig. 2.20 The transition diagram of the number of channels used by PUs in one time slot

successfully transmitted. Thus, we assume that the OFF period of a channel follows the geometric distribution, where the probability mass function (pmf) is given by

$$\Pr(N_{OFF} = n) = p(1-p)^n \tag{2.32}$$

where N_{OFF} is the number of time slots of an OFF period.

Let $\Omega(t)$ be the number of channels used by PUs at time slot t. The process $\{\Omega(t), t = 0, 1, 2, \cdots\}$ forms a Markov chain whose state transition diagram is given in Fig. 2.20, in which the self loops are omitted. To characterize the behavior of the PU channels, we define $\mathcal{D}_\alpha^l$ as the event that l PUs finish their transmissions given that there are α PUs in the network in a time slot. We also define $\mathcal{A}_\gamma^m$ as the event that m PUs start new transmissions given that there are γ idle PUs in a time slot. Thus, the probabilities of events $\mathcal{D}_\alpha^l$ and $\mathcal{A}_\gamma^m$ are:

$$\Pr\left(\mathcal{D}_\alpha^l\right) = \binom{\alpha}{l} v^l (1-v)^{\alpha-l} \tag{2.33}$$

$$\Pr\left(\mathcal{A}_\gamma^m\right) = \binom{\gamma}{m} p^m (1-p)^{\gamma-m} \tag{2.34}$$

where v is the probability that a PU finishes its transmission in a slot. If the average length of a PU packet is denoted as $\bar{L}$, then $v = 1/\bar{L}$. Therefore, the state transition probability from state $\{\Omega(t) = a\}$ to state $\{\Omega(t+1) = b\}$ can be written as

$$p_{ab} = \begin{cases} \sum_{l=0}^{a} \Pr\left(\mathcal{D}_a^l\right) \Pr\left(\mathcal{A}_{M-a+l}^{b-a+l}\right), & \text{for } b \geq a \\ \sum_{l=a-b}^{a} \Pr\left(\mathcal{D}_a^l\right) \Pr\left(\mathcal{A}_{M-a+l}^{b-a+l}\right), & \text{for } b < a \end{cases} \tag{2.35}$$

Therefore, we can obtain the steady-state probabilities of the number of busy channels in the band in a time slot, denoted as $\mathbf{g} = [g_0 \;\; g_1 \;\; g_2 \;\; \cdots \;\; g_M]^T$, where g_i denotes the steady-state probability that there are i busy channels in a time slot. Hence, $u = \sum_{i=0}^{M-1} g_i$.

2.6.4 Results Validation

In this section, we validate the numerical results obtained from our proposed Markov model using simulation. Note that we only consider two SUs in the network,

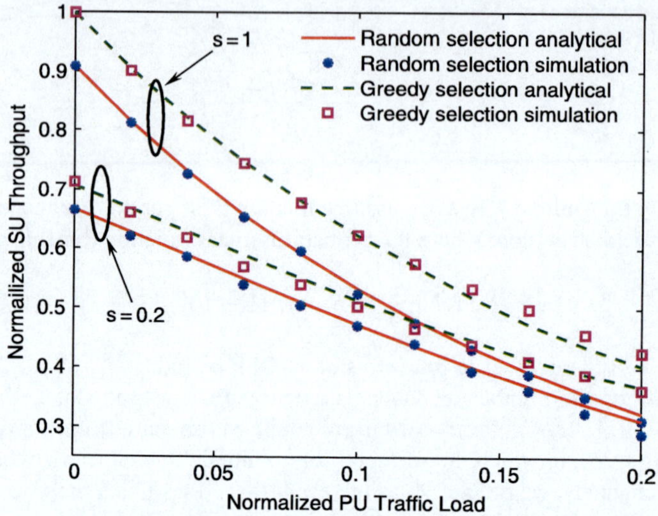

Fig. 2.21 Analytical and simulation results of the normalized SU throughput in a two-SU scenario

the probability of collision among SUs is always zero (i.e., $q = 0$). Thus, we validate our numerical results in a two-SU scenario, where the number of PU channels, $M = 10$. The number of frames in a SU packet, $h = 1$, and the number of slots in a frame, $c = 10$. We assume that the SU packets are of fixed length. Thus, $\sigma = 1/(ch)$. Figure 2.21 depicts the analytical and simulation results of the normalized SU throughput using the random channel selection scheme and the greedy channel selection scheme. It can be seen that the simulation results match extremely well with the numerical results in both schemes with the maximum difference only 3.84% for the random selection and 4.09% for the greedy selection. It is also shown that, under the same SU traffic load, the greedy channel selection scheme always outperforms the random channel selection scheme in terms of higher SU throughput.

2.6.5 The Impact of Spectrum Sensing Delay

In this section, we investigate the impact of the spectrum sensing delay on the performance of a spectrum handoff process. The spectrum sensing delay considered in this chapter is defined as the duration from the moment that a collision between a SU and PU happens to the moment that the SU detects the collision (i.e., the overlapping time between a SU and PU transmission). Let T_s be the spectrum sensing delay. Therefore, a SU does not need to wait till the last time slot of a frame to realize the collision. It only needs to wait for T_s to realize that a collision with a PU packet occurs and stops the current transmission immediately. In a recent work [9], the spectrum sensing time is considered as a part of the spectrum handoff delay. However, the definition of the spectrum sensing time in [9] is different from the

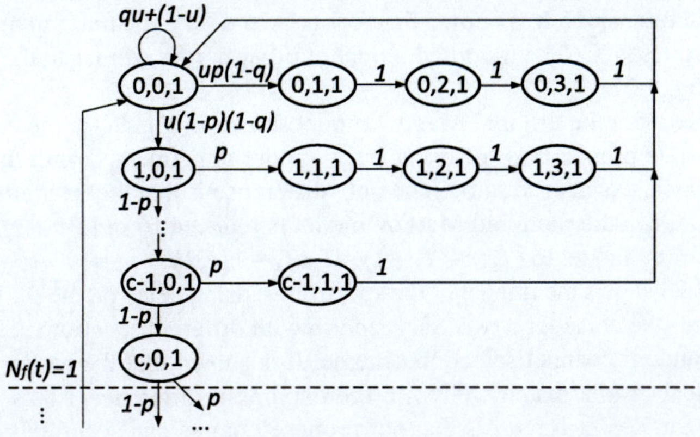

Fig. 2.22 The modified Markov model based on the spectrum sensing delay when T_s equals three time slots

definition considered in this chapter. In [9], the spectrum sensing time only refers to the duration that a SU finds an available channel for transmission after a collision occurs. Thus, the spectrum sensing time can be as low as zero in [9]. In addition, the overlapping time of a SU and PU collision is neglected in [9]. However, the spectrum sensing delay considered in this chapter is not negligible.

The spectrum sensing delay, T_s, can be easily implemented in our proposed three dimensional Markov model with minor modifications. Figure 2.22 shows the first tier of the modified three-dimensional discrete-time Markov chain when T_s

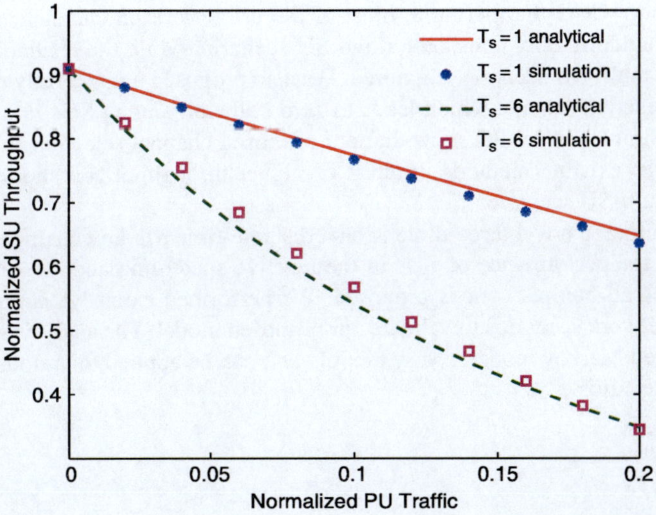

Fig. 2.23 Analytical and simulation results of the normalized SU throughput under different spectrum sensing delay

equals three time slots. It is shown that, for a fixed $N_t(t)$, the maximum number of *Collided* states is T_s. The modified model of other tiers is similar to the first tier as shown in Fig. 2.22.

Compared with the original Markov model shown in Fig. 2.19, the derivation of the steady-state probabilities of the Markov model implemented with the spectrum sensing delay is exactly the same. The only difference is that the total number of the *Collided* states in the modified Markov model is reduced from $[c(c+1)/2]h$ in the original Markov model to $[T_s(c-T_s+1)+T_s(T_s-1)/2]h$.

Figure 2.23 shows the impact of the spectrum sensing delay on the SU throughput performance. We consider a two-SU scenario with different spectrum sensing delay using the random channel selection scheme. It is shown that the numerical results and analytical results match well with the maximum difference 1.83% for $T_s = 1$ and 4.56% for $T_s = 6$. It reveals that our proposed model can accurately predict the SU throughput.

2.7 Conclusion

In this chapter, a proactive spectrum handoff framework in a CR ad hoc network scenario without the existence of a CCC is proposed. Compared with the sensing-based reactive spectrum handoff approach, the proposed framework can achieve fewer disruptions to primary transmissions by letting SUs proactively predict the future spectrum availability and perform spectrum handoffs before a PU occupies the current spectrum. We incorporated a single rendezvous and a multiple rendezvous network coordination scheme into the spectrum handoff protocol design, thus our proposed spectrum handoff framework is suitable for the network scenarios that do not need a CCC. Furthermore, most of the prior work on channel selection in spectrum handoffs only considers a two-SU scenario, while the channel selection issue for a multi-SU scenario is ignored. We also proposed a novel fully distributed channel selection scheme which leads to zero collision among SUs in a multi-SU scenario. Simulation results show that our proposed channel selection scheme outperforms the existing methods in terms of higher throughput and shorter handoff delay in multi-SU scenarios.

Furthermore, a novel three-dimensional discrete-time Markov chain is proposed to analyze the performance of SUs in the reactive spectrum handoff scenario in a two-SU CR ad hoc network is proposed. We performed extensive simulations in different network scenarios to validate our proposed model. The analysis shows that our proposed Markov model is very flexible and can be applied to various practical network scenarios.

References

1. FCC, "Notice of proposed rule making and order," no. 03-222, December 2003.
2. FCC, (2003, November) Et docket no. 03-237. [Online]. Available: http://hraunfoss.fcc.gov/edocs_public/attachmatch/FCC-03-289A1.pdf

3. J. Mitola, "Cognitive radio: an integrated agent architecture for software defined radio," Ph.D. dissertation, KTH Royal Institute of Technology, 2000.
4. I. F. Akyildiz, W.-Y. Lee, M. C. Vuran, and S. Mohanty, "NeXt generation/dynamic spectrum access/cognitive radio wireless networks: A survey," *Computer Networks (Elsevier)*, vol. 50, pp. 2127–2159, September 2006.
5. D. Willkomm, J. Gross, and A. Wolisz, "Reliable link maintenance in cognitive radio systems," in *Proc. IEEE DySPAN*, pp. 371–378, November 2005.
6. S. Mishra, A. Sahai, and R. Brodersen, "Cooperative sensing among cognitive radios," in *Proc. IEEE ICC*, pp. 1658–1663, June 2006.
7. C.-W. Wang and L.-C. Wang, "Modeling and analysis for proactive-decision spectrum handoff in cognitive radio networks," in *Proc. IEEE ICC*, pp. 1–6, June 2009.
8. Y. Zhang, "Spectrum handoff in cognitive radio networks: Opportunistic and negotiated situations," in *Proc. IEEE ICC*, pp. 1–6, June 2009.
9. C.-W. Wang and L.-C. Wang, "Modeling and analysis for reactive-decision spectrum handoff in cognitive radio networks," in *Proc. IEEE GlobeCom*, 2010.
10. L. Yang, L. Cao, and H. Zheng, "Proactive channel access in dynamic spectrum networks," *Physical Communication (Elsevier)*, vol. 1, pp. 103–111, June 2008.
11. T. Clancy and B. Walker, "Predictive dynamic spectrum access," in *Proc. SDR Forum Technical Conference*, Orlando, FL, November 2006.
12. S. Yarkan and H. Arslan, "Binary time series approach to spectrum prediction for cognitive radio," in *Proc. IEEE Vehicular Technology Conference (VTC)*, pp. 1563–1567, October 2007.
13. S.-U. Yoon and E. Ekici, "Voluntary spectrum handoff: A novel approach to spectrum management in CRNs," in *Proc. IEEE International Conference on Communications (ICC)*, 2010.
14. C. Song and Q. Zhang, "Intelligent dynamic spectrum access assisted by channel usage prediction," in *IEEE INFOCOM Conference on Computer Communications Workshops*, 2010, pp. 1–6.
15. S. Geirhofer, J. Z. Sun, L. Tong, and B. M. Sadler, "Cognitive frequency hopping based on interference prediction: Theory and experimental results," *ACM SIGMOBILE Mobile Computing and Communications Review*, vol. 13, no. 2, pp. 49–61, 2009.
16. I. F. Akyildiz, W.-Y. Lee, and K. R. Chowdhury, "CRAHNs: Cognitive radio ad hoc networks," *Ad Hoc Networks*, vol. 7, no. 5, pp. 810–836, July 2009.
17. Y. R. Kondareddy and P. Agrawal, "Synchronized MAC protocol for multi-hop cognitive radio networks," in *Proc. of IEEE International Conference on Communication (ICC)*, 2008, pp. 3198–3202.
18. C. R. Stevenson, G. Chouinard, Z. Lei, W. Hu, S. J. Shellhammer, and W. Caldwell, "IEEE 802.22: The first cognitive radio wireless regional area network standard," *IEEE Communications Magazine*, vol. 47, no. 1, pp. 130–138, January 2009.
19. K. Challapali, C. Cordeiro, and D. Birru, "Evolution of spectrum-agile cognitive radios: First wireless Internet standard and beyond," in *Proc. International Wireless Internet Conference (WICON)*, 2006.
20. J. Mo, H.-S. W. So, and J. Walrand, "Comparison of multichannel MAC protocols," *IEEE Transactions on Mobile Computing*, vol. 7, no. 1, pp. 50–65, January 2008.
21. A. Tzamaloukas and J. J. Garcia-Luna-Aceves, "Channel-hopping multiple access," in *Proc. IEEE International Conference on Communications (ICC)*, June 2000.
22. A. Tzamaloukas and J. J. Garcia-Luna-Aceves, "Channel-hopping multiple access with packet trains for ad hoc networks," in *Proc. IEEE Mobile Multimedia Communications (MoMuC)*, October 2000.
23. H. W. So and J. Walrand, "McMAC: A multi-channel MAC proposal for ad-hoc wireless networks," in *Proc. of IEEE Wireless Communication and Networking Conference (WCNC)*, 2007.
24. P. Bahl, R. Chandra, and J. Dunagan, "SSCH: Slotted seeded channel hopping for capacity improvement in IEEE 802.11 ad-hoc wireless networks," in *ACM MobiCom*, 2004, pp. 216–230.

25. N. C. Theis, R. W. Thomas, and L. A. DaSilva, "Rendezvous for cognitive radios," *IEEE Transactions on Mobile Computing*, vol. 10, no. 2, pp. 216–227, February 2010.
26. J. Neel, "Analysis and design of cognitive radio networks and distributed radio resource management algorithms," Ph.D. dissertation, Virginia Polytechnic Institute and State University, September 2006.
27. S. Sengupta and M. Chatterjee, "Designing auction mechanisms for dynamic spectrum access," *Mobile Network and Applications*, vol. 13, no. 5, pp. 498–515, 2008.
28. D. Niyato and E. Hossain, "Competitive pricing for spectrum sharing in cognitive radio networks: Dynamic game, inefficiency of nash equilibrium, and collusion," *IEEE Journal on Selected Areas in Communications*, vol. 26, no. 1, pp. 192–202, January 2008.
29. L. Cao and H. Zheng, "Distributed spectrum allocation via local bargaining," in *Proc. IEEE Conference on Sensor, Mesh and Ad Hoc Communications and Networks (SECON)*, pp. 475–486, September 2005.
30. E. Ferro and F. Potorti, "Bluetooth and Wi-Fi wireless protocols: A survey and a comparison," *IEEE Wireless Communications*, vol. 12, no. 1, pp. 12–26, February 2005.
31. F. Sivrikaya and B. Yener, "Time synchronization in sensor networks: A survey," *IEEE Network*, vol. 18, no. 4, pp. 45–50, July–August 2004.
32. H. Su and X. Zhang, "Cross-layer based opportunistic MAC protocols for QoS provisionings over cognitive radio wireless networks," *IEEE Journal on Selected Areas in Communications (JSAC)*, vol. 26, no. 1, pp. 118–129, January 2008.
33. H. Su and X. Zhang, "Channel-hopping based single transceiver MAC for cognitive radio networks," in *Proc. 42nd Annual Conference on Information Sciences and Systems (CISS)*, pp. 197–202, 2008.
34. L.-C. Wang and C.-W. Wang, "Spectrum handoff for cognitive radio networks: Reactive-sensing or proactive-sensing?" in *Proc. IEEE International in Performance, Computing and Communications Conference (IPCCC)*, pp. 343–348, December 2008.
35. F. Gebali, *Analysis of Computer and Communication Networks*. Springer, 2008.
36. S. Tang and B. L. Mark, "Modeling and analysis of opportunistic spectrum sharing with unreliable spectrum sensing," *IEEE Transactions on Wireless Communications*, vol. 8, no. 4, pp. 1934–1943, 2009.

Chapter 3
Environment–Mobility Interaction Mapping for Cognitive MANETs

Irene Macaluso, Timothy K. Forde, Oliver Holland, and Keith E. Nolan

Abstract Cognitive MANETs are likely to be complex radio systems. We already know that no single MANET solution can address all environments that may be encountered; such is the rationale of an ad hoc network that it must address the networking demands of unforeseen scenarios. Rather, a cognitive MANET should be viewed as a feature-rich radio system, i.e. one which has access to a range of radio and network components, each suited to different demands. Such a reconfigurable system requires cognitive functionality to self-architect the radios when they are deployed in addition to the cognitive functionality required for the various layers to self-organise. However, any cognitive decision-making process requires awareness of the world for which it is trying to optimise the system. This chapter introduces the concept of an *environment–mobility interaction* map, a persistent internal representation of the network which captures the presence of areas in the network's environment in which particular, sustained, mobility dynamics are observed. Such a self-generated map enables the cognitive MANET to plan a response to challenges brought about by these network dynamics.

3.1 Introduction

Cognitive mobile ad hoc networks (MANETs) will have an increasing role in the wireless communications realm as there is a strong emerging general trend that sees the world moving away from predominantly centrally pre-planned structures to more 'Wifi-like' approaches in which nodes are deployed randomly and have to 'fit in' to the environment in which they are deployed. This points to a self-planning or self-architecting wireless system consisting of a set of nodes that can form dense small cell networks, provide sparse regional area network coverage as well as form distributed structures. Nodes in such networks will need to be highly cognitive and hence capable of being *persistently aware of environmental constraints* through the many observations they can make.

In this chapter we attempt to ascertain if there is a way for a network to create a view of long-lasting network-wide mobility dynamics in spite of the fact that

I. Macaluso (✉)
CTVR, Trinity College, Dublin, Ireland
e-mail: macalusi@cs.tcd.ie

F.R. Yu (ed.), *Cognitive Radio Mobile Ad Hoc Networks*,
DOI 10.1007/978-1-4419-6172-3_3,

the cognitive MANET is transient and individual nodes are just *passing through*. In order to plan strategically, nodes (and the network) need access to information that tells them something about their likely future experiences. Otherwise they can only react to what they currently experience – and in the case of a MANET the local observations of a node are likely to be quite chaotic and random. Any sense of order or patterns can only be seen in a MANET by abstracting features of the network away from individual nodes, rather focussing on *all and any* nodes that might experience the same *persisting* and *orderly pattern* of events.

We explore the potential for a cognitive MANET to create a network-wide view of persistent features of the mobility dynamics that are experienced in various areas of the network, e.g. cross-paths, congregation areas, pinch-points and pathways. If each node, and thereby the network, can ascertain some understanding of the behaviour of a node's mobility in different areas, it may be able to plan the network in a better way, rather than each node simply reacting naively in each instance in ignorance of the behaviour having occurred persistently in the area heretofore. This chapter builds on some of the concepts presented by the authors [1].

The next section describes some pertinent features of MANETs and cognitive networks, drawing attention to the fact that cognition already exists in MANETs, although it is rarely labelled as such. Section 3.3 then looks at how mobility affects the operation of a MANET. In Section 3.4 we discuss the process of planning in a MANET, looking at how MANETs have coped with self-organisation heretofore and develop the concept of an *environment–mobility interaction* map. In Section 3.5, we present the details of our simulation of the *environment–mobility interaction* map concept. Section 3.6 concludes the chapter.

3.2 The MANET as a Cognitive Network

In this chapter we focus on the effects of mobility on a cognitive MANET and look to ways in which we can improve the network's ability to plan for mobility dynamics experienced in a given environment. Before exploring how mobility effects the operation of a cognitive MANET, we should understand what makes a cognitive MANET different to any other cognitive wireless network, and in understanding these differences we will see that such a network will require more self-generated and enhanced awareness of the world in which it operates.

MANETs are essentially distributed, peer-to-peer systems. They were initially proposed for situations in which the existing infrastructure-based networks either failed or did not exist. As such, the MANET is designed to be a robust, independent, relief network, resistant to failure [2, 3]. Decision making of any kind in a MANET is hard if not impossible; the Fischer–Lynch–Patterson impossibility result clearly delineates the limits of deterministic decision making in systems of the kind characterised by open mobile ad hoc networks [4].

All of the nodes in a MANET are autonomous with regard to their mobility. As such, no entity controls the mobility of any node other than the node itself. This autonomy makes the mobility of nodes somewhat unpredictable; any given node's mobility or immobility is at the whim of the entity hosting the radio, whether human, machine or other. The absence of any stationary or fixed network entities, as exist in most other network architectures, increases the complexity of organisation; fixed nodes act as natural cluster heads but their presence cannot be guaranteed in a MANET.

All nodes deal with the challenge of end-to-end network planning, e.g. multihop routing from source nodes to destination nodes via intermediate nodes. As such, a MANET requires more than the ability to create and maintain ad hoc links between adjacent nodes; it requires the ability to organise and maintain logical and physical connections across multiple nodes, at multiple layers of the stack, which may not have direct connectivity with, or awareness of, each other.

Given these challenges, the MANET is an inherently self-organising system, reliant on the ability of its distributed, autonomous, mobile nodes to support communication from end-to-end and from link-to-link in an adaptive and dynamic manner. That self-organisation can range from immediate, reactionary adaptions and fine-tuning, such as PHY-level transmit-power-control changes, to more involved, considered and longer-term network planning, such as multihop routing and traffic loading.

Cognition, whether using the definition used in the cognitive radio field [5] or definitions derived from the artificial intelligence community [6], implies that a radio or network can learn about its environment and make decisions based on what it learns and then act on these decisions. MANETs are already performing these actions in the form of multi-hop routing protocols which *learn* routes, then *plan* an end-to-end route and finally route date from end-to-end. In essence, the MANET already contains cognitive entities, although they are rarely labelled as such.

However, within the wireless networking domain we suggest that a cognitive network will exhibit a number of additional features that distinguish it from other wireless networks. A cognitive network, and its component nodes, should be feature-rich, i.e. the nodes should have access to a suite of components so that they can make choices about what techniques to use and when to use them. A cognitive MANET node should be able to reconfigure its various layers, the components within those layers and the parameters of those components. Software-defined radios are often cited as ideal platforms for cognitive radios as it is relatively easy to manipulate and reconfigure a complex radio system in software [7]. Choices, options and decisions arise *everywhere* in a cognitive MANET; at every layer of the conceptual communication stack and, if the cognitive MANET is also a dynamic spectrum access system, in the organisation of access to spectrum. At the network layer choices concern the appropriate routing and group management protocols. At the medium access and physical layer choices are even more complex and interconnected; everything from modulation to FEC, multiplexing, etc., is up for grabs. Then in the dynamic spectrum domain, choices may range between licenced and unlicenced spectrum, channelisation of said spectrum and tolerable adjacent channel interference levels.

Such a reconfigurable network both necessitates the use of cognitive functionality and motivates its exploration and development. Furthermore, when less structure, whether in the form of lighter radio standards or more technical and service neutral spectrum policy or etiquette is imposed on the network, then the network will have more degrees of freedom. In the absence of cognitive functionality to impose order, such freedom would likely lead to *chaos*, i.e. an environment in which radios fail to converge to a state that allows for communication between them.

To elaborate the point, we can look to the choices that a cognitive MANET could make if operating an IEEE 802.16-*like* MAC and PHY. We say *like* as IEEE 802.16 itself is not a cognitive radio standard [8]. IEEE 802.16 has many variants, both approved; IEEE 802.16, 802.16a, 802.16d and 802.16e, and pending approval or in development; IEEE 802.16h and IEEE 802.16j. Each of these variants represents a different component or building block of a radio system. IEEE 802.16 introduced a single carrier-based PHY. IEEE 802.16a introduced OFDM to the PHY. IEEE 802.16d introduced Fixed Mesh MAC capabilities in addition to the PMP MAC of 802.16. IEEE 802.16e introduced mobility to the PMP MAC and OFMDA at the PHY. Additionally, new enhancements for the IEEE 802.16 family of standards are under development. 802.16h addresses the use of WiMAX systems in unlicenced spectrum and, as such, investigates co-existence mechanisms for adjacent independent systems. Of more interesting note to the development of MANETs, 802.16j addresses the creation of so-called mobile multi-hop relay (MMR) systems. These networks would consist of basestation couple with fixed, nomadic and/or mobile relay stations.

Taken together, as a suite of components, these enhancements would give a network of enabled cognitive radios the capability to address the needs of a wide array of topologies that may be formed by a MANET; from fixed to nomadic to mobile, from licenced to unlicenced, from best effort traffic support to real-time support. But while vendors currently sell radio solutions that offer one solution (IEEE 802.16e) or another (IEEE 802.16d) as distinct products, a cognitive self-architecting ad hoc system would assemble the required radio components to address the challenges of a given deployment scenario.

So, from our perspective there are two broad categories of cognition within a MANET; cognition at the layer/component level and cognition at the system level. While all cognition in the network is concerned with making decisions, the timescale on which decision are made varies from immediate, and often unilateral, node-specific internal adaptations to longer-term network-wide decisions that require multilateral negotiation, consent and execution. Some decisions are internal to the radio, perhaps optimising the way it uses its computational resources, while others are external to the radio, optimising the way that it interacts with other radios. These decisions, which involve interaction with, and consideration of, other radios in the network can be classified as self-organising or self-architecting decisions.

We distinguish self-organising decisions as cognition manifesting itself at the layer/component level. Here, components or layers plan on a node-to-node link-level (PHY), neighbourhood level (MAC) and network level using the same protocols or algorithms to achieve their objectives.

This contrasts with self-architecting decisions which involve system-level cognition where the MANET network, through its nodes, is able to choose the right combination of building blocks, i.e. layers, components, protocols, algorithms, to make a functioning MANET. In this chapter we are interested in developing the awareness of a cognitive MANET towards enabling it to plan its architecture. Awareness of the world in which the reconfigurable cognitive MANET operates is a prerequisite to it being able to adapt itself.

3.3 Mobility Perturbs the MANET

Regardless of where the cognition lies in a MANET, whether at the PHY or the network-layer, one of the biggest hurdles that the cognitive MANET must overcome is the disruption caused by autonomous network mobility. The inherent autonomous mobility of a MANET clearly necessitates an ability to deal with unforeseen and changing network topologies by creating ad hoc networks that deal with ad hoc scenarios. Mobility and MANETs go hand-in-hand, the former necessitating the later. Whenever the MANET has reached a steady state, mobility has the effect of perturbing the network out of that steady state; the state of a wireless network is represented by both the internal state of the nodes and the state of the local links. A self-correcting, self-organising and self-architecting MANET has to overcome the various degrees of state perturbance caused by mobility.

The principle way in which node and network mobility upsets the network is that it causes changes in the link stability and link density characteristics of a network. Link stability describes the stability of links established between a given node and its immediate one-hop neighbours. Stability can be measured by the average duration for which a node-to-node link lasts or the quality of that link over time - a link may exist but the SNR experienced over it may change the capacity of the link. Neighbourhood links are the basic building block of a MANET. Without connection to neighbouring nodes, a node is not connected to the broader MANET network. When it comes to self-organising end-to-end protocols at the higher layers, i.e. the network, transport, session and application layers, neighbourhood link stability is a prerequisite. A lack of stability at the link-level inhibits certain types of higher layer transactions; if the MAC is constantly trying to establish links with new neighbours the creation of communication paths beyond the immediate neighbourhood, i.e. distant, multihop communication, is impossible. One of the factors affecting link stability is the speed of nodes relative to each other; if nodes are passing in and out of each others communication ranges at speed then a link will not endure.

Node density, or node degree, describes the number of one-hop neighbours that a node has or is aware of. The presence of neighbours can be established at the PHY layer, the MAC or the network layer. High-node densities and low-node densities characterise dense and sparsely connected networks; there are pros and cons to each. A dense network means that each node has many immediate one-hop neighbours and, if each of these neighbours is relatively stable, then the node has a more robust connection to the network and more redundancy and options when it comes to how it communicates with other nodes. On the other hand, unless the protocols and algorithms in use have been designed for a dense network, such that message flooding or broadcasting and relaying is managed to reduce duplicated transmissions within the same area, then the local links will become very congested and the network will not converge. A sparse network, whilst not suffering from problems of network congestion, will suffer from a paucity of local links which gives the network little choice in managing traffic load balancing and having alternative options in the case of failing routes.

So, a large array of ad hoc network topologies may be formed. The transience of these topologies will vary from place to place and from time to time. The cognitive MANET topology patterns which emerge may mimic known, or previously planned for, *standard* topologies, e.g. a topology may arise with a stationary node (basestation) and a group of mobile nodes that operate within its transmission radius (mobile subscribers). For such a network topology there are many proven ways to organise the PHY, the MAC and the upper layers. Cognitive protocols and algorithms within the network impose order and bring about a stable system in spite of mobility.

Given that mobility affects the operation of a MANET at such a basic level, through the effect it has on node-to-node links, mobility models are extensively used in the domain to develop and stress test novel network protocols and algorithms. The standard method of evaluation for MANET technologies is to impose a number of mobility models on a simulated MANET network in order to benchmark the efficiency or efficacy of these techniques. Mobility models range from random-waypoint models and group mobility models [9, 10] to more realistic models such as Manhattan models [11]. However, while mobility models are *useful* in development, they are *useless* in practice unless the network itself can create a representation (i.e. model) of the mobility dynamics it experiences.

But, when the mobility characteristics of a network change beyond the design scope of a particular radio or network component, then the system requires recalibration. MANET routing protocols in particular have been shown to have optimal ranges of operation with regard to the mobility characteristics and topologies of the underlying networks [1, 12]. In [1] the authors developed a system that enabled nodes to detect certain network dynamics and relate these observations to prior knowledge that was made available to each node such that it could begin a decision-making process regarding reconfiguration of the routing protocol. The process in [1] was a recognition-triggered decision-making process; once a node recognised the type of dynamics it was experiencing it triggered the decision-making process locally at that node. Nodes had no representation of the network environment, and no way to tell in what part of the network they were since no location-based services were used, so that if a node moved back and forth between two areas dominated by distinct and different network dynamics it acted as though it was encountering these dynamics for the first time. In [1] each node created an internal representation of its local world, i.e. the dynamics it observed and those of its neighbours. However, as this system did not use location-based services such as GPS, it could not detect its own movement from one area to another; a node could simply observe that its neighbours changed but that is not enough to infer whether it has moved to a new area or the other nodes have moved away. The node's internal representation of its world moved with it and could not be shared with nodes that came to the same area of the network after it had left.

In this regard, deliberative planning of changes to the network was not possible as each node could only act on immediate local and neighbourhood knowledge, knowledge which could appear to each node to oscillate without recognisable pattern as it passed from one area to another.

In this chapter, we relax the assumption in [1] that the network has no access to a system such as GPS, and advance that work by developing an internal representation of the network at each node that captures the persistence of certain mobility dynamics as they occur across a given environment. Using this approach, it does not matter that each node may observe different dynamics as it passes through the environment as it adds these observations to a representation that is shared by the network. When a node moves into a new area of the networking environment, it knows which part of the map represents that area, if the map has been created already, and the node can then update the map for the area while it too is present. With access to a map, nodes can see beyond local oscillations in observations allowing for decision-making techniques, such as those used in [1], to be used with greater certainty and for greater changes to the network architecture. When a network has a better internal representation of its world it can plan for change by weighing the cost of disruption caused by change against the benefits of using more suitable components, algorithms etc.

3.4 MANET Planning In Spite of Mobility

Looking back to the example of IEEE 802.16, a significant challenge for each of the IEEE 802.16 variants is the organisation of the schedule for the TDMA MAC; a task that can be approached using either centralised or distributed techniques [13]. In 802.16j the challenge is to develop an efficient MAC that allows for much more dynamic network topology changes and the timely organisation of schedules for the five classes of traffic supported by the standard which range from best-effort to real-time. The 802.16j MAC addresses the challenges of a TDMA MAC for an actively mobile MANET. So, given a particular networking environment, i.e. as characterised by the physical layout of the space, the spatial distribution of the nodes, their mobility characteristics and the spectrum in use, certain networking choices, e.g. choices within the IEEE 802.16 suite of standards, are more effective than others. But for a cognitive MANET, using an IEEE 802.16-*like* MAC/PHY, to plan the appropriate radio system for the network, it must be able to create a representation of those aspects of the networking environment that influence the efficient operation of its candidate MAC and PHY options.

Oftentimes in an ad hoc network mobility-related events happen too fast for any entity, whether at the local radio level or at the broader network level, to analyse the situation and plan an optimal response. More often than not the network simply reacts to changing events according to predetermined triggers, e.g. if the SNR experienced at the PHY deteriorates beyond a threshold then a lower-order modulation scheme is adopted. Such an immediate change requires no *thought*, no deliberation, simply reaction. To a MANET radio, constant, locally perceived change can seem quite random.

However, from a higher perspective, the network may have more order, i.e. there may be patterns of network behaviour or patterns in the underlying physical environment that, while undetectable locally by myopic nodes on their own, can be seen by stitching together local views to form a larger picture of the network's behaviour.

As we mentioned in Section 3.2, MANETs *do* exist already and *do* manage to function and organise themselves to some degree, notably at the network layer by use of advanced multi-hop routing protocols. These protocols allow multihop-distant, distributed, mobile nodes to communicate in such a way that they collectively build up a view of the logical links that exist throughout the network. But the routing information itself is as transient as the nodes are, and routing protocols are constantly reacting to the topology in a *just-in-time* fashion. The view that they assemble is almost always incomplete and its validity is immediately decaying, eventually being deleted from a node's memory if not refreshed. It should be noted that many routing protocols have been proposed for MANETs taking many forms; reactive source-routed [14], table-driven [15], GPS-assisted [16, 17]. Other routing protocols use cognitive techniques in terms of link management, e.g. OLSR link hysteresis [18] and others [19], while some use location data to predict the stability of links and routes [20].

Regardless of the actual method, routing protocols enable the network's radios to build an internal representation of their external world. From this representation an individual node is able to plan; to choose the best multihop route available to its destination based on its internalised understanding of the world beyond.

Referring to Fig. 3.1a, a network of nodes is depicted operating within a space; the gray blocks represent physical barriers. Corresponding to the actual distribution of nodes at time T_1 is a routing table, Fig. 3.1c created sometime thereafter at time T_2 ($>T_1$). As this is a MANET, by time T_3 ($>T_2$) the distribution of individual nodes may have changed, Fig. 3.1b, resulting in the routing table created at time T_2 being partially or completely out-of-date. But by time T_4 ($>T_2$) an updated routing table, as shown in Fig. 3.1d, is created.

But in spite of the fact that for any given physical environment nodes are *transient* and the routing information is *perishable*, underlying pertinent network characteristics, which are a function of the network's mobility and the environment it inhabits, may *persist*. Even though mobility is autonomous, the mobility actions taken by nodes are both constrained and induced by the environment in which the node operates. Physical barriers, e.g. walls, buildings, rivers, limit the range of movement of a node/network. Referring to Fig. 3.1a, it can be seen that the nodes are outside the buildings. Also, depending on the characteristics of the radio host, i.e. whether it is human-held, vehicle mounted or embedded in a stationary object, the environment will also promote or induce certain behaviours. Generally, humans walk on pavements and pass through doorways whereas cars travel along streets. The interaction of mobile nodes and their physical environment has a significant effect on the ability of a given algorithm (e.g. a PHY FEC scheme) or protocol (e.g. an IEEE 802.16

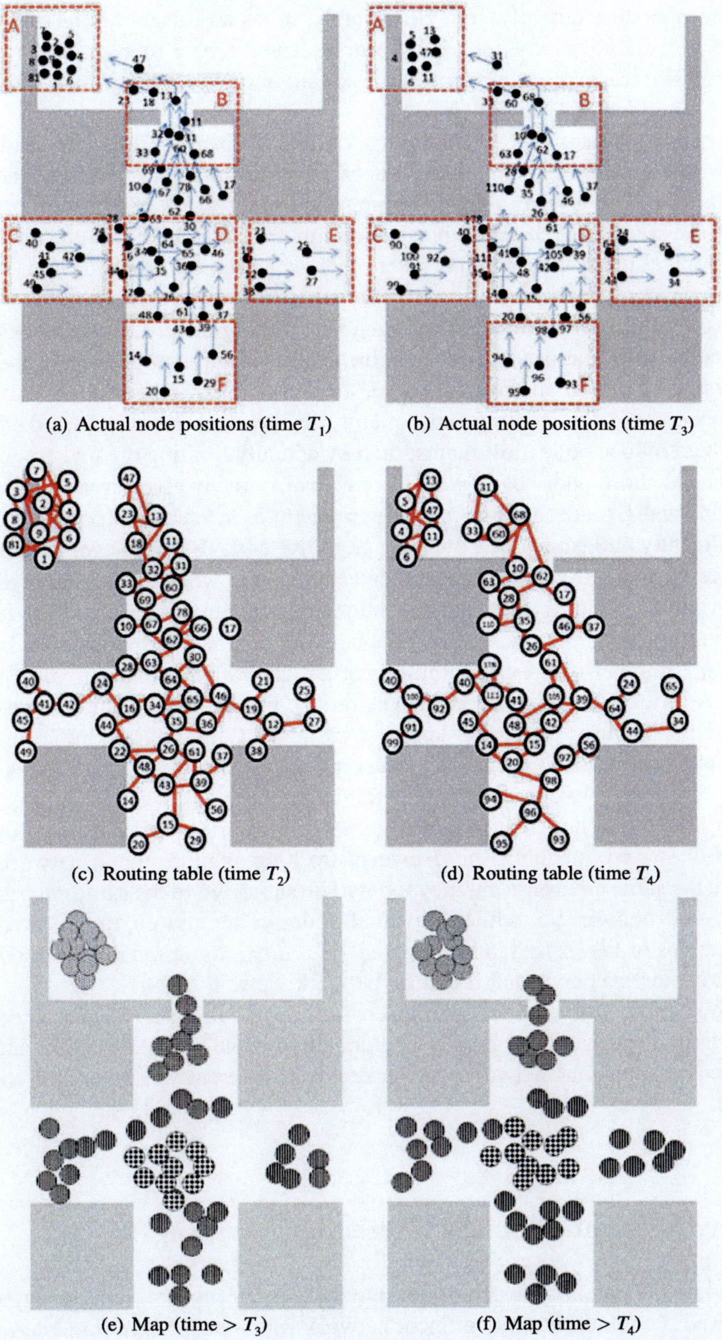

(a) Actual node positions (time T_1) (b) Actual node positions (time T_3)

(c) Routing table (time T_2) (d) Routing table (time T_4)

(e) Map (time > T_3) (f) Map (time > T_4)

Fig. 3.1 A network as viewed over time from 3 perspectives; actual node positions **(a)**, **(b)**; corresponding routing tables **(c)**, **(d)**; maps of pertinent features **(e)**, **(f)**

MAC) to operate efficiently, if at all. Notably, it has an impact on the node degree and the link stability experienced by a node. Capturing a representation of these persistent features, which we term *environment–mobility interactions*, can further enable a cognitive MANET to plan.

The question arises as to whether it is possible to abstract from the world as it is, i.e. from the actual unknown (from a network point-of-view) reality of individual nodes locations, to a fairly transient representation of the world in terms of routes through the creation of routing tables which capture a perishable, but detailed, snapshot of logical links, to a more persistent representation of the world in the form of a map of perceived and pertinent *environment–mobility interactions*.

Unlike the information that is captured in a routing table, i.e. precise node- and link-specific information, the information that we want to represent is specific to activity that occurs in an area. As noted, the physical environment drives a node's mobility behaviour by creating paths, plains, pinch-points, landmarks and other features which induce node movements, thereby actually limiting the real autonomy of the nodes. In short, nodes can't walk through walls. In any given area these features, e.g. doors, walls, roads, are likely to be permanent or at least their presence is likely to significantly outlast that of a transient cognitive MANET. However, they are only of interest to a cognitive MANET, and detectable by it, when the cognitive MANET exists in that area and its mobile nodes' interactions are influenced by those features.

Returning to Fig. 3.1a we see that the network can be divided, by visual inspection, into a number of zones exhibiting certain *environment–mobility interactions*. In zone A, a ticket-booth, or other such landmark, causes nodes to queue and cluster, i.e. good link stability, high node density. In zones B, C, E and F, nodes follow a path and in zone D the two paths intersect. A certain approach to MAC/PHY management may be appropriate in one zone, it may fail in another.

In Fig. 3.1b, the nodes have moved to new positions; continued down the path, finished or started queueing, etc. However, so long as more (new) network nodes populate the same environment, they too will be subjected to the same *environment–mobility interactions*. So, while individual nodes come and go, as reflected in the routing tables of Fig. 3.1e, f, a *environment–mobility interaction map* created by the network captures a persistent, if coarse, network view.

Of course, if no more nodes enter an environment then the ad hoc network ceases to exist in that space and the map is abandoned, or if the number of nodes increases or decreases then that network may take on a different characteristic changing the map.

3.5 Environment–Mobility Interaction Mapping

To elucidate the potential feasibility or intractability of this problem, we investigate how MANETs could identify pertinent network *environment–mobility interactions*.

The nodes' mobility pattern emerges from the interaction of a collection of radios and a man-made environment. As nodes' movements are constrained by the environment features, we classify radios according to the following definitions:

- *Way*: a radio that moves along with some of its neighbours in a certain direction.
- *Cross-way*: a mobile radio whose neighbours move in many different directions.
- *Passage*: a mobile radio that experiences an increase in the density of neighbours.
- *Landmark*: a static radio whose neighbours are also static.
- *Other*: a radio that does not belong to any of the previous definitions.

In the following we will describe the procedure to build a *environment–mobility interaction* map. Such representation should be flexible enough to account for sensing uncertainty, thus allowing it to be robustly modifiable as new information is added or the old one is updated. Furthermore the entire process should be autonomous and unsupervised. The problem of space representation has been the subject of research both in the robotics and in the cognitive science community [21]. Depending on the application, spatial knowledge can be represented at different levels, from a coarse topological map to a fine-grained metric map. Metric maps are usually accurate, but they do not scale well with the size of the environment. Topological maps represent the environment as a graph, where each node is an area or a place and the link between nodes implies a connection between places. Their structure allows them to naturally interface with symbolic planners [22]. As the accuracy provided by metric maps is not required in this context, the most suitable choice for our purposes is a topological map.

Radios are randomly distributed in a urban-like environment, which is composed by two orthogonal streets (see Fig. 3.1a). A congregation point, around which radios gather with a given probability, is also included. In order to build the *environment–mobility interaction* map, radios have to estimate their mobility state and communicate it to their neighbours: each radio periodically sends its position, its class (landmark, way, etc) and its direction of motion. Periodic beaconing is highly likely to already exist at some layer of the MANET, this data may piggy-back an existing beacon.

Each radio classifies itself based on its current direction of motion and its neighbors' direction of motion, according to the definition above. In the example in Fig. 3.2a, the radio highlighted as N1 moves in the same direction as the majority of its neighbours and it accordingly identifies itself as a way radio (cf. Fig. 3.2b). The movement information required to estimate the device's mobility state is provided by GPS measurements.

Several sources of error might hinder the self-identification process performance. For example, the GPS measurements are likely to be inaccurate or messages containing the neighbours' mobility state can be lost. However, we aim to get a representation which should not be precise at the level of a single radio, but that arises from the merging of the information provided by all the devices in the network. Therefore possible errors on the self-identification process are unlikely to deteriorate the pattern emerging from a set of radios.

Each device uses the location and the identity of all the radios in the network to segment the environment into areas. Each area, i.e. a portion of the environment occupied by radios presenting the same mobility pattern, constitutes a node in the topological map, as illustrated in Fig. 3.2a, b. The segmentation of the

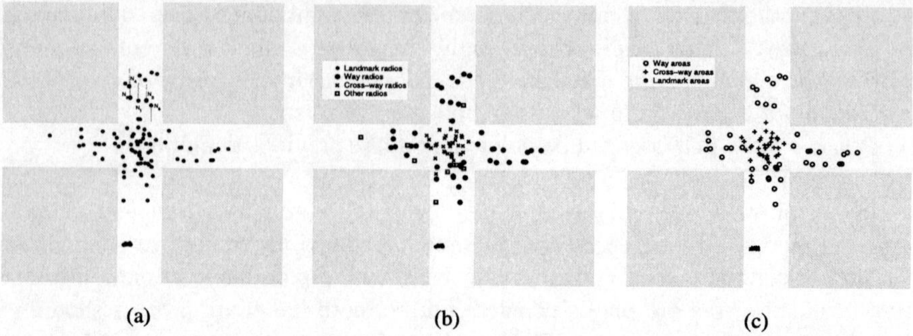

Fig. 3.2 **(a)** Radios in the simulated environment. Node N1 and the majority of its neigbours display the same mobility behaviour. The nearby N5 node moves in the opposite direction. **(b)** Nodes self-identification. Node N1 and nodes N2, N3 and N4 identify themselves as way radios; node N5 does not belong to any of the above definitions. **(c)** Segmentation of the environment into areas

environment in different areas takes into account the state of the nodes in the previous k periods and integrates all the observations performed during this temporal window. As soon as new information is available, the oldest one is discarded. This sliding window mechanism allows the representation to be robust to errors and adaptable to changes in the environment-radios situation. A link between two nodes i and j in the topological map exists if the radios previously located in the area corresponding to node i moved into the area corresponding to node j or vice versa. Each link may have a weight which depends on the number of devices that passed from node i to node j or vice versa.

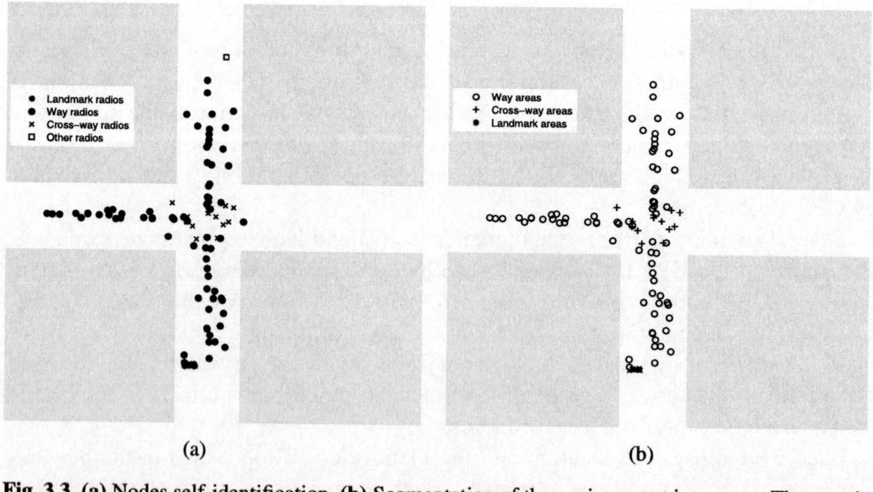

Fig. 3.3 **(a)** Nodes self-identification. **(b)** Segmentation of the environment into areas. The map is updated according to the radios operating in the environment: the area east of the intersection is removed (cf. Fig. 3.2c)

It should be highlighted that the map, or parts of it, is maintained as long as radios are operating on that environment, and it is eventually abandoned or altered according to the current situation. For example, a node is removed in the topological map shown in Fig. 3.3 as in the corresponding area radios are not longer operating.

3.6 Concluding Discussion

MANETs already exhibit some level of cognition. Nodes in a MANET are able to perceive and monitor their environment; to a certain extent they can plan activities to acquire additional knowledge; they are able to modify their behaviour in response to changes in the environment. For example, many different protocols have been proposed to estimate and update the network routing tables.

Currently, however, the MANETs use procedural knowledge to accomplish specific tasks. We believe that allowing a MANET to access different kinds of knowledge and to build a representation which is relatively stable over time is a key feature to extend its capabilities. This is the case when considering the self-configuration of single node or the self-architecting of the entire network. If *network maps* of the kind proposed in this chapter are developed for cognitive MANETs to address different aspects of the network environment then decision-making techniques, more deliberative than that described in [1], can be developed to manage a cognitive MANET. The environment–mobility interaction map discussed in the previous sections is to be considered as one example of the integration of different information into a single assessment of the environmental situation. Nonetheless, it exploits some cognitive capabilities [6], namely:

- Recognition of situations as instances of known patterns: radios recognize the mobility pattern displayed by themselves or other devices in the network. This process occurs in a single cycle and it operates on the output of the perceptual system and on the information conveyed by other agents.
- Categorization of situations: radios are able to assign the perceived situations (mobility patterns) to known categories (way, landmark, etc).
- Situation assessment: radios are able to combine the information coming from the perceptual system and the communication with other devices and create a model of the current environment-radios situation.

Although not explicitly addressed in the simulation, this representation also enables other cognitive capabilities, such as prediction and planning. Both functionalities require a model of the environment along with a model of the effects of the agent's actions.

Acknowledgments This material is based upon work supported by Science Foundation Ireland under Grant No. 03/CE3/I405.

References

1. Forde, T.K., Doyle, L.E., O'Mahony, D., "Ad hoc innovation: Distributed decision making in ad hoc networks", DOI: 10.1109/MCOM.2006.1632660, 2006, pp: 131–137.
2. Burbank, J.L., Chimento, P.F., Haberman, B.K., Kasch, W.T., "Challenges of military tactical networking and the elusive promise of MANET technology", Communications Magazine, IEEE, Vol. 44, No. 11 DOI: 10.1109/COMM.2006.248156, 2006, pp: 39–45.
3. Tsunemine, T., Kadokawa, E., Ueda, Y., Fukumoto, J., Wada, T., Ohtsuki, K., Okada, H., "Emergency urgent communications for searching evacuation route in a local disaster", Consumer Communications and Networking Conference, 2008. CCNC 2008. 5th IEEE, DOI: 10.1109/ccnc08.2007.267, 2008, pp: 1196–1200.
4. Fischer, M.J., Lynch, N.A., and Paterson, M.S., "Impossibility of distributed consensus with one faulty process," *Journal of the ACM*, Vol. 32, No. 2, Apr. 1985, pp. 374–82.
5. Doyle, L.E., "Essentials of Cognitive Radio (The Cambridge Wireless Essentials Series)", Cambridge University Press; 1 edition, 30 April 2009.
6. Langley, P., Laird, J., and Rogers, S., "Cognitive architectures: Research issues and challenges," Cognitive Systems Research, Vol. 10, no. 2, 2009, pp. 141–160.
7. Nolan, K.E., Sutton, P.D., Doyle, L.E., Rondeau, T.W., Le, B., Bostian, C.W., "Dynamic spectrum access and coexistence experiences involving two independently developed cognitive radio testbeds", New Frontiers in Dynamic Spectrum Access Networks, 2007. DySPAN 2007. 2nd IEEE International Symposium on, DOI: 10.1109/DYSPAN.2007.43, 2007, pp: 270–275.
8. Hossain, E., "IEEE802.16/WiMAX-based broadband wireless networks: Protocol engineering, applications, and services", Communication Networks and Services Research, 2007. CNSR '07. Fifth Annual Conference on, DOI: 10.1109/CNSR.2007.37, 2007, pp: 3–4.
9. Ariyakhajorn, J., Wannawilai, P., Sathitwiriyawong, C., "A comparative study of random waypoint and Gauss-Markov mobility models in the performance evaluation of MANET", Communications and Information Technologies, 2006. ISCIT '06. International Symposium on, DOI: 10.1109/ISCIT.2006.339866, 2006, pp: 894–899.
10. Biradar, S.R., Sarma, H.H.D., Sharma, K., Sarkar, S.K., Puttamadappa, C., "Performance comparison of reactive routing protocols of MANETs using group mobility model", 2009 International Conference on Signal Processing Systems, DOI: 10.1109/ICSPS.2009.56, 2009, pp: 192–195.
11. Gowrishankar, S., Sarkar, S., Basavaraju, T.G., "Simulation based performance comparison of community model, GFMM, RPGM, Manhattan Model and RWP-SS mobility models in MANET", Networks and Communications, 2009. NETCOM '09. First International Conference on, DOI: 10.1109/NetCoM.2009.31, 2009, pp: 408–413.
12. Doyle, L.E., Kokaram, A.C., Doyle, S.J., Forde, T.K., "Ad hoc networking, Markov random fields, and decision making", Signal Processing Magazine, IEEE, Vol. 23, No. 5, 2006 , pp: 63–73.
13. Abu Ali, N.A., Taha, A.-E.M., Hassanein, H.S., Mouftah, H.T., "IEEE 802.16 Mesh Schedulers: Issues and design challenges", Network, IEEE, Vol. 22, No. 1, DOI: 10.1109/MNET.2008.4435904, 2008, pp: 58–65.
14. Ahmad, S., Awan, I., Waqqas, A., Ahmad, B., "Performance analysis of DSR & extended DSR protocols", Modeling & Simulation, 2008. AICMS 08. Second Asia International Conference on, DOI: 10.1109/AMS.2008.72, 2008, pp: 191–196.
15. Rasheed, T., Javaid, U., Jerbi, M., Al Agha, K., "Scalable multi-hop ad hoc routing using modified OLSR routing protocol", Personal, Indoor and Mobile Radio Communications, 2007. PIMRC 2007. IEEE 18th International Symposium on, DOI: 10.1109/PIMRC.2007.4394079, 2007, pp: 1–6.
16. Blazevic, L., Le Boudec, J.-Y., Giordano, S., "A location-based routing method for mobile ad hoc networks", Mobile Computing, IEEE Transactions on, Vol. 4, No. 2, DOI: 10.1109/TMC.2005.16, 2005, pp: 97–110.

17. Blazevic, L., Buttyan, L., Capkun, S., Giordano, S., Hubaux, J.-P., Le Boudec, J.-Y., "Self organization in mobile ad hoc networks: the approach of Terminodes", Communications Magazine, IEEE, Vol. 39, No. 6, DOI: 10.1109/35.925685, 2001, pp: 166–174.
18. Ali, H.M., Busson, A., Veque, V., "Network layer link management using signal strength for ad-hoc networks", Computers and Communications, 2009. ISCC 2009. IEEE Symposium on, DOI:10.1109/ISCC.2009.5202353, 2009, pp: 141–146.
19. Hua, E.Y., Haas, Z.J., "An algorithm for prediction of link lifetime in MANET based on unscented kalman filter", Communications Letters, IEEE, Vol. 13, No. 10, DOI: 10.1109/LCOMM.2009.090974, 2009, pp: 782–784.
20. Nen-Chung Wang, Yu-Li Su, "A power-aware multicast routing protocol for mobile ad hoc networks with mobility prediction", Local Computer Networks, 2005. 30th Anniversary. The IEEE Conference on, DOI: 10.1109/LCN.2005.15, 2005.
21. Kuipers, B., "The spatial semantic hierarchy," Artificial Intelligence, Vol. 119, No. 1–2, 2000, pp: 191–233.
22. Thrun, S., "Learning metric-topological maps for indoor mobile robot navigation* 1," Artificial Intelligence, Vol. 99, No. 1, 1998, pp: 21–71.

Chapter 4
Spectrum Sharing in DS-CDMA/OFDM Wireless Mobile Networks

Keivan Navaie, Halim Yanikomeroglu, Mohammad G. Khoshkholgh, Ahmad R. Sharafat, and Hamidreza Nikoofar

Abstract In this chapter, we define DS-CDMA/OFDM spectrum sharing systems and investigate the impact of the primary service communication activity as well as other system parameters on the interference level at the secondary service receiver. The achieved capacity of the secondary service is directly related to the interference level at the secondary service receiver as well as the secondary service adopted sub-channel selection policy. The achievable capacity of the secondary service in such systems is obtained under different sub-channel selection policies in fading environments. Two general sub-channel selection policies are studied in this chapter: *uniform sub-channel selection* and *non-uniform sub-channel selection*. Uniform sub-channel selection fits into the cases where a priori knowledge on sub-channels state information is not available at the secondary transmitter. For cases with available a priori knowledge on sub-channels state information, a variety of non-uniform sub-channel selection policies are studied. We then present results on the scaling law of the opportunistic spectrum sharing in DS-CDMA/OFDM systems with multiple users. We provide numerical results to compare different sub-channel selection policies.

4.1 Introduction

Spectrum sharing was first proposed by the Federal Communications Commission (FCC) to improve the utilization of the allocated frequency bands [1]. In this method, under certain conditions, a *secondary service* is able to access a frequency band formally allocated to the *primary service* [2]. Various schemes are proposed in the literature for spectrum sharing (see e.g., [3] and references therein). Here our focus is on opportunistic spectrum access (OSA).

In this chapter, we consider a direct sequence code division multiple access/orthogonal frequency division multiplexing (DS-CDMA/OFDM) spectrum sharing system in which the spectrum of a DS-CDMA-based primary service is

K. Navaie (✉)
School of Electronic & Electrical Engineering, University of Leeds, Leeds LS2 9JT, UK
e-mail: keivan.navaie@ieee.org

F.R. Yu (ed.), *Cognitive Radio Mobile Ad Hoc Networks*,
DOI 10.1007/978-1-4419-6172-3_4,

shared with a secondary service that utilizes OFDM. DS-CDMA is the dominant air interface technique for the third-generation (3G) mobile communications and some wireless local area network (WLAN) technologies. Therefore, spectrum sharing over existing DS-CDMA-based networks is anticipated to be one of the spectrum sharing applications in the near future [4].

In practice, OFDM provides the required flexibility to the secondary service to access separate under-utilized portions of the spectrum band [2] and at the same time exploits the frequency selectiveness of the wireless channel. Furthermore, the spreading characteristic of DS-CDMA makes it more robust to the narrow-band interference which may be imposed by spectrum sharing. Therefore, DS-CDMA/OFDM combination provides a new degree of freedom by enabling the secondary service to adaptively select appropriate sub-channels for spectrum sharing. The pros-and-cons of DS-CDMA/OFDM spectrum sharing viz-a-vie the maximum achievable capacity as well as its implementation are elaborated in our previous works [4–6].

DS-CDMA systems have dynamic channel sharing and are naturally interference limited [7]. As a metric for recognizing an under-utilized portion of the primary spectrum, we consider a threshold on the acceptable level of the imposed interference at the primary receiver caused by secondary users. Therefore, an under utilized portion of spectrum is defined as a frequency band in which the received interference level is below the *interference threshold*.

The main subject of this chapter is to study the maximum achievable capacity of the secondary service. The maximum achievable capacity of the secondary service for the additive white Gaussian channel is obtained in [8] and [9]. For flat fading environments, the maximum achievable capacity of the secondary service is also obtained in [10] and [11]. In most of the related previous works in the literature, portions of the available primary spectrum are randomly selected for secondary access (see, e.g., [12] and [13]).

The achieved capacity of the secondary service, among other things, is a function of the level of interference imposed by the primary service transmission at the secondary service receiver. The temporal variations of the interference in DS-CDMA systems are elaborately studied in [14]. Here, we investigate the impact of the primary service activity on the corresponding interference level imposed at the secondary service receiver.

The problem of channel assignment to multiple secondary users is also considered in [15], in which algorithms are proposed for selecting appropriate portions of the available primary spectrum based on the interference threshold constraint. Furthermore, in [16] a game theoretic approach for channel selection problem for multiple secondary users in the spectrum sharing networks is investigated.

In this chapter, we present a framework for investigating the sub-channel selection policies with different objectives on the achievable capacity of the secondary service. The secondary service conducts sub-channel selection based on a selection criteria. The selection criteria is a function of the corresponding sub-channel gains including the channel between the secondary transmitter and the primary receiver, namely, the *cross-sub-channel* and the one between the secondary transmitter and receiver called the *secondary-sub-channel*. We divide the sub-channel selection policies into two categories: *uniform sub-channel selection* and *non-uniform sub-channel selection*.

The organization of this chapter is as follows. In Section 4.2, the system model is presented. Then in Section 4.3, we investigate the impact of the primary service activity on the received interference at the secondary user. Section 4.4 presents the basic definitions of the opportunistic spectrum access in DS-CDMA/OFDM. In Section 4.5, the achievable capacity of the secondary service is obtained for the case of one secondary user for which different sub-channel selection scenarios are investigated. In Section 4.6, the system's achievable capacity in case of multiple secondary service users is studied. Finally, numerical results are presented in Section 4.7 followed by conclusions in Section 4.8.

4.2 System Model

Two services access B Hz spectrum: the *primary service* and the *secondary service*. The frequency band has been licensed to the primary service. The secondary service does not have the spectrum license, but may access the spectrum by adopting OSA. In our notations, subscripts s and p refer to the secondary service and the primary service, respectively (Table 4.1). Hereafter, we simply refer to "primary spectrum" as "spectrum" unless otherwise stated.

The primary service is the uplink of a DS-CDMA-based system with processing gain G. Our focus on the uplink is due to the fact that most of the modern data applications are asymmetric, i.e., the amount of downlink communications is much higher than that of the uplink. Therefore, for spectrum sharing over 2G and/or 3G cellular communications, the uplink spectrum is most likely under-utilized, which makes it a good candidate for the OSA.

The wireless channel is a B Hz point-to-point frequency selective with additive white Gaussian noise (AWGN) whose power spectral density is N_0. The channel is divided into N Rayleigh fading B_c Hz sub-channels where B_c is the channel coherence bandwidth. Sub-channels are indexed by $i = 1, 2, \ldots, N$. The sub-channel gains are independent and identically distributed (i.i.d.) random variables.

For a large number of users in the primary network's coverage area, invoking the central limit theorem justifies a Gaussian approximation for the interference process. Using second-order statistics, it is also shown that the interference process is white [17]. Therefore, the average interference in the receiver of the secondary service user in each sub-channel is $(K - 1)N_0 B_c$, $K \geq 1$, where K is a system parameter related to the primary network characteristics [18].

In our model of the OSA system, the spectrum is divided into a number of sub-channels, and the secondary service access one (or more) of these sub-channels, subject to the interference threshold constraint. OFDM is utilized by the secondary network to access the spectrum. Let M, $0 \leq M \leq N$, be the number of accessible sub-channels by the secondary service indexed by $j = 1, 2, \ldots, M$. Sub-channel selection is discussed in Section 4.4. The system we consider in this chapter is time

Table 4.1 Summary of notations and parameters

Name	Description
Q_T	Interference threshold constraint
$\mathbf{Q} = (Q_1, Q_2, \ldots, Q_M)$	Narrow-band interference vector
B Hz	Total bandwidth of the primary service
α	Path-loss exponent
R	Primary service cell radius
P_s	Secondary service transmit power constraint
P_p	Primary service transmit power constraint
$\mathbf{P}_s = (P_{s1}, \ldots, P_{sM})$	Secondary service transmit power vector
g_{0i}	i^{th} Sub-channel power gain from the secondary transmitter to the primary receiver
g_{1i}	i^{th} Sub-channel power gain from the secondary transmitter to the secondary receiver
$f_0(g_{1i})$	Probability distribution function of g_{0i}
$f_1(g_{0i})$	Probability distribution function of g_{1i}
N_s	Number of secondary service active transmitter-receiver pairs
G	Processing gain of DS-CDMA primary network
f	Ratio of other cell's interference to the home cell interference
N_0	White noise power spectral density
B_c	Sub-channel bandwidth
N	Number of narrow-band sub-channels
M	Number of accessible sub-channels
E_b/I_0	Bit energy to the interference spectral density
ρ_y	Required E_b/I_0 for primary user y
R_y	Average data rate of primary user y
η_y	Load factor of the primary user y
P_p^y	Transmitted power of the primary user y
Q_y	Average received power of primary user y at the base station
D_p^y	Distance between the primary mobile user y and the base station
$D_{s\|p}^y$	Distance between the primary mobile user y and the secondary receiver
D	Distance between the secondary service transmitter and receiver
η_{UP}	Primary service uplink load factor
Q_{total}	Total interference caused by primary users to the base-station
Q_{home}	Home-cell interference caused by primary users to the base-station
Q_{other}	Other-cell interference caused by primary users to the base-station
$I_{p\|s}$	Interference caused by secondary users to the primary service receiver
$I_{s\|p}$	Interference caused by the primary users to the secondary service receiver
h_1^y	Channel power gain between primary user y and the base station
$h_{0,i}^y$	Channel power gain between the primary user y and the secondary service receiver on sub-channel i

slotted. The interference threshold Q_T is the maximum allowable temporal interference in the receiver of the primary service caused by concurrent operation of the secondary service at the same frequency band. Therefore, the secondary network access to spectrum should be managed in each time slot in such a way that the interference threshold constraint is held.

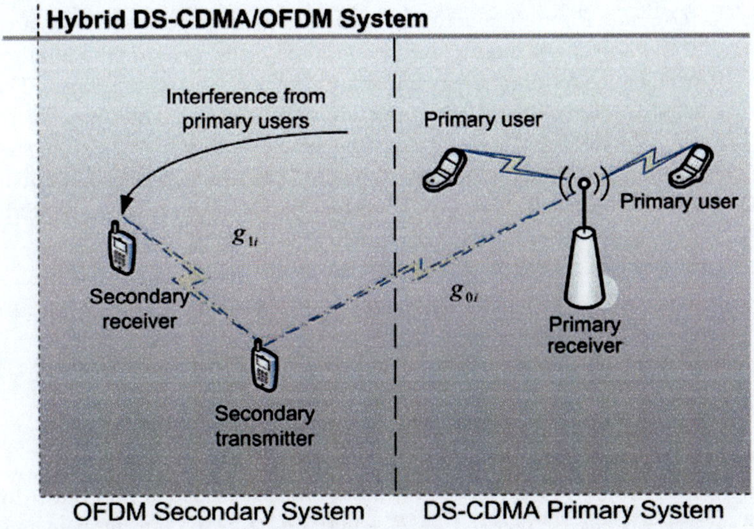

Fig. 4.1 The spectrum sharing structure for sub-channel i

A schematic of the considered system is depicted in Fig. 4.1. For sub-channel i, g_{0i} and g_{1i} in Fig. 4.1 denote the instantaneous gains of sub-channel i from the secondary transmitter to the primary receiver, i.e., *cross-sub-channel*, and the secondary receiver, i.e., *secondary-sub-channel*, respectively. Both g_{0i} and g_{1i} are assumed to be stationary and ergodic independent random variables with unit-mean and probability density functions (pdf) $f_{0i}(g_{0i})$ and $f_{1i}(g_{1i})$, respectively. Channel gains g_{0i} and g_{1i} are i.i.d. for $i = 1, 2, \ldots, N$. In our analysis, we assume that perfect channel state information (CSI) pair (g_{0i}, g_{1i}), $i = 1, \ldots, N$ is available at transmitters. Note that the results derived based on this assumption are the upper bound for the cases with perfect CSI pair.

4.3 Impact of Primary Service Activity

The secondary service opportunistically accesses the spectrum; therefore, its achieved capacity, among other things, is also a function of the level of interference caused by the primary service transmission to the secondary service receiver. The temporal variations of the interference in DS-CDMA systems are elaborately studied in [14]. Here, we investigate the impact of the primary service activity on the corresponding interference level caused to the secondary service receiver.

For the primary mobile user y, the average received bit energy to the interference-plus-noise power spectral density (E_b/I_0) is

$$\rho_y = \frac{B}{\upsilon_y R_y} \frac{Q_y}{Q_{\text{total}} - Q_y} \tag{4.1}$$

where υ_y, Q_y, and R_y are the activity factor, the average received power, and the bit rate of the yth primary mobile user, respectively, and Q_{total} is the total received power of primary users plus noise and interference at the base station. The ratio B/R_y is usually referred to as the processing gain corresponding to the primary user y. Let $I_{p|s}$ be the interference caused by secondary users at the base station of primary service. Due to the interference threshold constraint, we set $I_{p|s}$ equal to the worst case situation, i.e., $I_{p|s} = Q_T$. Therefore, the effective noise and interference spectral power at the base station is $\bar{N}_0 B$, which is equal to $N_0 B + Q_T$.

The load factor of mobile user y is denoted by η_y, where $Q_y = \eta_y Q_{\text{total}}$, and η_y is the contribution of user y in the total interference at the base station, Q_{total}. Using η_y in (4.1), it is easy to show that

$$\eta_y = \left(1 + \frac{B}{\rho_y \upsilon_y R_y}\right)^{-1} \tag{4.2}$$

At the base station, the total received interference is divided into three different parts as $Q_{\text{total}} = Q_{\text{home}} + Q_{\text{other}} + \bar{N}_0 B$, where Q_{home} is the interference generated by home cell users and Q_{other} is the interference caused by users in other cells. Similar to the related literature (see, e.g., [19]) we also assume that $Q_{\text{other}} = f Q_{\text{home}}$. The parameter f is the ratio of other cell's interference to the home cell interference and is referred to as the network load factor. We define η_{UP} by

$$\eta_{\text{UP}} = \frac{Q_{\text{home}} + Q_{\text{other}}}{Q_{\text{total}}} \tag{4.3}$$

as the uplink load factor which shows the contribution of the home cell and other cells' interference into Q_{total}. We note that $\eta_{\text{UP}} = (1 + f) \sum \eta_y$, then noting (4.2) for $\mathcal{Y}$ different services, $\varsigma = 1, \ldots, \mathcal{Y}$ each with L_ς users, η_{UP} is obtained by

$$\eta_{\text{UP}} = (1 + f) \sum_{\varsigma=1}^{\mathcal{Y}} L_\varsigma \eta_\varsigma = (1 + f) \sum_{\varsigma=1}^{\mathcal{Y}} L_\varsigma \left(1 + \frac{B}{\rho_\varsigma \upsilon_\varsigma R_\varsigma}\right)^{-1} \tag{4.4}$$

Rewriting Q_{total} as $Q_{\text{total}} = \eta_{\text{UP}} Q_{\text{total}} + \bar{N}_0 B$, and replacing η_{UP} from (4.4), Q_{total} is obtained by

$$Q_{\text{total}} = \left(1 - (1 + f) \sum_{\varsigma=1}^{\mathcal{Y}} L_\varsigma \left(1 + \frac{B}{\rho_\varsigma \upsilon_\varsigma R_\varsigma}\right)^{-1}\right)^{-1} \bar{N}_0 B \tag{4.5}$$

Equation (4.5) indicates that in DS-CDMA cellular networks with a large number of mobile users, the received interference at the base station in the uplink channel behaves as an additive white Gaussian noise whose power spectral density is Q_{total}. We denote $Q_{\text{total}} = \kappa \bar{N}_0 B$, where parameter κ is obtained from

$$\kappa = \left(1 - (1+f)\sum_{\varsigma=1}^{\mathcal{Y}} L_\varsigma \left(1 + \frac{B}{\rho_\varsigma \upsilon_\varsigma R_\varsigma}\right)^{-1}\right)^{-1} \tag{4.6}$$

Parameter κ is evaluated at the base station. This parameter reflects the impact of primary mobile users on the base station's received interference. However, we are also interested to study the impact of primary users on secondary users, especially on their achievable capacity. One may also assume that in this case, the impact of primary users is reflected in κ, as it is assumed in [4, 5, 18].

In what follows, we obtain a closed form for the interference caused by primary users to secondary users. In cell zero, the received power of the yth primary service user at the base station located in the center of the cell is

$$Q_y = \eta_y Q_{\text{total}} = \frac{\kappa N_0 B + \kappa Q_T}{1 + \frac{B}{\rho_\varsigma \upsilon_\varsigma R_\varsigma}}, \qquad y \in \{1, \ldots, L_\varsigma\} \tag{4.7}$$

The truncated power control mechanism is adopted in the primary service DS-CDMA-based technology to mitigate fast fading fluctuation [20]. Indeed, noticing the allowed maximum transmission power of each primary user, the inversion power control mechanism is implemented with the objective of enforcing (4.7). Consequently, taking the truncated power control mechanism into account, the transmitted power of the primary service user y is

$$P_p^y = \min\left\{\frac{Q_y}{\left(D_p^y\right)^{-\alpha} h_1^y}, P_p\right\} \tag{4.8}$$

where P_p is the maximum transmit power of primary mobile users. Here, we assume that all primary mobile users have the same maximum transmit power. Moreover, D_p^y is the distance between the primary mobile user y and its base station, α is the path-loss exponent ($\alpha > 2$), and h_1^y is the channel power gain between this user and its base station. Generally, distance-dependent path-loss and fast fading are independent.

Since secondary users access the spectrum utilizing OFDM technology, the generated interference at the secondary service receiver corresponding to the primary user's transmits power P_p^y on accessible sub-channel j is

$$\begin{aligned} I_{s|p}^{(j)} &= \frac{1}{N} P_p^y \left(D_{s|p}^y\right)^{-\alpha} h_{0,j}^y \\ &= \frac{1}{N} P_p \left(D_{s|p}^y\right)^{-\alpha} h_{0,j}^y \mathbf{1}_{\left\{h_1^y < \frac{Q_y}{P_p}\left(D_p^y\right)^\alpha\right\}} \\ &= \frac{Q_y}{N}\left(\frac{D_{s|p}^y}{D_p^y}\right)^{-\alpha} \frac{h_{0,j}^y}{h_1^y} \mathbf{1}_{\left\{h_1^y \geq \frac{Q_y}{P_p}\left(D_p^y\right)^\alpha\right\}} \end{aligned}$$

where $D^y_{s|p}$ is the distance between the yth primary service user and the secondary service receiver, and $\mathbf{1}_{\{\cdot\}}$ is the indicator operator.

Let $h^y_{0,j}$ be the channel power gain between the primary mobile user y and the secondary service receiver on sub-channel j. The value of $I^{(j)}_{s|p}$ is only sensitive to a $1/N$ portion of the transmission power P^y_p which generates interference over the corresponding $B_c = B/N$ Hz sub-channel. Consider

$$\mathcal{A}_\varsigma = \left\{ y : h^y_1 < \frac{Q_y}{P_p} (D^y_p)^\alpha \right\}$$

as a set that contains all primary mobile users $y = 1, \ldots, L_\varsigma$ with transmit power $P^y_p = P_p$, i.e., those primary users that cannot consistently increase their transmit powers proportional to fading. The set $\mathcal{A}_\varsigma{}^c$ (the complement set of $\mathcal{A}_\varsigma$) contains those primary mobile users with transmission power $P^y_p = \frac{Q_y}{(D^y_p)^{-\alpha} h^y_1}$.

The total received interference at the secondary service receiver generated by primary service transmissions is

$$\begin{aligned} I_{s|p} &= \tfrac{1+f}{N} \sum_{\varsigma=1}^{\mathcal{Y}} \sum_{y=1}^{L_\varsigma} P^y_p \left(D^y_{s|p}\right)^{-\alpha} h^y_{0,j} \\ &= \tfrac{1+f}{N} P_p \sum_{\varsigma=1}^{\mathcal{Y}} \sum_{y \in \mathcal{A}_\varsigma} \left(D^y_{s|p}\right)^{-\alpha} h^y_{0,j} + \tfrac{1+f}{N} \sum_{\varsigma=1}^{\mathcal{Y}} \sum_{y \in \mathcal{A}^c_\varsigma} Q_y \left(\frac{D^y_{s|p}}{D^y_p}\right)^{-\alpha} \frac{h^y_{0,j}}{h^y_1} \end{aligned} \tag{4.9}$$

Considering (4.7), the total average received interference at the secondary service receiver is

$$\begin{aligned} \mathbf{E}\left[I_{s|p}\right] &= \frac{1+f}{N} P_p \mathbf{E}\left[\sum_{\varsigma=1}^{\mathcal{Y}} \sum_{y \in \mathcal{A}_\varsigma} \left(D^y_{s|p}\right)^{-\alpha} h^y_{0,j} \right] \\ &\quad + (1+f)\kappa \frac{N_0 B + Q_T}{N} \mathbf{E}\left[\sum_{\varsigma=1}^{\mathcal{Y}} \sum_{y \in \mathcal{A}^c_\varsigma} \frac{\left(\frac{D^y_{s|p}}{D^y_p}\right)^{-\alpha}}{1 + \frac{B}{\rho_\varsigma \upsilon_\varsigma R_\varsigma}} \frac{h^y_{0,j}}{h^y_1} \right] \\ &= \frac{1+f}{N} P_p \sum_{\varsigma=1}^{\mathcal{Y}} \mathbf{E}\left[\sum_{y \in \mathcal{A}_\varsigma} \left(D^y_{s|p}\right)^{-\alpha} h^y_{0,j} \right] \\ &\quad + (1+f)\kappa \frac{N_0 B + Q_T}{N} \sum_{\varsigma=1}^{\mathcal{Y}} \mathbf{E}\left[\sum_{y \in \mathcal{A}^c_\varsigma} \frac{\left(\frac{D^y_{s|p}}{D^y_p}\right)^{-\alpha}}{1 + \frac{B}{\rho_\varsigma \upsilon_\varsigma R_\varsigma}} \frac{h^y_{0,j}}{h^y_1} \right] \end{aligned}$$

We denote the followings

$$I_\varsigma^1 = \sum_{y \in \mathcal{A}_\varsigma} \left(D_{s|p}^y\right)^{-\alpha} h_{0,j}^y, \qquad I_\varsigma^2 = \sum_{y \in \mathcal{A}_\varsigma^c} \frac{\left(\frac{D_{s|p}^y}{D_p^y}\right)^{-\alpha}}{1 + \frac{B}{\rho_\varsigma \upsilon_\varsigma R_\varsigma}} \frac{h_{0,j}^y}{h_1^y}$$

Considering I_ς^1, for the case of Rayleigh fading environment, it is easy to verify that

$$\mathbf{E}\left[I_\varsigma^1\right] = \sum_{y=1}^{L_\varsigma} \mathbf{E}\left[\left(D_{s|p}^y\right)^{-\alpha} \left(1 - \exp\left(-\frac{Q_y}{P_p}\left(D_p^y\right)^\alpha\right)\right)\right] \mathbf{E}\left[h_{0,j}^y\right] \tag{4.10}$$

Note that D_p^y is a uniform random variable in the interval $[1, R]$, where R is the cell radius. Similarly, $D_{s|p}^y$ is also a uniform random variable in the interval $[1, 2R]$ for each y. Consequently,

$$\mathbf{E}\left[\left(D_{s|p}^y\right)^{-\alpha}\right] = \frac{1}{2R} \frac{1 - (2R)^{-\alpha+1}}{\alpha - 1} \tag{4.11}$$

Noting that $\mathbf{E}\left[h_{0,j}^y\right] = 1$, and utilizing (4.7) and (4.11), (4.10) is reduced to

$$\mathbf{E}\left[I_\varsigma^1\right] = \frac{1}{2R^2} \frac{1 - (2R)^{-\alpha+1}}{\alpha - 1} \sum_{y=1}^{L_\varsigma} \int_1^R \left(1 - \exp\left(-\frac{\kappa}{P_p} \frac{N_0 B + Q_T}{1 + \frac{B}{\rho_\varsigma \upsilon_\varsigma R_\varsigma}} r^\alpha\right)\right) dr \tag{4.12}$$

Similarly, $\mathbf{E}\left[I_\varsigma^2\right]$ is following

$$\mathbf{E}\left[I_\varsigma^2\right] = \frac{1}{2R^2} \frac{1-(2R)^{-\alpha+1}}{\alpha-1} \sum_{y=1}^{L_\varsigma} \frac{1}{1+\frac{B}{\rho_\varsigma \upsilon_\varsigma R_\varsigma}} \int_1^R r^\alpha E_1\left(\frac{\kappa}{P_p} \frac{N_0 B+Q_T}{1+\frac{B}{\rho_\varsigma \upsilon_\varsigma R_\varsigma}} r^\alpha\right) dr \tag{4.13}$$

where $E_n(x)$ is the exponential integral of order n defined by

$$E_n(x) \triangleq \int_1^\infty t^{-n} e^{-xt} dt, \ x \geq 0$$

Therefore, a closed-form expression for $\mathbf{E}\left[I_{s|p}\right]$ is

$$\mathbf{E}\left[I_{s|p}\right] = (1+f)\frac{1}{2R^2}\frac{1-(2R)^{-\alpha+1}}{\alpha-1}\sum_{\varsigma=1}^{\mathcal{Y}} L_\varsigma$$

$$\int_1^R \left[\frac{P_p}{N}\left(1-\exp\left(-\frac{\kappa}{P_p}\frac{N_0B+Q_T}{1+\frac{B}{\rho_\varsigma\upsilon_\varsigma R_\varsigma}}r^\alpha\right)\right)\right.$$

$$\left.+\frac{\kappa}{N}\frac{N_0B+Q_T}{1+\frac{B}{\rho_\varsigma\upsilon_\varsigma R_\varsigma}}r^\alpha E_1\left(\frac{\kappa}{P_p}\frac{N_0B+Q_T}{1+\frac{B}{\rho_\varsigma\upsilon_\varsigma R_\varsigma}}r^\alpha\right)\right]dr \qquad (4.14)$$

Note that the value of the average interference at the secondary service receiver caused by primary service transmissions increases consistently as κ is increased.

Note also that $\mathbf{E}\left[I_{s|p}\right]$ is a function of the interference threshold Q_T. When $\mathbf{E}\left[I_{s|p}\right] = (K-1)N_0B_c$, the value of K is obtained by

$$K = 1+(1+f)\frac{1}{2R^2}\frac{1-(2R)^{-\alpha+1}}{\alpha-1}\sum_{\varsigma=1}^{\mathcal{Y}} L_\varsigma$$

$$\int_1^R \left[\frac{P_p}{N_0B}\left(1-\exp\left(-\frac{\kappa}{P_p}\frac{N_0B+Q_T}{1+\frac{B}{\rho_\varsigma\upsilon_\varsigma R_\varsigma}}r^\alpha\right)\right)\right.$$

$$\left.+\kappa\frac{1+\frac{Q_T}{N_0B}}{1+\frac{B}{\rho_\varsigma\upsilon_\varsigma R_\varsigma}}r^\alpha E_1\left(\frac{\kappa}{P_p}\frac{N_0B+Q_T}{1+\frac{B}{\rho_\varsigma\upsilon_\varsigma R_\varsigma}}r^\alpha\right)\right]dr \qquad (4.15)$$

4.4 Opportunistic Spectrum Sharing in DS-CDMA/OFDM Systems: Basic Definitions

As mentioned earlier, our main objective is to obtain the maximum secondary service achievable capacity; therefore, similar to [9, 10], we start our analysis considering one secondary user.

In cases with more than one secondary user, the total secondary service achievable capacity is upper-bounded by the case with only one secondary user. This is due to the fact that secondary users also cause interference to each other. Interference is caused due to imperfect multiple access techniques utilized by secondary users. We later extend our analysis into multiple secondary service users in Section 4.6. Here we define the required concepts.

At a given time instant, we define a policy $\mathcal{P}_{\Phi}$, based on a deterministic selection criteria $\Phi(\cdot,\cdot)$ and set

$$\mu_i = \Phi(g_{0i}, g_{1i}) \tag{4.16}$$

For the observed random variables $\mu_i, i = 1, \ldots, N$, we also define the selection sequence

$$\Upsilon_N = \left(\mu_{r_1}, \mu_{r_2}, \ldots, \mu_{r_N}\right) = \mathcal{P}_{\Phi}\left(\mu_1, \mu_2, \ldots, \mu_N\right) \tag{4.17}$$

The N*-tuple selection sequence* is arranged so that its first element, indexed by r_1, is *the most suitable* sub-channel for spectrum sharing based on the selection criteria in (4.16). Adopting a new indexing, we also define the M*-tuple selected sequence*

$$\Theta_M = (\theta_1, \theta_2, \ldots, \theta_M) = \left(\mu_{r_1}, \mu_{r_2}, \ldots, \mu_{r_M}\right) \tag{4.18}$$

We assume that the probability density function (pdf) of random variable θ_j is $k_j(\theta)$, $j = 1, \ldots, M$. Note that based on such a policy, if θ_{j_1} and θ_{j_2} are entities in the selected sequence and $j_1 < j_2$, then the corresponding sub-channel with index j_1 is considered *more suitable* for spectrum sharing as compared to that of j_2.

Since g_{0i} and g_{1i} $\forall i$ are i.i.d. random variables, for constant $\Phi(g_{0i}, g_{1i})$, μ_i and consequently θ_j $j = 1, \ldots, M$ are i.i.d.. In other words, sub-channels are considered uniform and M out of N sub-channels are selected randomly without any a priori knowledge on their status. We call this selection strategy *uniform sub-channel selection*. For cases with variable $\Phi(g_{0i}, g_{1i})$, different sub-channels based on the corresponding values of $\Phi(g_{0i}, g_{1i})$ are treated non-uniformly. We call this selection strategy as *non-uniform sub-channel selection*.

The instantaneous transmit power of the secondary service over the jth sub-channel is $P_{sj}(g_{0j})$, which we refer to as P_{sj}. We set $\mathbf{P}_s = \left(P_{s1}, \ldots, P_{sM}\right)$ as the secondary service transmit power vector over M sub-channels. Assume that the secondary service communicates over the selected sub-channel j with transmit power P_{sj}. Narrow-band interference denoted by Q_j is correspondingly caused to the front-end of the primary service receiver, where

$$Q_j = g_{0r_j} P_{sj} \tag{4.19}$$

Since the air interface in the primary network is DS-CDMA, the narrow-band interference Q_j is then spread out over the whole B Hz bandwidth and manifests itself as an equivalent wide-band interference at the primary receiver.

For M accessible sub-channels, the secondary service transmits with the transmit power vector $\mathbf{P}_s = (P_{s1}, P_{s2}, \ldots, P_{sM})$. Therefore, the equivalent narrow-band interference $\mathbf{Q} = (Q_1, Q_2, \ldots, Q_M)$ is implied at the primary receiver. Consequently, to comply with the interference threshold Q_T, we should have

$$\sum_{j=1}^{M} g_{0r_j} P_{sj} \leq Q_T \tag{4.20}$$

Note that here we assume that due to the E_b/I_0 requirement, the primary service receiver assigns the interference threshold constraint; however, in [5, 18], we assumed that this parameter is assigned based on the signal-to-interference-plus-noise ratio (SINR) requirement.

The ergodic capacity of the secondary service for the transmit power vector $\mathbf{P}_s$ over M accessible sub-channels is

$$C_s = \sum_{j=1}^{M} B_c \int_{g_{1r_j}, g_{0r_j}} \mathbf{E}\left[\log\left(1 + \frac{D^{-\alpha} g_{1r_j} P_{sj}}{N_0 B_c + I_{s|p}}\right) f_{1j}\left(g_{1r_j}\right) f_{0j}\left(g_{0r_j}\right) dg_{0r_j} dg_{1r_j} \right] \tag{4.21}$$

where $\mathbf{E}[\cdot]$ is the expectation operator on the random variable $I_{s|p}$. In (4.21), D denotes the distance between the secondary service transceiver. Here, we assume that D is a given parameter. In what follows, we present a lower bound for C_s. Due to convexity of function $\log\left(1 + \frac{a}{b+x}\right)$ for $x \geq 0$ and given positive parameters a and b, and employing Jensen's inequality [21], we have

$$\begin{aligned} C_s &\geq \sum_{j=1}^{M} B_c \int_{g_{1r_j}, g_{0r_j}} \log\left(1 + \frac{D^{-\alpha} g_{1r_j} P_{sj}}{N_0 B_c + \mathbf{E}[I_{s|p}]}\right) f_{1j}\left(g_{1r_j}\right) f_{0j}\left(g_{0r_j}\right) dg_{0r_j} dg_{1r_j} \\ &= \sum_{j=1}^{M} B_c \int_{g_{1r_j}, g_{0r_j}} \log\left(1 + \frac{g_{1r_j} P_{sj}}{D^{\alpha} K N_0 B_c}\right) f_{1j}\left(g_{1r_j}\right) f_{0j}\left(g_{0r_j}\right) dg_{0r_j} dg_{1r_j} \\ &\triangleq C^{\Phi}_{s|M} \end{aligned}$$

For a given Q_T, the maximum achievable capacity of the secondary service, for M selected sub-channels based on policy $\mathcal{P}_{\Phi}$, denoted by $C^{\Phi}_{s|M}$, is the solution to the following problem.

Problem $\mathcal{O}_1$:

$$C^{\Phi}_{s|M} = \max_{\mathbf{P}_s} \sum_{j=1}^{M} B_c \int_{g_{1r_j}, g_{0r_j}} \log\left(1 + \frac{g_{1r_j} P_{sj}}{D^{\alpha} K N_0 B_c}\right) f_{1j}\left(g_{1r_j}\right) f_{0j}\left(g_{0r_j}\right) dg_{0r_j} dg_{1r_j}, \tag{4.22}$$

$$\text{subject to} \sum_{j=1}^{M} g_{0r_j} P_{sj} \leq Q_T \tag{4.23}$$

$$\sum_{j=1}^{M} P_{sj} \leq P_s \tag{4.24}$$

In $\mathcal{O}_1$, (4.22) is the Shannon's Capacity (ergodic capacity) formula for the power vector $\mathbf{P}_s$, (4.23) is the interference threshold, (4.24) is the secondary service maximum transmit power constraint, and P_s is the secondary service maximum transmit power. In practice, DS-CDMA cellular systems are single frequency; therefore, the operation of the secondary service in the primary band may impose unexpected interference on the base stations of the adjacent cells. To deal with this issue, one may either add new constraints to Problem $\mathcal{O}_1$ or consider a conservative value for Q_T. Hereafter, for brevity we consider the latter.

In practice, the interference threshold constraint is usually tight enough so that the transmit power constraint for the secondary service is not satisfied; therefore, for clarity of expositions, similar to the related literature (see e.g., [10]), we do not consider the transmit power constraint for the secondary service in (4.24). In cases where the transmit power constraint is the dominant constraint as compared to the interference threshold, it is shown in [4] that the achieved capacity without considering the transmit power constraint serves as an upper bound. The optimization problem in $\mathcal{O}_1$ is an instance of water-filling problem (for water-filling problem see, e.g., [17]).

4.5 Single Secondary Service User

4.5.1 Uniform Sub-channel Selection

In uniform sub-channel selection, sub-channels are considered uniformly and M out of N sub-channels are selected randomly by $\mathcal{P}_1$ without any a priori knowledge on their status. Therefore, $\Phi(g_{0i}, g_{1i}) = 1$, thus $\mathcal{P}_1$ returns θ_j, $j = 1, \ldots M$ which are i.i.d. The probability of selecting a sub-channel in uniform sub-channel selection scenario is $1/N$. Substituting $P_{sj} = \frac{Q_j}{g_{0r_j}}$, $j = 1, 2, \ldots, M$ and defining

$$\gamma_{Q_j} = \frac{Q_j}{D^\alpha K N_0 B_c} \tag{4.25}$$

$\mathcal{O}_1$ is converted into the following problem.

Problem $\mathcal{O}_2$:

$$C^1_{s|M} = \max_{\mathbf{Q}} \sum_{j=1}^{M} B_c \int_{\nu_{r_j}} \log\left(1 + \nu_{r_j}\gamma_{Q_j}\right) h_j\left(\nu_{r_j}\right) d\nu_{r_j} \tag{4.26}$$

$$\text{subject to} \sum_{j=1}^{M} Q_j = Q_T, \;\; 0 \le Q_j \le Q_T \tag{4.27}$$

where we define *reward factor* ν_{r_j} as

$$\nu_{r_j} = \frac{g_{1r_j}}{g_{0r_j}}, \qquad 0 < \nu_{r_j} < \infty$$

In cases where sub-channel gains $\sqrt{g_{0i}}$ and $\sqrt{g_{1i}}$ are i.i.d, Rayleigh random variables g_{0i} and g_{1i} are exponentially distributed random variables, thus, the pdf of ν_{r_j}, denoted by $h_j(\nu_j)$, is

$$\begin{aligned} h_j\left(\nu_{r_j}\right) &= \frac{d}{d\nu_{r_j}} \int_0^{\infty} \int_0^{g_{0r_j}\nu_{r_j}} e^{-g_{0r_j}} e^{-g_{1r_j}} dg_{0r_j} dg_{1r_j} \\ &= \int_0^{\infty} g_{0r_j} e^{-g_{0r_j}\left(1+\nu_{r_j}\right)} dg_{0r_j} \\ &= \frac{1}{\left(1+\nu_{r_j}\right)^2}, \qquad 0 < \nu_{r_j} < \infty \end{aligned} \tag{4.28}$$

By substituting (4.28) into (4.26), and integrating by part, $\mathcal{O}_2$ is simplified to the following problem.

Problem $\mathcal{O}_3$:

$$C^1_{s|M} = \max_{\underline{\gamma_Q}} \sum_{j=1}^{M} B_c \frac{\gamma_{Q_j}}{\gamma_{Q_j} - 1} \log\left(\gamma_{Q_j}\right) \tag{4.29}$$

$$\text{subject to} \sum_{j=1}^{M} \gamma_{Q_j} = N\gamma_Q, \quad 0 \le \gamma_{Q_j} \le N\gamma_Q \tag{4.30}$$

where γ_Q is the *spectrum sharing load factor* defined by

$$\gamma_Q = \frac{Q_T}{D^{\alpha} K N_0 B} \tag{4.31}$$

and $\underline{\gamma_Q} = \left(\gamma_{Q_1}, \gamma_{Q_2}, \ldots, \gamma_{Q_M}\right)$ is the *spectrum sharing load vector*.

An approximate solution to $\mathcal{O}_3$ is obtained by substituting (4.29) with the following pseudo-linear approximation

$$\frac{x}{x-1}\log(x) \approx A_1 + A_2 x + A_3 \log\left(A_4 x + A_5\right) \tag{4.32}$$

Utilizing this approximation, the Lagrangian function corresponding to $\mathcal{O}_3$ is

$$L\left(\underline{\gamma_Q}, \lambda\right) = \sum_{j=1}^{M} A_1 + A_2\gamma_{Q_j} + A_3 \log\left(A_4\gamma_{Q_j} + A_5\right) - \lambda\left(\sum_{j=1}^{M} \gamma_{Q_j} - N\gamma_Q\right)$$

where λ is the Lagrangian coefficient. Taking the derivative of $L(\underline{\gamma_Q}, \lambda)$ with respect to γ_{Q_j}, and setting this derivative to zero, we have

$$\gamma_{Q_j}^* = \frac{A_3}{\lambda^* - A_2} - \frac{A_5}{A_4} \tag{4.33}$$

Substituting (4.33) into (4.30) yields

$$\sum_{j=1}^{M} \left[\frac{A_3}{\lambda^* - A_2} - \frac{A_5}{A_4}\right] = N\gamma_Q \tag{4.34}$$

hence,

$$\lambda^* = A_2 + \frac{A_3}{\frac{N\gamma_Q}{M} + \frac{A_5}{A_4}} \tag{4.35}$$

Consequently, substituting (4.35) into (4.33) yields

$$\gamma_{Q_j}^* = \frac{N\gamma_Q}{M}, \qquad j = 1, 2, ..., M \tag{4.36}$$

Note that (4.36) suggests that for given M and Q_T, the maximum achievable capacity is obtained by dividing the total acceptable interference $N\gamma_Q$ into M equal portions for each sub-channel. This is, in fact, a direct consequence of selecting M out of N sub-channels without a priori knowledge on their status.

The optimal transmit power vector $\mathbf{P}_s^*$ is then obtained using (4.19) and (4.25)

$$\mathbf{P}_s^* = \left(\frac{1}{g_{0r_1}}\frac{Q_T}{M}, \frac{1}{g_{0r_2}}\frac{Q_T}{M}, \ldots, \frac{1}{g_{0r_M}}\frac{Q_T}{M}\right) \tag{4.37}$$

Equation (4.37) indicates that the interference share for each selected sub-channel j, i.e., $\gamma_{Q_j}^*$, is mapped into the corresponding transmit power, P_{sj}^*, proportional to $1/g_{0r_j}$. Therefore, for a large g_{0r_j}, the secondary user's transmission creates a large interference at the primary service receiver. In this case, (4.37) suggests a lower secondary transmit power in selected sub-channel j.

The maximum achievable capacity of the secondary service is approximated by substituting (4.36) into (4.29) as

$$C_{s|M}^1 \approx MB_c \frac{N\gamma_Q}{N\gamma_Q - M} \log\left(\frac{N\gamma_Q}{M}\right) \tag{4.38}$$

4.5.2 Non-uniform Sub-channel Selection

Basically, uniform sub-channel selection ignores the fact that some sub-channels are actually more appropriate for OSA because of their corresponding CSIs, i.e., a larger capacity is achieved and/or a smaller interference on primary users is caused. Therefore, we expect that non-uniform sub-channel selection based on a priori knowledge of the secondary-sub-channel and/or cross-sub-channel CSIs results in a higher secondary service achieved capacity.

Intuitively, an appropriate policy could consider the interference that secondary users cause at the primary receiver. Such policy may select those sub-channel(s) with the lowest cross-sub-channel gain(s) g_{0i} for secondary users that cause a lower interference on the primary receiver. Potentially, a lower g_{0i} may also provide secondary users with the flexibility of allocating a higher power to achieve a higher capacity. This policy is referred to as the *cross-sub-channel based selection policy*.

To implement cross-sub-channel-based selection policy, the secondary service requires g_{0i} during each time slot. To obtain g_{0i}, a direct or indirect (i.e., through a third party such as spectrum manager) signaling channel is required between the primary service receiver (i.e., the base-station) and the secondary service transmitter.

Another option is the one that selects those sub-channels that achieve the highest capacity for allocating the secondary service transmit power. Such policy selects the sub-channel(s) with the highest secondary sub-channel gain g_{1i} for the secondary transmission. This policy is referred to as the *secondary sub-channel-based selection policy*.

An alternative approach can also be envisaged in which the sub-channel selection policy employs a combination of the aforementioned two strategies in some sense. For example, the objective can be to achieve the highest possible capacity while imposing the lowest possible interference. As an instance of such a combination we define g_{1i}/g_{0i} as the reward factor of sub-channel i and call the corresponding sub-channel selection policy as the *reward factor-based sub-channel selection policy*. In this section, we obtain the maximum achievable capacity for the three aforementioned sub-channel selection policies.

4.5.2.1 Cross-Sub-channel-Based Selection Policy

In cross-sub-channel based selection policy the selection criteria is

$$\Phi(g_{0i}, g_{1i}) = g_{0i} \tag{4.39}$$

and correspondingly, $\mu_i = g_{0i}$. Policy $\mathcal{P}_{g_0}$ in then defined so that in the N-tuple selection sequence Υ_N

$$\begin{aligned} \mu_{r_1} \leq \mu_{r_2} \leq \ldots \leq \mu_{r_N} \\ \mu_{r_1} = \min_i\{\mu_i\} \end{aligned} \tag{4.40}$$

The M-tuple selected sequence Θ_M is

$$\theta_1 = \mu_{r_1} \leq \theta_2 = \mu_{r_2} \leq \ldots \leq \theta_M = \mu_{r_M}$$

The main objective of $\mathcal{P}_{g_0}$ is to select the sub-channel(s) that cause(s) the lowest interference at the primary receiver. The pdf of θ_j, $\forall j$, is obtained using order statistics,

$$k_j(\theta) = N_j F_\mu^{j-1}(\theta) \left[1 - F_\mu(\theta)\right]^{N-j} f_\mu(\theta) \tag{4.41}$$

$$N_j = \frac{N!}{(j-1)!(N-j)!} \tag{4.42}$$

and $f_\mu(\theta)$ and $F_\mu(\theta)$ are the pdf and probability distribution function (PDF) of the random variable μ, respectively. By following the same argument as in Section 4.5.1 for Rayleigh fading, $f_\mu(\theta) = e^{-\theta}$, and $F_\mu(\theta) = 1 - e^{-\theta}$. Thus, $k_j(\theta)$ is

$$k_j(\theta) = N_j \left(1 - e^{-\theta}\right)^{j-1} e^{-\theta(N-j+1)} \tag{4.43}$$

Replacing the binomial expansion of $\left(1 - e^{-\theta}\right)^{j-1}$ in (4.43) yields

$$k_j(\theta) = N_j \sum_{l=0}^{j-1} \Theta_l^{j-1} e^{-\theta(N-l)} \tag{4.44}$$

where Θ_l^{j-1} is

$$\Theta_l^{j-1} = \binom{j-1}{l} (-1)^{j-1-l} \tag{4.45}$$

Proposition 1 *The maximum achievable capacity of the secondary service based on policy $\mathcal{P}_{g_0}$, $C_{s|M}^{g_0}$, is obtained from the following optimization problem.*

Problem $\mathcal{O}_4$:

$$C_{s|M}^{g_0} = \max_{\underline{\gamma_Q}} \sum_{j=1}^{M} \sum_{l=0}^{j-1} B_c N_j \Theta_l^{j-1} \frac{\gamma_{Q_j} \log\left((N-l)\gamma_{Q_j}\right)}{(N-l)\gamma_{Q_j} - 1} \tag{4.46}$$

$$\text{subject to} \sum_{j=1}^{M} \gamma_{Q_j} = N\gamma_Q, \quad 0 \leq \gamma_{Q_j} \leq N\gamma_Q \tag{4.47}$$

Proof Substituting $P_{sj} = \frac{Q_j}{g_{0r_j}}$ into (4.22), and setting $\nu_{r_j} = g_{1r_j}/g_{0r_j}$, the maximum achievable capacity $C^{g_0}_{s|M}$ based on policy $\mathcal{P}_{g_0}$ is obtained from $\mathcal{O}_1$ as

$$C^{g_0}_{s|M} = \max_{\mathbf{Q}} \sum_{j=1}^{M} B_c \int_{\nu_{r_j}} \log\left(1 + \nu_{r_j}\gamma_{Q_j}\right) h_j\left(\nu_{r_j}\right) d\nu_{r_j} \tag{4.48}$$

$$\text{subject to} \sum_{j=1}^{M} \gamma_{Q_j} = N\gamma_Q,\ 0 \le \gamma_{Q_j} \le N\gamma_Q$$

In cases where the sub-channel gains $\sqrt{g_{0r_j}}$ and $\sqrt{g_{1r_j}}$ are i.i.d., Rayleigh random variables g_{0r_j} and g_{1r_j} are exponentially distributed random variables, therefore, using (4.44), the pdf of ν_{r_j}, denoted by $h_j(\nu_j)$, is

$$h_j\left(\nu_{r_j}\right) = \frac{d}{d\nu_{r_j}} \int_0^{\infty} \int_0^{\nu_{r_j} g_{0r_j}} N_j \sum_{l=0}^{j-1} \Theta_l^{j-1} e^{-(N-l)g_{0r_j}} e^{-g_{1r_j}} dg_{1r_j} dg_{0r_j}$$

or equivalently

$$\begin{aligned} h_j\left(\nu_{r_j}\right) &= \frac{d}{d\nu_{r_j}} \sum_{l=0}^{j-1} N_j \Theta_l^{j-1} \int_0^{\infty} \left(1 - e^{-g_{0r_j}\nu_{r_j}}\right) e^{-(N-l)g_{0r_j}} dg_{0r_j} \\ &= \sum_{l=0}^{j-1} N_j \Theta_l^{j-1} \int_0^{\infty} g_{0r_j} e^{-(N-l+\nu_{r_j})g_{0r_j}} dg_{0r_j} \\ &= \sum_{l=0}^{j-1} N_j \Theta_l^{j-1} \frac{1}{(N-l+\nu_{r_j})^2} \end{aligned} \tag{4.49}$$

Substituting (4.49) into (4.48), and calculating the integral completes the proof. □
Note that in practice $M << N$, thus $N\gamma_{Q_j} >> 1$. Therefore, $\mathcal{O}_4$ is approximated by the following optimization problem.

Problem $\mathcal{O}_5$:

$$C^{g_0}_{s|M} \approx \max_{\underline{\gamma_Q}} \sum_{j=1}^{M} \sum_{l=0}^{j-1} B_c \frac{N_j \Theta_l^{j-1}}{N-l} \log\left((N-l)\gamma_{Q_j}\right) \tag{4.50}$$

$$\text{subject to} \sum_{j=1}^{M} \gamma_{Q_j} = N\gamma_Q,\ 0 \le \gamma_{Q_j} \le N\gamma_Q \tag{4.51}$$

The Lagrangian function for $\mathcal{O}_5$ is

$$L\left(\underline{\gamma_Q},\lambda\right) = \sum_{j=1}^{M}\sum_{l=0}^{j-1}\frac{N_j\Theta_l^{j-1}}{N-l}\log\left((N-l)\gamma_{Q_j}\right) - \lambda\left(\sum_{j=1}^{M}\gamma_{Q_j} - N\gamma_Q\right)$$

where λ is the Lagrangian coefficient. By differentiating $L(\underline{\gamma_Q},\lambda)$ with respect to γ_{Q_j} and setting this derivative equal to zero, we have

$$\gamma_{Q_j}^* = \frac{1}{\lambda^*}\vartheta_j \tag{4.52}$$

where

$$\vartheta_j \triangleq \sum_{l=0}^{j-1} N_j \frac{\Theta_l^{j-1}}{N-l}$$

Substituting (4.52) into (4.51) yields

$$\lambda^* = \frac{1}{N\gamma_Q}\sum_{j=1}^{M}\vartheta_j \tag{4.53}$$

The optimal spectrum sharing load factor $\gamma_{Q_j}^*$ is obtained by substituting (4.53) into (4.52),

$$\gamma_{Q_j}^* = N\gamma_Q\frac{\vartheta_j}{\sum_{j=1}^{M}\vartheta_j} \tag{4.54}$$

Also, using (4.19) and (4.25), the optimal transmit power vector $\mathbf{P}_s^*$ is

$$\mathbf{P}_s^* = \frac{Q_T}{\sum_{j=1}^{M}\vartheta_j}\left(\frac{\vartheta_1}{g_{0r_1}}, \frac{\vartheta_2}{g_{0r_2}}, \ldots, \frac{\vartheta_M}{g_{0r_M}}\right) \tag{4.55}$$

The maximum achievable capacity of the secondary service is approximated by substituting (4.54) into (4.50),

$$C_{s|M}^{g_0} \approx \sum_{j=1}^{M}\sum_{l=0}^{j-1}\frac{B_cN_j\Theta_l^{j-1}}{N-l}\log\left((N-l)N\gamma_Q\frac{\vartheta_j}{\sum_{j=1}^{M}\vartheta_j}\right) \tag{4.56}$$

For $M = 1$, the approximated achievable capacity in (4.56) reduces to

$$C_{s|1}^{g0} \approx B_c \log(N^2 \gamma_Q) \tag{4.57}$$

which is very close to the exact solution of $\mathcal{O}_5$ for $M = 1$, which is

$$C_{s|1}^{g0} = B_c \frac{N^2 \gamma_Q}{N^2 \gamma_Q - 1} \log(N^2 \gamma_Q) \tag{4.58}$$

4.5.2.2 Secondary Sub-channel-Based Selection Policy

For secondary sub-channel based selection policy, the selection criteria is

$$\Phi(g_{0i}, g_{1i}) = g_{1i} \tag{4.59}$$

and $\mu_i = g_{1i}$. The policy $\mathcal{P}_{g_1}$ is defined so that in the N-tuple selection sequence Υ_N

$$\begin{aligned} \mu_{r_N} \le \mu_{r_{N-1}} \le \ldots \le \mu_{r_1} \\ \mu_{r_1} = \max_i \{\mu_i\} \end{aligned} \tag{4.60}$$

The M-tuple selected sequence Θ_M is

$$\theta_M = \mu_{r_M} \le \theta_{M-1} = \mu_{r_{M-1}} \le \ldots \le \theta_1 = \mu_{r_1}$$

Here, $\mathcal{P}_{g_1}$ selects those sub-channels that result in the highest achieved capacity for the secondary service. Using order statistics, the pdf of θ_j is

$$k_j(\theta) = N_j F_\mu^{N-j}(\theta) \left[1 - F_\mu(\theta)\right]^{j-1} f_\mu(\theta) \tag{4.61}$$

where $f_\mu(\theta)$ and $F_\mu(\theta)$ are probability density function (pdf) and probability distribution function (PDF) of the random variable μ, respectively. Following the same line of argument as in Section 4.5.1, for Rayleigh fading, $k_j(\theta)$ is obtained from

$$k_j(\theta) = N_j \left(1 - e^{-\theta}\right)^{N-j} e^{-\theta j} \tag{4.62}$$

Replacing the binomial expansion of $\left(1 - e^{-\theta}\right)^{N-j}$ in (4.62) yields

$$k_j(\theta) = N_j \sum_{l=0}^{N-j} \Theta_l^{N-j} e^{-\theta(l+j)} \tag{4.63}$$

where N_j is obtained from (4.42) and

$$\Theta_l^{N-j} = \binom{N-j}{l}(-1)^l \tag{4.64}$$

Proposition 2 *The maximum achievable capacity of the secondary service based on policy $\mathcal{P}_{g_1}$, denoted by $C_{s|M}^{g_1}$, is obtained by solving the following optimization problem.*

Problem $\mathcal{O}_6$:

$$C_{s|M}^{g_1} = \max_{\underline{\gamma_{\mathbf{Q}}}} \sum_{j=1}^{M} \sum_{l=0}^{N-j} B_c \frac{N_j \Theta_l^{N-j}}{l+j} \frac{\frac{\gamma_{Q_j}}{l+j}}{\frac{\gamma_{Q_j}}{l+j} - 1} \log\left(\frac{\gamma_{Q_j}}{l+j}\right) \tag{4.65}$$

$$\text{subject to} \sum_{j=1}^{M} \gamma_{Q_j} = N\gamma_Q, \quad 0 \le \gamma_{Q_j} \le N\gamma_Q \tag{4.66}$$

Proof Substituting $P_{sj} = \frac{Q_j}{g_{0r_j}}$ into (4.22), and setting $\nu_{r_j} = g_{1r_j}/g_{0r_j}$, the maximum achievable capacity $C_{s|M}^{g_1}$ based on policy $\mathcal{P}_{g_1}$ is obtained from $\mathcal{O}_1$ as

$$C_{s|M}^{g_1} = \max_{\mathbf{Q}} \sum_{j=1}^{M} B_c \int_{\nu_{r_j}} \log\left(1 + \nu_{r_j}\gamma_{Q_j}\right) h_j\left(\nu_{r_j}\right) d\nu_{r_j} \tag{4.67}$$

$$\text{subject to} \sum_{j=1}^{M} \gamma_{Q_j} = N\gamma_Q, \quad 0 \le \gamma_{Q_j} \le N\gamma_Q$$

In cases where the sub-channel gains $\sqrt{g_{0r_j}}$ and $\sqrt{g_{1r_j}}$ are i.i.d., Rayleigh random variables g_{0r_j} and g_{1r_j} are exponentially distributed random variables, therefore, using (4.63) and some straightforward mathematical derivations, the pdf of ν_{r_j}, denoted by $h_j(\nu_j)$, is

$$\begin{aligned}
h_j\left(\nu_{r_j}\right) &= \frac{d}{d\nu_{r_j}} \int_0^{\infty} \int_0^{\nu_{r_j} g_{0r_j}} N_j \sum_{l=0}^{N-j} \Theta_l^{N-j} e^{-(l+j)g_{1r_j}} e^{-g_{0r_j}} dg_{1r_j} dg_{0r_j} \\
&= \frac{d}{d\nu_{r_j}} \sum_{l=0}^{N-j} \frac{N_j \Theta_l^{N-j}}{l+j} \int_0^{\infty} \left(1 - e^{-g_{0r_j}\nu_{r_j}(l+j)}\right) e^{-g_{0r_j}} dg_{0r_j} \\
&= \sum_{l=0}^{N-j} N_j \Theta_l^{N-j} \int_0^{\infty} g_{0r_j} e^{-\left(1+\nu_{r_j}(l+j)\right)g_{0r_j}} dg_{0r_j}
\end{aligned}$$

$$= \sum_{l=0}^{N-j} N_j \Theta_l^{N-j} \frac{1}{((l+j)\nu_{r_j}+1)^2} \tag{4.68}$$

Substituting (4.68) into (4.67), and obtaining the integral completes the proof. □

For small values of $\gamma_{Q_j}/(l+j)$, $l = 0, 1, \ldots, N-j$, we have

$$\frac{x}{x-1}\log x \approx \sqrt{x} \tag{4.69}$$

which reduces the Problem $\mathcal{O}_6$ into the following problem.

Problem $\mathcal{O}_7$:

$$C_{s|M}^{g_1} \approx \max_{\underline{\gamma_Q}} \sum_{j=1}^{M} \sum_{l=0}^{N-j} B_c \frac{N_j \Theta_l^{N-j}}{l+j} \left(\frac{\gamma_{Q_j}}{l+j}\right)^{\frac{1}{2}} \tag{4.70}$$

$$\text{subject to} \sum_{j=1}^{M} \gamma_{Q_j} = N\gamma_Q, \ \ 0 \le \gamma_{Q_j} \le N\gamma_Q \tag{4.71}$$

Similar to $\mathcal{O}_5$, by utilizing Lagrange multipliers for solving $\mathcal{O}_7$, the optimal spectrum sharing load factor is

$$\gamma_{Q_j}^* = N\gamma_Q \frac{\chi_j^2}{\sum_{j=1}^{M} \chi_j^2} \tag{4.72}$$

where

$$\chi_j = \sum_{l=0}^{N-j} N_j \frac{\Theta_l^{N-j}}{2(l+j)^{1.5}}$$

Also, using (4.25) and (4.19), the optimal transmit power vector $\mathbf{P}_s^*$ is

$$\mathbf{P}_s^* = \frac{Q_T}{\sum_{j=1}^{M} \chi_j^2} \left(\frac{\chi_1^2}{g_{0r_1}}, \frac{\chi_2^2}{g_{0r_2}}, \ldots, \frac{\chi_M^2}{g_{0r_M}}\right) \tag{4.73}$$

The maximum achievable capacity of the secondary service is approximated by substituting (4.72) into (4.70),

$$C_{s|M}^{g_1} \approx \sum_{j=1}^{M} \sum_{l=0}^{N-j} B_c \frac{N_j \Theta_l^{N-j}}{(l+j)^{1.5}} \left(N\gamma_Q \frac{\chi_j^2}{\sum_{j=1}^{M} \chi_j^2} \right)^{\frac{1}{2}} \tag{4.74}$$

For $M = 1$, the approximated achievable capacity in (4.74) is reduced to

$$C_{s|1}^{g_1} \approx B_c \sum_{l=0}^{N-1} N \frac{\Theta_l^{N-1}}{l+1} \left(\frac{N\gamma_Q}{l+1}\right)^{\frac{1}{2}} \tag{4.75}$$

Considering (4.69), note that (4.75) is similar to the exact solution of Problem $\mathcal{O}_8$, which is

$$C_{s|1}^{g_1} = B_c \sum_{l=0}^{N-1} N \frac{\Theta_l^{N-1}}{l+1} \frac{\frac{N\gamma_Q}{l+1}}{\frac{N\gamma_Q}{l+1} - 1} \log\left(\frac{N\gamma_Q}{l+1}\right) \tag{4.76}$$

4.5.2.3 Reward Factor-Based Sub-channel Selection Policy

Assume that the selection criteria is

$$\Phi(g_{0i}, g_{1i}) = \nu_i = \frac{g_{1i}}{g_{0i}} \tag{4.77}$$

and $\mu_i = \nu_i$. Policy $\mathcal{P}_\nu$ in then defined so that in the N-tuple selection sequence Υ_N

$$\begin{aligned} \mu_{r_N} \leq \mu_{r_{N-1}} \leq \ldots \leq \mu_{r_1} \\ \mu_{r_1} = \max_i\{\mu_i\} \end{aligned} \tag{4.78}$$

The M-tuple selected sequence Θ_M is

$$\theta_M = \mu_{r_M} \leq \theta_{M-1} = \mu_{r_{M-1}} \leq \ldots \leq \theta_1 = \mu_{r_1}$$

Using order statistics, pdf $k_j(0)$ is

$$k_j(\theta) = N_j H_\mu^{N-j}(\theta) \left[1 - H_\mu(\theta)\right]^{j-1} h_\mu(\theta) \tag{4.79}$$

where $H_\mu(\theta)$, $h_\mu(\theta)$ are the pdf and the PDF of the random variable μ, respectively.

For cases in which $\sqrt{g_{0j}}$ and $\sqrt{g_{1j}}$ are i.i.d. with Rayleigh distribution $\forall j$, it was already shown that $h_\mu(\theta)$ is (see Section 4.5.1)

$$h_\mu(\theta) = \frac{1}{(1+\theta)^2}, \quad 0 < \theta < \infty \tag{4.80}$$

and $H_\mu(\theta)$ is

$$H_\mu(\theta) = \frac{\theta}{(1+\theta)}, \quad 0 < \theta < \infty \tag{4.81}$$

Substituting (4.81) and (4.80) into (4.79), we have

$$k_j(\theta) = N_j \frac{\theta^{N-j}}{(1+\theta)^{N+1}}, \quad 0 < \theta < \infty \tag{4.82}$$

where N_j is obtained from (4.42).

Substituting $P_{sj} = \frac{Q_j}{g_{0r_j}}$ into (4.22), and setting $\theta_j = \frac{g_{1r_j}}{g_{0r_j}}$, the maximum achievable capacity $C^{\nu}_{s|M}$ for the policy $\mathcal{P}_\nu$ is obtained from $\mathcal{O}_1$ by solving the following optimization problem.

Problem $\mathcal{O}_8$:

$$C^{\nu}_{s|M} = \max_{\underline{\gamma_Q}} \sum_{j=1}^{M} B_c \int_0^{\infty} \log\left(1 + \theta_j \gamma_{Q_j}\right) k_j\left(\theta_j\right) d\theta_j \tag{4.83}$$

$$\text{subject to} \sum_{j=1}^{M} \gamma_{Q_j} = N\gamma_Q, \; 0 \le \gamma_{Q_j} \le N\gamma_Q \tag{4.84}$$

Obtaining a closed form solution for $C^{\nu}_{s|M}$ in $\mathcal{O}_8$ is complicated. This is mainly due to the form of pdf $k_j(\theta)$ in (4.83). Here, we obtain $C^{\nu}_{s|M}$ utilizing numerical results. Note that for $M = 1$, (4.83) is reduced to

$$C^{\nu}_{s|1} = B_c \int_0^{\infty} \log\left(1 + \theta N\gamma_Q\right) k_1(\theta)\, d\theta \tag{4.85}$$

A summary of the results of Section 4.5 is presented in Table 4.2.

Table 4.2 Achievable capacity in Rayleigh fading for different policies: single secondary user

Sub-channel selection policy	$\Phi(g_{0i}, g_{1i})$	Maximum achievable capacity
Uniform	1	$C^{1}_{s\|M} \approx MB_c \frac{N\gamma_Q}{N\gamma_Q - M} \log\left(\frac{N\gamma_Q}{M}\right)$
Cross-sub-channel based	g_{0i}	$C^{g_0}_{s\|M} \approx \sum_{j=1}^{M}\sum_{l=0}^{j-1} \frac{B_c N_j \Theta_l^{j-1}}{N-l} \log\left((N-l)N\gamma_Q \frac{\vartheta_j}{\sum_{j=1}^{M}\vartheta_j}\right)$
Secondary-sub-channel based	g_{1i}	$C^{g_1}_{s\|M} \approx \sum_{j=1}^{M}\sum_{l=0}^{N-j} B_c \frac{N_j \Theta_l^{N-j}}{(l+j)^{1.5}} \left(N\gamma_Q \frac{\chi_j^2}{\sum_{j=1}^{M}\chi_j^2}\right)^{\frac{1}{2}}$
Reward factor based	g_{1i}/g_{0i}	Solution of problem $\mathcal{O}_8$

4.6 Multiple Secondary Service Users

Here, we consider the case where more than one secondary service transmitter–receiver pair are communicating. The *secondary network* consists of a number of secondary service users that employ OSA to access the spectrum. Let N_s be the number of secondary service active transmitter–receiver pairs, indexed by s, each with their corresponding spectrum sharing load factor γ_s, $s = 1, \ldots, N_s$. In what follows, we first assume that each secondary service transmitter–receiver pair selects only one sub-channel by utilizing a sub-channel selection scenario. We then extend our analysis to the case where each secondary service transmitter–receiver pair selects multiple sub-channels.

The wide-band interference caused by each secondary service transmitter–receiver pair is $Q_s = KN_0B\gamma_s$. Since our main objective is to obtain the maximum achievable capacity, for brevity we ignore the interfering effect of different secondary service transmitter–receiver pairs on each other. The interfering effect, if any, reduces the achieved capacity of the secondary network, therefore, the presented results are the upper-bound for the maximum achievable capacity.

4.6.1 Uniform Sub-channel Selection

Consider the case where only one sub-channel is selected corresponding to each secondary service transmitter–receiver pair. Utilizing uniform sub-channel selection, as in (4.38), the maximum achievable capacity for the secondary service s is

$$C_s^1 = B_c \frac{N\gamma_s}{N\gamma_s - 1} \log(N\gamma_s). \tag{4.86}$$

The total achievable capacity of the secondary network C^1 is obtained by solving the following optimization problem.

Problem $\mathcal{O}_9$:

$$C^1 = \max_{\underline{\underline{\gamma_S}}} \sum_{s=1}^{N_s} B_c \frac{N\gamma_s}{N\gamma_s - 1} \log(N\gamma_s) \tag{4.87}$$

$$\text{subject to} \sum_{s=1}^{N_s} \gamma_s = \gamma_Q, \quad 0 \le \gamma_s \le \gamma_Q \tag{4.88}$$

where $\underline{\underline{\gamma_S}} = (\gamma_1, \ldots, \gamma_{N_s})$. Following the same line of arguments as in Section 4.5.1, the optimal spectrum sharing load factor γ_s^* is

$$\gamma_s^* = \frac{\gamma_Q}{N_s} \tag{4.89}$$

The total achievable capacity of the secondary network is obtained by substituting (4.89) into (4.87),

$$C^1 = B_c N_s \frac{N\gamma_Q}{N\gamma_Q - N_s} \log\left(\frac{N\gamma_Q}{N_s}\right) \tag{4.90}$$

4.6.2 Non-uniform Sub-channel Selection

We also consider the case in which non-uniform sub-channel selection based on policy $\mathcal{P}_{g_0}$ is employed for the secondary service. In this case the maximum achievable capacity of the secondary service transmitter–receiver pair s with one sub-channel selection is obtained from (4.58),

$$C_{s|1}^{g_0} = B_c \frac{N^2\gamma_s}{N^2\gamma_s - 1} \log(N^2\gamma_s) \tag{4.91}$$

The total achievable capacity of the secondary network C^{g_0} is obtained by solving the following optimization problem.

Problem $\mathcal{O}_{10}$:

$$C^{g_0} = \max_{\underline{\underline{\gamma_S}}} \sum_{s=1}^{N_s} B_c \frac{N^2\gamma_s}{N^2\gamma_s - 1} \log(N^2\gamma_s) \tag{4.92}$$

$$\text{subject to} \sum_{s=1}^{N_s} \gamma_s = \gamma_Q, \; 0 \le \gamma_s \le \gamma_Q \tag{4.93}$$

In this case, similar to Section 4.6.1 the optimal spectrum sharing load factor γ_s^* is also obtained from (4.89). Intuitively, from the secondary network's point of view, each user shares an equal spectrum sharing load factor because each secondary service transmitter–receiver pair selects one sub-cannel based on policy $\mathcal{P}_{g_0}$. Substituting (4.89) into (4.92), the secondary network's total achievable capacity is

$$C^{g_0} = B_c N_s \frac{N^2\gamma_Q}{N^2\gamma_Q - N_s} \log\left(\frac{N^2\gamma_Q}{N_s}\right) \tag{4.94}$$

4.6.3 Impact of Intersecondary Service Interference

In practice, in the secondary network, the transmission made by a secondary service transmitter–receiver pair, also causes interference to other active secondary service transmitter–receiver pairs. To understand the scaling effect of the achievable capacity of the secondary network, we assume $N_s \gg N$.

Suppose that the secondary service transmitter–receiver pairs are fixed communications entities. In such a case, adopting the result from [22], the total achievable capacity of the secondary network tends to zero with increasing N_s by the rate of $1/\sqrt{N_s}$. Therefore, for a uniform sub-channel selection, we have

$$\lim_{N_s \to \infty} C^1 \propto B_c \sqrt{N_s} \frac{N\gamma_Q}{N\gamma_Q - N_s} \log\left(\frac{N\gamma_Q}{N_s}\right) \tag{4.95}$$

and for non-uniform sub-channel selection,

$$\lim_{N_s \to \infty} C^{g_0} \propto B_c \sqrt{N_s} \frac{N^2\gamma_Q}{N^2\gamma_Q - N_s} \log\left(\frac{N^2\gamma_Q}{N_s}\right) \tag{4.96}$$

Therefore,

$$\lim_{N_s \to \infty} \frac{C^{g_0}}{C^1} \propto N \tag{4.97}$$

Equation (4.97) shows that by increasing the number of secondary service users, the total achievable capacity by policy $\mathcal{P}_{g_0}$ is N time higher than that of $\mathcal{P}_1$.

For the case where the secondary service users are mobile and delay tolerant, from adopting the results in [23], we note that the decreasing rate of the total achievable capacity of the secondary network can be kept constant. With assuming mobility along with the infinite delay tolerance, both uniform and non-uniform sub-channel selections are able to achieve the corresponding achievable capacity in (4.90), and (4.94), respectively.

4.6.4 Multiple Sub-channel Selection

We extend our analysis to the case where each secondary service transmitter–receiver pair selects multiple sub-channels. We further assume that the inter-secondary service interference is ignorable. Let each secondary service transmitter–receiver pair select M sub-channels. For uniform sub-channel selection policy, using (4.38), the problem $\mathcal{O}_9$, is converted to the following problem.

Problem $\mathcal{O}_{11}$:

$$C_M^1 = \max_{\underline{\underline{\gamma_S}}} \sum_{s=1}^{N_s} M B_c \frac{N\gamma_s}{N\gamma_s - M} \log\left(\frac{N\gamma_s}{M}\right) \tag{4.98}$$

$$\text{subject to} \sum_{s=1}^{N_s} \gamma_s = \gamma_Q, \quad 0 \le \gamma_s \le \gamma_Q$$

By defining $\hat{\gamma}_s \triangleq \gamma_s/M$, the above optimization problem can be written as

Problem $\mathcal{O}_{12}$:

$$C_M^1 = \max_{\underline{\hat{\gamma}_S}} M \sum_{s=1}^{N_s} B_c \frac{N\hat{\gamma}_s}{N\hat{\gamma}_s - 1} \log\left(N\hat{\gamma}_s\right)$$

$$\text{subject to} \sum_{s=1}^{N_s} \hat{\gamma}_s = \frac{\gamma_Q}{M}, \quad 0 \leq \hat{\gamma}_s \leq \gamma_Q$$

Similar to $\mathcal{O}_9$, the optimal value of $\hat{\gamma}_s^*$ is

$$\hat{\gamma}_s^* = \frac{\gamma_Q}{MN_s} \tag{4.99}$$

which is similar to the optimal spectrum sharing load factor $\gamma_s^* = \gamma_Q {N_s}^{-1}$ in (4.89). Substituting γ_s^* into (4.98), the total achievable capacity of the secondary network is

$$C_M^1 = B_c M N_s \frac{N\gamma_Q}{N\gamma_Q - MN_s} \log\left(\frac{N\gamma_Q}{MN_s}\right) \tag{4.100}$$

For a non-uniform selection of multiple sub-channels in $\mathcal{P}_{g_0}$, by using (4.56), the optimization problem $\mathcal{O}_{10}$ is converted to the following problem.

Problem $\mathcal{O}_{13}$:

$$C_M^{g_0} = \max_{\underline{\gamma_S}} \sum_{s=1}^{N_s} \sum_{j=1}^{M} \sum_{l=0}^{j-1} \frac{B_c N_j \Theta_l^{j-1}}{N-l} \log\left((N-l)N\gamma_s \frac{\vartheta_j}{\sum_{j=1}^{M} \vartheta_j}\right) \tag{4.101}$$

$$\text{subject to} \sum_{s=1}^{N_s} \gamma_s = \gamma_Q, \quad 0 \leq \gamma_s \leq \gamma_Q \tag{4.102}$$

where N_j and Θ_l^{j-1} are obtained from (4.42) and (4.45), respectively, and

$$\vartheta_j \triangleq \sum_{l=0}^{j-1} N_j \frac{\Theta_l^{j-1}}{N-l}$$

Utilizing Lagrange multipliers for solving $\mathcal{O}_{13}$, the optimal spectrum sharing load factor, γ_s^*, is also obtained from (4.89). Consequently, the total achievable capacity of the secondary network is obtained by substituting γ_s^* into (4.101),

$$C_M^{g_0} = \sum_{j=1}^{M} \sum_{l=0}^{j-1} \frac{N_s B_c N_j \Theta_l^{j-1}}{N-l} \log\left((N-l) N \frac{\gamma_Q}{N_s} \frac{\vartheta_j}{\sum_{j=1}^{M} \vartheta_j} \right) \tag{4.103}$$

4.7 Numerical Studies

Here we compare the achievable capacity of different sub-channel selection policies under different scenarios. Main system parameters are given in Table 4.3. The primary service is a cellular DS-CDMA network with a single service, i.e., $\mathcal{Y} = 1$. For simplicity we assume that the number of the primary service users in each cell is the same and denoted by L.

4.7.1 Comparing Sub-channel Selection Policies

First we compare the achieved spectral efficiency of the sub-channel selection policies versus M. For easy reference, here we repeat our definitions for the four sub-channel selection policies; $\mathcal{P}_1$: uniform sub-channel selection policy, $\mathcal{P}_{g_0}$: sub-channel selection policy based on the cross-sub-channel, $\mathcal{P}_{g_1}$: sub-channel selection policy based on the secondary-sub-channel, and $\mathcal{P}_\nu$: reward factor-based sub-channel selection policy.

As can be seen in Fig. 4.2, the achieved spectral efficiency of uniform sub-channel selection $C^1_{s|M}/MB_c$ is lower than that of non-uniform case in most cases. For $M = 1$, the gap in the achieved spectral efficiency between $C^1_{s|M}/MB_c$ and $C^\nu_{s|M}/MB_c$ is very large. However, by increasing M, this gap is significantly reduced. This gap is related to the ratio M/N, and we note that for larger values of this ratio, the gap is smaller. This is mainly due to the fact that for higher values M/N, the set of M selected sub-channels by $\mathcal{P}_\nu$ and $\mathcal{P}_1$ may overlap to a large extent.

It is also seen that $\mathcal{P}_{g_1}$ performs very similar to $\mathcal{P}_\nu$ for larger values of M. Comparing the rate of decreasing the achieved spectral efficiency by increasing M

Table 4.3 Parameters in the numerical studies

Parameter	Value
Noise power, N_0B	0.1 Watt
Maximum transmit power of the primary user, P_p	1 Watt
Cell radios, R	100 m
Number of users, L	30
Required E_b/I_0, ρ	4 dB
Activity factor, υ	0.3
Distance between the secondary service transmitter and receiver, D	6 m
Processing gain, G	128

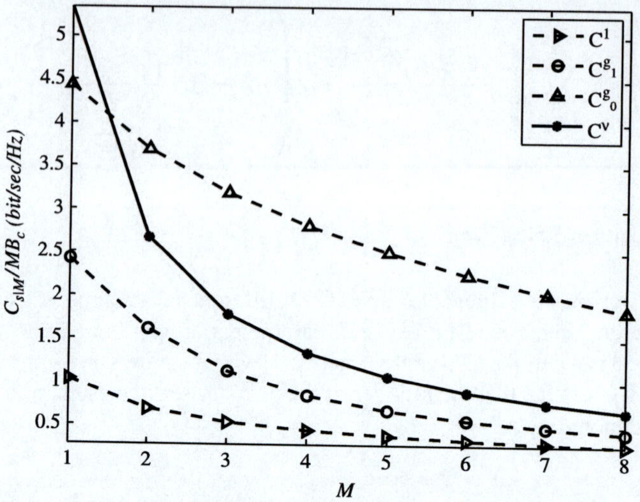

Fig. 4.2 Achieved spectral efficiency of the secondary service for various sub-channel selection policies versus M, for $Q_T = N_0B$, $N = 20$, $\upsilon = 0.3$, and $M = 1$

indicates that $C^{g0}_{s|M}/MB_c$ is decreased with a slower rate than that of the others. The slower decay rate is mainly due to the fact that considering the cross-sub-channel gain g_{0i} allows the secondary transmitter to transmit at maximum power, causing less interference on primary users, while satisfying the interference threshold constraint. In most cases, the maximum transmit power by $\mathcal{P}_{g0}$ can be higher than that of the others, thus the corresponding achievable capacity is higher.

Figure 4.2 indicates that in terms of the achieved spectral efficiency for a given number of accessible sub-channels to the secondary service, $\mathcal{P}_{g0}$ outperforms the other sub-channel selection policies.

To study the impact of the primary service activity factor υ, in Fig. 4.3 we compare the achieved spectral efficiency of the secondary service for $M = 1$ versus υ for different sub-channel selection policies. As expected, for both uniform and non-uniform sub-channel selection policies, the achieved spectral efficiency is decreased by increasing υ. This is due to the fact that increasing υ, increases the interference caused by primary users to secondary users. The decrease in the rate of the achieved spectral efficiency for all sub-channel selection policies looks similar.

To study the impact of N, in Fig. 4.4 we compare the achieved spectral efficiency of the secondary service for $M = 1$ versus N for different sub-channel selection policies. As expected, for both uniform and non-uniform sub-channel selection policies, the achieved spectral efficiency is increased by increasing N. This is due to the fact that the probability of selecting proper sub-channel for OSA is increased by increasing N. The trend in the achieved spectral efficiency for $\mathcal{P}_\upsilon$ and $\mathcal{P}_{g0}$ is the same, and is higher as compared to those of $\mathcal{P}_1$ and $\mathcal{P}_{g1}$. It is also interesting to note that by increasing N, the gap between the achieved spectral efficiency of $\mathcal{P}_{g0}$ and $\mathcal{P}_1$ widens.

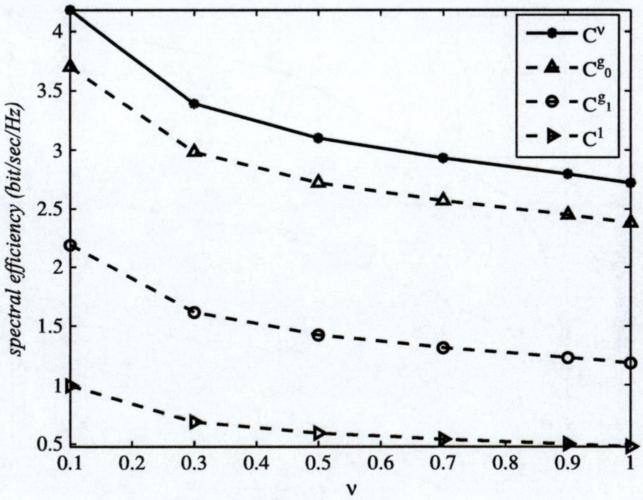

Fig. 4.3 Achieved spectral efficiency of the secondary service for various sub-channel selection policies versus υ, for $N = 40$, $M = 1$, and $Q_T = N_0 B$

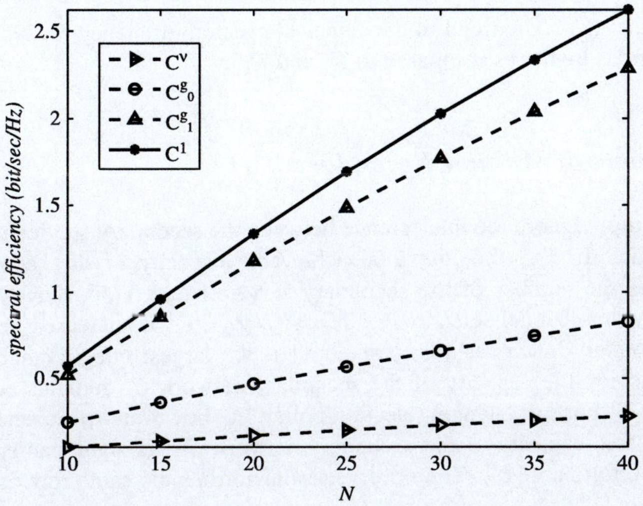

Fig. 4.4 Achieved spectral efficiency of the secondary service for various sub-channel selection policies versus N, for $Q_T = 0.1 N_0 B$, $M = 1$, and $\upsilon = 0.3$

Figure 4.5 shows the impact of the interference threshold constraint on the achieved spectral efficiency of the secondary service for $M = 1$ versus $\frac{Q_T}{N_0 B}$ for different sub-channel selection policies. As can be seen, for all sub-channel selection policies, the achieved spectral efficiency of the secondary service is increased

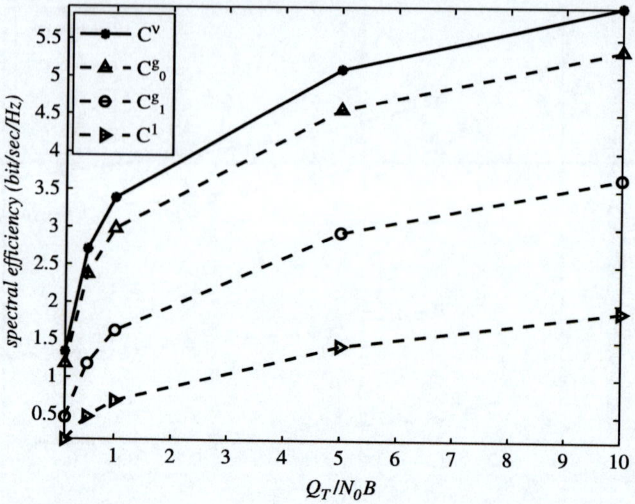

Fig. 4.5 Achieved spectral efficiency of the secondary service for various sub-channel selection policies, versus $\frac{Q_T}{N_0 B}$, for $M = 1$, $N = 20$, and $\upsilon = 0.3$

by increasing $\frac{Q_T}{N_0 B}$. The trend in the achieved spectral efficiency for $\mathcal{P}_\nu$ and $\mathcal{P}_{g_0}$ is the same and is higher as compared to $\mathcal{P}_1$ and $\mathcal{P}_{g_1}$.

4.7.2 Impact of Multiple Secondary Users

First, we simply ignore the interference between the secondary service transmitter-receiver pairs. In Fig. 4.6, the total achievable capacity of the secondary network versus the number of the secondary service users N_s is shown. Note that for uniform sub-channel selection for $N_s < \lfloor N\gamma_Q \rfloor$, C^1 is increased versus N_s in an approximately linear fashion, where $\lfloor x \rfloor$ is the largest integer smaller than x. The total achievable capacity of the secondary network C^1 remains constant for $N_s \gg \lfloor N\gamma_Q \rfloor$. For sub-channel selection policy $\mathcal{P}_{g_0}$ however, by increasing N_s, the total achievable capacity of the secondary network C^{g_0} is significantly increased as compared to that of C^1. The observed behavior has the same root as *multiuser diversity gain* [17].

In Fig. 4.7, we plot the asymptote of the achieved capacity obtained in Section 4.6.3. As can be seen, for large values of N_s, the value of C^1 is very close to zero. In non-uniform sub-channel selection for $0 \leq N_s \leq \lfloor N^2\gamma_Q \rfloor$, the value of C^{g_0} is increased versus N_s. For $N_s = \lfloor N^2\gamma_Q \rfloor$, the maximum achievable capacity in the secondary network is $C^{g_0} = \sqrt{N^2\gamma_Q}$, which can also be obtained by setting the derivative of (4.96) to zero. Therefore, although the number of the secondary service users in the network is increased, the total achievable capacity of the secondary network is constant in non-uniform sub-channel selection for $N_s >> \lfloor N^2\gamma_Q \rfloor$.

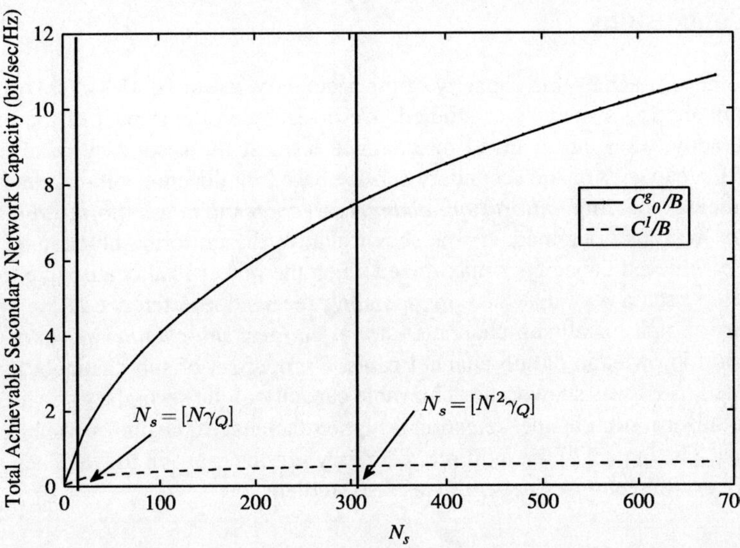

Fig. 4.6 Total achievable capacity of the secondary network versus N_s, for uniform and non-uniform sub-channel selection without considering cross interference among the secondary service users ($Q_T = 10N_0B$, $\upsilon = 0.3$, and $N = 60$)

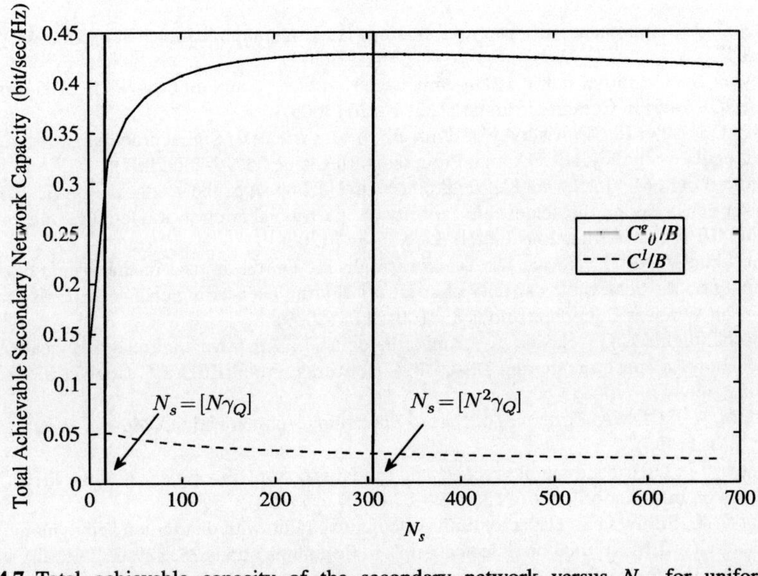

Fig. 4.7 Total achievable capacity of the secondary network versus N_s, for uniform and non-uniform sub-channel selection with considering cross interference among the secondary service users ($Q_T = 10N_0B$, $\upsilon = 0.3$, and $N = 60$)

4.8 Conclusions

In this chapter, achievable capacity of the secondary service in DS-CDMA/OFDM spectrum sharing systems was studied. We investigated the impact of the primary service activity on the received interference level at the secondary receiver. The achievable capacity of the secondary service based on different sub-channel selection policies including *uniform sub-channel selection* and *non-uniform sub-channel selection* was also obtained. It was shown that in the uniform sub-channel selection, the achieved capacity is maximized when the power is allocated to each sub-channel in such a way that the corresponding received interference at the primary receiver is equal for all sub-channels. In *non-uniform sub-channel selection* based on a priori knowledge of sub-channel gains, a proper set of sub-channels for OSA is selected. As it was shown, the achievable capacity of the secondary service based on non-uniform sub-channel selection is higher than that of the uniform sub-channel selection. The impact of the multiple secondary service users on the total achievable capacity of the secondary network was also studied.

Acknowledgments Part of the work presented in this chapter was supported by the Mobile Communication Company of Iran (MCI).

References

1. Peha J. M: Approaches to spectrum sharing. IEEE Communications Magazine **43**, 10–12 (2005)
2. Haykin S.: Cognitive radio: Brain-empowered wireless communications. IEEE Journal on Selected Areas in Communications **23**, 201–220 (2005)
3. Zhao O., Sadler B. M.: A survey of dynamic spectrum access: Signal processing, networking, and regulatory policy. IEEE Signal Processing Magazine **24**, 79–89 (2007)
4. Khoshkholgh M. G., Navaie K., Yanikomeroglu H.: Impact of the secondary service transmit power constraint on the achievable capacity of spectrum sharing in Rayleigh fading environment. IEEE Communications Letters **12**, 865–867 (2008)
5. Khoshkholgh M. G., Navaie K., Yanikomeroglu H.: On the impact of the primary network activity on the achievable capacity of spectrum sharing over fading channels. IEEE Transactions on Wireless Communications **8**, 2100–2111 (2009)
6. Khoshkholgh M. G., Navaie K., Yanikomeroglu H.: Adaptive multiple time-scale power allocation for spectrum sharing DS-CDMA Networks. Proc. IEEE ICC CogNet'08, Beijing, China, 466–470 (2008)
7. Viterbi, A. J: CDMA: Principles of Spread Spectrum Communication. Addison-Wesley, Reading Mass (1995)
8. Gastpar M.: On capacity under receive and spatial spectrum-sharing constraints. IEEE Transactions on Information Theory **53**, 471–487 (2007)
9. Jafar S. A., Srinivasa S.: Capacity limits of cognitive radio with distributed and dynamic spectral activity. IEEE Journal on Selected Areas in Communications **25**, 529–537 (2007)
10. Ghasemi A., Sousa E. S.: Fundamental limits of spectrum-sharing in fading environments. IEEE Transactions on Wireless Communications **6**, 649–658 (2007)
11. Zhang L., Liang Y. C., Xin Y.: Joint beamforming and power allocation for multiple access channels in cognitive radio networks. IEEE Journal on Selected Areas in Communications **26**, 38–51 (2008)

12. Zhang H. Su, X.: Cross-layer based opportunistic MAC protocols for QoS provisionings over cognitive radio wireless networks. IEEE Journal on Selected Areas in Communications **26**, 118–129 (2008)
13. Geirhofer S., Tong L., Sadler L.: Cognitive medium access: Constraining interference based on experimental models. IEEE Journal on Selected Areas in Communications **26**, 95–105 (2008)
14. Navaie K., Valaee S., Sharafat A. R., Sousa E. S.: On the downlink interference in heterogeneous wireless DS-CDMA networks. IEEE Transactions on Wireless Communications **5**, 384–393 (2006)
15. Sharma M., Sahoo A., Nayak K. D.: Channel selection under interference temperature model in multi-hop cognitive mesh networks. Proc. IEEE DySPAN'07, Dublin, Ireland, 133–136 (2007)
16. Nie N., Comaniciu C.: Adaptive channel allocation spectrum etiquette for cognitive radio networks. Proc. IEEE DySPAN'05, Baltimore, MD, 331–335 (2005)
17. Tse D., Viswanath P.: Fundamentals of Wireless Communication. Cambridge University Press, New York, NY (2004)
18. Khoshkholgh M. G., Navaic K., Yanikomeroglu H.: Achievable capacity in hybrid DS-CDMA/OFDM spectrum-sharing. IEEE Transactions on Mobile Computing **9**, 756–777 (2010)
19. Holma H., Toskala A.: WCDMA for UMTS: Radio Access for Third Generation Mobile Communications. Wiley, New York, NY (2000)
20. Goldsmith A. J.: Wireless Communications. Cambridge University Press, New York, NY (2005)
21. Cover T. M., Thomas J. A.: Elements of Information Theory. Wiley, Hoboken, NJ (2006)
22. Gupta P., Kumar P. R.: The capacity of wireless networks. IEEE Transactions on Information Theory **46**, 388–404 (2000)
23. Grossglauser M., Tse D. N. C.: Mobility increases the capacity of ad hoc wireless networks. IEEE/ACM Transactions on Networking **10**, 477–486 (2002)

Chapter 5
CREAM-MAC: Cognitive Radio-Enabled Multi-channel MAC for Wireless Networks

Xi Zhang and Hang Su

Abstract As the novel and effective approach to improve the utilization of the precious radio spectrum, cognitive radio technology is the key to realize the dynamic spectrum access (DSA) networks, where the *secondary* (unlicensed) users can opportunistically utilize the unused licensed spectrum in a way that confines the level of interference to the range the *primary* (licensed) users can tolerate. However, there are many new challenges associated with cognitive radio networks, such as the multi-channel hidden terminal problem and the fact that the time-varying channel availability differs for different secondary users, in the medium access control (MAC) layer. To overcome these challenges, we propose an efficient Cognitive Radio-EnAbled Multi-channel MAC (CREAM-MAC) protocol, which integrates the cooperative sequential spectrum sensing at physical layer and the packet scheduling at MAC layer, over the wireless cognitive radio networks. Under the proposed CREAM-MAC protocol, each secondary user is equipped with a cognitive radio-enabled transceiver and multiple channel sensors. Our cooperative sequential spectrum sensing scheme improves the accuracy of spectrum sensing and further protects the primary users. The proposed CREAM-MAC enables the secondary users to best utilize the unused frequency spectrum while avoiding the collisions among secondary users and between secondary users and primary users. We develop the Markov chain model and $M/G^Y/1$ queueing model to rigorously study our proposed CREAM-MAC protocol for both the saturation networks and the non-saturation networks. We also conduct extensive simulations to validate our developed protocol and analytical models.

5.1 Introduction

The rapid growth in the ubiquitous wireless services has imposed increasing stress on the fixed and limited radio spectrum. Allocating a fixed frequency band to each wireless service, which is the current frequency allocation policies, is an easy and natural approach to eliminate interference between different wireless services.

X. Zhang (✉)
Networking and Information Systems Laboratory, Department of Electrical and Computer Engineering, Texas A&M University, College Station, TX 77843, USA
e-mail: xizhang@ece.tamu.edu

F.R. Yu (ed.), *Cognitive Radio Mobile Ad Hoc Networks*,
DOI 10.1007/978-1-4419-6172-3_5,

However, extensive measurements reported indicate that the static frequency allocation results in a low utilization of the licensed radio spectrum in most of the time [1, 2]. Even when a channel is actively used, the bursty nature of most data traffics still implies that a great amount of opportunities exist in using the spare spectrum.

In order to better utilize the licensed spectrum, the Federal Communication Committee (FCC) has recently suggested a new concept/policy for *dynamically* allocating the spectrum [3, 4]. Consequently, a promising implementation technique of this concept, called the *cognitive radio* [5], is proposed to take advantage of this more open spectrum policy for alleviating the severe scarcity of spectrum bandwidth. Cognitive radio is typically built on top of the software-defined radio (SDR) technology, in which the transmitter's operating parameters, such as the frequency range, modulation type, and maximum transmission power can be dynamically adjusted by software [6, 7]. In the cognitive radio networks, the *secondary* (unlicensed) users (SUs) can periodically scan and identify the vacant channels in the spectrum. Based on the scanned results, the SUs dynamically tune their transceivers to the identified spare channel spectrum to communicate among themselves while limiting their interference imposed onto the *primary* (licensed) users (PUs) to an acceptable low and harmless level.

The cognitive radio technology has received intensive attention since it was first coined by Dr. J. Mitola III in 1999 [8]. The cognitive radio wireless networks can be broadly categorized into the following two types: synchronous cognitive radio wireless networks [9–14] and asynchronous cognitive radio wireless networks [15–19]. In the synchronous cognitive radio networks, the time axis is divided into slots. The SU networks are synchronized with the PU networks. In other words, the SUs have the same knowledge on the boundary of time slots as the PUs. In the synchronous cognitive radio networks, a given PU starts utilizing the licensed spectrum only at the beginning of a time slot. If the PU uses the licensed spectrum at the beginning of the time slot, it continues to utilize the spectrum for the rest of this time slot. Since the SUs are synchronized with the PUs, the SUs only need to sense the licensed spectrum at the beginning of a time slot to determine whether this time slot is available to be used. In the synchronous cognitive radio networks, the SUs repeat the sensing-transmission cycle for every time slot. If the spectrum sensing returns perfect sensing results, there is no interference caused to the PUs.

On the other hand, unlike the counterpart in synchronous cognitive radio networks, the SUs in asynchronous cognitive radio networks cannot be synchronized with the PUs. For example, the PU networks are based on the carrier sensing multiple access (CSMA)-like random access. In this chapter, we consider the asynchronous cognitive radio networks. Besides all the problems encountered in the synchronous cognitive radio networks, in the asynchronous cognitive radio networks, when transmitting/receiving the SUs are not aware whether/when the PUs become active due to the half-duplex nature of the wireless spectrum medium. Thus, the SUs in the asynchronous cognitive radio networks may inevitably cause interference to the PUs. How the SUs limit the interference caused to the PUs to an acceptable level is the critical problem in the design of MAC protocols for asynchronous or non-time-slotted cognitive radio networks. Moreover, the problems become more complicated when there are no centralized controllers and each SU is equipped with only a single transceiver.

To tackle the aforementioned problems, in this chapter we propose an efficient Cognitive Radio-EnAbled Multi-channel MAC protocol, called CREAM-MAC protocol, which integrates the cooperative sequential spectrum sensing at physical layer and packet scheduling at MAC layer, over the wireless cognitive radio networks. Under the CREAM-MAC protocol, each SU is equipped with a SDR-based transceiver that can dynamically utilize one or multiple licensed channels to receive/transmit the SUs' packets, and multiple sensors that can detect multiple licensed channels simultaneously. The proposed cooperative sequential spectrum sensing scheme aims at improving the accuracy of spectrum sensing to decrease the interference imposed to the PUs. The CREAM-MAC protocol enables the SUs to dynamically utilize the unused licensed frequency spectrum in a way that confines the level of interference to the range the PUs can tolerate. With the help of the four-way handshakes of control packets, the CREAM-MAC protocol with a single transceiver can efficiently handle the traditional hidden terminals and the multi-channel hidden terminals.

The rest of this chapter is organized as follows. Section 5.2 presents the related works. Section 5.3 describes the system models. Section 5.4 develops the CREAM-MAC protocol with the cooperative sequential spectrum sensing scheme. Section 5.5 develops the analytical model to study the CREAM-MAC protocol with the cooperative spectrum sensing scheme for the saturation network case. Applying the $M/G^Y/1$ queuing model, Section 5.6 analyzes the packet transmission delay and throughput of the proposed CREAM-MAC protocol in the non-saturation network case. Section 5.7 evaluates our proposed multi-channel MAC protocol by using our developed analytical models and simulation experiments. The chapter concludes with Section 5.8.

5.2 Related Works

Several decentralized cognitive MAC protocols have been proposed recently. The authors of [20] proposed a cognitive MAC with statistical channel allocation, in which the secondary users select the channel that has the highest successful transmission probability to send packets based on the channel statistics. However, the computational complexity determining the successful transmission probabilities increases quickly with the number of licensed channels. The author of [21] proposed a multi-channel opportunistic MAC protocol, which, however, targets only at the Global System for Mobile Communications (GSM) cellular networks. The authors of [10] developed a cognitive MAC protocol based on the partially observable Markov decision processes (POMDPs) framework. In [11], we proposed opportunistic MAC protocols with random and negotiation-based sensing policies for the time-slotted wireless networks. In [14], we developed a channel hopping-based cognitive MAC protocol which enables the secondary users to conduct channel negotiations at multiple rendezvous. However, the schemes in [10, 11, 14] require global synchronization between primary and secondary users, which is not easy to implement. The authors of [15] proposed a cognitive MAC protocol aiming to opportunistically utilize the TV broadcast bands. However, the proposed protocol is costly and complicated as it requires not only a cognitive radio-based transceiver but also a regular radio receiver, which operate on the unused TV channels and the control channel, respectively. Our proposed CREAM-MAC is designed to tackle the aforementioned problems in the existing works.

5.3 The System Models

We consider the scenario where there are two non-cooperating types of users, namely PUs and SUs. The PUs, for example, TVs, cellular phones, or wireless microphones, are those to which an amount of wireless spectrum is licensed. On the other hand, the SUs are referred to those without pre-assigned wireless spectrum. However, the SUs equipped with the cognitive radios can transmit their own packets by seizing the opportunities that arise when the PUs do not use the licensed wireless spectrum. In this chapter, the wireless spectrum accessible to the SUs is further divided into a number of channels, each with a fixed amount of frequency bandwidth.

5.3.1 Primary Users' Behaviors

Our system model focuses on the asynchronous cognitive radio networks, which impose more challenges as compared with the synchronous cognitive radio networks. In particular, we consider a scenario where a spectrum licensed to the PUs consists of M channels, as depicted in Fig. 5.1, in which the PUs operating at different channels are unsynchronized. We assume that for each channel, the channel usage pattern of the PUs follows independent and identically distributed (i.i.d.) ON/OFF renewal process, as shown in Fig. 5.2. An ON state represents that the channel is occupied by the PUs. An OFF state represents that the channel is vacant

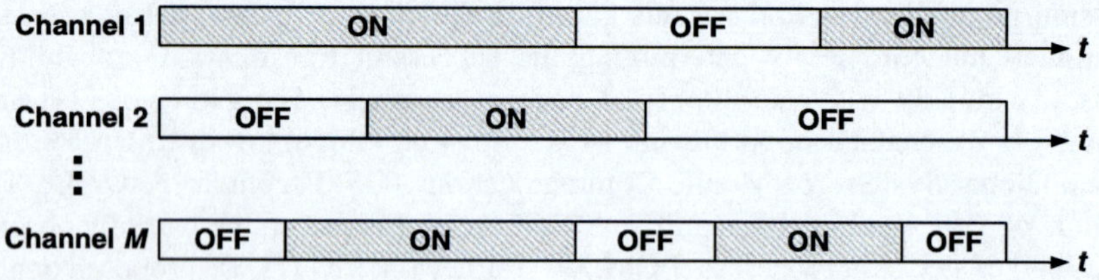

Fig. 5.1 Illustration of PUs' channel utilization in cognitive radio networks. There are M licensed channels. The ON and OFF states of PUs operating at different licensed channels are unsynchronized

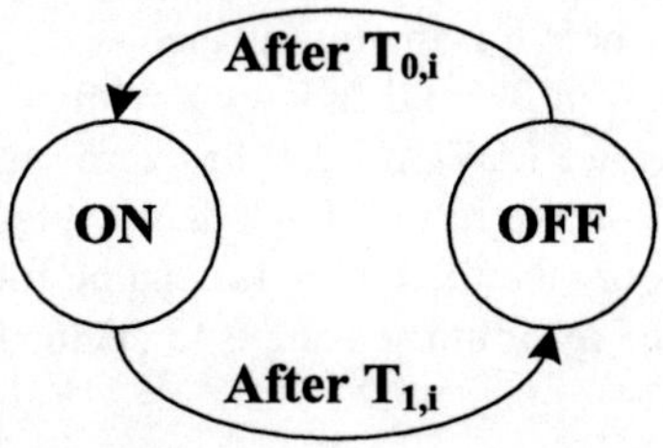

Fig. 5.2 The ON/OFF channel model for the ith licensed data channel

and thus can be opportunistically used by the SUs. Note that the average ON- and OFF-periods depend on the channel usage pattern of the PUs.

In the viewpoint of the SUs, the channel is alternating between ON and OFF states. We refer a spectrum access cycle for the SUs as a ON state followed by an OFF state. For an alternating renewal channel, let random variables $T_{1,i}$ and $T_{0,i}$ represent the sojourn times of ON and OFF states, respectively, for the ith licensed channel. Without loss of generality, we assume that $T_{1,i}$ is independent of $T_{0,i}$. Denote $f_{T_{1,i}}(s)$ and $f_{T_{0,i}}(s)$ as the probability density functions (pdf) for the durations of the ith licensed channel's ON state and OFF state, respectively. Then, we can derive the probability, denoted by γ_i, that the channel is in its ON state at an arbitrary time instance as follows:

$$\gamma_i = \frac{\int_0^\infty s f_{T_{1,i}}(s)ds}{\int_0^\infty s f_{T_{1,i}}(s)ds + \int_0^\infty s f_{T_{0,i}}(s)ds} = \frac{\overline{T}_{1,i}}{\overline{T}_{1,i} + \overline{T}_{0,i}}, \tag{5.1}$$

where $\overline{T}_{1,i}$ and $\overline{T}_{0,i}$ are the mean sojourn times of ON and OFF states, respectively, for ith channel utilization. Note that in fact γ_i is the ith channel utilization w.r.t. PUs. We assume that the PUs' channel utilization pattern is homogeneous, i.e., $\overline{T}_{1,i} = \overline{T}_{1,j}$ and $\overline{T}_{0,i} = \overline{T}_{0,j}, \forall i \neq j$.

5.3.2 The Spectrum Sensing Model

The spectrum sensing scheme plays an important role in cogitative radio networks. There are several signal detection techniques used for spectrum sensing, such as the energy detection, feature detection, and matched filter, for the SUs to detect the presence of the PUs [22]. We mainly focus on the energy detection approach in this chapter because the energy detection approach is efficient and simple to be implemented in hardware and more importantly, it does not require the knowledge of signal features of the PUs, which typically may not be known by the SUs. We assume that all channels experience Rayleigh fading. If a PU is active and its sent signal is s with the transmit power E_s, then the received signal at a given SU's side, denoted by r, is

$$r = hs + \omega, \tag{5.2}$$

where h is the instantaneous amplitude gain of the channel between the PU and the given SU and follows Rayleigh distribution, and ω is the additive white Gaussian noise (AWGN) that has zero mean and variance of σ^2. The instantaneous signal-to-noise ratio (SNR), denoted by υ, is equal to $(E_s|h|^2/\sigma^2)$. If the PU is idle, then only the thermal noise can be found at the receiver of the SUs. Thus, the objective of spectrum sensing is to decide between the following two hypotheses:

$$r = \begin{cases} hs + \omega, & \mathcal{H}_1 \\ \omega, & \mathcal{H}_0 \end{cases} \tag{5.3}$$

where $\mathcal{H}_1$ is the hypothesis stating that the given licensed channel is in ON state, and $\mathcal{H}_0$ is the hypothesis stating that the given licensed channel is in OFF state.

When the SUs perform the spectrum sensing, all of the SUs need to keep their radios silent for τ_s time units in order to obtain an accurate sensing outcome. During this silent period, no transmissions can be made. Since the sender and the receiver may have different sensing outcomes due to the relatively different locations to the PUs and the wireless fading channels, the SUs cooperatively exchange the sensing outcomes to make the more accurate decisions based on the overall sensing results.

5.3.3 Channel Aggregating Technique

After the SUs sense the licensed channels for a period of time, they have the information of the licensed channel conditions. By using this information, the SUs can opportunistically utilize multiple unused channels simultaneously. However, in most cases, the unused channels are discontinuous. Fortunately, the orthogonal frequency division multiplexing access (OFDMA) has been introduced to help the SUs aggregate the discontinuous channels. In particular, the cognitive radios with OFDMA can enable or suppress the corresponding subcarriers based on the channel availability and thus access the multiple continuous/discontinuous unused channels simultaneously.

5.4 The Proposed CREAM-MAC Protocol

5.4.1 Protocol Overview

There are many challenges imposed on the design of MAC protocols for the cognitive radio networks. Among them, the following three problems are most important: (i) the problem when to transmit data packets in a way that limits the interference on the PUs, (ii) synchronization between the SU sender and the SU receiver due to the difference of the channel availability between them, and (iii) the traditional hidden terminal problem and the multi-channel hidden terminal problem.[1] Keeping these in mind, we start to develop the CREAM-MAC protocol under which the SUs can dynamically utilize the vacant licensed channels.

The CREAM-MAC protocol employs a *common control channel* as the rendezvous where the SUs exchange the control packets for multi-channel resource reservation. The control channel can be either statically assigned or dynamically selected. Under the statistical case, the control channel can use either the dedicated channel licensed to the SUs or the unlicensed spectrum band (e.g., 2.4 GHz spectrum for IEEE 802.11 b/g). On the other hand, for the dynamical case, the control channel can select the most reliable one from the unused channels which are licensed to the PUs [23]. In this chapter, we do not delve into which way the control channel is selected. Instead, we assume that control channel is *always* reliable and available.

Under the CREAM-MAC protocol, each SU is equipped with n sensors, such that at most n licensed channel can be sensed simultaneously. After sensing the licensed spectrum for a period of time, each SU has the information of the channel states in these spectrum bands. Then, the SUs can opportunistically access the vacant channels which are not being occupied by the PUs. Since the interference from SUs' transmission must be constrained to a modest level the PUs can tolerate, we limit each channel access time of SUs to be no more than the maximum tolerable interference period, denoted by $T_d^{\max}$. Thus, the constraint that each opportunistic access of the SUs does not exceed $T_d^{\max}$ time units ensures that the PUs only experience the acceptable and limited interference imposed by the SUs. Moreover, the CREAM-MAC protocol employs the cooperative sequential spectrum sensing scheme for the CREAM-MAC protocol to improve the spectrum sensing accuracy, and thus further protect the PUs by reducing the interference caused by SUs.

One of the important components for the CREAM-MAC protocol is to employ the four types (two pairs) of control packets, namely, Ready-to-Send/Clear-to-Send (RTS/CTS) and Channel-State-Transmitter/Channel-State-Receiver (CST/CSR) packets, to implement the *channel negotiation* which is a process for multiple SUs to compete for the vacant licensed channels. All of the above four types of control packets are exchanged over the control channel. First, the functions of the RTS/CTS control packets include (i) reserving the control channel and (ii) solving the hidden terminal problem. The SU sender sends the RTS packet over the control channel based on the contention-based mechanism. Without loss of generality, we adopt the binary exponential backoff-based IEEE 802.11 Distributed Coordination Function (DCF) [24] as the contention algorithm. The RTS/CTS handshakes can prevent the neighboring SUs from selecting the same channels to transmit data, guaranteeing no collisions between the SUs. Thus, exchanging the RTS/CTS control packets can efficiently solve the hidden terminal problem. Second, the function of the CST/CSR handshakes aims to synchronize the vacant channel information between the SU

[1] In multi-channel systems, especially those with only one single transceiver, the multi-channel hidden terminal problem emerges. The reason is that a single transceiver may operate on only one channel, which makes it difficult to use virtual carrier sensing to handle the hidden terminals [32].

sender and the SU receiver, and thus to prevent the collisions between the SUs and the PUs. The CST packet includes the lists of the vacant channels at the transmitter's side, while the CSR packet includes the lists at the receiver's side. The exchange of the CST/CSR packets ensures that the SU sender and the SU receiver select the set of the vacant channels, which are shared by both of them. In summary, the objective of the RTS/CTS control packets is to prevent the collisions among the SUs, while the objective of the CST/CSR control packets is to avoid the collisions between the SUs and the PUs. We will detail the process on how the control packets are exchanged over the control channel in Section 5.4.4 and Section 5.4.5.

To better understand the CREAM-MAC protocol, Fig. 5.3 illustrates an example case with 1 control channel and 3 data channels (CH 1, CH 2, CH 3) which form a Channel Group (to be detailed in Section 5.4.4). In this particular example, when an SU sender wants to communicate with an SU receiver through the cognitive radio network, the SU sender contends for the data channels via the control channel by going through the binary exponential backoff algorithm as described in the above. After the successful *backoff* stage, the SU sender conducts the *channel negotiation* with the SU receiver by exchanging RTS/CTS/CST/CSR control packets over the control channel. Since the common idle licensed channels for the SU sender and SU receiver are CH 2 and CH 3 (CH 1 is being occupied by PUs) at this time point, after the successful channel negotiation, the SU sender utilizes CH 2 and CH 3 simultaneously to send the data to the SU receiver. Upon successfully receiving the data, the SU receiver responds with two ACK packets to the SU sender over CH 2 and CH 3, respectively. Note that the channel negotiation over the control channel and the opportunistic data transmission over the licensed data channels can be performed simultaneously by different pairs of SU senders and SU receivers. As shown in Fig. 5.3, because the licensed CH 1 becomes idle during the time when the SU sender and SU receiver exchange data over the licensed CH 2 and CH 3, another SU sender can start its own channel negotiation by sending the RTS packet.

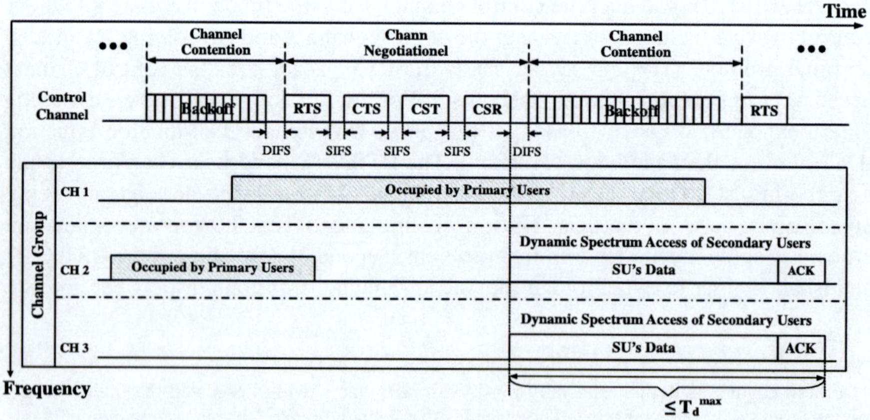

Fig. 5.3 Illustrations of the CREAM-MAC protocol for an example case with 1 control channel and 3 data channels (CH 1, CH 2, CH 3) which form a Channel Group

5.4.2 The Maximum Allowable Transmission Duration for SUs

Before transmitting packets, the SUs need to sense the licensed channel and detect the presence of the PUs. Due to the half-duplex characteristic of the wireless radio, SUs cannot sense the channel while transmitting their own signals, which implies that the SUs cannot accurately know when the PUs become active again, especially when the SUs are not synchronized with PUs. The SUs may inevitably cause harmful interference to PUs. We have to limit the duration of interference that is caused by SUs. Particularly, we apply the interference constraint in time domain by limiting the amount of time when the SUs and PUs transmit simultaneously.

In this chapter, we assume the duration of each OFF state ($T_{0,i}$) of the ith data channel follows the exponential distribution. Let t_{sp} and t_{ls} be time that the channel switches from an ON state to an OFF state and the most recent time that the channel state is sensed, respectively. If the channel is in an OFF state at t_{ls}, then there is still a positive probability that the channel remains OFF for a certain period, and thus it is possible for the SUs to opportunistically access the spectrum. Let τ_t be the transmit duration of SUs. Given that the channel is in an OFF state at t_{ls}, we can derive the probability that the SUs will interfere with the PUs in a given channel during the transmission as follows:

$$I_1(\tau_t) = \Pr\{T_{0,i} < t_{ls} - t_{sp} + \tau_t | T_{0,i} > t_{ls} - t_{sp}\} = 1 - e^{-\frac{\tau_t}{\overline{T}_{0,i}}}. \tag{5.4}$$

In our proposed scheme, the SUs are allowed to utilize up to n licensed channels during a transmission. Then, we obtain the probability that at least a channel is interfered by the SUs as follows:

$$I_n(\tau_t) = 1 - (1 - I_1)^n = 1 - e^{-\frac{\tau_t}{n\overline{T}_{0,i}}}. \tag{5.5}$$

It is clear that the probability $I_n(\tau_t)$ is a monotonically increasing function of τ_t. For the given predefined interference constraint, denoted by I_{th}, we can derive the maximum allowable transmission duration $T_{d,i}^{\max}$ for the ith data channel as follows:

$$T_{d,i}^{\max} = \operatorname*{argmax}_{\tau_t > 0}\{I_n(\tau_t) < I_{th}\}. \tag{5.6}$$

Since we assume that the PUs' channel utilization pattern is homogeneous, i.e., $\overline{T}_{0,i} = \overline{T}_{0,j}, \forall i \neq j$, we have $T_d^{\max} = T_{d,i}^{\max}, \forall i$. In other words, the maximum allowable transmission durations for all channels are the same, and thus in the rest of chapter we use $T_d^{\max}$ only.

5.4.3 The Selection of Licensed Channels

Since the licensed channels are sometimes utilized by the PUs unevenly, some licensed channels may be utilized more than the others. Because the SUs can only sense a limited number of licensed channels simultaneously, to fully utilize the licensed channels, the SUs need to select the licensed channels which are used less intensively to sense. At the beginning, the SUs randomly select a number ($\leq$ n) of channels to sense. They can update the statistical utilization information of licensed channels in either the non-cooperation way or the cooperation way. In the non-cooperation way (e.g., the POMDP scheme [10]), the SUs update its channel by themselves without exchanging information. On the other hand, the cooperation-based channel selection allows the SUs to exchange information such that the SUs can learn the global channel states. Compared with the non-cooperation-based schemes, the cooperation-based schemes allow the SUs to obtain the updated channel states more accurately and quickly, but need more communication overheads. We adopt the cooperation-based scheme for our CREAM-MAC protocol. In particular, the channel state information is embedded in the control packets. Thus, the SUs can obtain the neighbors' channel state information by overhearing the control packets. After obtaining the statistical utilization information of the licensed channels, the SUs select n of the licensed channels, which form a *channel group* (see Fig. 5.3 for an example), to sense by using the n sensors.

5.4.4 Channel Contention

Under the CREAM-MAC protocol, to decrease the collision probability of the control packets, the SU sender, which attempts to send an RTS packet, selects a backoff counter within a contention window and maintains the contention window size. At the initial state, the contention window size is set to be equal to a predefined value, denoted by CW_{min}. The counter is deducted by one after a time slot during which both the control channel and at least one data channel in the channel group are idle. Otherwise, the counter remains the same. When the backoff counter reaches zero, the SU sender tries to reserve the control channel by sending RTS to the SU receiver. The binary exponential backoff algorithm is employed when the collisions occur. In other words, the collided SUs double their contention window size to lower the probability of further collision. Fig. 5.3 demonstrates when/where the channel contention occurs.

5.4.5 Channel Negotiation

The CREAM-MAC does not require global synchronization among all the PUs and SUs. Under the CREAM-MAC, the contention mechanism over the control channel is similar to IEEE 802.11 DCF. The SU sender reserves time for the following transmission operations within the neighborhood through the control channel by exchanging RTS/CTS control packets with the SU receiver. In particular, after the backoff counter of the SU sender reaches zero, the SU sender senses the control channel for a duration equal to the DCF interframe space (DIFS). If the control channel remains idle after the DIFS duration, the SU sender sends RTS packet including its channel group list to the SU receiver through the control channel. Upon receiving the RTS packet, if at least one data channel in the channel group is currently not used by its neighboring SUs, the SU receiver replies to the SU sender with a CTS packet after a duration equal to the short interframe space (SIFS). In the meantime, the SU receiver uses its sensors to detect the channel group indicated in the RTS packet. The other neighboring SUs overhear the RTS/CTS control packets to update the available channel list.

After SU sender and SU receiver reserve the control channel by successfully exchanging RTS/CTS packets, they negotiate on the licensed channels which are vacant for both the transmitter and the receiver. More precisely, the SU sender first sends the CST packet which includes the vacant channel list at the transmitter's side. Upon receiving the CST packet, the SU receiver replies with the CSR packet telling the SU sender which common channels are vacant and how long the communication will last over these common channels. Since the communication interval can be less than or equal to $T_d^{\max}$, the other neighboring SUs can overhear the CST/CSR packets to precisely predict when the channels used by this pair of SUs will be released. The above channel negotiations operations are also illustrated in an example given in Fig. 5.3.

The handshakes of RTS/CTS can only solve the traditional hidden terminal problem, but not the multi-channel hidden terminal problem. Specially, the SUs which just finished the data transmission over the licensed channels may miss their neighbor's control packets while their transceivers worked over the licensed data channels. They will probably win the control channel contention and then enter the licensed channels over which its neighbors are receiving data. Consequently, these SUs become the hidden terminals interrupting their neighbors' ongoing communications. To prevent this from happening, we need to put additional rules on CREAM-MAC. In particular, the SUs which just finished the data transmission can only select the same channel group which they just released within a *waiting* interval of $T_d^{\max}$. After the waiting period, these secondary can select any other channel groups to use. It allows these SUs to have enough time to observe the current spectrum activities before they start packet transmissions and to prevent them from interfering the neighbors' ongoing communications, since the maximum time interval that the SUs can occupy the licensed channels each time is $T_d^{\max}$. During the waiting period, if these SUs receive any control packets, they can obtain the updated channel state from the control packets. Otherwise, it is safe for these SUs to assume that all the licensed channels are not being used by any SUs after the waiting period, which can efficiently solve the multi-channel hidden terminal problem. Note that the same rules also apply to the new SUs which first join the network.

5.4.6 Data Transmissions

Totally, there are six-way handshakes in a successful data exchange between the SU sender and the SU receiver. Besides the four-way handshakes of the control packets over the control channel, there are another two-way handshakes of Data/ACK over the licensed data channels. In particular, after the successful four-way handshakes of the control packets over the control channel, the SU sender starts transmitting data to the SU receiver over the channel group's idle channels. The SU receiver sends ACK to the SU sender after successful receiving the data packets from the SU sender. The above data transmissions operations are also illustrated in Fig. 5.3 for an example. Data transmissions over multiple idle channels in a channel group (see CH 2 and CH 3 in Fig. 5.3) can be implemented by using the OFDMA-based channel aggregating technique as described in Section 5.3.2. Also, each SU data transmission only lasts for a variable duration less than or equal to $T_d^{\max}$, as shown in Fig. 5.3.

5.4.7 The Distributed Spectrum Sensing Scheme

As described in Section 5.3, we adopt the energy detection approach. A typical *energy detector* consists of a bandpass filter which chooses the center frequency and bandwidth of interest, a squaring device which calculates the energy of the signal samples, and an integrator which controls the observation intervals. Let Y be the output of the integrator in the energy detector. Following the work in [25], given that the PUs are present (i.e., $\mathcal{H}_1$) and instantaneous SNR is υ, we have the conditional pdf, denoted by $f_{Y|\upsilon,\mathcal{H}_1}(y)$, of Y, which follows the non-central chi-square distribution, i.e.,

$$f_{Y|\upsilon,\mathcal{H}_1}(y) = \frac{1}{2}\left(\frac{y}{2\upsilon}\right)^{\frac{m-1}{2}} e^{-\frac{2\upsilon+y}{2}} I_{m-1}\left(\sqrt{2\upsilon y}\right), \tag{5.7}$$

where m denotes the integer number of samples measured and $I_\upsilon(\cdot)$ denotes the υth order modified Bessel function of the first kind. Since the channel gain is assumed to follow the Rayleigh distribution, the SNR υ follows the exponential distribution with the mean SNR equal to $\overline{\upsilon}$. Thus, taking into account the fading factor, we have

$$\begin{aligned} f_{Y|\mathcal{H}_1}(y) &= \int_0^\infty f_{Y|\upsilon,\mathcal{H}_1}(y)\frac{1}{\overline{\upsilon}}e^{-\frac{\upsilon}{\overline{\upsilon}}}d\upsilon \\ &= \frac{(1+\overline{\upsilon})^m e^{-\frac{y}{2(1+\overline{\upsilon})}}}{2(1+\overline{\upsilon})^2\overline{\upsilon}^{m-1}}\left[1-\frac{\Gamma\left(m-1,\frac{\overline{\upsilon}y}{2(1+\overline{\upsilon})}\right)}{\Gamma(m-1)}\right] \end{aligned} \tag{5.8}$$

where $\Gamma(\cdot)$ is the complete gamma function, and $\Gamma(a,z) = \int_z^\infty t^{a-1}e^{-t}dt$ is the upper incomplete gamma function.

On the other hand, suppose that the PUs do not occupy the spectrum (i.e., $\mathcal{H}_0$), then the output of the integrator in the energy detector can be characterized by the central chi-square distribution as follows:

$$f_{Y|\mathcal{H}_0}(y) = \frac{1}{2^m\Gamma(m)}y^{m-1}e^{-\frac{y}{2}} \tag{5.9}$$

Then, we obtain its cumulative distribution function (cdf), denoted by $F_{Y|\mathcal{H}_1}(y)$, of Y given $\mathcal{H}_1$ as follows:

$$\begin{aligned} F_{Y|\mathcal{H}_1}(y) &= \int_0^y f_{Y|\mathcal{H}_1}(t)dt \\ &= \frac{\Gamma(m-1,0) - \Gamma\left(m-1, \frac{y}{2}\right)}{\Gamma(m-1)} + \left(\frac{1+\overline{\upsilon}}{\overline{\upsilon}}\right)^{m-1} \left[1 - e^{-\frac{y}{2(1+\overline{\upsilon})}} \right. \\ &\quad \left. - \frac{\Gamma(m-1,0)}{\Gamma(m-1)} + e^{-\frac{y}{2(1+\overline{\upsilon})}} \frac{\Gamma\left(m-1, \frac{\overline{\upsilon}y}{2(1+\overline{\upsilon})}\right)}{\Gamma(m-1)}\right] \end{aligned} \tag{5.10}$$

Similarly, we can get the cdf, denoted by $F_{Y|\mathcal{H}_0}(y)$, of Y given $\mathcal{H}_0$ as follows:

$$F_{Y|\mathcal{H}_0}(y) = \int_0^y f_{Y|\mathcal{H}_0}(t)dt = 1 - \frac{\Gamma\left(m, \frac{y}{2}\right)}{\Gamma(m)} \tag{5.11}$$

Because in the cognitive systems the PUs have the higher priority in spectrum access than the SUs, the missed detection probability of the PUs' presence should be limited to a small value. However, based on the traditional single-threshold energy-detection method [26], decreasing the missed detection probability is equivalent to increasing the false alarm probability, which consequently decreases the spectrum access opportunities for SUs. It is contradicted to decrease the missed detection probability while decreasing the false alarm probability. To overcome the contradiction, we propose to use the *two-threshold-based sequential sensing policy*, which can decrease the false alarm probability while confining the missed detection probability to a predefined threshold, denoted by P_{MD}^{th}. The basic idea of our proposed spectrum sensing policy is to collect PUs' signal samples sequentially in multiple uncorrelated sensing rounds to enhance the decision process. In particular, we design two thresholds, denoted by β_1 and β_2, with $\beta_1 < \beta_2$, for the two hypotheses, $\mathcal{H}_1$ and $\mathcal{H}_0$, respectively. Our proposed policy can be described as follows: (1) When the energy (Y) of the detected signal is larger than β_2, then we claim that the PUs are active; (2) When Y is less than β_1, we claim that the PUs are inactive; (3) Otherwise, if $\beta_1 < Y < \beta_2$, the SUs need to take one more sensing round to collect PUs' signal samples after a duration equaling the channel's coherence time until $Y > \beta_2$ or $Y < \beta_1$.

Then, we define and derive the *missed detection probability*, denoted by P_{MD}, as follows:

$$P_{\text{MD}} \triangleq \Pr\{Y < \beta_1 | \mathcal{H}_1\} = F_{Y|\mathcal{H}_1}(\beta_1) \tag{5.12}$$

where $F_{Y|\mathcal{H}_1}(\cdot)$ is specified in (5.10). Similarly, we define and derive the *false alarm probability*, denoted by P_{FA}, as follows:

$$P_{\text{FA}} \triangleq \Pr\{Y > \beta_2 | \mathcal{H}_0\} = 1 - F_{Y|\mathcal{H}_0}(\beta_2) \tag{5.13}$$

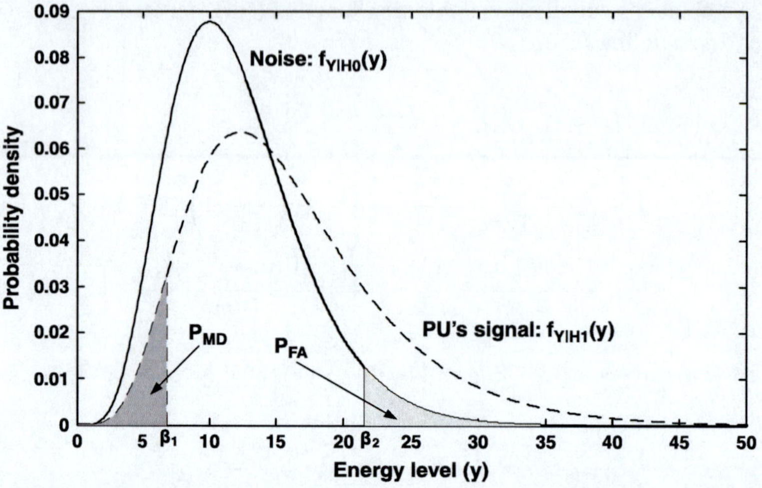

Fig. 5.4 The illustration of the relationship between the P_{FA} and P_{MD} in the two-threshold sequential sensing. The two thresholds β_1 and β_2 control P_{MD} and P_{FA}, respectively

where $F_{Y|\mathcal{H}_0}(\cdot)$ is specified in (5.11). Fig. 5.4 illustrates $f_{Y|\mathcal{H}_0}(y)$, $f_{Y|\mathcal{H}_1}(y)$, P_{MD}, P_{FA}, β_1, and β_2. Figure 5.4 also shows how β_1 and β_2 impact the missed detection probability P_{MD} and the false alarm probability P_{FA}, respectively. In the two-threshold sensing policy, the missed detection and false alarm probabilities are controlled by two separated parameters, namely, β_1 and β_2, respectively. As shown in Fig. 5.4, decreasing β_1 and increasing β_2 at the same time can decrease both the missed detection and false alarm probabilities, which, however, may also increase the number of required uncorrelated sensing rounds because the probability that the sensed signal's energy level falls between the two thresholds becomes larger. Hence, we proceed to study the impact of β_1 and β_2 on the average number of required sensing rounds by deriving the probability, denoted by q, that the detected signal energy falls between the two thresholds as follows:

$$q = \left[F_{Y|\mathcal{H}_1}(\beta_2) - F_{Y|\mathcal{H}_1}(\beta_1)\right]\gamma + \left[F_{Y|\mathcal{H}_0}(\beta_2) - F_{Y|\mathcal{H}_0}(\beta_1)\right](1-\gamma), \quad (5.14)$$

where γ is the PUs' channel utilization of any given licensed channel. The number of sensing rounds that need to be taken during the ith spectrum sensing follows the geometry distribution, i.e.,

$$\Pr\{N_s = n\} = q^{n-1}(1-q).$$

Thus, the average number, denoted by $\overline{N}_s$, of the sensing rounds that need to be taken during the ith spectrum sensing is given by

$$\overline{N}_s = \frac{1}{1-q}. \quad (5.15)$$

It is clear that there is a tradeoff between the number of uncorrelated sensing rounds and false alarm probability while the missed detection probability is upper-bounded. When the missed detection probability is given, β_1 is determined. The larger the value of $\overline{N}_s$, the smaller the value of β_2, and thus the smaller the false alarm probability. It is interesting to note that the single-threshold sensing policy is a special case of our sequential sensing policy. Specifically, when setting $\overline{N}_s = 1$, we have $\beta_1 = \beta_2$, which implies that our proposed sequential sensing policy reduces to the single-threshold based sensing policy.

5.5 Throughput Analysis for the Saturation Network Case

In this section, we develop an analytical model to analyze the aggregate throughput of our proposed CREAM-MAC protocol under the saturation network case, where each SU has always an infinite amount of data packets to send.

5.5.1 The Analysis for the Licensed Data Channels

Suppose that there are M licensed data channels and each SU is equipped with n sensors. Based on the CREAM-MAC protocol, the SUs can reserve at most n licensed channels and utilize all of them with the help of channel aggregating technique if n licensed channels are free. There are $\lfloor M/n \rfloor$ channel groups that can be utilized by the SUs.

Denote a discrete random variable H as the number of vacant channels in a specified channel group with n licensed channels. To make the model tractable, we assume that each channel is evenly utilized by the PUs. Thus, we apply the same γ to all licensed channels, i.e., $\gamma = \gamma_i = \gamma_j$ for $1 \leq i, j \leq M$, where γ_i is given in (5.1). Since the channel states among different channels are independent with each other, we get the probability that the number H of vacant channels in a specified channel group is equal to i following the binomial distribution as follows:

$$\Pr\{H = i\} = \binom{n}{i} \left[1 - \gamma(1 - P_{FA})^2\right]^i \left[\gamma(1 - P_{FA})^2\right]^{n-i}, \tag{5.16}$$

where the term $(1 - P_{FA})^2$ is due to the available-channel list synchronizing between the SU sender and the SU receiver. Thus, we can derive the average number, denoted by the expectation $\mathbb{E}[H]$, of vacant channels as follows:

$$\mathbb{E}[H] = \sum_{i=0}^{n} i \Pr\{H = i\} = n\left[1 - (1 - P_{FA})^2 \gamma\right]. \tag{5.17}$$

5.5.2 The Analysis for the Control Channels

In order to analyze the saturation network throughput of the proposed CREAM-MAC, we need to study the contention behavior over the control channel where the control packets are transmitted based on the IEEE 802.11 DCF. We develop the analytical model based on the works of [27, 28], which uses a Markov chain model to analyze the backoff operations for IEEE 802.11 DCF. Following the previous works, if we denote the probability that a given SU transmits in a randomly chosen time slot by τ, and the probability that a transmitted packet collides by p, respectively, then we obtain the following equations:

$$\begin{cases} \tau = \frac{2(1-2p)}{(1-2p)(CW_{min}+1)+CW_{min}\,p[1-(2p)^m]} \\ p = 1-(1-\tau)^{u-1} \end{cases} \tag{5.18}$$

where m is the maximum backoff stage, u is the number of the contending SUs, $CW_{\min}$ is the initial contention backoff window size. Note that τ is the function of p while p is also the function of τ. Solving simultaneously the two equations in (5.18), we can obtain the numerical solutions for τ and p. Obviously, $\tau, p \in (0, 1)$. Observing (5.18), we can learn that p only depends on the number of the contending SUs (u), the maximum backoff stage (m), and the initial contention backoff window size ($CW_{\min}$).

Let P_{tr} be the probability that there is at least one transmission in a given time. Since each contending SU transmits with probability τ at any given time, given that there are u contending SUs, we get that P_{tr} can be expressed as:

$$P_{\text{tr}} = 1-(1-\tau)^u. \tag{5.19}$$

Then, we can derive the probability, denoted by P_s, that an SU transmits successfully without collisions, given that at least one SU transmits, as follows:

$$P_s = \frac{u\tau(1-\tau)^{u-1}}{P_{\text{tr}}} = \frac{u\tau(1-\tau)^{u-1}}{1-(1-\tau)^u}. \tag{5.20}$$

Denote the duration of a time slot by σ. Under the IEEE 802.11 DCF, the backoff counter of the contention node decreases by 1 when the sensed channel is idle in a time slot. However, it should be noted that under the CREAM-MAC protocol, only when both the control channel and at least one data channel in the channel group are idle, the backoff counters of the contending SUs decrease by 1. In other words, if all of channels in the channel group are busy, the backoff counter should remain the same until the time slot in which control channel and at least one data channel in channel group are idle, which is different from the backoff mechanism in traditional IEEE 802.11 DCF. Therefore, we introduce a new parameter, namely the effective duration of a time slot, denoted by σ', which represents the average duration of a time slot after taking the above descriptions into account. The effective duration of a time slot includes the duration of the normal time slot and the average duration of time slots where the backoff counter remains the same due to all the channels in the channel group being busy. Thus, we can derive σ' as follows:

$$\sigma' = \sigma(1 + \mathbb{E}[N_{\text{busy}}]), \tag{5.21}$$

where N_{busy} is the random number of time slots in which all the channels in a channel group are busy between the two consecutive time slots where the backoff counter decreases, and $\mathbb{E}[N_{\text{busy}}]$ is the mathematical expectation of N_{busy}. Since the channel states in different time slots are independent, N_{busy} follows the geometric distribution, and thus we can obtain its probability mass function (pmf) as follows:

$$\Pr\{N_{\text{busy}} = i\} = P_{\text{busy}}^{i}(1 - P_{\text{busy}}), \tag{5.22}$$

where $P_{\text{busy}} = \Pr\{H = 0\}$ is the probability that all the channels in a channel group are busy. According to (5.16), we have $\Pr\{H = 0\} = [\gamma(1 - P_{FA})^2]^n$. Then, we can get $\mathbb{E}[N_{\text{busy}}]$ by

$$\mathbb{E}[N_{\text{busy}}] = \sum_{i=0}^{\infty} i \Pr\{N_{\text{busy}} = i\} = \frac{[\gamma(1 - P_{FA})^2]^n}{1 - [\gamma(1 - P_{FA})^2]^n}. \tag{5.23}$$

Hence, substituting (5.23) into (5.21), we can calculate the effective duration (σ') of a time slot.

Let T_{succ} and T_{coll} be the time used for successful transmission and the time spent when collisions happen, respectively. Then, T_{succ} and T_{coll} can be expressed as:

$$\begin{cases} T_{\text{succ}} = \frac{\text{RTS}+\text{CTS}+\text{CST}+\text{CSR}}{R_c} + 3 \times \text{SIFS} + \text{DIFS}, \\ T_{\text{coll}} = \frac{\text{RTS}}{R_c} + \text{DIFS}, \end{cases} \tag{5.24}$$

where R_c is the transmission rate of control channel, SIFS is the duration of the short interframe space, DIFS is the duration of DCF interframe space, RTS, CTS, CST, and CSR are the sizes of RTS, CTS, CST, CSR control packets, respectively. Then, we can derive the average time, denoted by $\mathbb{E}[T_c]$, spent for the successful four-way handshakes of RTS/CTS/CST/CSR (i.e., the channel negotiation) as follows:

$$\begin{aligned} \mathbb{E}[T_c] &= \frac{(1 - P_{\text{tr}})\sigma' + P_s P_{\text{tr}} T_{\text{succ}} + P_{\text{tr}}(1 - P_s) T_{\text{coll}}}{P_s P_{\text{tr}}} \\ &= T_{\text{succ}} + \frac{1 - P_{\text{tr}}}{P_s P_{\text{tr}}}\sigma' + \frac{1 - P_s}{P_s} T_{\text{coll}}. \end{aligned} \tag{5.25}$$

5.5.3 The Aggregate Throughput

For convenience of presentation, Table 5.1 lists the important parameters for the design and analysis of our proposed CREAM-MAC protocol. Let N_c be the maximum number of SUs that successfully reserve the licensed channel groups during the length of $T_d^{\max}$ on average. Clearly N_c is inversely proportional to $\mathbb{E}[T_c]$, and thus we obtain:

$$N_c = \frac{T_d^{\max}}{\mathbb{E}[T_c]}.$$

Note that there are at most $\lceil N_c \rceil$ SUs that can opportunistically transmit data over the licensed data channels at the same time from the global viewpoint. Comparing the value of $(N_c + 1)$ and the number $\lfloor M/n \rfloor$ of channel groups, we can determine whether the control channel gets saturated.

On one hand, if $(N_c + 1) \leq \lfloor M/n \rfloor$, then there are always vacant channel groups that can accommodate the SUs successfully conducting channel negotiation. As a result, the control channel gets saturated and is the bottleneck to the aggregate throughput. Figure 5.5a shows an example of the saturated control channel case where the number $\lfloor M/n \rfloor$ of channel groups is equal to 4 and $3 < N_c + 1 < 4$. As shown in Fig. 5.5a, the SUs can always find the idle channel group to transmit data whenever the SUs successfully complete the channel negotiation over the control channel. On the other hand, if $(N_c+1) > \lfloor M/n \rfloor$, then there is always an idle period between the two consecutive channel negotiations on the control channel because the data channels in each channel group become saturated. Figure 5.5b shows an example of this case where the number $\lfloor M/n \rfloor$ of channel groups is equal to 2 and

Table 5.1 The parameters for design and analysis of the CREAM-MAC protocol

RTS	20 Bytes	The size of RTS packet
CTS	20 Bytes	The size of CTS packet
CST	20 Bytes	The size of CST packet
CSR	20 Bytes	The size of CSR packet
σ	9 μs	Mini-slot interval
SIFS	15 μs	Short interframe space
DIFS	34 μs	DCF interframe space
R_c	1 Mbps	Transmission rate of the control channel
R_d	1 Mbps	Transmission rate of a licensed channel
n		The number of sensors each SU has
u		The number of contending SUs
γ		Channel utilization of PUs
M		The number of licensed channels
$\mathbb{E}[T_c]$		Avg. time for successful four-way handshakes
$T_d^{\max}$		Max. tolerable interference-time of PUs
$CW_{\min}$		The minimum size of contention window
β_1		Threshold that determines the missed detection probability
β_2		Threshold that determines the false alarm probability
υ		Instantaneous SNR of PUs' signal at the side of the SU

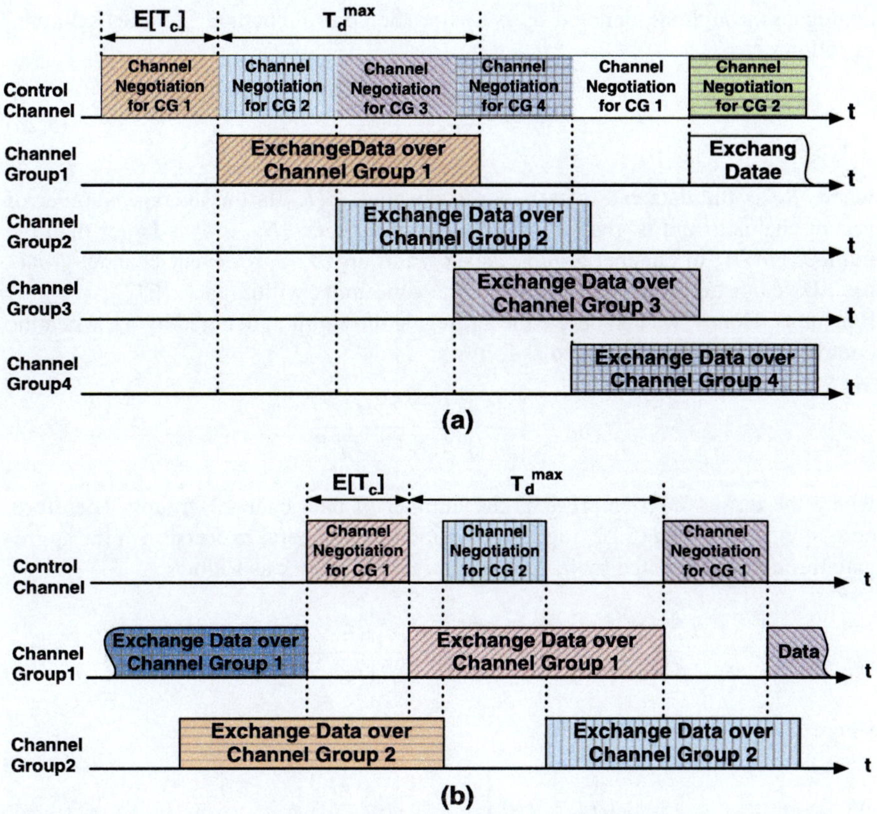

Fig. 5.5 Illustrations of the CREAM-MAC protocol for the saturation network case. (**a**) The case where the number of channel group is 4 and the *control channel gets saturated.* (**b**) The case where the number of channel group is 2 and the *control channel does not get saturated*, implying that the data channels get saturated. Here CG i represents Channel Group i. The channel occupation blocks drawn with the *same* filling pattern and color in either (**a**) or (**b**) represent that the control-channel and data-channel resources are occupied by the *same* pair of SU sender and SU receiver. The average duration of the channel negotiation is equal to $\mathbb{E}[T_c]$ and the duration for the exchange data block is equal to $T_d^{\max}$

$N_c + 1 > 3$. As shown in Fig. 5.5b, due to the requirements of the CREAM-MAC protocol, the SUs cannot start channel negotiation until at least one channel group becomes idle, which results in an idle period between two consecutive channel negotiations.

Based on whether the control channel gets saturated or not, we derive the aggregate throughput in two different cases, respectively. First, for the case where the control channel gets saturated, as shown in Fig. 5.5a, on average, the SUs can transmit data for $T_d^{\max}$ time units at the cost of $\mathbb{E}[T_c]$ time units. Note that in the saturation network case, all the SUs that win the channel reservation use up all of the transmission time $T_d^{\max}$ to transmit packets. Consequently, we derive the aggregate throughput, denoted by η_c, when the control channel becomes saturated as follows:

$$\eta_c = \frac{T_d^{\max}\mathbb{E}[H]R_d}{\mathbb{E}[T_c]} \tag{5.26}$$

where R_d is the data rate of a licensed channel, $\mathbb{E}[H]$ is the average number of vacant channels and is given by (5.17). Second, when $(N_c + 1)$ is larger than the number $\lfloor M/n \rfloor$ of channel groups, as shown in Fig. 5.5b, for every channel group, the SUs can effectively transmit for $T_d^{\max}$ time units within each $\left(\mathbb{E}[T_c] + T_d^{\max}\right)$ time units. Hence, we can derive the aggregate throughput, denoted by η_d, when the control channel is not saturated as follows:

$$\eta_d = \left\lfloor \frac{M}{n} \right\rfloor \frac{T_d^{\max}\mathbb{E}[H]R_d}{\mathbb{E}[T_c] + T_d^{\max}} \tag{5.27}$$

where the term $\lfloor M/n \rfloor$ represents the number of data channel groups. Therefore, combining (5.26) and (5.27) together, we obtain the general expression of the aggregate throughput, denoted by η, for the saturation network as follows:

$$\eta = \frac{T_d^{\max}N_d\mathbb{E}[H]R_d}{\mathbb{E}[T_c] + T_d^{\max}} = \frac{T_d^{\max}N_d n[1-\gamma(1-P_{FA})^2]R_d}{\mathbb{E}[T_c] + T_d^{\max}}, \tag{5.28}$$

where

$$N_d = \min\left\{(N_c + 1), \left\lfloor \frac{M}{n} \right\rfloor\right\} \tag{5.29}$$

which distinguishes between the control-channel saturation (if $N_d = N_c + 1$) and data-channel saturation (if $N_d = \lfloor M/n \rfloor$).

5.6 Performance Analysis for the Special Non-saturation Network Case

In this section, we analyze the performance of the CREAM-MAC protocol, including the network throughput, queueing delay, and service delay (or packet transmission delay), for the non-saturation-network case, where the SUs may have the empty queues. To make the model tractable, we consider the case where the number of licensed channels is equal to the number of equipped sensors, i.e., $M = n$. As a result, there is only one channel group for the SUs. Without loss of generality, we suppose that the packets of the SUs arrive according to the Poisson process with a mean arrival rate λ and the size of each data packet is equal to $R_d T_d^{\max}$. Every time the SUs successfully reserve the licensed data channels, they occupy the licensed channels for a period of $T_d^{\max}$ to transmit their data packets. For convenience of presentation, we call the procedure, during which an SU successfully reserves the data channels and transmits data packets, a *service procedure*, or simply, a *service*, in the following discussions.

Utilizing the channel-aggregating technique, the SUs can send multiple data packets simultaneously after they successfully exchanged the control packets. Note that the number of data packets that an SU can send depends on the number of common unused channels, which is a random variable, between the SU sender and the SU receiver. This implies that the service capacity for the SUs *varies* from time to time. Therefore, for this non-saturation-network case, we use the single-server bulk-service queueing model, $M/G^Y/1$, to characterize the network throughput, queueing delay, and service delay, where Y stands for the variable service capacity.

We first obtain the equilibrium-state distribution of the number of buffered packets in the queue for any given SU at any random points in order to derive the queueing delay and network throughput. We start with studying the random number, $N_\alpha^+, \alpha = 0, 1, 2, \cdots$, where α is used to index the services, of packets in the system for a given SU immediately after the α-th service. The probability, denoted by P_j^+, that the system has j packets in the equilibrium state can be expressed as:

$$P_j^+ = \lim_{\alpha \to \infty} \Pr\left\{N_\alpha^+ = j\right\}. \tag{5.30}$$

Denote by $Y_\alpha, \alpha = 0, 1, 2, \cdots$ the service capacity during the α-th service, i.e, the maximum number of data packets the given SU can send during the α-th service. Since each data packet is transmitted over an unused channel and the size of each data packet is equal to $R_d T_d^{\max}$, the service capacity is equal to the maximum number of unused channels in the channel group. Thus, we obtain Y_α's distribution as follows:

$$\begin{aligned}\Pr\{Y_\alpha = i\} &= \Pr\{H = i\} \\ &= \binom{n}{i}\left[1 - \gamma(1 - P_{\text{FA}})^2\right]^i \left[\gamma(1 - P_{\text{FA}})^2\right]^{n-i},\end{aligned} \tag{5.31}$$

where n is the number of sensors equipped in each SU and equal to the number of licensed channels. Then, the pmf, denoted by y_i, that the given SU sends i data packets during a service at the equilibrium state, can be determined by

$$y_i = \lim_{\alpha \to \infty} \Pr\{Y_\alpha = i\}. \tag{5.32}$$

Note that the sequence of $\{y_i\}$ is independent of the arrival process of the packets for a given SU. Then, we get average number, denoted by $\overline{y}$, of packets that an SU can send during a service as follows:

$$\overline{y} = \mathbb{E}[\Pr\{Y_\alpha = i\}] = n\left[1 - \gamma(1 - P_{\text{FA}})^2\right]. \tag{5.33}$$

Let us define the following two equations:

$$\begin{cases} \phi_j \triangleq \sum_{m=j}^{n} y_m, \\ \Phi_j(z) \triangleq \sum_{m=j}^{n} y_m z^m, \end{cases} \tag{5.34}$$

where $j = 0, 1, 2, \cdots$, and note that $\Phi_0(z)$ is the probability generating function (PGF) for $\{y_i\}$ and $\Phi_0(1) = \phi_0$.

For a given SU, a service period consists of four components: (i) the successful transmission time of packets sent by itself, (ii) the successful transmission time of packets sent by other SUs, (iii) the backoff time, and (iv) the time spent due to collisions. Since the sum of components (iii) and (iv) is much smaller than the components (i) and (ii), we only focus on the components (i) and (ii) to derive the service period. The successful transmission time, denoted by D, of a node includes the time spent by transmitting control packets and data packets, i.e.,

$$D = T_{\text{succ}} + T_d^{\max}. \tag{5.35}$$

Let p_s be the probability that a given SU successfully reserves the data channels given that a successful reservation occurs during a contention period, and V be the random number of other nodes' transmissions between two successfully transmissions of a given node plus its own successful transmission (i.e., the service period in terms of the number of successful transmissions), respectively. Then, we obtain [29]

$$p_s = \frac{1}{\rho(u-1)+1}, \tag{5.36}$$

where ρ is the system utilization. Then, the service period in terms of the number of successful transmissions (V) follows the geometric distribution, which has the following pmf:

$$\Pr\{V = v\} = p_s(1-p_s)^{v-1}, \tag{5.37}$$

where $v = 1, 2, \cdots$. Thus, we can get the average service period in terms of the number of successful transmissions, denoted by $\overline{v}$ for a given SU as follows:

$$\overline{v} \triangleq \mathbb{E}[V] = \frac{1}{p_s} = \rho(u-1)+1. \tag{5.38}$$

According to the definition of the system utilization, ρ can also be written as

$$\rho \triangleq \frac{\lambda D \overline{v}}{\overline{y}}, \tag{5.39}$$

where $\overline{y}$ is given by (5.33). Consequently, solving (5.38) and (5.39) simultaneously, we can obtain

$$\rho = \frac{\lambda D}{\overline{y} - \lambda(u-1)D}. \tag{5.40}$$

For the equilibrium-state probability distribution to exist, ρ should be less than 1.

Denote ψ as the random number of arrived packets for a given SU during the α-th service. Note that the packet arrivals follow the Poisson process. Given that the length of the service period is vD, we can obtain the probability that the number of arrived packets is j as follows:

$$\Pr\{\psi = j | V = v\} = \frac{e^{-\lambda D v}(\lambda D v)^j}{j!}. \tag{5.41}$$

By removing the conditioning on variable V in (5.41), we get the probability that the number of arrived packets is j as follows:

$$\begin{aligned} \Pr\{\psi = j\} &= \sum_{v=1}^{\infty} \Pr\{\psi = j | V = v\} \Pr\{V = v\} \\ &= \sum_{v=1}^{\infty} \frac{e^{-\lambda D v}(\lambda D v)^j}{j!} \left[p_s (1 - p_s)^{v-1} \right]. \end{aligned} \tag{5.42}$$

Then, we get the PGF, denoted by $\Psi(z)$, of $\Pr\{\psi = j\}$ as follows:

$$\Psi(z) = \sum_{j=0}^{\infty} \Pr\{\psi = j\} z^j = \frac{p_s e^{\lambda D(z-1)}}{1 - (1 - p_s) e^{\lambda D(z-1)}}. \tag{5.43}$$

Let us denote N_s as the average number of packets that SUs send during a service period at the equilibrium state. We can obtain:

$$N_s = \min\left\{ N_q, \lim_{\alpha \to \infty} Y_\alpha \right\}, \tag{5.44}$$

where Y_α is the service capacity at α-th service. Thus, we obtain the network throughput, denoted by θ, as follows:

$$\theta = \mathbb{E}[N_s] T_d^{\max} R_d. \tag{5.45}$$

Following the work in [30, 31], we can obtain the average queue length, denoted by L_q, of the packets for any given SU is the first moment of N_q as follows

$$L_q = \frac{(\lambda D)^2 \Psi^{(2)}(1) + \Phi_0^{(2)}(1)}{2\overline{y}(1-\rho)} + \frac{1-n+\rho(n-\rho\overline{y})}{1-\rho} + \sum_{i=1}^{n-1} (1-z_i)^{-1} - \overline{y}\rho + \frac{(\lambda D)^2 \Psi^{(2)}(1)}{2\overline{y}\rho}, \tag{5.46}$$

where $f^{(i)}(\cdot)$ indicates the i-th derivative of $f(\cdot)$, and z_i with $1 \leq i \leq n-1$ are the $(n-1)$ roots of $\Psi(z)\Phi_0(z^{-1}) = 1$, which are located inside the unit circle [30]. Using the Little's law, we can derive the queueing delay, denoted by W_q, for a given SU as follows:

$$W_q = \frac{L_q}{\lambda}. \tag{5.47}$$

Finally, we can derive the service delay, or the average packet transmission delay, denoted by W_s, for a given SU as follows:

$$W_s = W_q + \overline{v}D = \frac{L_q}{\lambda} + [\rho(u-1)+1]D. \tag{5.48}$$

5.7 Performance Evaluations

The parameters used to evaluate the CREAM-MAC protocol are summarized in Table 5.1. We first investigate the aggregate throughput for the saturation network case. Let the number n of sensors of each SU be 4, the channel utilization γ of PUs be fixed at 0.5, and R_c be equal to 1 Mbps. By tuning the thresholds β_1 and β_2 in the spectrum sensing scheme, the false alarm probability and missed detection probability are set to 10^{-3} and 10^{-4}, respectively. Using Eq. (5.28), we plot the aggregate throughput η against the size of the contention window $CW_{\min}$ in Fig. 5.6. In Fig. 5.6, we observe that the optimal $CW_{\min}$, denoted by $CW^*_{\min}$, which achieves the highest aggregate throughput, changes with the different number (u) of contending SUs. This is expected because given that there are sufficient licensed channels, the aggregate throughput only depends on the time spent to accomplish the RTS/CTS/CST/CSR four-way handshakes over the control channel, which is ultimately determined by the IEEE 802.11 DCF parameters, such as $CW_{\min}$ and R_c. If we can obtain the number of contending SUs in advance, we can pre-select the optimal $CW_{\min}$ which achieves the highest aggregate throughput. On the other hand, if the number of contending SUs dynamically fluctuates, we can adopt the algorithms proposed in [28] to dynamically adjust the value of $CW_{\min}$. In the rest of our chapter, we assume that in each scenario the number of contending SUs is fixed, and thus we can pre-select $CW^*_{\min}$ for different scenarios with different u's.

After setting the optimal $CW^*_{\min}$ to be 256 for the case where $u = 30$, we use (5.28) to get numerical results of the aggregate throughput against the channel utilization of PUs as shown in Fig. 5.7. The aggregate throughput (η) decreases as the

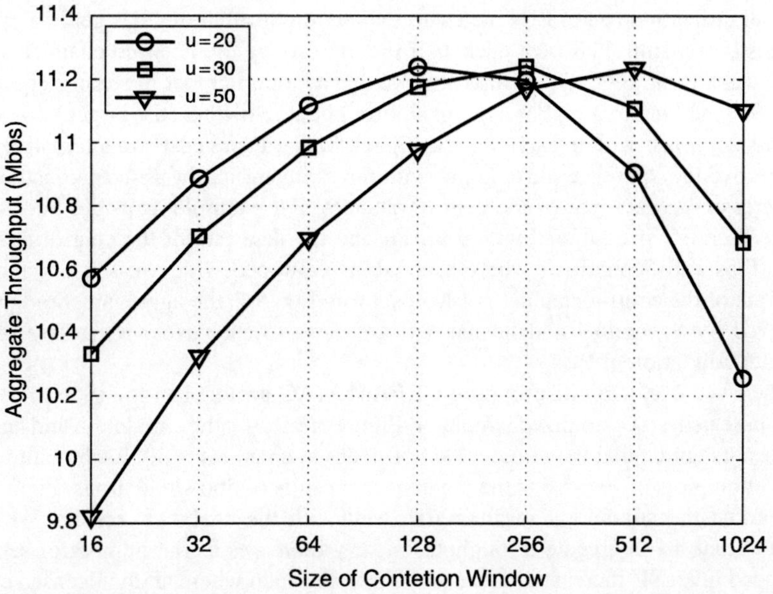

Fig. 5.6 The aggregate throughput against the size of contention window ($CW_{\min}$). The number (n) of sensors is set to be 4. The channel utilization (γ) of PUs is set to be 0.5. R_c and R_d are set to be 1 Mbps

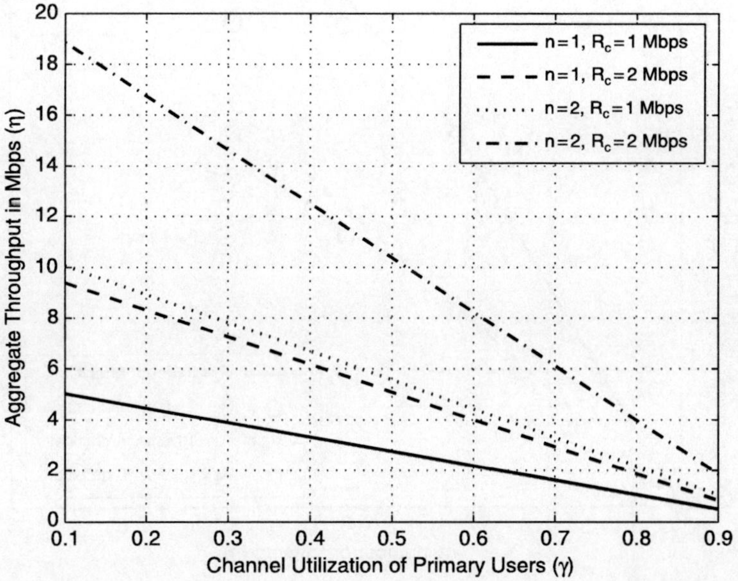

Fig. 5.7 The aggregate throughput against the channel utilization of PUs, when the number (M) of licensed channels is set to be 30, the number (u) of contending SUs is set to be 30, and $CW_{\min} = 256$

channel utilization (γ) of PUs increases, which implies that the SUs get less opportunities to transmit their own packets if the PUs utilize the licensed channels more intensively. From Fig. 5.7, we also observe that a larger number of sensors equipped in an SU can lead to a higher aggregate throughput. However, the larger number of sensors equipped with in each SU, the higher the hardware cost. An alternative way to improve the aggregate throughput with the stringent hardware-cost constraint is to increase the data rate of the control channel. For example, consider Scenario I where each SU is equipped with a sensor and the data rate of the control channel is 2 Mbps and Scenario II where each SU is equipped with two sensors and the data rate of the control channel is 1 Mbps. From Fig. 5.7, the aggregate throughputs achieved by Scenario I and Scenario II are close to each other regardless of the channel utilization of PUs.

Then, we evaluate our proposed CREAM-MAC protocol in the saturation network case using a customized simulator. Figure 5.8 shows the simulation and analytical results, given that the number (M) of licensed channels is 30. Each point in the simulation plots of Fig. 5.8 is the mean of the results of 500 simulations. Figure 5.8 shows that the simulation results agree well with the analytical results. We also observe that the aggregate throughput linearly increases as the number of sensors equipped in an SU increases before they reach the state where all the licensed channels are saturated. This is expected by (5.29) and (5.28). More precisely, when the SUs can only access a small number of channels simultaneously with fewer sensors,

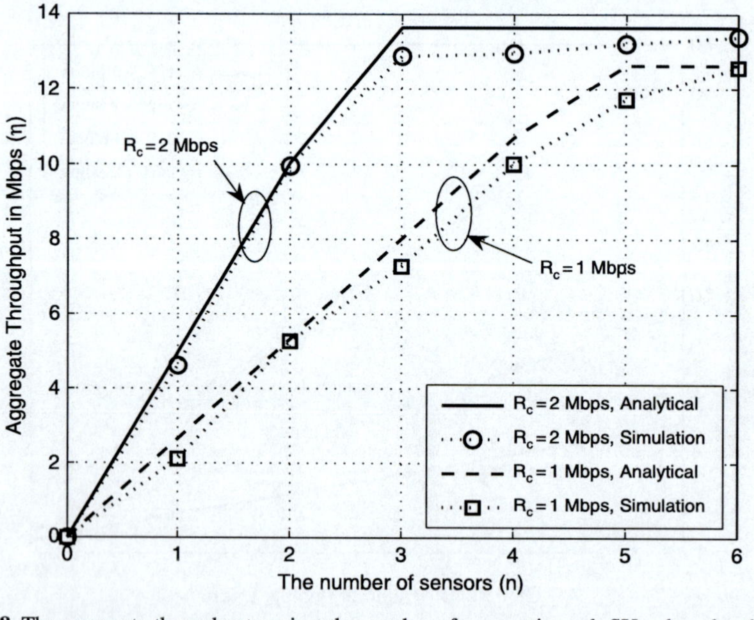

Fig. 5.8 The aggregate throughput against the number of sensors in each SU, when the channel utilization (γ) of the PUs is set to be 0.5, the number (M) of licensed channels is set to be 30, the number (u) of contending secondary is set to be 30, and $CW_{\min} = 256$

there are sufficient channel groups for the SUs to access. As a result, the control channel becomes saturated, which implies that the average number of winning SUs during a fixed amount of time is constant. Thus, increasing the number of sensors can efficiently increase the aggregate throughput. On the other hand, when the number of sensors equipped within each SU is large enough, the licensed data channels become saturated before the control channel gets saturated, which implies that the control channel is not the bottleneck any more. Thus, further increasing the number of sensors cannot increase the aggregate throughput any more. However, as shown in Fig. 5.8, we can still increase the aggregate throughput by increasing the data rate of the control channel because the higher control-channel data rate implies the less time spent to accomplish the RTS/CTS/CST/CSR four-way handshakes.

We further proceed to the non-saturation network case. Using (5.48), we plot the packet transmission delay against the channel utilization of PUs when the packet arrival rate is set to be a constant of 0.049, as shown in Fig. 5.9. The packet transmission delay increases as the channel utilization (γ) of PUs gets larger. In particular, when γ is larger than 0.83, the packet transmission delay for $u = 10$ increases much faster than that for $u = 5$. This is because when γ is larger than 0.83, the system utilization ρ for $u = 10$ approaches to 1, implying that the packet transmission delay goes to infinity. Figure 5.9 shows that when the number of contending SUs is fixed, the packet transmission delay decreases by utilizing more channels.

Also, we conduct simulations to verify the analytical model for the non-saturation network case. Figure 5.10 shows the analytical and simulation results

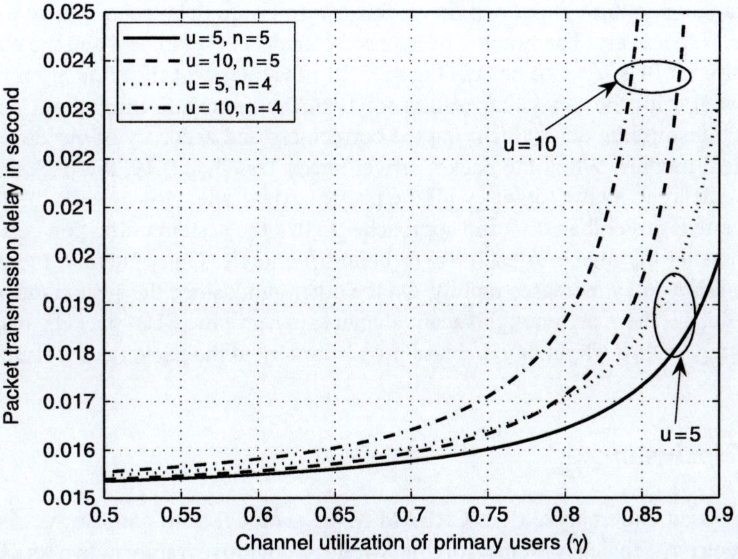

Fig. 5.9 The packet transmission delay against the channel utilization of PUs with the combinations of the number (u) of contending SUs being set to be 5 or 10 and the number (n) of licensed channels being set to be 4 or 5

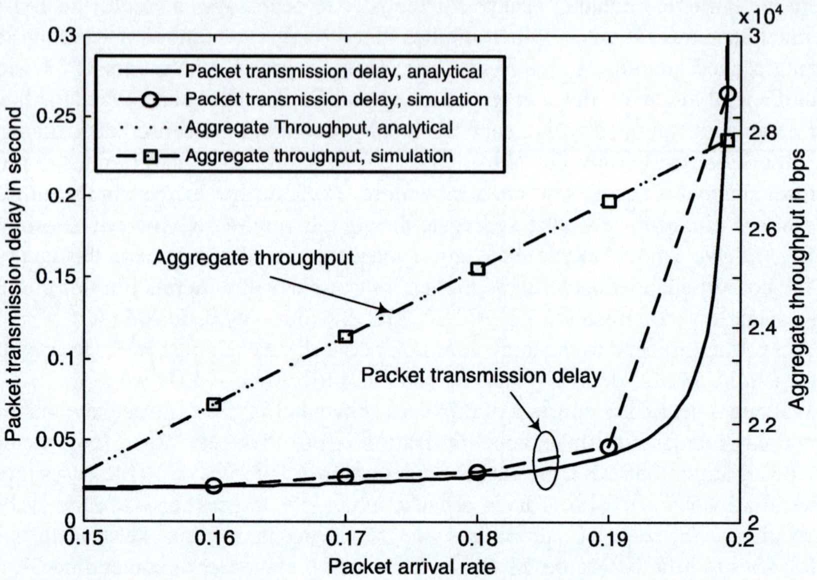

Fig. 5.10 Analytical and simulation results for the non-saturation network case when the packet arrival rate varies. The number (u) of contending SUs is set to be 10, the number (n) of licensed channels is set to be 4, the channel utilization (γ) of PUs is set to be 0.5

of the aggregate throughput and the packet transmission delay, denoted by the two Y-axes, respectively. The number of licensed channels is set to be 4 and the channel utilization of PUs is set to be 0.5. Figure 5.10 illustrates that the simulation results agree well with the analytical results for both the aggregate throughput and the packet transmission delay, verifying the correctness and accuracy of our developed analytical models. When the packet arrival rate is less than 0.19, the packet transmission delay increases slowly with the packet arrival rate. However, as the packet arrival rate is larger than 0.19 and approaches to 0.2, the system utilization gets close to 1, causing the queue of each SU to build up quickly. Consequently, the packet transmission delay increases rapidly. On the other hand, since the system utilization is less than 1, there are enough vacant channels to serve the SUs' packets, and thus the aggregate throughput can increase proportionally to the packet arrival rate.

5.8 Conclusions

We proposed and analyzed the CREAM-MAC protocol with cooperative sequential spectrum sensing scheme for the wireless cognitive radio networks. Under the CREAM-MAC protocol, each SU, which is equipped with a cognitive radio and multiple sensors, seizes the opportunity whenever the vacant licensed channels are available to exchange their own packets while only causing the acceptable interference to the PUs. Without using any centralized controllers, our proposed CREAM-MAC protocol can still solve both the traditional and multi-channel hidden terminal problems by introducing the four-way handshakes of control packets over the control channel. Our developed cooperative sequential spectrum sensing scheme can enhance the accuracy of spectrum sensing to reduce the interference imposed to the PUs. Applying the IEEE 802.11 DCF-based model and the $M/G^Y/1$ queueing model, we developed analytical models to accurately evaluate the performance of our proposed CREAM-MAC protocol for both the saturation network case and non-saturation network case. The simulation results also verify the analytical models and analyses.

References

1. McHenry, M. (Shared Spectrum Company, 2005). NSF spectrum occupancy measurements project summary.
2. McHenry, M., Tenhula, P., McCloskey, D., Roberson, D., and Hood, C. (2006). Chicago spectrum occupancy measurements & analysis and a long-term studies proposal. In *Proc. the First International Workshop on Technology and Policy for Accessing Spectrum (TAPAS)*, Boston, MA.
3. FCC Spectrum Policy Task Force (ET Docket No. 02-135, 2002). Spectrum policy task force report.
4. FCC Spectrum Policy Task Force (ET Docket No. 03-237, 2003). Establishment of an interference temperature metric to quantify and manage interference and to expand available unlicensed operation in certain fixed, mobile and satellite frequency band.
5. J. Mitola III (Ph.D. Thesis, KTH Royal Inst. Technology, Stockholm, Sweden, 2000). Cognitive radio: an integrated agent architecture for software defined radio.
6. Haykin, S. (2005). Cognitive radio: brain-empowered wireless communications. *IEEE Journal on Selected Areas in Communications*, 23(2):201–220.
7. Akyildiz, I., Lee, W., Vuran, M., and Mohanty, S. (2006). Next generation/dynamic spectrum access/cognitive radio wireless networks: a survey. *Elsevier Computer Networks*, 50: 2127–2159.
8. Mitola, J. I. and Maguire, G. Q. J. (1999). Cognitive radio: making software radios more personal. *IEEE Personal Communications*, 6(4):13–18.
9. Zhao, Q., Tong, L., and Swami, A. (2005). Decentralized cognitive MAC for dynamic spectrum access. In *Proc. First IEEE International Symposium on New Frontiers in Dynamic Spectrum Access Networks (DySPAN)*, Baltimore, MD, pages 224–232.
10. Zhao, Q., Tong, L., Swami, A., and Chen, Y. (2007). Decentralized cognitive MAC for opportunistic spectrum access in ad hoc networks: a POMDP framework. *IEEE Journal on Selected Areas in Communications*, 25(3):589–600.
11. Su, H. and Zhang, X. (2008b). Cross-layer based opportunistic MAC protocols for QoS provisionings over cognitive radio wireless networks. *IEEE Journal on Selected Areas in Communications*, 26(1):118–129.
12. Su, H. and Zhang, X. (2007b). Opportunistic MAC protocols for cognitive radio based wireless networks. In *Proc. 41st Annual Conference on Information Sciences and Systems (CISS)*, Baltimore, MD, pages 363–368.
13. Su, H. and Zhang, X. (2007a). Cognitive radio based multi-channel mac protocols for wireless ad hoc networks. In *Proc. IEEE Global Telecommunications Conference (GLOBECOM)*, pages 4857–4861.
14. Su, H. and Zhang, X. (2008a). Channel-hopping based single transceiver MAC for cognitive radio networks. In *Proc. 42nd Annual Conference on Information Sciences and Systems (CISS)*, Princeton, NJ, pages 197–202.

15. Yuan, Y., Bahl, P., Chandra, R., Chou, P. A., Ferrell, J. I., Moscibroda, T., Narlanka, S., and Wu, Y. (2007). KNOWS: cognitive radio networks over white spaces. In *Proc. 2nd IEEE International Symposium on New Frontiers in Dynamic Spectrum Access Networks (DySPAN)*, Dublin, Ireland, pages 416–427.
16. Huang, S., Liu, X., and Ding, Z. (2009). Optimal transmission strategies for dynamic spectrum access in cognitive radio networks. *IEEE Transactions on Mobile Computing*, 8(12): 1636–1648.
17. Zhang, X. and Su, H. (2011a). CREAM-MAC: Cognitive radio-enabled multi-channel MAC protocol over dynamic spectrum access networks. *IEEE Journal of Selected Topics in Signal Processing*, 5(1):110–123.
18. Zhang, X. and Su, H. (2011b). Opportunistic spectrum sharing schemes for CDMA-based uplink MAC in cognitive radio networks. *IEEE Journal on Selected Areas in Communications*, 29(4):716–730.
19. Su, H. and Zhang, X. (2009). Adaptive uplink MAC for CDMA-based cognitive radio networks. In *Proc. IEEE MILCOM*, Boston, MD.
20. Chia-Chun Hsu, A., Weit, D. S. L., and Kuo, C.-C. J. (2007). A cognitive mac protocol using statistical channel allocation for wireless ad-hoc networks. In *Proc. IEEE Wireless Communications and Networking Conference WCNC 2007*, pages 105–110, Hong Kong, China.
21. Mishra, A. (2006). A multi-channel MAC for opportunistic spectrum sharing in cognitive networks. In *Proc. Military Communications Conference MILCOM 2006*, pages 1–6.
22. Cabric, D., Mishra, S. M., and Brodersen, R. W. (2004). Implementation issues in spectrum sensing for cognitive radios. In *Proc. Thirty-Eighth Asilomar Conference on Signals, Systems and Computers*, Pacific Grove, CA, volume 1, pages 772–776.
23. Cordeiro, C. and Challapali, K. (2007). C-MAC: A cognitive MAC protocol for multi-channel wireless networks. In *Proc. 2nd IEEE International Symposium on New Frontiers in Dynamic Spectrum Access Networks (DySPAN)*, Dublin, Ireland, pages 147–157.
24. Institute of Electrical and Electronics Engineers. *IEEE Standard 802.11 - 1999, Wireless LAN Medium Access Control (MAC) and Physical Layer (PHY) Specifications, Nov 1999.*
25. Digham, F. F., Alouini, M.-S., and Simon, M. K. (2007). On the energy detection of unknown signals over fading channels. *IEEE Transactions on Communications*, 55(1):21–24.
26. Ghasemi, A. and Sousa, E. S. (2005). Collaborative spectrum sensing for opportunistic access in fading environments. In *Proc. First IEEE International Symposium on New Frontiers in Dynamic Spectrum Access Networks (DySPAN)*, Baltimore, MD, pages 131–136.
27. Bianchi, G. (2000). Performance analysis of the IEEE 802.11 distributed coordination function. *IEEE Journal on Selected Areas in Communications*, 18(3):535–547.
28. Su, H., Zhang, X., Qiu, P., and Guizani, M. (2006). An enhanced IEEE 802.11 MAC algorithm for tradeoff between delay and energy-consumption. In *Proc. IEEE International Conference on Communications (ICC)*, Istanbul, Turkey, volume 9, pages 3935–3940.
29. Medepalli, K. and Tobagi, F. A. (2006). Towards performance modeling of IEEE 802.11 based wireless networks: a unified framework and its applications. In *Proc. 25th IEEE INFOCOM*, Barcelona, Spain, pages 1–12.
30. Briere, G. and Chaudhry, M. (1989). Computational analysis of single-server bulk-service queues, M/G Y/1. *Advances in Applied Probability*, 21(1):207–225.
31. Chaudhry, M. and Templeton, J. (1983). *First course in bulk queues.* Wiley, New York, NY.
32. So, J. and Vaidya, N. H. (2004). Multi-channel MAC for ad hoc networks: handling multi-channel hidden terminals using a single transceiver. In *Proc. ACM MobiHoc*, Tokyo, Japan, pages 222–233.

Chapter 6
Cognitive MAC Protocol with Transmission Tax: Probabilistic Analysis and Performance Improvements

Vojislav B. Mišić and Jelena Mišić

Abstract We investigate the performance of a cognitive personal area network (CPAN) in which spectrum sensing is linked to packet transmissions. Efficient CPAN operation may be achieved if each data transmission is taxed by requiring the transmitting node to participate in cooperative sensing for a prescribed time period. In this approach, each node is allowed to transmit a single packet in one transmission cycle, but must then 'pay' for it by spectrum sensing, which not only ensures fairness with respect to transmission but also distributes the sensing burden to all nodes. We describe a probabilistic model of the integrated system and evaluate its performance with respect to packet transmissions and spectrum sensing. We discuss two modifications that involve centralized and distributed selection of the channels to be sensed. We also propose an adaptive algorithm to determine the tax coefficient and show that it offers superior data transmission performance while not affecting the sensing accuracy.

6.1 Introduction

The cognitive communications paradigm (also known as opportunistic spectrum access, or OSA) allows efficient use of existing wireless spectrum opportunities [1, 2] in many wireless networks. In case of wireless personal area networks (PANs), opportunistic spectrum access may be combined with frequency hopping spread spectrum (FHSS) [3] to offer increased resilience to interference from primary users' activity. Unlike earlier FHSS technologies such as Bluetooth [4], the hopping sequence of a frequency-hopping cognitive personal area network (CPAN) should not only be random, but also dynamically adapted to primary users' activity patterns, which in turn necessitates accurate and timely sensing of all working channels in the chosen band. Fig. 6.1 schematically shows frequency hopping in a CPAN.

Efficient operation of a frequency-hopping OSA network is critically dependent on accurate and timely sensing of all working channels in the chosen band. Ideally, all channels should be sensed all the time [5], but in reality there is an error which is

V.B. Mišić (✉)
Department of Computer Science, Ryerson University, Toronto, ON M5B 2K3, Canada
e-mail: vmisic@scs.ryerson.ca

F.R. Yu (ed.), *Cognitive Radio Mobile Ad Hoc Networks*,
DOI 10.1007/978-1-4419-6172-3_6,

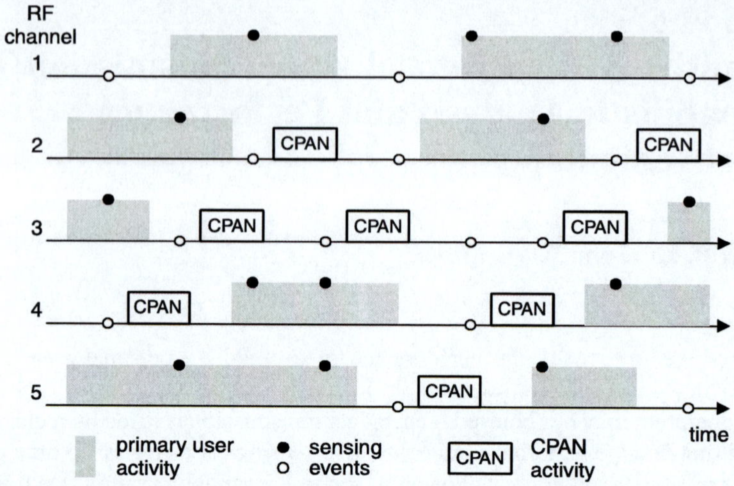

Fig. 6.1 Frequency hopping in a CPAN

caused by two factors. First, sensing is performed in discrete intervals, and changes in channel status (which can occur at any time) are not detected until the next sensing event [6]. Second, not all channels are sensed in every sensing cycle, and the delay in detecting the channel status may be longer than the sensing cycle.

However, if the CPAN coordinator is the only node to perform the sensing, sensing error may be unacceptably high [7] since each sensing takes a finite time, which limits the number of channels that can be sensed. This problem may be overcome by requiring that ordinary nodes in the CPAN help with sensing: this is commonly referred to as *cooperative sensing* [6]. The coordinator, then, coordinates the sensing process through scheduling sensing events, collecting the results, and combining them to form a coherent map of busy and idle channels [6, 8].

At the same time, CPAN nodes have data traffic to send to and receive from each other, which obviously interferes with sensing. As a result, a high-level balance must be struck between communication and sensing activities for each node, and a suitable protocol to schedule their individual activities must also be devised. It is the duty of the CPAN coordinator to ensure fairness among nodes, which means not only equal opportunities for transmission but also equal contribution to smooth operation of the CPAN through sensing. As transmission and sensing activities cannot be done at the same time, an effective balance between the two must be achieved and maintained, which necessitates a flexible transmission scheduling protocol.

In this chapter, we present one such protocol at the MAC level in which equal opportunity to transmit a single packet is given to all nodes through round robin scheduling. At the same time, nodes are required to 'pay' for transmissions by conducting sensing after transmission; this mechanism will be referred to as the 'transmission tax.' For each packet transmitted, the node has to perform spectrum

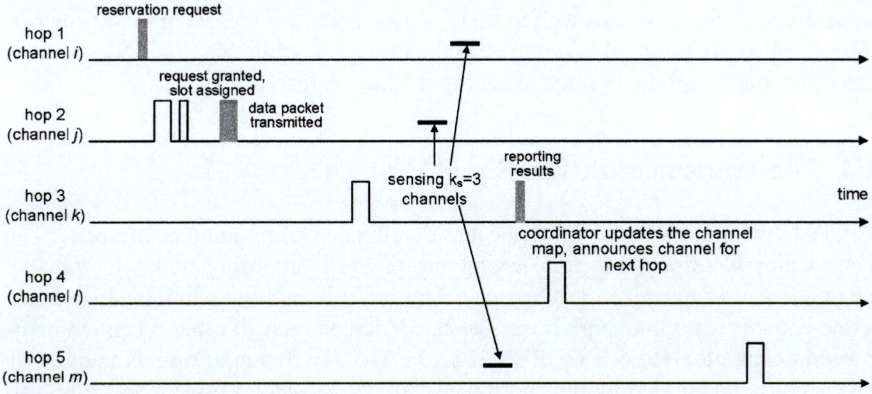

Fig. 6.2 CPAN operation: node activity for a single data packet to be transmitted

sensing on some k_s channels [9]. A schematic view of node activity related to a single packet to be transmitted is shown in Fig. 6.2. (For clarity, we have aligned the different channels from top to bottom, even though the actual channels may occur in any order, depending on the activity of primary sources.) In this manner, all nodes in the CPAN will be given equal opportunity to transmit data, but at the same time, all of them are expected to equally contribute to sensing, which reduces sensing errors and, thus, ensures smooth operation of the CPAN. The value of k_s is determined by the coordinator which monitors the CPAN traffic and maintains the channel map, with the goal of keeping the total sensing error below some pre-defined limit. Reception is not penalized, as will be seen below, and it does not affect the ratio of the number of transmitted packets to the sensing time. We then present a detailed probabilistic analysis of the integrated protocol and discuss its performance.

Several factors may affect the performance of the CPAN in this setting. Obviously, the tax rate will be the most important among them, as it limits the amount of time a node can spend in transmission, and its value has to be carefully chosen for the best tradeoff between transmission and sensing. The accuracy of sensing may be affected by the initial selection of channels to be sensed, and it may be done in a centralized fashion (i.e., by the coordinator) or it in a distributed fashion, locally by each node. We will investigate the performance of both approaches. Finally, we propose a simple adaptive algorithm for determining the tax rate so as to maintain a steady influx of sensing information.

The chapter is organized as follows: Section 6.2 gives more details about the operation of a frequency hopping cognitive personal area network, while Section 6.3 models the durations of transmission, sensing, and waiting times. Packet access delay is derived in Section 6.4. In Section 6.5 we present sensing model with control of sensing error. Section 6.6 presents the performance of the original protocol with respect to both data transmission and sensing. Modifications regarding the selection of channels to be sensed, as well as their performance, are presented in Section 6.7. The adaptive tax protocol is described in Section 6.8, while Section 6.9 concludes the chapter and highlights some avenues for future research.

6.2 The Transmission Tax-Based Protocol

A CPAN piconet consists of a dedicated coordinator and a number of nodes. The coordinator is responsible for starting the piconet, admitting nodes to join the piconet, monitoring and controlling its operation, and for other administrative tasks. Time is partitioned into superframes of fixed size, similar to other recent communication technologies such as IEEE 802.15.3 [10]. Each superframe is marked by a beacon frame emitted by the coordinator; activities such as node association and disassociation, bandwidth allocation requests and announcements, sensing allocation and reporting, and actual data transfers take part in dedicated sub-frames within the superframe, as shown in Fig. 6.3.

Each superframe uses a different channel from the working channel set, and the hopping sequence is random as well as adaptive, in the sense that the next working channel is chosen in a pseudo-random manner from the set of channels which are currently free of primary source activity. The working frequency band consists of N channels, each of which is used by a distinct primary source which operates in an ON-OFF regime. The durations of active and inactive periods of these sources may be characterized with suitable probability distributions. For simplicity, we assume that all primary sources have the same mean durations of active and inactive periods and, consequently, the same value of the activity factor $p_{a,r}$.

When the node has data to transmit, it requests a suitable bandwidth allocation, expressed as a certain number of time slots, during the reservation sub-frame in hop 1. Each node can request transmission of at most one packet in a single request. The coordinator receives the requests and allocates time slots to nodes in a

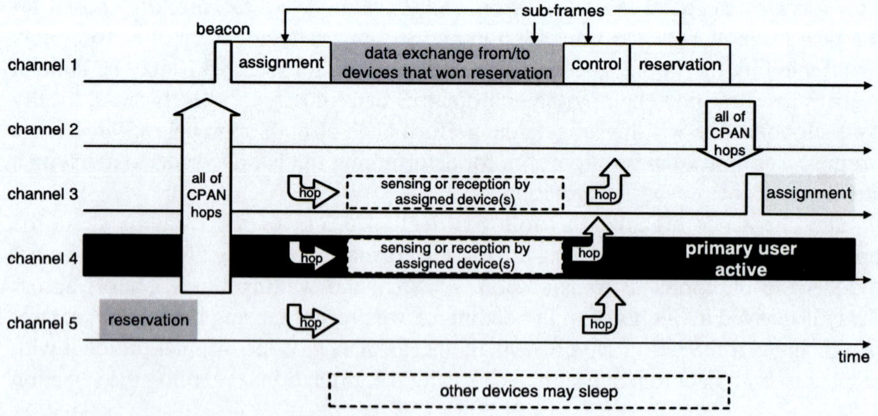

Fig. 6.3 Superframe format and node activities

round-robin fashion, as follows. All requests received during the reservation slot are sorted according to node addresses. The address of the last node serviced in the previous superframe is recorded, so that the allocations are assigned beginning from the next higher address. (Service policy in which nodes are served in a round-robin fashion and allowed to transmit at most one packet at a time is known as one-limited policy [11, 12]; it offers best performance at high traffic loads and ensures both short- and long-term fairness among the nodes.)

The allocations are announced in the assignment sub-frame which immediately follows the beacon. This packet is sent in the assignment sub-frame after the beacon, in the superframe at hop 2. (The allocation packet is shown in white, since it is received, rather than transmitted, by the current node.) The node then transmits its data packet to the designated receiver in that same superframe.

Allocation includes as many requests as can fit into one superframe, given that each node is granted transmission of at most one packet; requests that cannot fit in the current superframe are deferred to the next one(s). Appropriate announcements are made during the assignment sub-frame which immediately follows the beacon. The announcement also includes the channel which the coordinator would like to be sensed, but the node can choose to ignore that and perform the selection itself.

Upon transmitting the packet, the node has to perform sensing duty for k_s subsequent superframes, where k_s is the tax coefficient, before requesting to transmit another packet, as shown in Fig. 6.2. Sensing is performed by switching to the selected channel and listening to it for some time to decide whether there is a primary user activity. It may be done through energy detection, carrier detection, or feature identification, with an increasing accuracy but also with an increasing detection time [1].

Nodes that are currently engaged in sensing on account of an earlier transmission cannot request bandwidth again until they sense k_s primary channels and report the results of sensing to the coordinator.

Since sensing involves the overhead of switching channels back and forth, it is most efficient if performed throughout the entire data sub-frame; therefore the tax coefficient p will denote the number of sub-frames to spend in sensing. In this manner, each node given bandwidth to transmit actually pays for it by contributing to sensing. During the control sub-frame, the nodes that did sensing send the results back to the coordinator, which then updates the channel map and decides on the next working channel.

All nodes, regarding of whether they are currently doing the sensing duty or not, must listen to the beacon of the next superframe and subsequent assignment sub-frame. The reasoning behind this is simple: if a node that is currently sensing learns from the beacon that it is to receive a packet, it suspends sensing during that superframe, receives the packet, and resumes sensing after the next beacon frame. (Suspension may again be needed if another packet is to be received.) In theory a node could interrupt its sensing to switch back to the working channel, receive a packet, and then go back to finish sensing, the overhead of channel switching makes this prohibitively expensive, which is why we have adopted the simpler approach in which sensing is preempted by reception.

Transmission need not be suspended because of reception, as the node only has to switch its radio from one mode to another within the same superframe.

Packets that arrive during the transmission of an earlier packet and/or sensing are queued in the node buffer but not transmitted; bandwidth requests are prohibited until the node has spent at least p superframes in sensing. Sensing duty is thus discharged at the expense of extending the packet service cycle. Once the sensing is done, the node may immediately apply for bandwidth allocation in the next reservation sub-frame – if there are packets queued for transmission. This is the rationale for placing the control sub-frame before the reservation sub-frame; otherwise, a node would have to wait idle for an entire superframe before it could apply for bandwidth again.

6.3 Modeling the Protocol

Let us denote the time interval between two consecutive services of a given node, hereafter referred to as the piconet cycle, with C. Packet service cycle, denoted with Y, represents the time needed to service a single packet from the moment when the node applies for bandwidth. It consists of several components: waiting time for the beginning of round-robin packet service; packet service time (packet reception may occur in the same superframe); waiting time for the next beacon; potential reception of the packet; and spectrum sensing time.

From the viewpoint of queuing theory, interaction between transmission and sensing can be modeled as 1-limited M/G/1 system with vacations, in which packet transmission corresponds to the service period, while everything else – the sensing process, waiting for the transmission after requesting bandwidth from the coordinator, synchronization with the beacon, and potential reception of packet – comprises the total vacation period. These interactions are schematically depicted in Fig. 6.4 for the case with and without the preemption of the sensing process.

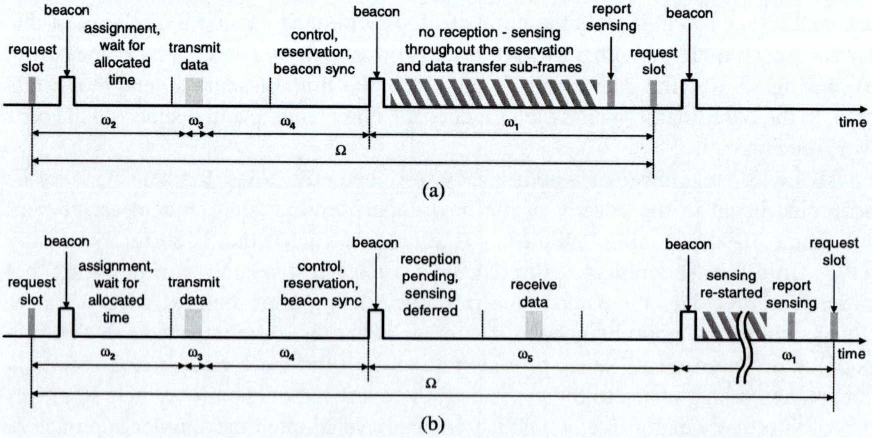

Fig. 6.4 Timing of node operation and distribution of arrivals during a service cycle. **(a)** Case without the preemption of sensing process; **(b)** Case with the preemption of sensing process

We assume that time is slotted and the total superframe length is s_f basic slots, while the packet size is constant and equal to k_d basic slots. Every data packet will be acknowledged immediately after the transmission, and the acknowledgment packet takes exactly one basic slot. Let the probability generating function (PGF) for the packet size with fixed size acknowledgment be $b(z) = z^{k_d+1}$ with a mean value of $\overline{b} = k_d + 1$. The Laplace–Stieltjes transform (LST) of the packet time (including the acknowledgment) is obtained by replacing the variable z with e^{-s}, i.e., $b^*(s) = \mathrm{e}^{-s(k_d+1)}$. Packets arrive to a node according to a Poisson process with the arrival rate of λ, so that the offered load to the node transmission interface is $\rho = \lambda\overline{Y}$. Nodes are assumed to have buffers of infinite capacity.

Vacation due to sensing activity – Each node has to perform k_s sensing events within the next superframe or several of them (if the node buffer is empty when sensing starts). From the aspect of queuing theory, this is a system with multiple vacations where the basic vacation duration is equal to one superframe. The PGF for the basic time spent in sensing is $V(z) = z^{s_f}$, with a mean value of $\overline{V} = s_f$, while the LST of a single vacation (sensing) period is $V^*(s) = \mathrm{e}^{-s_f s}$.

According to [12], the number of packet arrivals to the node during a single vacation period has the PGF of $\omega_1(z) = V^*(\lambda - \lambda z) = \sum_{k=0}^{\infty} f_k z^k$, and the probability of zero packet arrivals is $f_0 = V^*(\lambda)$.

Vacation due to waiting for round robin service – A node has to wait until all nodes with addresses lower than its own that have packets have been served. Assuming the CPAN has M nodes, the LST for the cycle time is $C^*(s) = ((1-\rho) + \rho S^*(s))^M$, where $S^*(s)$ is defined below. Then, the mean duration of the CPAN cycle is $\overline{C} = -C^{*'}(0) = -\left.\frac{dC^*(s)}{ds}\right|_{s=0} = \rho M\overline{S}$.

In terms of renewal theory [13], the time between the node request and (random) beginning of the service within the cycle is denoted as 'elapsed' or backward recurrence time in the discrete-time renewal process, where the distribution of the renewal interval is given by the cycle time C. Namely, the target node observes a cycle which starts with the node being served immediately after the beacon in the frame following its request for the bandwidth. Backward recurrence time will be denoted as $C_{\_}$ and its LST is $C_{\_}^*(s) = \frac{1-C^*(s)}{s\overline{C}}$.

The number of packet arrivals while waiting for the round-robin service has the PGF of $\omega_2(z) = C_{\_}^*(\lambda - \lambda z)$.

Duration of service period – Since only one packet can be transmitted in each service cycle, the PGF for the service time is $S(z) = b(z)$. The LST of the duration of node service period is $S^*(s) = b(e^{-s}) = b^*(s)$, and its mean value is $\overline{S} = \overline{b}$. Finally, the number of packet arrivals to the node during the transmission time has the PGF of $\omega_3(z) = S^*(\lambda - \lambda z)$.

Vacation due to synchronization – After packet transmission, the node needs to wait for the next control sub-frame in order to report sensing results to the coordinator. This waiting time presents 'residual' or forward recurrence time in the renewal period presented by the superframe. Its LST has the form $R_{\_}^*(s) = \left(1 - e^{-s_f s}\right)/(s_f s)$, and the number of packet arrivals to the node buffer during that time has the PGF of $\omega_4(z) = R_{\_}^*(\lambda - \lambda z)$.

Effect of packet reception – To model the impact of packet reception, we need the probability distribution of the time interval between transmission and reception by the given node. We will look at the piconet cycle started by the transmission by the target node $i \in 1 \ldots M$, where M is the number of nodes in the piconet. Reception by the target node will be triggered by some other node j, where j is between $i+1$ and $(i+M-1) \bmod M$ and we assume that the node address is located uniformly in that range.

Given that packet size is k_d slots and one slot is needed for acknowledgment, the PGF for the number of slots between transmission and reception by the node i is

$$\Theta(z) = \sum_{k=1}^{M-1} \frac{((1-\rho)+\rho S(z))^k}{M-1} = \sum_{i=1}^{(M-1)(k_d+1)} \theta_i z^i \tag{6.1}$$

and the probability that the node will transmit and receive in the same superframe is $P_\theta = P(\Theta < s_f - \Delta) = \sum_{i=1}^{s_f-\Delta} \theta_i$, where Δ represents the portion of the superframe dedicated to the reporting of sensing results and bandwidth reservation. The PGF for the number of packet arrivals during the superframe in which the node is engaged in reception is $\omega_5(z) = \mathrm{e}^{-s_f(\lambda-\lambda z)}$.

6.4 Packet Service Cycle and Access Delay

A packet service cycle begins when the node applies for bandwidth and ends when the node returns from sensing associated with the packet which was transmitted in this cycle. Each packet needs a total service time with a LST of $Y^*(s) = S^*(s)C^*_-(s)R^*_-(s)\left(P_\theta + (1-P_\theta)e^{-s_f s}\right)V^*(s)$, and an average value of $\overline{Y} = \overline{S} + \overline{C_-} + \overline{R_-} + (1-P_\theta)s_f + \overline{V}$. The offered load then becomes $\rho = \lambda\overline{Y}$, instead of $\rho' = \lambda\overline{S}$ as is the case for exhaustive service. As both cycle time and probability of reception in the same superframe are functions of offered load, the equation for $\overline{Y}$ can actually be solved for ρ. The PGF for the number of packet arrivals during the packet service cycle is $\Omega(z) = \omega_1(z)\omega_2(z)\omega_3(z)\omega_4(z)\left(P_\theta + (1-P_\theta)\omega_5(z)\right)$.

Let us now consider the node packet queue length at the moments of end of packet service cycle and at the end of the node's vacation. The probability that there are k packets in the queue after packet's service cycle is denoted with π_k, and the PGF for the queue length after a packet service cycle is denoted with $\Pi(z) = \sum_{k=0}^{\infty} \pi_k z^k$. Probability that there are k packets in device queue at the end of a single vacation is denoted with q_k, while the PGF for the queue length after the sensing period is denoted with $Q(z) = \sum_{k=0}^{\infty} q_k z^k$. Then, the state transition equations for these two kinds of Markov points are

$$\begin{aligned} q_k &= (q_0 + \pi_0) f_k & k \geq 0 \\ \pi_k &= \sum_{j=1}^{k+1} (q_j + \pi_j) a_{k-j+1} & k \geq 0 \end{aligned} \tag{6.2}$$

subject to condition $\sum_{k=0}^{\infty} q_k + \sum_{k=0}^{\infty} \pi_k = 1$. The PGFs $\Pi(z)$ and $Q(z)$ then become

$$\Pi(z) = \frac{1 - \lambda\overline{Y}}{1 - \lambda\overline{Y} + \lambda\overline{V}} \cdot \frac{Y^*(\lambda - \lambda z)(1 - V^*(\lambda - \lambda z))}{Y^*(\lambda - \lambda z) - z}$$
$$Q(z) = \frac{1 - \lambda\overline{Y}}{1 - \lambda\overline{Y} + \lambda\overline{V}} V^*(\lambda - \lambda z) \tag{6.3}$$

Knowing $Q(z)$, we can find $q_0 = \frac{f_0(1-\rho)}{1-\rho+\lambda\overline{V}}$. Moreover, we are able to find the probability distribution of the entire node inactive period (which occurs when the node finds an empty buffer after sensing in the current superframe) as $I(z) = V(z)(1 - q_0)\sum_{k=0}^{\infty}(V(z)q_0)^k$, with an average value of $\overline{I} = \overline{V}/(1 - q_0)$.

The PGF for the number of packets in the device queue immediately after a packet departure is

$$\Pi_d(z) = \frac{\Pi(z) - \pi_0 + Q(z) - q_0}{(1 - \pi_0 - q_0)z} \omega_2(z)\omega_3(z) \tag{6.4}$$

and the average number of packets left after a packet departure is $\overline{L} = \Pi_d'(1) = \overline{\omega_2} + \overline{\omega_3} - \frac{1}{2} + \frac{\omega_1^{(2)}}{2\overline{\alpha}} + \frac{\Omega^{(2)}}{2(1-\rho)} - \frac{\rho}{2(1-\rho)}$, where $\omega_1^{(2)}$ denotes the second moment of the number of packet arrivals during the vacation time and $\Omega^{(2)}$ denotes the second moment of the number of packet arrivals during the packet service cycle.

Probability distribution of the waiting time in the node buffer can be found from the observation that, under the FIFO serving discipline, the number of packets left after a packet departure is equal to the number of packets which have arrived to the queue while the target packet was in the system. Therefore, $\Pi_d(z) = T_a^*(\lambda - \lambda z) = W^*(\lambda - \lambda z)Y^*(\lambda - \lambda z)$.

6.5 Model of the Sensing Process

Upon completing their packet transmissions, ordinary nodes revert to sensing of k_s channels out of N working channels in the RF band. Given the results of the previous section, let us denote the probability that a node is available for sensing as $P_{\text{sec}} = k_s\overline{V}/\left(\overline{Y} - \overline{V} + \overline{I}\right)$. Then, the number of secondary nodes X_{sec} available for sensing is a random variable; under constant packet arrival rate (per node) of λ and piconet size M, this number has a binomial distribution with the PGF of

$$X_{\text{sec}}(y) = \sum_{j=0}^{M} \binom{M}{j} P_{\text{sec}}^j (1 - P_{\text{sec}})^{M-j} y^j \tag{6.5}$$

with the average value of $\overline{X_{\text{sec}}} = P_{\text{sec}} M$. Note that we use variable y in the PGF to carry mass probabilities related to the number of sensors active at any given time.

Since sensing is performed in discrete intervals, and the number of nodes available for sensing is typically lower than the number of channels, $X_{\text{sec}} < N$, the information available to the coordinator is only partially correct at any given time. The magnitude of the error (i.e., the mean number of channels with incorrect information) and the delay in detecting changes in channel status will determine the success rate of CPAN transmissions and, ultimately, its QoS.

Without loss of generality, we may also assume that each of the N working channels is used by a distinct primary user or source. For simplicity, we will assume that active and idle times on each channel follow the probability distributions with cumulative density functions $T_{a,r}(x)$ and $T_{i,r}(x)$, and mean values of $\overline{T_{a,r}}$ and $\overline{T_{i,r}}$, respectively. The mean value of real cycle time will be denoted as $\overline{T_{\text{cyc},r}} = \overline{T_{a,r}} + \overline{T_{i,r}}$. For clarity, we will denote relevant variables in the active and idle (free) periods of the channel state, with subscripts a and i, respectively, while the subscripts r and o will denote real and observed values of respective network parameters.

Sensing is performed in a frequency hopping manner with the sensing period regulated by the occupancy of the node's data queue. As long as the node buffer is empty, the period between consecutive sensing actions is equal to the superframe duration s_f. As explained above, if there is a packet in the node buffer at any time during sensing, the node will accelerate sensing to one sensing per basic slot, so that the remaining sensing activity can be completed within the current superframe. Therefore, the sensing period is a random variable with the PGF of

$$T_s(z) = (1-\pi_0)z + \pi_0 \left((1-q_0) \sum_{j=1}^{k_s-1} q_0^{j-1} \left(\frac{j}{k_s} z^{s_f} + \frac{k_s - j}{k_s} z \right) + q_0^{k_s} z^{s_f} \right) \quad (6.6)$$

where the variable z carries the mass probabilities of time periods between successive sensing events.

Assuming that $T_s \ll \overline{T_{a,r}}, \overline{T_{i,r}}$, the probability distributions of active and inactive times of primary users may be considered discrete, and their state changes can occur at the boundaries of the sensing period T_s. Let us denote the PGFs for active and idle times of primary user activity as $T_{a,r}(z) = \sum_{k=0}^{\infty} p_a(k) z^k$ and $T_{i,r}(z) = \sum_{k=0}^{\infty} p_i(k) z^k$, where the variable z corresponds to the sensing period T_s. For example, if $T_{a,r}$ and $T_{i,r}$ are geometrically distributed with parameters α and β respectively, then $p_a(k) = \alpha(1-\alpha)^{k-1}$ and $p_i(k) = \beta(1-\beta)^{k-1}$.

At the beginning of the superframe, CPAN coordinator announces that X_{sec} nodes will sense primary channels, some of which are marked as idle in the channel map while others are marked as active. Due to random selection of channels to sense and random allocation of these channels to sensing nodes, a given sensing node need not be tied to sensing only idle or only active channels. Since the average interval between successive sensing events depends on the ratio of the number of sensing nodes and the number of channels, the sensing intervals for idle and active channels will be equal, as will be the delays in detecting the beginning and end of channel activity.

Let the mean observed durations of active and inactive period be $\overline{T_{i,o}}$ and $\overline{T_{a,o}}$, respectively. Then, the probability that a channel is idle (as observed via the channel map) is $p_{i,o} = \frac{\overline{T_{i,o}}}{\overline{T_{a,o}+T_{i,o}}}$, while the probability that a channel is active is $p_{a,o} = 1 - p_{i,o}$. The mean observed numbers of idle and active channels are $\overline{N_{i,o}} = p_{i,o}N$ and $\overline{N_{a,o}} = p_{a,o}N = N - \overline{N_{i,o}}$, respectively. On the other hand, the probability that a channel is actually idle is $p_{i,r} = \frac{\overline{T_{i,r}}}{\overline{T_{a,r}+T_{i,r}}} \neq p_{i,o}$, and by the same token $p_{a,r} \neq p_{a,o}$.

The choice of channels to be sensed is made randomly, but no channel is sensed by more than one sensing node as long as the number of sensors is smaller than the number of channels.

Then, the PGF for the single-channel sensing rate becomes

$$P_w(y) = \sum_{j=0}^{M} \binom{X}{j} P_{\text{sec}}^{j}(1 - P_{\text{sec}})^{M-j} y^{\frac{j}{N}} \tag{6.7}$$

when $X_{\text{sec}} < N$, and $P_w(y) = y$, otherwise. This polynomial does not qualify as a PGF, since the exponents of y are not integers. However, it satisfies the condition that $P_w(1) = 1$ and its mean value is finite, $\overline{P_w} = P_w'(1)$, and therefore it does represent a probability distribution.

The time between two consecutive sensing events on the same channel is a random variable which depends on the number of sensors and on the sensing period of a particular sensor. It has geometric distribution with respect to the sensing period and binomial distribution with respect to the number of sensors available for sensing, i.e.,

$$\begin{aligned} H(y,z) &= \sum_{j=0}^{M} \frac{T_s(z)j}{(1 \quad T_s(z)(1 \quad \frac{j}{N}))} y^{\frac{j}{N}} \cdot \binom{M}{j} P_{\text{sec}}^{j}(1 - P_{\text{sec}})^{M-j}, && X_{\text{sec}} < N \\ H(y,z) &= yT_s(z) && X_{\text{sec}} \geq N \end{aligned} \tag{6.8}$$

Using the results of renewal theory [13] it is possible to calculate probability distribution between the change of channel state and next sensing event. The sensing process for a single channel may be considered as a discrete-time renewal process where sensing events correspond to renewal points, while the renewal time corresponds to the period between two consecutive sensing events. In terms of renewal theory, the time between the channel state change and the next sensing event is denoted as the residual life(time) or forward recurrence time; Then, the probability generating function for the residual sensing time can be derived as

$$R(y,z) = \sum_{j=0}^{M} \sum_{k=0}^{\infty} R_{j,k} z^k y^{\frac{j}{N}} = \frac{1 - H(y,z)}{\overline{h}(1-z)} \tag{6.9}$$

and the mean value of the residual sensing time is

$$\overline{R} = \frac{\lim\limits_{z\to 1}\lim\limits_{y\to 1} \dfrac{\partial^2 H(y,z)}{\partial z^2}}{2\lim\limits_{z\to 1}\lim\limits_{y\to 1} \dfrac{\partial H(y,z)}{\partial z}} \tag{6.10}$$

From the above system of equations we can find the values of $p_{i,o}$, $p_{i,a}$, $P_{w,i}$, $P_{w,a}$, and $\overline{R}$. The average delays in detecting inactivity and activity of the channel are $(1 - p_{s,i})\overline{R}$ and $(1 - p_{s,a})\overline{R}$, respectively. Given that the primary source is idle or active with the probabilities $p_{i,r} = \frac{\overline{T_{i,r}}}{\overline{T_{a,r}+T_{i,r}}}$ and $p_{a,r} = 1 - p_{i,r}$, respectively, the number of channels with obsolete information will be $E_a = a_1 p_{a,r} N$, for active channels, and $E_i = b_1(1 - p_{a,r})N$, for idle ones.

We note that skipping an entire active or inactive period will also affect the duration of incorrect information about channel state, which will now differ from the residual sensing time. More details about the impact of this can be found in [8].

6.6 Performance of the Original Protocol

In order to evaluate the performance of the proposed scheme we had to evaluate the performance of transmission scheduling as well as that of channel sensing. Our evaluation environment used the following basic assumptions:

- Time is partitioned into basic time slots of equal duration; to ensure generality, all calculations are done in units of time slots.
- There are 30 primary channels, each with an independent ON-OFF source that has a negative exponential distribution of active and inactive periods. The mean activity period is 1000 unit time slots, with the mean duty cycle of 1/3.
- Each superframe has a duration of 100 unit time slots; the duration of the data/sensing sub-frame is set to 85 time slots.
- The CPAN consists of a variable number of identical nodes and a dedicated coordinator. The coordinator neither generates nor receives any data traffic.
- Each node generates packets at a variable rate according to Poisson distribution. The destinations are uniformly distributed over all other nodes. Packet duration was set to 10 unit time slots, with an optional acknowledgment.
- Each node has a buffer with 20 packets capacity; packets that arrive to a full buffer are simply dropped.
- Each node is allowed to request a single packet transmission slot from the coordinator; the requests are serviced in the round-robin fashion. The allocation announcement includes the number of the channel that the coordinator would like to be sensed.
- Once it has transmitted its packet, the node performs sensing on a certain number of channels. We have assumed that the sensing of a single channel, including the

channel switching overhead, lasts for 8 unit time slots (which is rather conservative). The node can therefore sense up to 10 channels in a single superframe; for simplicity, we assume that sensing is perfect and there is no discrepancy between sensing results obtained by different nodes. (This will be a topic for future research, since the focus of this chapter is the tradeoff between data transmission and sensing.)

We have then solved the system of equations outlined above using Maple 11 by Maplesoft, Inc. [14]. The results are as follows.

To evaluate the performance of the transmission-tax based MAC protocol, we have performed two experiments, first by fixing the number of CPAN nodes to 15 and varying the packet arrival rate between 0.0005 and 0.003 packets per unit time slot per node and second by fixing the packet arrival rate to 0.0015 packets per unit time slot per node and varying the number of CPAN nodes between 15 and 40; in both cases, the other variable parameter was the tax coefficient, i.e., the number of superframes to be spent in sensing for each packet transmitted.

Performance of data transmission in the CPAN, as indicated by mean packet access delay, offered load, and mean blocking probability at the device buffer, is illustrated through diagrams in Fig. 6.5. As could be expected, the CPAN enters saturation at higher packet arrival rates and/or higher values of the tax coefficient, as both lead to more sensing. However, when the nodes spend too much time in sensing, they do not transmit data as often, their input buffers overflow and too many packets are dropped; hence the blocking probability exhibits a sharp increase. As a result, the offered load, increases in an almost linear fashion when the tax coefficient is 1, but begins to flatten out at higher tax rates; for tax coefficient of 5, offered load cannot exceed 0.26 due to extensive sensing. Note that the offered load was calculated with reference to the duration of the data/sense sub-frame, i.e., 85 unit time slots, rather than relative to the entire superframe duration of 100 unit time slots.

The sensing performance, expressed as mean number of nodes reporting sensing results per superframe, mean number of sensing reports received per superframe, mean number of channels for which the information in the channel map is incorrect, and mean delay in detecting the beginning or end of primary user activity, is shown in Fig. 6.6. As can be seen, the onset in saturation is also reflected in distinct flattening of the number of reporting nodes per superframe in Fig. 6.6a, as well as in an increase of sensing error, Fig. 6.6c. Interestingly enough, another measure of sensing accuracy – the detection delay – flattens out in the saturation regime. Note that the number of reports received does include duplicate reports; as mentioned above, we assume that sensing is perfect, and leave the issue of reconciling different sensing results for future work.

The number of sensing nodes and the number of sensing reports increase with both the tax coefficient and the number of CPAN nodes. Sensing accuracy, however, does not improve in the same manner: the sensing error decreases linearly with the number of nodes, but shows a floor when the tax coefficient increases, even though the number of nodes exceeds the number of channels in part of the observed range.

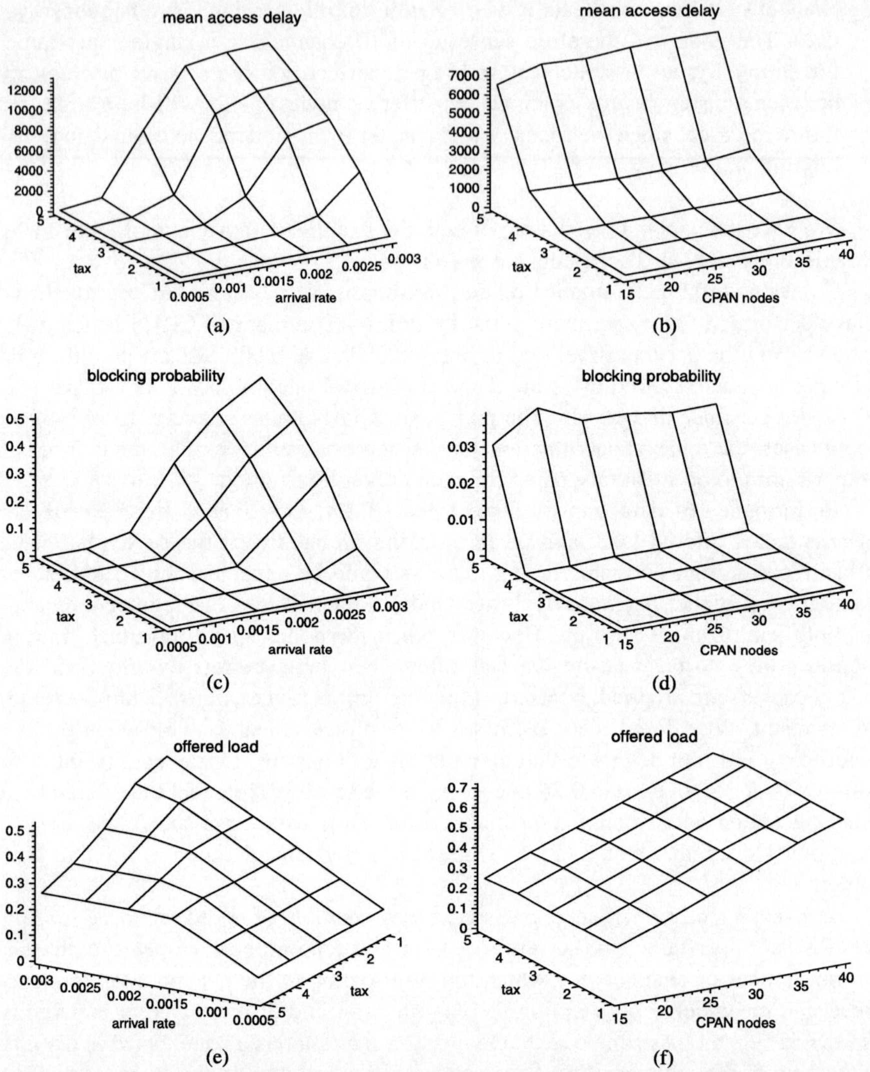

Fig. 6.5 Original transmission tax protocol: data communication performance. On the *left*, variable packet arrival rate and tax coefficient; on the *right*, variable number of CPAN nodes and tax coefficient. (**a**) Mean packet access delay; (**b**) Mean packet access delay; (**c**) Mean blocking probability at the device buffer; (**d**) Mean blocking probability at the device buffer; (**e**) Offered load; (**f**) Offered load (note the different orientation of the plot)

This is the consequence of primary user activity following a negative exponential distribution.

Finally, to evaluate the impact of sensing on the operation of the CPAN, we have also measured the probability that a channel which is considered to be free at the

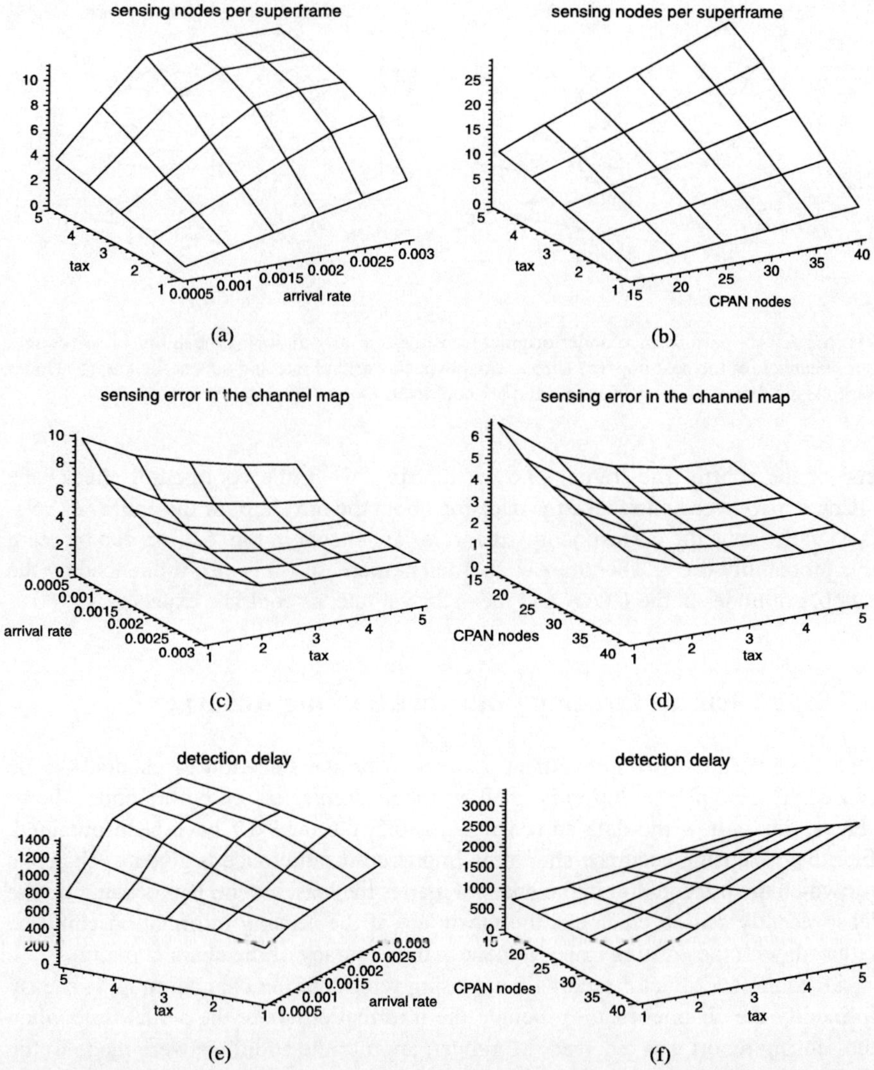

Fig. 6.6 Original transmission tax protocol: sensing performance. On the left, variable packet arrival rate and tax coefficient; on the right, variable number of CPAN nodes and tax coefficient. (**a**) Mean number of reporting nodes per superframe; (**b**) Mean number of reporting nodes per superframe; (**c**) Mean number of incorrect entries in the channel map; (**d**) Mean number of incorrect entries in the channel map; (**e**) Mean detection delay for primary user activity; (**f**) Mean detection delay for primary user activity

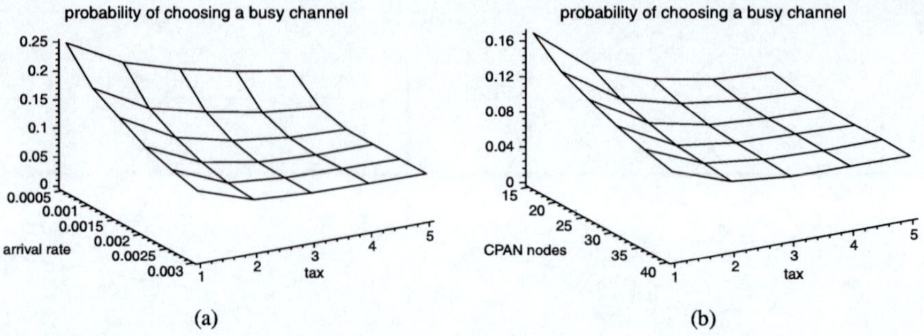

Fig. 6.7 CPAN performance under original transmission tax protocol: probability of selecting a busy channel for the next hop. (**a**) Under variable packet arrival rate and tax coefficient; (**b**) Under variable number of primary channels and tax coefficient

end of the control sub-frame – i.e., at the time when the coordinator checks the channel map in order to make the decision about the next hop for the entire CPAN – is actually busy; the corresponding diagrams are shown in Fig. 6.7. As can be seen, this probability drops when the tax coefficient increases; it is also influenced by the number of nodes in the CPAN and mean arrival rate, as could be expected.

6.7 Can Channel Ordering Improve Sensing Accuracy?

One possible area of improvement appears to be the selection of channels to be sensed. The simplest solution is to allow for random selection by the nodes themselves; this is how the data shown in Figs. 6.5, 6.6, and 6.7 have been obtained. Intuitively, sensing accuracy should be improved if preference is given to channels for which the information in the channel map is the least recent. By focusing on the least recently sensed channels, the mean age of the sensing information could be reduced, with the ensuing improvement in the accuracy of the channel map.

Since the coordinator records the sensing reports in the channel map, it is easy to modify the channel map to include the information about the actual time when the sensing report was received. Moreover, appropriate additions were made to the design of ordinary nodes so that they can also keep track of the sensing time. In one experiment, the coordinator instructed the nodes to begin sensing from the channels with least recent information. Since the packet format includes provisions for only one channel (the coordinator does not know for sure how many channels will the node be able to sense in one superframe), the remaining channels were assigned at random by the node itself. In the second experiment, the nodes themselves performed sensing on the channels for which their local information was least recent. The sensing performance obtained in these two experiments, referred to as centralized and distributed solution, respectively, are shown in Figs. 6.8 and 6.9. As can be seen, local selection (the second set of diagrams) gives somewhat better sensing accuracy, although the difference is not too big.

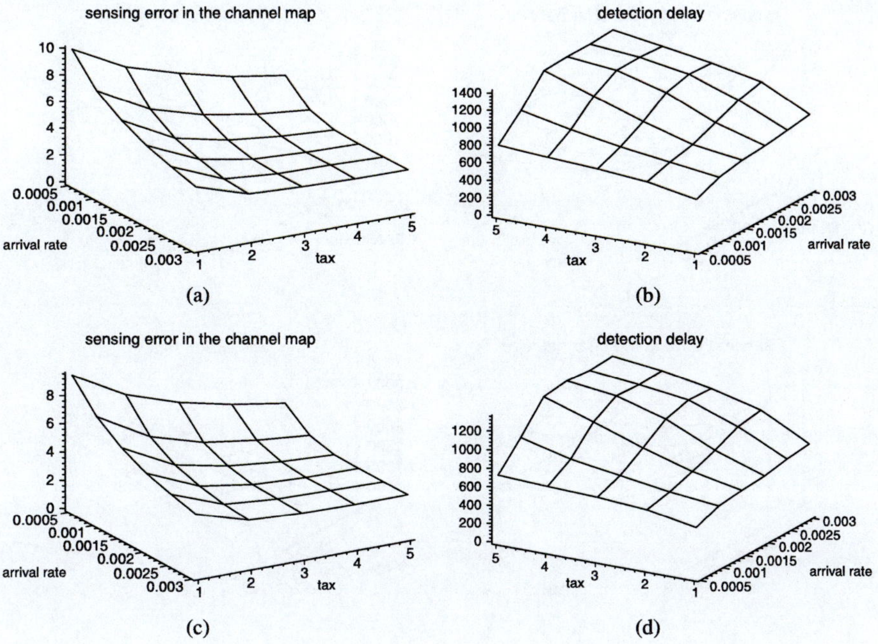

Fig. 6.8 Sensing performance of the protocol modified to account for least recently sensed channels, part I: centralized selection of channels to sense. (**a**) Mean number of incorrect entries in the channel map; (**b**) Mean detection delay for primary user activity; (**c**) Mean number of incorrect entries in the channel map; (**d**) Mean detection delay for primary user activity

Fortunately, the improvement in sensing accuracy does not come at the expense of data transmission performance, shown in Fig. 6.10 for the case of local selection of channels to be sensed; as can be seen, mean access delay and blocking probability are virtually unaffected by this modification.

6.8 Adapting the Transmission Tax

One of the problems with the transmission tax is that the actual amount of sensing information is highly dependent on the data traffic. Namely, when several nodes apply for bandwidth, we may expect abundant sensing information in the next superframes – with a lot of overlap; but when only one node applies, or none at all, there will be a dearth of sensing reports in the next few superframes. This imbalance might be remedied by making the tax coefficient vary in inverse proportion to data traffic intensity: it can be low when there are many transmission requests, but must be high when there are only few of them.

To this end, we have built a simulator of the protocol, using the object-oriented Petri net-based simulation engine Artifex [15], and included adaptive calculation of the tax coefficient using the algorithm below. Namely, in each superframe, the

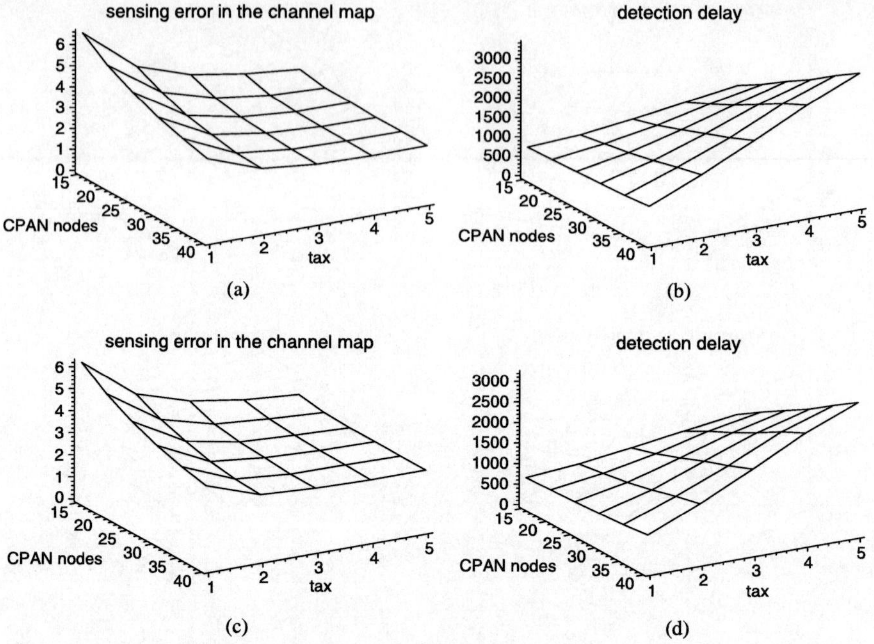

Fig. 6.9 Sensing performance of the protocol modified to account for least recently sensed channels, part II: local selection of channels to sense. **(a)** Mean number of incorrect entries in the channel map; **(b)** Mean detection delay for primary user activity; **(c)** Mean number of incorrect entries in the channel map; **(d)** Mean detection delay for primary user activity

coordinator updates an exponential moving average (EWMA) of the number of distinct channels sensed per superframe, denoted with S; this number is compared to the total number of channels N and the new value of the tax coefficient is calculated on the basis of this comparison. The updated value of the tax coefficient is then broadcast to all the nodes in the beacon frame; nodes that have to perform sensing will use this value throughout their sensing duty, even though it may go up or down in subsequent superframes.

Algorithm 1: Adaptive calculation of the tax coefficient.

Input: number of sensing reports in the last superframe s;
number of incoming transmission requests r;
maximum number of channels that can be sensed in a superframe σ;
smoothing coefficient α
Result: tax coefficient t

calculate $S(i+1) = \alpha * S(i) + (1-\alpha) * s$;
calculate raw tax coefficient as $\tau = (N/r) * \sigma$;
if $S(i+1) > M/2$ **then** $t = \tau$;
else if $S(i+1) > M/3$ **then** $t = 1.0 + \tau$;
else if $S(i+1) > M/4$ **then** $t = 2.0 + \tau$;
else $t = 3.0 + \tau$;

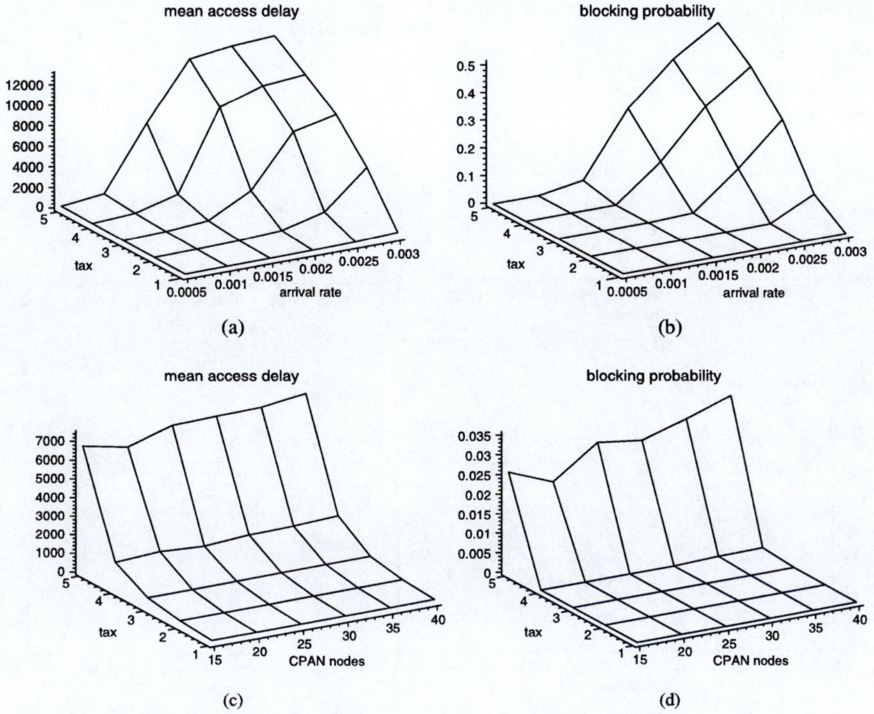

Fig. 6.10 Protocol modified to include local selection of the channel to sense: data communication performance. *Top row*: variable packet arrival rate and tax coefficient; *bottom row*: variable number of CPAN nodes and tax coefficient. **(a)** Mean packet access delay; **(b)** Mean blocking probability at the device buffer; **(c)** Mean packet access delay; **(d)** Mean blocking probability at the device buffer

The results obtained with the adaptive algorithm are shown in Figs. 6.11 and 6.12, for data transmission and sensing performance, respectively. As can be seen, the improvement in data transmission performance is drastic: access delays are low and offered load increases almost linearly with the arrival rate, reaching the value of approx. 0.7 when the CPAN has 40 nodes. At the same time, the CPAN does not come close to saturation in the observed range of arrival rates – blocking probability does not exceed 10^{-4}, which is why it is not shown. Sensing performance is also improved, esp. the detection delay which stays below 900 unit time slots throughout the observed range of independent variables. In all the diagrams, solid lines correspond to the case without least-recently-sensed modification, while diamonds and asterisks correspond to the centralized and local least-recently-sensed channel selection, respectively; however, the differences are minuscule, even though the sensing error is lower under local selection of least-recently-sensed channels.

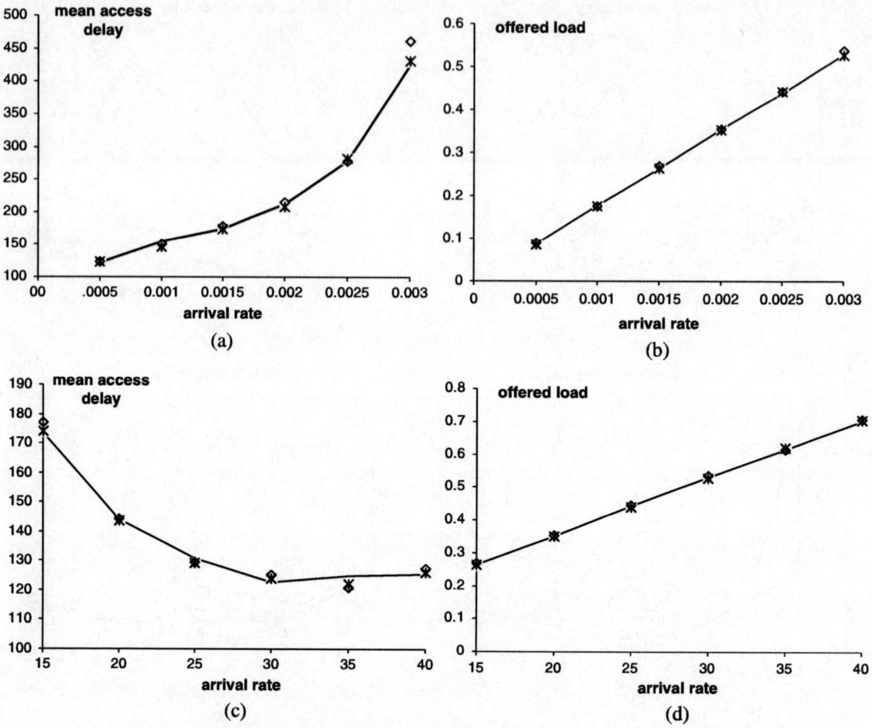

Fig. 6.11 Adaptive protocol: data transmission performance. (**a**) Mean access delay vs. packet arrival rate; (**b**) Offered load vs. packet arrival rate; (**c**) Mean access delay vs. number of CPAN nodes; (**d**) Offered load vs. number of CPAN nodes

6.9 Conclusion

We have presented one possible scheme of interaction between transmission and sensing in cognitive personal area networks and investigated its performance through analytical modeling and simulation. The scheme conforms to the 1-limited round robin service policy, where nodes can transmit only single packet they've had at the moment of applying for the bandwidth using the first-come, first-served policy. In addition, nodes need to perform sensing duty after each transmitted packet in order to control the sensing error. Reception is not penalized because it is associated with transmission, and because it may have to be performed at the expense of extending the sensing cycle.

We have modeled the effect of packet reception of the node which might preempt, but not reduce sensing activity. In this manner, total sensing time is not affected; instead, any reception actually contributes to the packet delay. The 1-limited policy is simple to implement and fair to all nodes. However, it suffers from the overhead associated with each transmitted packet, which limits the stable operation region.

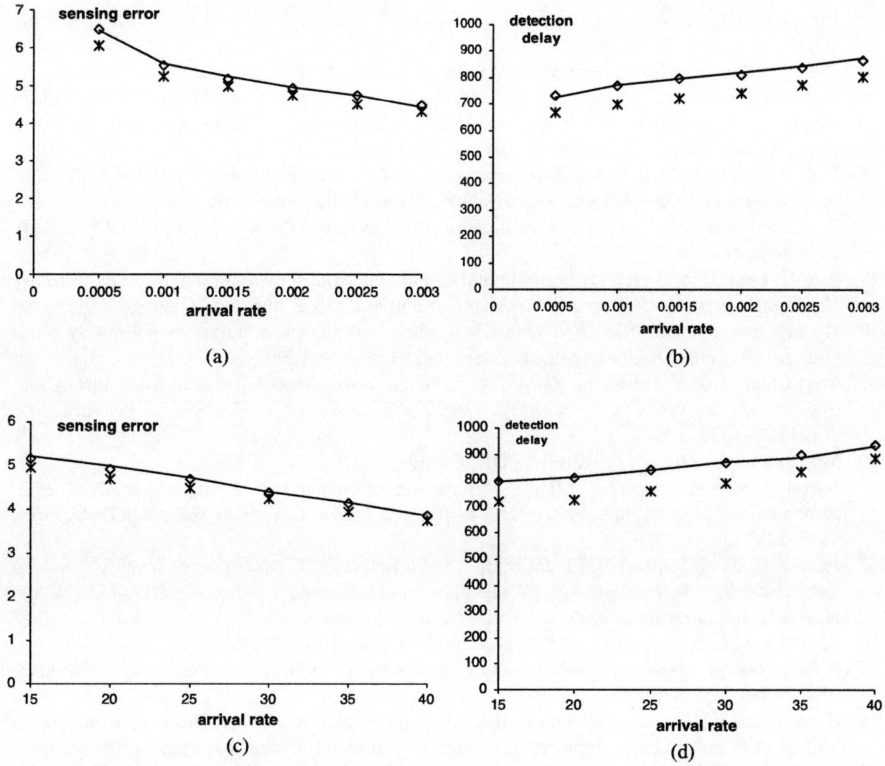

Fig. 6.12 Adaptive protocol: sensing performance. (**a**) Sensing error vs. packet arrival rate; (**b**) Detection delay vs. packet arrival rate; (**c**) Sensing error vs. number of CPAN nodes; (**d**) Detection delay vs. number of CPAN nodes

We have also integrated the sensing policy which controls the total sensing error with the packet transmission policy. Control policy limits the total error on all channels and calculates the necessary number of sensing events per each transmitted packet. Our results indicate that future refinement of the integrated sensing-for-transmission policy is needed in order to separately control the sensing error on active and idle channels.

Finally, we have shown that the performance of the MAC with respect to both data transmission and sensing is vastly improved if the tax coefficient is dynamically adapted to traffic variations, and that slight improvements in sensing accuracy may be obtained when the selection of channels to be sensed is performed locally by each node.

References

1. Ian F. Akyildiz, Won-Yeol Lee, Mehmet C. Vuran, and Shantidev Mohanty. NeXt generation/dynamic spectrum access/cognitive radio wireless networks: A survey. *Computer Networks*, 50:2127–2159, 2006.

2. Bluetooth SIG. *Core Specification of the Bluetooth System.* Version 2.0 + EDR, November 2004.
3. DARPA. The XG vision. Request for comments, January 2004.
4. Stefan Geirhofer, Lang Tong, and Brian M. Sadler. Cognitive medium access: A protocol for enhancing coexistence in WLAN bands. In *Proceedings Global Telecommunications Conference GLOBECOM'07*, Washington, DC, November 2007.
5. D. P. Heyman and M. J. Sobel. *Stochastic Models in Operations Research, Volume 1: Stochastic Processes and Operating Characteristics.* McGraw-Hill, New York, 1982.
6. IEEE. Wireless MAC and PHY specifications for high rate WPAN. IEEE Std 802.15.3, IEEE, New York, NY, 2003.
7. Paul K. Lee. Joint frequency hopping and adaptive spectrum exploitation. In *IEEE Military Communications Conference MILCOM2001*, volume 1, pages 566–570, October 2001.
8. Hanoch Levy, Moshe Sidi, and Onno J. Boxma. Dominance relations in polling systems. *Queueing Systems Theory and Applications*, 6(2):155–171, 1990.
9. Ying-Chang Liang, Yonghong Zeng, E.C.Y. Peh, and Anh Tuan Hoang. Sensing-throughput tradeoff for cognitive radio networks. *IEEE Transactions on Wireless Communications*, 7(4):1326–1337, 2008.
10. Maplesoft, Inc. *Maple 11.* Waterloo, ON, Canada, 2007.
11. Jelena Mišić and Vojislav B. Mišić. Performance of cooperative sensing at the MAC level: Error minimization through differential sensing. *IEEE Transactions on Vehicular Technology*, 58(5):2457–2470, 2009.
12. Jelena Mišić and Vojislav B. Mišić. Simple and efficient MAC for cognitive wireless personal area networks. In *Proceedings Global Telecommunications Conference GLOBECOM'09*, Honolulu, HI, November 2009.
13. RSoft Design. *Artifex v.4.4.2.* RSoft Design Group, San Jose, CA, 2003.
14. Hideaki Takagi. *Queueing Analysis*, volume 1: Vacation and Priority Systems. North-Holland, Amsterdam, The Netherlands, 1991.
15. Danijela Čabrić, Shridhar Mubaraq Mishra, Daniel Willkomm, Robert Brodersen, and Adam Wolisz. A cognitive radio approach for usage of virtual unlicensed spectrum. In *Proceedings of the 14th IST Mobile Wireless Communications Summit*, Dresden, Germany, June 2005.

Chapter 7
Control Channel Management in Dynamic Spectrum Access-Based Ad Hoc Networks

Tao Chen, Honggang Zhang, and Zhifeng Zhao

Abstract In this chapter we introduce the concept of the control channel cloud to solve the control channel problem in dynamic spectrum access (DSA)-based ad hoc networks. A DSA-based ad hoc network is an infrastructure-less wireless network based on DSA and featured by self-organization, self-configuration, and self-healing. One of the challenges in such a network is the common control channel problem, which is caused by the opportunistic spectrum sharing nature of secondary users (SU) in the network. Without a common control channel, it is a challenge to coordinate the behaviors of SU nodes in a DSA-based ad hoc network. The control channel cloud approach, which relies only on the local information exchange to function, aligns the control channel of SU nodes to the same channel in a distributed way if a common control channel exists. It provides a simple but scalable way to synchronize the control channel in a dynamically changed radio environment. The convergence of the proposed approach is proved. The performance of proposed algorithms is studied by simulation.

7.1 Introduction

Powered by *cognitive radio* (CR) [10], which is capable of sharing spectrum resource flexibly and efficiently, dynamic spectrum access (DSA) [17] is regarded as a promising remedy to solve the emerging spectrum scarcity problem. The paradigm shift of the radio spectrum from the conventional command and control allocation to DSA will have fundamental impact on the whole wireless ecosystem. To enable DSA, the adaptation has to be introduced into most layers of the network protocol stack in a system [1], leading to significant changes on the system and network design [3, 5, 14].

Designed for different purposes, wireless systems show great diversity. Obviously, when applying DSA, different types of wireless systems have their own requirements on the system design. Roughly dividing wireless systems into infrastructure based and ad hoc systems, we study the DSA-based ad hoc networks in this chapter. A DSA based ad hoc network is an infrastructure-less wireless network

T. Chen (✉)
VTT, Oulu, Finland
e-mail: tao.chen@vtt.fi

F.R. Yu (ed.), *Cognitive Radio Mobile Ad Hoc Networks*,
DOI 10.1007/978-1-4419-6172-3_7,

which uses DSA for spectrum access. Like a conventional wireless ad hoc network, such a network is featured by self-organization, self-configuration, and self-healing. However, it is more flexible on spectrum, energy, and network resource usage, therefore being superior to wireless ad hoc networks on performance and resource efficiency.

At present the study on DSA-based ad hoc networks is still on its early phase. Many open problems are remaining. We identify the control channel problem as one of the main challenges in DSA-based ad hoc networks. Indeed, in a DSA environment where primary users (PU) and secondary users (SU) of spectrum are assumed, because SUs opportunistically share spectrum with PUs, the network cannot rely on a global common control channel for operation, which distinguishes itself from conventional wireless networks.

In this chapter, we analyze the common control channel problem in DSA-based ad hoc networks and propose feasible solutions. We propose a control channel cloud concept for the control channel selection. A *control channel cloud* is formed by a group of connected SUs using a common control channel. The main reason to introduce the control channel cloud concept is to help the control channels chosen by individual SU nodes aggregate to the same control channel as possible in a dynamic changed radio environment. We introduce four basic operations to manage clouds in the network and provide details to merge clouds in different network dynamics. The algorithms based on four basic operations are presented and their correctness is proven. The performance of the proposed approach is studied by simulation.

The chapter is structured as follows: DSA and impacts on DSA-based ad hoc networks are introduced in Section 7.2; the control channel problem in DSA-based ad hoc networks is examined in Section 7.3; the studied problem and the corresponding system model are presented in Section 7.4; the requirements on the control channel in DSA-based ad hoc networks are summarized in Section 7.5; after in Section 7.6 we propose a new concept on the control channel management, i.e., the control channel cloud concept; the cloud approach under the dynamics of the network is described in Section 7.7; the algorithms and correction of the algorithms are provided in Sections 7.8 and 7.9, respectively; in Section 7.10 the advantages of the cloud approach are emphasized; the performance of the proposed approach is studied by simulation in Section 7.11; finally, the conclusion is drawn in Section 7.12.

7.2 Dynamic Spectrum Access and Impacts on Ad Hoc Networks

DSA is a general term to a new set of spectrum access principles and methods as compared to the conventional command and control spectrum access model [17]. It allows the previously exclusively allocated spectrum resource to be reused among heterogenous systems under predefined constraints on time, space, and interference conditions, and thus significantly improves the usage of precious spectrum resource. The enormous success of Wi-Fi access networks built upon the unlicensed

Industrial, Scientific and Medical (ISM) band and Unlicensed National Information Infrastructure (U-NII) band indicates a bright future of DSA.

According to the use right on the spectrum, DSA can be roughly categorized to three models: dynamic exclusive use model, open sharing model, and hierarchical access model. In the dynamic exclusive use model, spectrum bands are dynamically allocated to spectrum users for exclusive use. It can be seen as a flexible extension of the command and control access model, which enables spectrum trading. The open sharing model, as the name suggested, allows spectrum being equally shared among spectrum users. The spectrum in this context is common and no license is required to access this part of spectrum. The hierarchical access model is the most widely used model in the CR research. It allows an SU of a spectrum to opportunistically share the licensed spectrum with PU of that spectrum when the SU only generates tolerant interference to PUs. The definitions of PU and SU can be found in [1]. Underlay or overlay spectrum share is used in this model.

There is certainly no limitation on which DSA model a wireless ad hoc network should be built upon. However, it is more interesting to study wireless ad hoc networks under the hierarchical access model. As we can see, networks under the dynamic exclusive use model have no big difference from networks under the command and control access model. Moreover, the open sharing model can be regarded as a special case of hierarchical access model without PUs. Therefore in this chapter we assume the use of the hierarchical access model.

7.2.1 DSA-Based Ad Hoc Networks

A wireless ad hoc network by a general definition is a self-organized wireless network without the support of infrastructure. It is normally a multi-hop network where data are routed from the source to the destination through multiple intermediate nodes by distributed algorithms. Wireless ad hoc networks have been studied intensively in the last two decades and the focus is on the routing problem in a single channel radio environment [15]. As shown by Gupta and Kumar in their seminal paper [9], the capacity of a wireless ad hoc network is limited by the number of nodes in the network. This conclusion is drawn based on the assumption that all nodes in the network share limited amount of spectrum. Ad hoc networks based on DSA have potential to increase the network capacity as DSA allows more spectrum to be used by networks. But introducing DSA into wireless networks is more than the capacity improvement. DSA provides a more flexible way for wireless networks to use the radio resource and therefore enabling new applications and use scenarios. DSA-based wireless ad hoc networks are able to share spectrum with infrastructure-based wireless networks. Even better, they can act as bridges between heterogenous networks and help the penetration of conventional infrastructure-based services to a wider area.

As aforementioned, in this chapter the hierarchical access model is applied. Under this model, a DSA-based ad hoc network is more like a multi-channel network where the availability of channels depends on the radio environment. The network coordination in multi-channel systems has been well investigated [4, 12, 13]. Proposals for conventional multi-channel wireless systems usually assume channels are available all the time. This assumption greatly simplifies efforts for coordination. However, it may not always hold in DSA-based ad hoc networks. DSA-based ad hoc networks use the radio resource in an opportunistic nature. They have the following common features and associated challenges:

- Distributed spectrum sensing is required in those networks in order to support DSA. Due to the ad hoc nature, each node should be able to sense the radio environment and determine its available spectrum resource independently. The basic requirement is that the spectrum sensing should be accurate enough to avoid unnecessary interference to PUs while minimizing false alarms on the spectrum hole detection.
- In a DSA-based ad hoc network, a common control channel may not be always available in the network. Network coordination relies on the local share channels of neighboring nodes. This imposes a prominent challenge on the network management.
- Connectivity in a DSA-based ad hoc network is not only influenced by mobility and power control of SU nodes but also by the activities of PUs. Spectrum management becomes a necessary function in a DSA-based ad hoc network to enable adaptive medium access and routing.
- New hidden terminal problems occur in DSA based ad hoc networks. On the one hand, PUs are a new type of hidden terminals to SUs, which cannot be avoided by using explicit signalings like Request to Send (RTS)/Clear to Send (CTS). On the other hand, SUs should not be the hidden terminals of PUs. This is normally solved by spectrum sensing.

7.3 Control Channel Problems in DSA-Based Ad Hoc Networks

Due to dynamics introduced by self-coordination activities, the control problem is critical in distributed networks. Until now most of proposed spectrum control protocols designed for the DSA scenario assume the availability of a common control channel [1]. For instance, Jing et al. [11] used common spectrum coordination channel (CSCC) etiquette protocol for coexistence of IEEE 802.11b and 802.16a networks. Moreover, the cognitive pilot channel (CPC) concept was proposed in [7] for the exchange of the spectrum information among nodes.

As aforementioned, a common control channel may not always exist in DSA-based ad hoc networks. Correspondingly, the topology management of DSA-based ad hoc networks is affected by two main factors: the absence of a common control channel in the network and the frequent topology changes according to the presence of PUs and SUs. In a DSA-based ad hoc network, SUs normally rely on local share channels for the network coordination.

However, until now only few proposals are made under the non-common control channel assumption. Zhao et al. observed that though very limited number of global common channels exist in a network, local neighbors may share numerous channels with others [16]. They proposed a distributed grouping scheme to solve the common control channel problem [16]. Bian et al. [2] used the concept of the segment, which is a group of nodes sharing common channels along a routing path, to organize control channels. In [6], this problem was tackled by a cluster-based approach, where the local users sharing common channels form a dynamic one-hop cluster and the spectrum is managed by cluster heads. DaSilva et al. (chapter 19 in [8]) used the sequence-based rendezvous mode in cognitive radio networks to synchronize SUs. The idea is to assign a well-designed channel access sequence to each SU so that two SUs can meet on certain channels following their channel hopping sequences.

In DSA-based ad hoc networks, the control channel problem is related to physical, medium access control (MAC), and network layer. The air interface structure, signal processing, and spectrum sensing ability in the physical layer determine the reach range and coexisting method of control signals. The MAC layer relies on those control signals to fulfill its channel access functions. The network layer, which builds its routing functions on top of the date delivery service of the link layer, uses network layer control messages to establish and maintain routing. It is easy to see the reliability of control channel and efficiency of control mechanism determine the performance of the network.

We focus in this chapter on the non-common control channel problem in DSA-based ad hoc networks. The key challenges in this problem include the synchronization of control channels, the reliability and scalability of control channel management mechanisms, and the trade-off between performance and control overhead on control channel management. In the following, we will use the control channel cloud concept and cluster-based network formation to deal with the common control channel problem.

7.4 Control Channel Management in DSA-Based Ad Hoc Networks

We give the system model before proceeding to the details of the control channel management method. As aforementioned, we study a DSA-based ad hoc network under a hierarchical spectrum access model. The nodes in studied network are SUs of spectrum, which opportunistically share the spectrum with PUs of the spectrum. The spectrum sharing of SUs can be spectrum overlay and underlay [1]. The principle is that SUs should not violate the tolerant interference level of PUs. We do not study the spectrum sensing here but assume the perfect spectrum sensing in each SU. That assumes synchronization is achieved among SUs to perform the spectrum sensing and avoid false alarm.

SUs coexist with PUs in a given range of spectrum. For SUs, the spectrum is divided into a series of non-overlapped channels. Without losing generality, we assume each channel has equal size. Every channel can be used to transmit control and data messages. A unique channel ID is assigned to each channel, ordered by its frequency range. After spectrum sensing, an SU gets a list of available channels, identified by the channel IDs. A channel is available to a pair of SUs only when the transmitter in the pair does not violate the interference limit of any neighboring PU receiver. When considering the interference at a PU receiver, the sources from multiple SU transmitters must be taken into account. We assume the spectrum sensing function of the SU will handle this problem.

No global common channel is assumed for the control purpose. An SU in the network should be ready to vacate any channel once a PU is detected on the channel. Consequently, SUs must rely on the local common channels for control and coordination.

For the studied DSA-based ad hoc network to operate, the link layer and network layer functions, for instance, neighbor discovery, MAC, and routing, are needed. In this chapter, we only focus on the control channel-specific problems, e.g., control channel selection, mobility, and maintenance problems, at the link layer.

To support the control message exchange, we extend the reference link layer model from [6]. Following the idea in [6], a DSA-based ad hoc network is organized by clusters. As shown in Fig. 7.1, a cluster is a group of nodes managed by a centric node called the cluster head. A cluster has the following properties: the cluster head is selected from the members of the cluster which is one-hop away from all other nodes in the cluster; all nodes in a cluster share at least one common channel; the cluster uses one common channel as the control channel; all control messages of the cluster are transmitted over this channel. We call the control channel of the cluster as the master channel of the cluster. The reasons to organize the DSA-based ad hoc network by clusters have been explained in [6].

As shown in Fig. 7.2, each cluster organizes the channel access time by superframes. In each superframe, there is a period for neighborhood broadcast, in which each node in the cluster is assigned a slot to broadcast the status information about the cluster and the node itself. The status information includes the master channel of the cluster, the available channels of the cluster, the available channels of the node, and the available channels and their associated clusters of neighboring nodes.

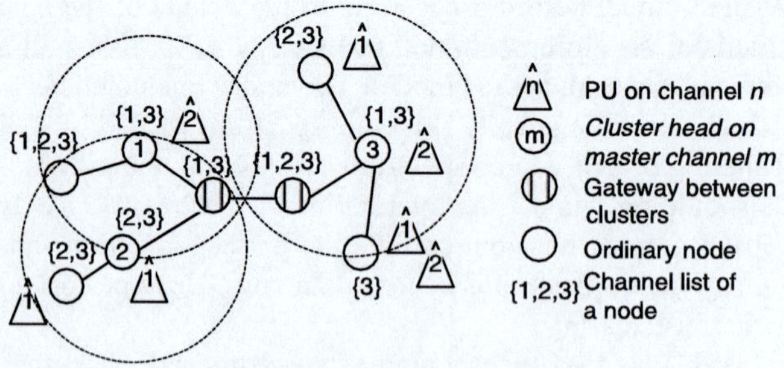

Fig. 7.1 DSA-based ad hoc network formed by clusters

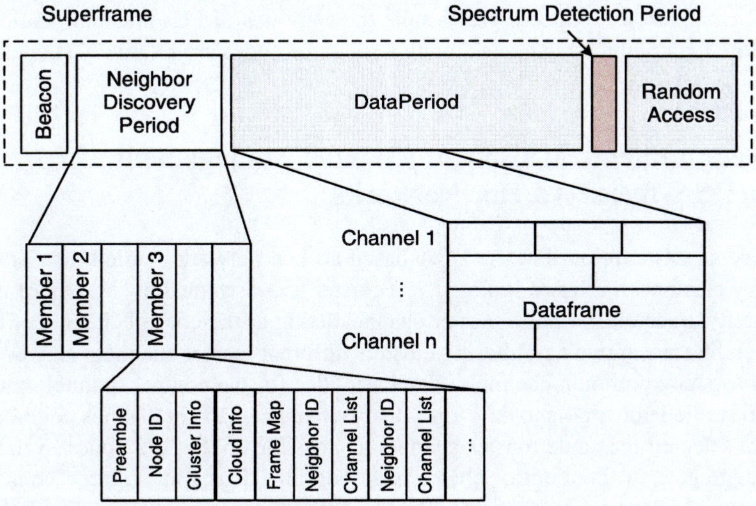

Fig. 7.2 Superframe structure of studied network

Using the status information, nodes are able to establish and maintain clusters. The detail information regarding the clustering in a DSA-based ad hoc network is referred to [6].

Inside a cluster, the collision can be avoided by the coordination of the cluster head. Among clusters, collisions may happen and harm the transmission of control messages. Although it is still an open question, we assume the mechanism like Carrier Sense Multiple Access/Collision Avoidance (CSMA/CA) is applied to reduce the collision among clusters. Therefore a node can listen to different channels and decode neighborhood broadcast messages correctly. Each node will gradually know its surrounding. According to the update of the neighbor information, nodes interact with each other to organize the network in a better way, for instance, using better cluster structure and selecting better control channels.

Depending on the size, purpose, and architecture of the network, there are various criteria to select control channels in multi-channel ad hoc networks [4, 12, 13]. In a DSA-based ad hoc network, it is reasonable to maintain the control channels of a group of nodes at the same channel as possible. The reasons are following: first, if a group of nodes can contact each other, using the same control channel provides a reliable and timely way to exchange control information; second, the common control channel approach simplifies the neighbor discovery process; third, the common control channel approach eases the way of routing and reduce overhead; fourth, the common control channel approach can help to manage the channel variation. If a stable common control channel is used by most of nodes, it is easy for them to deal with channel variation occurring in other channels.

The idea in this chapter is to make the nodes of the whole network select common control channels as possible so that efforts for communication cross different control channels are reduced. We assume the activities of PUs are semi-stationary. Therefore the common channels among a group of nodes are relatively stable.

7.5 Requirements of Control Channel Management in DSA-Based Ad Hoc Networks

We have an assumption that the DSA-based ad hoc network studied here is organized by clusters. As shown in Fig. 7.1, a cluster is constructed by a group of nodes voluntarily. Each cluster has a master channel acting as the control channel of all its members. Between two neighboring clusters different master channels may be used even they share common channels. The basic idea for the control channel management in studied network is to develop a distributive mechanism that has neighboring clusters selected the same master channel as possible so that the nodes in the network aggregate to few control channels as possible. It is a distributed consensus process in which nodes by clusters agree to use the same control channel only based on the local information exchange. In an extreme case if there is a global common channel in a fully connected network, using the mechanism all nodes will tune to the same control channel in a long run.

Several requirements need to be considered in the development of such a mechanism:

- The mechanism should work in a distributive way. It should only rely on the exchange of local information to function.
- The mechanism should be reliable. If a stable common channel is available in an area including multiple connected clusters, those clusters will finally converge to a common control channel.
- The mechanism must adapt to the channel changes. After channel changes due to the presence or absence of PUs, the mechanism is able to re-organize the common control channels so that less common control channels are used.
- The mechanism should be robust to the fluctuation of channels. The fluctuation of channels in short time intervals should not trigger the frequent change of control channels in the whole network.
- The mechanism should avoid the oscillation of control channels. A cluster should not bounce among two or more control channels frequently if those channels are always available.

We use a the concept of *cloud* to manage the control channels of the clusters in the network. We define the cloud as following. In a cluster, in addition to the master channel, all members may share other channels. Let us define the share channels of the neighboring cluster are the share channels of all members in a cluster, and the share channels of two neighboring clusters are the intersected share channels of two clusters. If two clusters share the same master channel and other channels, we put those two clusters into a cloud, identified by their master channel and share channels. We include the cloud identity in the neighborhood broadcast messages.

Since a neighborhood broadcast message defined in [6] has included the master channel, we only need to add a new field to include the share channel list of the cloud. Two clusters are in the same cloud if they are connected, not necessary in neighborhood, and have the same master channel and share channel list in neighborhood broadcast messages. Two connected clusters sharing the same master channel but different share channel lists are not in the same cloud. We describe the details of the cloud-based control channel management in the next section.

7.6 Cloud-Based Control Channel Management

We introduce the cloud concept into the DSA-based ad hoc network for the control channel management. The purpose is to have nodes in the network synchronized on the same control channel as possible in a distributive and self-organizing way. A cloud starts from one cluster and grows as more neighboring clusters joining. If two clouds sharing common channels meet, they can be merged to a single one. By the growing and merging processes, the clouds in the network continue evolving. If the channels in network are stable enough, they finally reach the stable state. It is an optimization process cross the network with the objective to put nodes on few common control channel as possible. In this process, the cloud becomes a method to share common channel information beyond one-hop neighbors and makes it possible at the network wide level to synchronize the control channel. All those are done by simply exchanging the cloud information among neighbors.

The cloud idea can be illustrated by a simple example. When a new cluster is formed, it first broadcasts its master channel and share channels in neighborhood broadcast messages. As it learns neighboring clusters from neighborhood broadcast messages of other clusters, it will know the cloud information of other clusters. There may be several clouds around it. It makes a decision to join another cloud or insist on it own, according to the factors like the choices of neighboring clusters, the connections with neighboring clusters, the need to shift the master channel, and others. If it decides to join another cloud working on a different master channel, the cluster head will configure the cluster to shift the master channel starting from the next superframe. The updated cloud information is broadcasted in the next round of neighborhood broadcast messages. As a result the cluster move from it own cloud to other cloud. It continues monitoring the cloud state by collecting the cloud information from neighboring clusters. Once detecting neighboring clusters change their clouds, the cluster will take action to keep staying in a desired cloud. Moreover, the cluster will update its cloud information if the available channels of the cluster are affected by the activities of PUs. In this case, the cluster creates a new cloud and interact with other clusters to deal with the changes.

As we can see, the key idea in the cloud based management is to merge clouds when possible. There are several basic situations where a mergence may occurs. We describe them in details before developing the whole idea of the cloud-based management.

7.6.1 Basic Cloud Operations

We say two clouds meet when a cluster of one cloud detects that there is a neighboring cluster belonging to another cloud. Two clusters are in different clouds if either the master channel or share channels of the cloud are different. The merging process happens if two clouds meet. There are four basic situations: two clouds are built upon the same master channel but different share channels; two clouds relies on different master channels and at least one master channel in their share channels; two clouds use different master channels and none of the master channels are in their share channels; cloud self-refresh. The fourth situation is a special case of the first situation. We introduce it separately since it is an additional approach for self-healing of clouds. Before describing the basic operations, several rules are defined to make the merging process move in one direction.

- The rule of the master channel. It includes several cases: when there are multiple options on the master channel for a cloud, the channel with the lowest frequency wins the game; if two clouds merge and the master channel of one cloud is in their intersected channels but the other is not, the master channel in the share channels is inherited by the new cloud.
- The rule of keeping previous cloud information. For the basic operations of cloud, it is required that each cluster remembers the previous cloud information of its own and neighboring clusters, which is used in the merging process. This rule sets that the previous cloud information of a cluster will be only kept for a given time period and then replaced by the state of the current cloud. This time period is a system parameter. We call a cloud of which the previous cloud information is equal to the current cloud information as a *persistent cloud.*
- The rule for back-off timer to change cloud. A rule is set that difference back-off timer are used to make changes to clouds. The use of back-off timers can be referred to Section 7.8. The back-off timer to change the master channel is longer than that to only update the share channel list.
- The rule of subset cloud overriding. We said one cloud is a *subset* of the other cloud if two clouds share the same master channel, and the share channels of one cloud are the subset of the share channels of the other cloud. We also define the cloud broadcasting as a cluster receives a neighborhood broadcast message from another cluster. The rule is that if a cloud receives a cloud broadcasting from a cloud which is its subset but is not its previous cloud, the cloud will convert to the new cloud.
- The rule of holding in a cloud. If a cluster joins a new cloud but receiving cloud broadcasting messages from its previous cloud, it will keep staying in the new cloud.
- The rule of overriding persistent cloud. It is that a persistent cloud will convert to a cloud with more share channels than it only when the cloud's previous cloud is a persistent cloud.

7.6.1.1 Operation 1: Same Master Channel But Different Share Channels

This situation can be divided into two cases: two clouds meet as they grow, or a new cloud is created inside a cloud due to the activities of PUs. Since a PU can be active or inactive, the new cloud can have less or more channels than the original cloud.

A cluster can distinguish if a new cloud is created and growing inside another cloud or two clouds meet as they grow based on the records of previous cloud information. If a new cloud is created from an old one, a cluster belonging to the new cloud will keep staying in the new one even it receives cloud broadcasting messages from the old cloud since it knows its previous cloud is the old one. For a cluster in the old cloud, it will join the new cloud as it treats the new cloud as a chance to adapt to channel changes. At the end, the old cloud will evolve to the new one. In a case a cluster finds there are more than one new cloud to join, it will join the cloud with most neighboring clusters on it. Otherwise it will join the first one who sends the cloud broadcasting message. We have to emphasize that the new cloud here is a cloud uses the same master channel but different channel lists from the original cloud, and therefore having no impact on the delivery of control messages.

It also worth noting that a new cloud may have more share channels than the old cloud. This provides an opportunity for a cloud to increase its share channels. As we can get from the common sense, if two clouds are merged, the share channels of the new cloud are normally reduced. At the end, if there is no mechanism to refresh the share channel lists of clouds at all according to the changes of the radio environment, clouds will become stale and finally not function correctly. By allowing a new cloud increases the share channels of an old cloud, we can introduce a cloud refreshing mechanism in which the clusters of a cloud can split from the cloud, create a new cloud with their local share channels, and refresh the old cloud by mergence.

If two clouds grow and finally meet, two neighboring clusters belonging to each cloud will detect this happening. Those two clusters will negotiate and create a new cloud, which uses the intersected share channels of two clouds as its share channels. If one cloud is a subset of another cloud, no new cloud will be created but the one cloud with less share channels will gradually convert the other one. We can use the previous cloud information to guarantee the mergence of two clouds occur correctly. Without lose generality, let us say a new cloud is created among two clouds. The clusters of the new cloud remember that its previous cloud is one of those old clouds. If a cluster in the new cloud receives a cloud broadcasting from its previous cloud, it will ignore this message. On the other hand, if it receives a cloud broadcasting from the other old cloud, it finds that the new cloud is the subset of this old cloud and then does nothing. Therefore it will take no action when receiving a cloud broadcasting from old clouds. The clusters in old clouds, while receiving a cloud broadcasting from the new cloud, will join the new cloud as the new cloud is not their previous clouds. As a result, two clouds will finally be merged to one cloud.

7.6.1.2 Operation 2: Different Master Channels But at Least One Master Channel in Share Channels

In this situation if two clouds meet, a new cloud will be created using the master channel of one cloud. There are two cases in this situation: two clouds have the same share channels or two clouds share some of their share channels. According to the rule of the master channel, in the former the cloud with the master channel of a lower frequency will convert the other cloud. For the latter, a new cloud using the intersected share channels of two clouds is created and grows among two clouds. We need to decide the master channel in the new cloud. It depends on the share channels and master channels of two clouds. If the share channels of one cloud are included in the other cloud, the master channel of the cloud will be inherited by the new cloud. Otherwise, the rule of the master channel is applied.

We can show that in this situation two clouds will finally be merged. If two clouds have equal share channels, the one with the master channel of a lower frequency remains untouched but the clusters in the other one will gradually move the first one. If two clouds have no equal share channels, according to the rule of subset cloud overriding, the clusters in two clouds will join the new cloud. For the clusters in the new cloud, assuming they previously belong to the cloud one, if they receive cloud broadcasting from the cluster one, it will take no action since their previous cloud is the cloud one; if they receive cloud broadcasting from the cluster two, since from their previous cloud information they will find the share channels of the current cloud is a subset of those in the cloud two, they conclude that their current cloud is derived from the cluster two and thus take no action. Following the same deduction, the clusters will take no action if they previously belong to the cluster two. Therefore two clouds will be merged.

7.6.1.3 Operation 3: Different Master Channels and No Master Channel in Share Channels

If no master channel of two clouds is in their share channels, a new master channel has to be selected for the new created cloud. According to the rule of the master channel, the share channel with the lowest frequency becomes the new master channel of the new cloud. Finally the new cloud will merge two old clouds.

It is easy to show that the merging process happens in one direction. For a cluster in any of two clouds, if it finds the new cloud, it will move to it since the new cloud is its subset. For a cluster in the new cloud, if it receives cloud broadcasting from any of two old clouds, it will take no action as its previous cloud information shows it just moves from another cloud.

7.6.1.4 Operation 4: Cloud Self-Refreshing

The aforementioned operations only occur when two external clouds meet. That means a cloud can change itself only when meeting with another cloud. If the clouds of a network are evolved to a stable situation, the evolution of clouds will be stopped even there is a chance to move to a more optimized cloud configuration. That chance happens when PUs release their spectrum so corresponding channels returns to the network. A self-refreshing operation continuously makes the network find new opportunities for better cloud configuration.

This operation works as follows. According to the rule of keeping previous cloud information, after a given time of staying in a cloud, a cluster will replace its previous cloud information with the current cloud information. After a given time when a cluster's previous cloud is its current cloud, the cluster will include all its available channels into the share channel list of its cloud. A new cloud is created if the available channels are more than the share channels of the cloud. According to the rule of overriding persistent cloud, the old cloud will convert to the new cloud. For the new cloud, if it receives cloud broadcasting from the old cloud, it knows it is from its previous cloud and takes no action. If a new channel is available to a cloud, according to this operation, the new cloud will include that new channel. When the new cloud meets other external clouds, the other three operations are applied.

7.7 Dynamics of Network

We show in this section how the operations of cloud are applied in a network under different network conditions. Those conditions include the initiation of the network, the loss and return of channels due to the PU activities, and the topology change due to nodes joining or leaving the network. We also introduce how clouds refresh themselves in this section.

7.7.1 Initiation

A DSA-based ad hoc network is organized in an autonomous way. Every node in the network is an equal entity from the communication viewpoint. A network is formed when messages can be delivered from one node to any other nodes in the network. As the channel availability, topology and nodes of the network undergo dynamic changes during the life time the network, the network is expected to experience frequent network partition and re-union. Therefore it is hard to define the initiation phase of the network since it may happen all the time during the life time of the network. To illustrate how a cloud is formed at the beginning, we show here a special initiation phase of the network where all nodes in the network start joining the network at the same time.

As described in [6], a node will first detect available channels and listen to those channels for available clusters. If it cannot detect any working clusters on its available channels, it will create the cluster of its own by choosing a proper master channel, which can be the channel of the lowest frequency, the channel of the best quality, or a randomly chosen channel. Otherwise, it will join the first cluster that it detects. A new cluster forms its own cloud at the beginning by using its share channels as the share channels of the cloud. The cloud information is broadcasted in the neighborhood discovery process. As the new cluster continuously monitors the radio environment, it will gradually know neighboring clusters on different channels. The cluster optimization algorithm, for instance, minimum dominating set (MDS) algorithm in [6], is run to optimize the cluster structure of the network. Meanwhile, the cloud operations 1–3 are performed among clusters to align nodes on the same control channel if possible. The initiation phase of the network can be described as the initiation and optimization of clusters and clouds to establish optimized cluster structure on few control channel as possible.

7.7.2 Loss and Return of Channels

We assume the loss and return of channels in the network are mainly affected by the activities of PUs. Nodes in the area have to vacate a channel when it is occupied by PUs in that area. On the other hand, the channel can be re-used by SU nodes if nodes sense the channel and make a conclusion that no surrounding PU is in its active state.

If nodes have to vacate a channel, the available channels of their clusters may be reduced. If the master channel of a cluster has to be vacated, the cluster will be dismissed. Nodes of the cluster join the network by forming new clusters. Clouds will be created in new clusters and interact with old clouds using the operation 1–3 to find a new optimized cloud distribution. If the vacated channel is not the master channel but a channel in the share channel list of a cloud, the affected clusters will update the share channel list of their clouds, and the operation 1 is applied to update other clusters in their clouds.

The return of a channel to a node will not immediately affect the cloud distribution of the network. However, if the new channel increases the share channel of a cluster, in self-refreshing process, the operation 4 will bring the new channel to the cloud. The network then re-optimizes the cloud distribution using the operation 1–3. The best result is that more nodes synchronize on the same master channel. However the worst case is that the return of a channel will have no impact on the cloud distribution.

7.7.3 Node Joins or Leaves Network

A node follows the procedure described in Section 7.7.1 to join an established network. That is it will join a working cluster or establish the cluster of its own. The impact on the cloud dynamics when a node join the network depends on the contribution of the node to the connectivity of the network and its available channels. If the node does not create new neighboring clusters for its cluster, it will have no impact on the master channel of the cloud, but it may reduce the share channels of the cloud if it changes the share channels of the cluster. The operation 1 will be applied in this case. If the new node connects two previous separated networks, it may have a profound impact on the cloud distribution of the whole network. If two previous separated networks use different master channels but now it is possible to move to a common master channel, the operation 1–3 will be applied to drive the cloud evolution.

In our network, a node can be a cluster head or a normal node in a cluster. The impact on the cloud when a node leaving network relies on the role of the node in the cluster. If a cluster head leaves the network, the cluster will be dismissed and the remaining nodes of the cluster will follow the initiation process to join the network. In this process, the operation 1–3 will be used to re-organize clouds in the local area. If the leaving of the node makes it possible to merge clouds of the network, the operation 4 will initialize this process and then the operation 1–3 will be invoked to complete the process. If a normal node in a cluster leave the network, there is no impact on the cloud at the beginning. But later on the operation 4 will be applied to refresh the cloud of the cluster. If after the self-refreshing, it is possible to merge more clouds in the network, the operation 1–3 will be applied. Note that the leaving of a node may create network partition. The use of the operation 4 will drive the evolution of two separated networks independently.

7.7.4 Refresh Cloud

As a cloud merges with others, it will normally reduce its share channel list. As a result, a cluster may have more share channels than its cloud. The refreshing of a cloud refers to a cluster replacing the share channels of its cloud with its own share channels. It is the way a cloud can increase its share channels when more channels are available to the cloud. For instance, the share channels of a cloud may increase if PUs release channels to SUs or nodes leave the network. The refreshing of cloud provides a way to make cloud adapt to channel changes.

The refreshing process can be regarded as two steps: first, a new cloud is created by a cluster using the self-refreshing operation; second, the new cloud start merging surrounding clouds by applying the rule of overriding persistent cloud. The introduction of the persistent cloud concept and mechanism makes the refreshing happen only when clouds are stable for enough long. It gives the self-refreshing cloud merging a lower priority than other merging. The reason is straightforward: if other merging happens, it will override the self-refreshing merging.

7.8 Algorithms for Control Channel Management

The whole algorithms for control channel management include the algorithms on the ordinary nodes of a cluster and those on the cluster head. All algorithms are run in a distributed way. The algorithms in an ordinary node of a cluster performs neighboring cluster discovery and cloud broadcasting. The cloud operations are done in the cluster head. We only describe the algorithms on the cluster head in this section. The basic functions of the algorithms on the cluster head consist of three parts: initiation of the cloud, cloud merging, and cloud self-refreshing. The initiation of the cloud in a cluster is rather simple. It starts by building the own cloud of the cluster and then merging with other surrounding clouds. The own cloud of a cluster is built by using the share channels of the cluster as the share channels of the cloud.

In this section, we provide pseudo-codes for cloud merging and cloud self-refreshing, listed in Algorithms 1, 2. The output of those two algorithms is the updated cloud information, which will be broadcasted in the beacon period of the next superframe. The functions for broadcasting and receiving of cloud information

---

Algorithm 1: MergeCloud run on CH_i

Data: Cloud information on CH_i and neighboring CH_j
Result: Updated Cloud information of CH_i
begin
  if $mCh_i = mCh_j$ **and** $SC_i \neq SC_j$ **then**
    if $T_s = 0$ **then**
      if $SC_j \in SC_i$ **then**
        if $SC_i' \neq SC_j$ **then**
          $SC_i' = SC_i$;
          $SC_i = SC_j$;
      else if $SC_i \in SC_j$ **then**
        if $SC_i' = SC_i$ **and** $mCh_i' = mCh_i$ **then**
          $SC_i' = SC_i$;
          $SC_i = SC_j$;
      else
        $SC_i' = SC_i$;
        $SC_i = SC_i \cap SC_j$;
      Reset T_s;
    else
      $T_s = T_s - 1$
  else if $mCh_i \neq mCh_j$ **then**
    if $T_d = 0$ **then**
      if $mCh_i \in SC_j$ **and** $mCh_j \in SC_i$ **then**
        $mCh_i' = mCh_i$;
        $mCh_i = \min(mCh_i, mCh_j)$;
      else if $mCh_j \in SC_i$ **then**
        $mCh_i' = mCh_i$;
        $mCh_i = mCh_j$;
      else if $mCh_i \notin SC_j$ **then**
        $mCh_i' = mCh_i$;
        $mCh_i = \min(SC_i \cap SC_j)$;
      $SC_i' = SC_i$;
      $SC_i = SC_i \cap SC_j$;
      Reset T_d;
    else
      $T_d = T_d - 1$

Algorithm 2: SelfRefresh run on CH_i

```
Data: Cloud information on CH_i
Result: Updated Cloud information of CH_i
begin
    if SC'_i ≠ SC_i or mCh'_i ≠ mCh_i then
        if T_p = 0 then
            SC'_i = SC_i;
            mCh'_i = mCh_i
            Reset T_p;
        else
            T_p = T_p − 1
    else
        if T_r = 0 then
            SC_i = CC_i;
            Reset T_r;
        else
            T_r = T_r − 1
```

Table 7.1 Notation

CH_i	Cluster Head *i*
mCh_i	Master channel of CH_i
mCh'_i	Previous master channel of CH_i
CC_i	Share channel list of cluster CH_i
SC_i	Share channel list of cloud stored at CH_i
SC'_i	Previous share channel list of cloud stored at CH_i
T_s	Back-off timer for cloud merging on same master channel
T_d	Back-off timer for cloud merging on different master channels
T_p	Back-off timer to change to persistent cloud
T_r	Back-off timer for cloud self-refreshing

are not included in the algorithms. The notations used in the algorithms are given in Table 7.1.

Each cluster is identified by its cluster ID i. In the algorithms the host cluster running the algorithms is identified as CH_i, while neighboring clusters are identified as CH_j. Four back-off timers are introduced to postpone the change of a cloud. T_s is used to update the share channels of a cloud but keeping the master channel the same; T_d is used to move a cloud to a different master channel; T_p is used to turn a cloud to a persistent cloud after it becomes a non-persistent cloud; T_r is used to self-refresh a persistent cloud. The time unit for those timers is one superframe period. Note that those timers are set that $T_s < T_d << T_p < T_r$.

7.9 Correctness of Cloud Formation Algorithm

Definition 1 A connected component of a network is a group of network nodes in which any two nodes can be reached by a direct link or links from other nodes in the group and no other nodes of the network can be added in the group.

Proposition 1 *In a radio environment where the availability of channels is fixed, the operation 1–3 finally create a stable cloud distribution.*

Proof First we prove that after the operation 1–3, a cluster will never return to the previous clouds it has stayed. As we can see, those three operations will never increase the share channels of a cloud. If the share channels of the new cloud and the previous cloud is the same, which happens in the operation 2, the new cloud will use a lower frequency master channel. Otherwise the new cloud will reduce its share channels. As a cluster continuously moves between different clouds, the new cloud to which it jumps will either use lower frequency channel or reduce share channels. Since the number of channels is limited, a cluster will finally be fixed at one cloud. Therefore for the whole network, the clouds of all nodes will finally be fixed. It implies the operation 1–3 evolve clouds in one direction.

Lemma 1 *In a radio environment where the availability of channels is fixed, if a connected component has common channels, by the operation 1–3 and without self-refreshing it will finally converge to a single control channel.*

Proof The operation 1–3 guarantee the common channels remaining in the share channel lists of clouds. The operation 2 makes two clouds with the same share channel list using the same control channel. The Proposition 1 makes sure the final cloud distribution stable. As a result, the connected component will converge to a single control channel.

Lemma 2 *If a connected component has common channels, the self-refreshing operation does not affect the connected component to select its common control channel.*

Proof If a connected component has already use a common control channel, since the self-refreshing operation does not change the master channel, the common control channel of the connected component is not affected.

If a connected component has common channels but has not converge to a common control channel, the operation 1–3 will merge clouds. Let us say a new cloud created by the operation 1–3 merges a old cloud. The self-refreshing can be considered in two cases: the old cloud and new cloud start self-refreshing, respectively. For the former, since self-refreshing only increases the share channels of the cloud, the new cloud is still the subset of the refreshed cloud. Consequently, it will not change the result of merging. For the latter, we can show that it does not affect the merging process either. As described in the operation 4, the self-refreshing operation includes a cluster into a new cloud only when the cluster is in a persistent cloud. New clouds created by the operation 1–3 are not persistent clouds. They will immune from self-refreshing as long as they are not persistent clouds. Since the time to become a persistent cloud is much longer than that to convert the cloud of a cluster, the edge of a new cloud will not be affected by its self-refreshing and will continue converting the old cloud.

Since the self-refreshing operation does not affect the operation 1–3, combining with the Lemma 1, the conclusion holds.

Lemma 3 *If common channels exist in a connected component after the PU activities, and at least one of them keeps available enough long, the connected component will finally converge to one common control channel.*

Proof Note that the PU activities refer to a PU appears or disappears in/from the network. We only discuss the case a PU occupies the control channel since other cases have no direct impact on the control channel of connected component.

If a PU shows up and occupies the control channel in a region, new clouds will be formed at that region. Those clouds will interact with the old cloud in the unaffected area. According to the Lemma 1 and 2, a new cloud based on a new common control channel will finally be formed in the connected component. This also holds if there are many PUs in different areas of the connected component occupying common channels.

If PUs release their occupied channels, according to the operation 4, the clusters in the network will self-refresh their clouds. The Proposition 1 states if the availability of channels is stable enough, the cloud distribution will be stable. According to the rule of keeping previous cloud information, those clouds will become persistent clouds. The self-refreshing operation will finally return released channels to the cloud. If there are common channels again in the connected component, according to the Lemma 1 and 2, a new cloud based on a common control channel will eventually be formed.

Lemma 4 *If two connected components become connected and there exists at least one common channel available enough long, the new connected component will finally use a single common control channel.*

Proof If two connected components become connected, they are merged into a single connected components. According to the Lemma 3, the new connected component will finally use a single common control channel.

Theorem 1 *If common channels exist and they are available in the network for time enough long, a common control channel will finally be selected by the network.*

Proof A network is composed by one or several connected components. According to the Lemma 4, when it becomes a single connected component, and meanwhile common channels are available for time enough long, the network will converge to a common control channel.

7.10 Advantages of Cloud Approach

The cloud-based control channel management approach is specifically proposed for DSA-based ad hoc networks. It fulfills the needs of DSA and has the follow advantages to deal with the control channel problem in similar networks.

- There is no need to assign a unique id to each cloud. Instead, a cloud is identified by the master channel and its share channel list. This significantly eases the cloud management. Clouds on the same master channel but with different share channel lists are different clouds. Merging happens if two of such clouds meet.

- Including a share channel list in a cloud makes it possible to align the control channel multiple hops away only relying on local information exchange. The proposed algorithms are therefore scalable.
- The cloud is managed in a loose way based on the cluster. A cluster has its own freedom to create and merge a cloud. It enables robust and fast control channel management and makes it possible for clouds quickly adapting to environment changes.
- Managing the cloud at the cluster level simplifies the control channel management. The cluster is responsible for the channel management in a local group and thus reduces the overhead for cloud interaction at the node level.
- The combination of cloud merging and self-refreshing mechanisms balances the adaptive ability for rapid changes of channels and long-term selection of control channel.

7.11 Simulation Studies

The behaviors and performance of the proposed algorithms are studied by simulation. The simulation environment is settled in a 600 m $\times$ 600 m 2-dimensional playground, in which PUs and SUs are randomly distributed according to Poisson processes. The maximin reach range of a PU is set to 200 m and that of an SU is 150 m. A number of channels are shared by PUs and SUs. A PU randomly chooses a channel to operate. During its active time, the surrounding SUs which may cause intolerable interference to the PU have to vacate that channel. We assume the transmission of cloud broadcasting message is collision free. The neighboring SUs listening on that channel can decode that message successfully. Cloud broadcasting messages are deliver to cluster heads by member nodes in the following beacon period. The algorithms are run on each cluster head to decide its cloud configuration.

The simulation is run sufficient times and the average values of the statistic data are analyzed. In each simulation run, PUs fix their operating channels. SUs initialize clusters and evolve their cluster structure according to the MDS algorithm proposed in [6]. The cloud algorithms are run based on the established clusters. We compare the control channel and cluster distributions without and with the cloud algorithm. In the network without the use of the cloud concept, the cluster head decides its own master channel without referring to the choices of neighboring clusters. The simulation results are presented and analyzed as follows.

The number of connected components on all master channels is a good performance indicator of the algorithm. A connected component on a master channel refers to a group of connected SUs in which all of them share the same master channel and no other SUs on the same master channel are connected to this group. Figure 7.3 shows the connected components on all master channels as a function of SU number. It is the result based on the cloud algorithms under different available

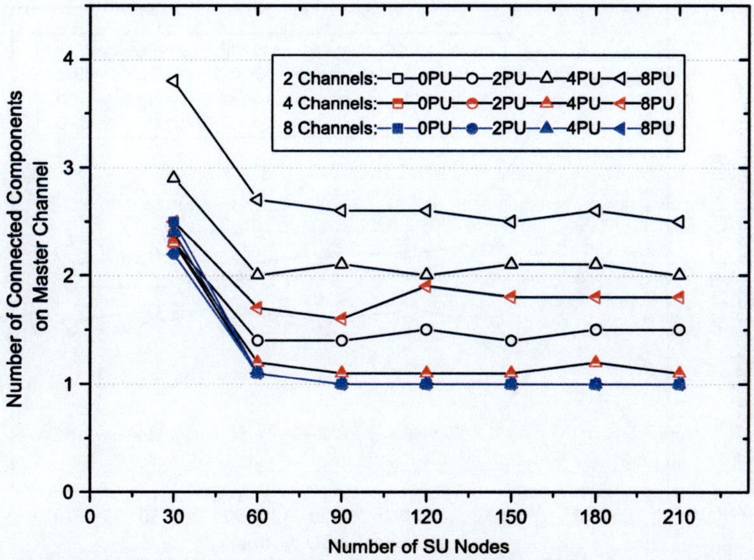

Fig. 7.3 Number of connected components on master channels as a function of number of SUs

channels and PUs setup. From the figure, it is easy to observe that more available channels lead to less connected components. In the case where the channel number is 8, there is only one connected component for the SU number greater than 90. That means only one control channel is used. We also notice the impact of PUs on the control channels. That is a larger PU number resulting in more connected components, which implies that more control channels have to be used. The figure shows the SU number of 30 has significantly more connected components than other larger SU numbers. It is because all nodes in the network cannot be fully connected when the SU number is 30 or less.

Figure 7.4 gives the comparison of the cloud and no cloud approach on the connected components on all master channels. We have already analyzed the performance of the cloud approach on connected components in Fig. 7.3. From Fig. 7.4, we can see when the PU number increases, the no cloud approach creates more connected components, therefore using more control channels. In the case of 8 channels, the cloud approach converges to a single control channel when the SU number is greater than 90, while for the no cloud approach, more than one control channels are used when the PU number is 4 and 8. That shows the effort of the cloud approach to drive SUs to the same control channel.

The other performance indicator of the cloud algorithms is the number of SUs on the dominant master channel. The dominant master channel refers to a channel selected by most of SUs as the control channel. Figure 7.5 provides the percentage of SUs on the dominant master channel as a function of SU number. The percentage is defined as the number of SUs on the master channel to the total number of SUs

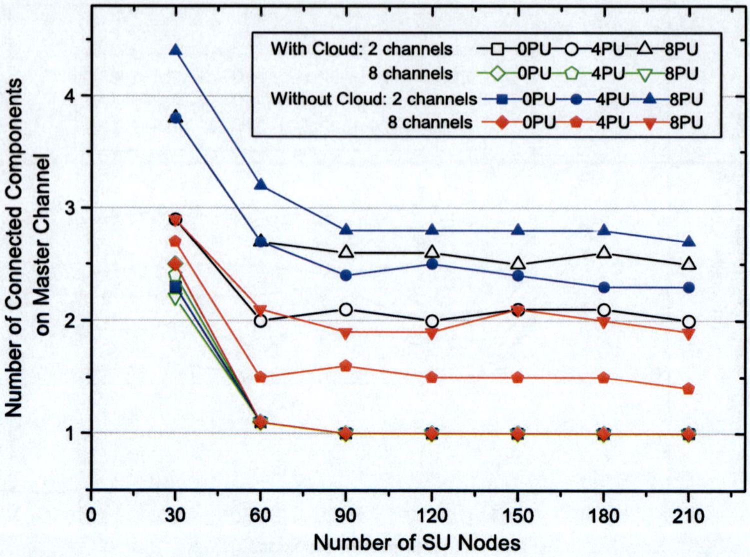

Fig. 7.4 Number of connected components on master channels as a function of number of SUs. Cloud approach vs. no cloud approach

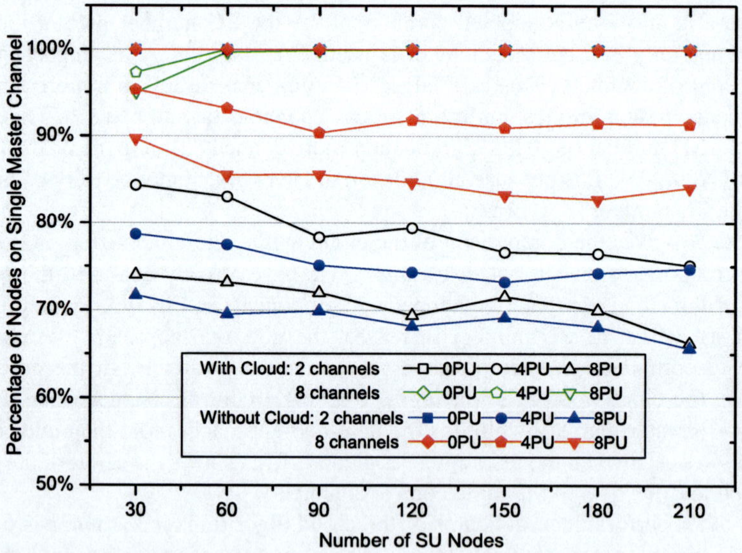

Fig. 7.5 Percentage of SUs on master channel selected by maximum number of SUs as a function of number of SUs. Cloud approach vs. no cloud approach

in the network. In line with Figs. 7.4, 7.5 shows that more PUs lead to less SUs on the dominant master channel. Moreover, in the no cloud approach, less SUs are on the dominant master channel than the cloud approach when the PU number is large. Since when the PU number is 0, the no cloud approach almost has the same results on control channels as the cloud approach, while in other cases their results are significantly different, it illustrates the cloud approach has advantage on the control channel selection.

Figure 7.6 reveals the impact of the cloud approach on the cluster formation. It shows that as the SU number increases, the number of clusters increases in both approaches. The increasing rates in both approaches are not linear and slow down when the SU number increases. Compared to the no cloud approach, the cloud approach slightly increases the number of clusters in the networks. That difference is prominent when there is only two channels and it becomes dimmish as available channels increase.

The final figure shows how the number of PUs impacts the control channel selection of SUs. As seen from Fig. 7.7, the increasing of the PU number leads to the decreasing of percentage of SUs on the dominant master channel. When there are more channels, more SUs are aggregated on the dominant master channel. When the channel number is 8, the cloud approach results in almost all SUs on the dominant master channel. It also shows that the no cloud approach has less SUs on the dominant master channel in all cases when PUs are in active.

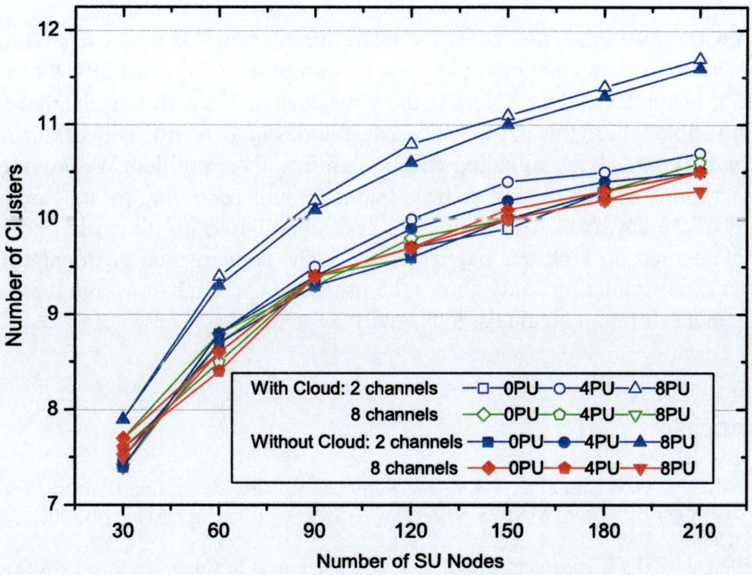

Fig. 7.6 Number of clusters as a function of number of SUs. Cloud approach vs. no cloud approach

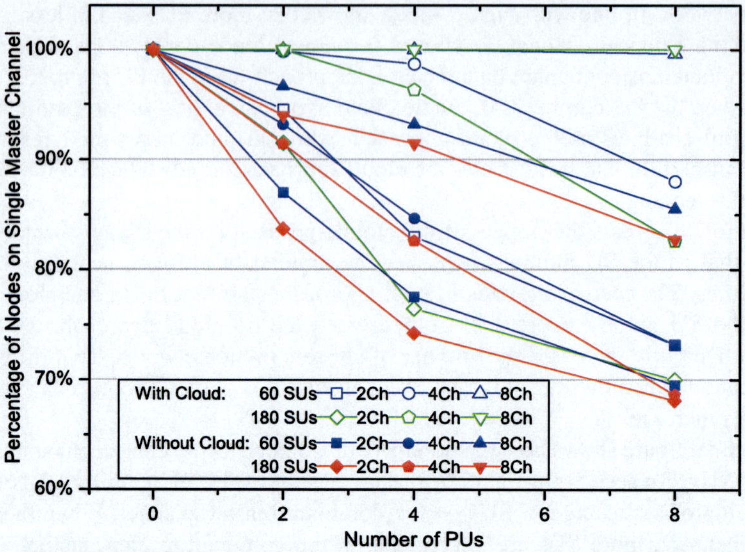

Fig. 7.7 Percentage of SUs on master channel with most SUs as a function of PUs. Cloud approach vs. no cloud approach

7.12 Conclusion

In this chapter, we study the control channel problem in DSA-based ad hoc networks. The control channel cloud approach is proposed to synchronize the control channel of SU nodes in the network to the same channel if a common channel exists. The motivation behind this approach is that if nodes in an ad hoc network using the same control channel, the signaling will be significantly simplified. We provide four basic operations to manipulate control channel clouds according to the cloud state of nodes in the network. Algorithms are developed based on the basic operations and implemented on a cluster based network. The convergence of the algorithms is proved. The simulation study shows the proposed approach converge the control channel under different channels, SUs, and PUs setups.

References

1. I.F. Akyildiz, W.Y. Lee, M.C. Vuran, and S. Mohanty. Next Generation/Dynamic Spectrum Access/Cognitive Radio Wireless Networks: A Survey. *Computer Networks*, 50(13):2127–2159, 2006.
2. K. Bian and J. Park. Segment-Based Channel Assignment in Cognitive Radio Ad Hoc Networks. *CROWNCOM 2007, Orlando, August*, 2007.
3. D. Bourse, K. El-Khazen, A. Delautre, T. Wiebke, M. Dillinger, J. Brakensiek, K. Moessner, G. Vivier, and N. Alonistioti. European Research in End-to-End Reconfigurability. *Proc. of IST Mobile & Wireless Communications Summit, Lyon, France*, pages 27–30, June 2004.

4. S.P. Chaudhuri, R. Kumar, and A.K. Saha. A MAC protocol for Multi frequency Physical layer. Technical report, Technical Report, Rice University, Houston, TX, January 2003.
5. T. Chen, H. Zhang, M. Matinmikko, and M.D. Katz. CogMesh: Cognitive Wireless Mesh Networks. In *WMSN Workshop in conjunction with IEEE GLOBECOM 2008, New Orleans*, pages 1–6, 2008.
6. T. Chen, H. Zhang, X. Zhou, G.M. Maggio, and I. Chlamtac. CogMesh: A Cluster Based Cognitive Radio Mesh Network. F. Fitzek et al ed., Cognitive Wireless Networks Concepts, Methodologies and Visions: Inspiring the Age of Enlightenment of Wireless Communications, pages 657–678, Springer, Dordrecht, Netherlands, 2007, ISBN: 978-1-4020-5978-0.
7. P. Cordier, D. Bourse, K. Moessner, et al. E2R Cognitive Pilot Channel concept. *Proc. of IST Mobile & Wireless Communications Summit 2006, Myconos, Greece*, June 2006.
8. B.A. Fette. *Cognitive Radio Technology*. Academic, Burlington, MA, 2009.
9. P. Gupta and P.R. Kumar. The capacity of wireless networks. *IEEE Transactions on information theory*, 46(2):388–404, 2000.
10. S. Haykin. Cognitive Radio: Brain-Empowered Wireless Communications. *Selected Areas in Communications, IEEE Journal on*, 23(2):201–220, 2005.
11. X. Jing and D. Raychaudhuri. Spectrum Co-existence of IEEE 802.11 b and 802.16 a Networks Using the CSCC Etiquette Protocol. *DySPAN 2005, Baltimore*, MD, November 2005.
12. J. So and N.H. Vaidya. Multi-Channel MAC for Ad Hoc Networks: Handling Multi-Channel Hidden Terminals Using a Single Transceiver. *MobiHoc 2004, Tokyo, Japan*, pages 222–233, May 2004.
13. C.Z. Tang and JJ Garcia-Luna-Aceves. Hop-Reservation Multiple Access (HRMA) for Ad hoc Networks. *IEEE INFOCOM 1999, New York*, NY, March 1999.
14. RW Thomas, DH Friend, LA Dasilva, and AB Mackenzie. Cognitive Networks: Adaptation and Learning to Achieve End-to-End Performance Objectives. *Communications Magazine, IEEE*, 44(12):51–57, 2006.
15. C.K.K. Toh. *Ad Hoc Wireless Networks: Protocols and Systems*. Prentice Hall PTR Upper Saddle River, NJ, 2001.
16. J. Zhao, H. Zheng, and G. Yang. Distributed Coordination In Dynamic Spectrum Allocation Networks. *IEEE DySPAN 2005, Baltimore*, MD, November, 2005.
17. Q. Zhao and B.M. Sadler. A Survey of Dynamic Spectrum Access. *Signal Processing Magazine, IEEE*, 24(3):79–89, 2007.

Part III
Topology Control and Routing

Chapter 8
Topology Control and Routing in Cognitive Radio Mobile Ad Hoc Networks

Quansheng Guan, F. Richard Yu, and Shengming Jiang

Abstract Recent research activities about cognitive radio (CR) are mainly focusing on opportunistic spectrum access and spectrum utilization. However, CR technology will have significant impacts on upper layer performance such as topology control and routing in wireless networks, especially in mobile ad hoc networks (MANETs). The dynamic spectrum availability issue imposes more challenges on routing in CR-MANETs. Since the spectrum availability is affected by primary user activities and the mobility of cognitive users, cognitive routing is required to be forward looking rather than reactive. To this end, a topology control and routing framework is presented in this chapter, where cognitive routing is enabled by topology control. In the framework, topology control serves as a middleware and a cross-layer module residing between routing and CR module. Prediction techniques can be used to construct a smart network topology, which provisions cognition capability to routing. Particularly, we present a distributed prediction-based cognitive topology control (PCTC) scheme to demonstrate the framework and verify its feasibility.

8.1 Introduction

Cognitive radio (CR) [35] is an enabling technology to allow cognitive users (CUs, i.e., unlicensed users or secondary users) to operate on the vacant parts of the spectrum allocated to primary users (PUs, i.e., licensed users). CR is widely considered as a promising technology to deal with the spectrum shortage problem caused by the current inflexible spectrum allocation policy. It is capable of sensing its radio environment and adaptively choosing transmission parameters according to sensing outcomes, which improves cognitive radio system performance and avoids interfering with PUs [21].

Recent research activities conducted in CR are mainly focusing on opportunistic spectrum access and spectrum utilization [25, 33, 34, 44, 57]. However, CR technology will have significant impacts on upper layer performance in wireless networks, especially in mobile ad hoc networks (MANETs), which enable wireless

Q. Guan (✉)
School of Electrical and Information Engineering, South China University of Technology, Guangzhou, China 510640
e-mail: guan.quansheng@mail.scut.edu.cn

F.R. Yu (ed.), *Cognitive Radio Mobile Ad Hoc Networks*,
DOI 10.1007/978-1-4419-6172-3_8,

CR devices to dynamically establish networks without using a fixed infrastructure. The end-to-end multi-hop transmissions in CR-MANETs may experience different spectrum transmissions according to the spectrum availability. In addition to the dynamic spectrum/channel availability introduced by CR, a key distinguishing issue in CR-MANETs is the importance of protecting the PU transmissions. In this sense, re-routing in CR-MANETs is more frequent due to the emergence of PU activities compared to classical MANETs.

Certainly, issues in non-cognitive MANETs in general are still of interest in the CR paradigm. However, some distinct characteristics of CRs introduce new non-trivial issues to CR-MANETs [2]. Though some efforts have been done to the MAC layer issues [45, 53], routing is still one of the particularly important networking issues in CR-MANETs. From routing perspective, it is expected that data packets are routed via a stable and reliable path to avoid frequent re-routing problem, since frequent re-routing may induce broadcast storms to the network, waste the scarce radio resources, and degrade end-to-end network performance such as throughput and delay [49]. Compared to classical MANETs, a path in CR-MANETs is especially unstable due to the highly dynamic spectrum availability, since it is affected not only by the mobility of CUs but also by the interference to PUs. In this sense, learning and forecasting abilities are demanded for all the protocol stack components in CR-MANETs. However, in the current protocol stack, the lower layer information regarding the CR environment is transparent to the routing layer. In this sense, topology control, working as a middleware, is adopted to provision cognitive capability to routing [20].

This chapter will address a topology control and routing framework for CR-MANETs, where cognitive routing is enabled by topology control. The rest of this chapter is structured as follows. The overview of topology control and routing is presented in Section 8.2. Section 8.3 introduces a prediction-based topology control to provide cognition capability to routing. Some author's pre-mature opinions on cognitive routing are summarized and this chapter is concluded in Section 8.4.

8.2 Topology Control and Routing

Topology control and routing work closely with each other. As routing is usually run over a given network topology, its behaviors and performance depend on topology control to a certain extent.

8.2.1 Topology Control

As we know, the network topology in a CR-MANET is changing in constant due to user mobility, the emergence of PU activities, barriers, etc. Meanwhile, the topology in a CR-MANET is controllable by adjusting some parameters such as the transmission power, channel assignment. In general, topology control is such a scheme to determines where to deploy the links and how the links work in wireless networks to form a *good* network topology, which will optimize the energy consumption, the capacity of the network, or end-to-end routing performance. Topology control is originally developed for wireless sensor networks (WSNs) [40], MANETs [12] and wireless mesh networks [46] to reduce energy consumption and interference. It usually results in a simpler network topology with small node degree and short transmission radius, which will have high-quality links and less contention in MAC layer. Spatial/spectrum reuse will become possible due to the smaller radio coverage. Other properties like symmetry and planarity are expected to obtain in the resultant topology. Symmetry can facilitate wireless communication and two-way handshake schemes for link acknowledgement while planarity increases the possibility for parallel transmissions and space reuse [17].

The area of energy-saving topology control in MANETs has attracted a great deal of attention [41]. Under the constraint of network connectivity, topology control adjusts the transmission range of each mobile node in order to save energy. The approach of some approximate graph based algorithms [32, 38, 51] is to remove long links while preserve network connectivity in order to force nodes to use multiple short hops, which saves the energy and prolongs the network lifetime. Network capacity is another concern of topology control [19, 24]. Thus interference-aware topology control schemes emerge [5, 27]. Since distributed topology control handles only the local settings, some research extends it to enhance the end-to-end paths [17].

Topology control focuses on network connectivity with the link information provided by medium access control (MAC) and physical layers, which both belong to cognitive radio module in CR-MANETs. The intermittently connected behavior of CR-MANETs is eager for efficient and opportunistic link management and routing due to its dynamic topology and then maximum transmission opportunities [53]. When constructing network topology, topology control takes care about the interference to PUs and link availability in CR-MANETs.

Distributed topology control suits the dynamic CR-MANETs well due to the high cost of maintaining network knowledge for highly dynamic networks [31]. While the distributed algorithm usually copes with merely local information, other schemes also propose topology control to tune with the decisions of routing. Rather than forming a topology for routing, topology control assigns channels based on the routing decisions [23]. By doing this, the network topology can be self-organized to support the current offered traffic, which also provides some end-to-end performance gain (e.g., delay or throughput) or network-wide gain (e.g., network capacity).

The power control and channel control issues are coupled in CR-MANETs while they are treated separately traditionally. Although CR module can sense the using opportunities of available spectrums, it lacks of the scope to make network-wide decisions. It therefore makes more sense to conduct power control and channel control via the topological viewpoint [47]. The goal of topology control is then to set up interference-free connections to minimizes the maximum transmission power and the number of required channels. It is also desirable to construct a reliable network topology since it will result in some benefits for the network performance. A prediction-based topology control is proposed to reduce the re-routing frequency in CR-MANETs so as to improve the end-to-end throughput [20]. The detailed introduction of this scheme will be presented in Section 8.3.

8.2.2 Routing

Most of the state-of-the-art routing schemes in CR-MANETs are focus on the joint consideration of dynamic spectrum allocation and path establishment by monitoring the PU activity pattern [10]. Taking the PU avoidance as the first consideration, SEARCH routing protocol [11], which is somewhat a geographic routing, leverages the geographic information for greedy forwarding and PU avoidance. Based on the interference model, a joint routing, opportunistic spectrum scheduling, and time sharing scheduling algorithm in [58] is proposed to minimize the aggregate interference form CUs to a PU. With the observation that the spectrum availability may change from time to time and hop by hop, [14] proposed a routing and dynamic spectrum allocation (ROSA) algorithm, which is a joint spectrum allocation, scheduling, and transmission power control scheme. ROSA is formulated as a optimization problem, which aims to maximize the network throughput and limit the physical interference to other users.

The robustness of a path, which refers to the probability that no PU activity appears during a transmission period, is considered in [43]. It first forms a skeleton set that meets the robustness requirement for a source–destination pair. A joint routing and channel assignment is use to determined the skeleton, which is formulated as an integer linear programming model to maximize the system throughput. It is argued that a basic level of robustness should be guaranteed or the path will be disconnected though it has high transmission throughput. [29] exploits the result that the probability distribution of the PU-to-CU interference at a given CU node over a given channel approximately follows a lognormal distribution [39] to establish a probabilistic path that satisfy a given bandwidth demand while avoiding the interference to PUs. While the available spectrum dynamic caused by PU activities is the main concern in the existing research, the terminal mobility issues are rarely explored. The mobile terminal may enter the interference region of active PUs, making the route unavailable. In this case, CR-MANETs experience not only the interference limitation to PUs but also the handover issues [13].

Butun et al. [6] utilizes prediction techniques for routing reliability, efficiency, and scalability performance. Prediction methods can used to predict the time of the next movement, the next location of mobile terminal, or both of them. Three fundamentally predictors, i.e., Markov predictors, moving average predictors, and CDF predictors [8], are exercised and their performance in improving routing protocols for CR-MANETs are evaluated. Simulation results show that predictions improve the overall performance significantly.

The routing problems are usually formulated into optimization problems [14, 22, 58] or described via graph theory. [22] introduces a mixed integer non-linear programming to minimize the spectrum usage, i.e., to increase spectrum reuse. The objective of [22] is to minimize the aggregate interference from CUs to a PU. Optimization routing often involves joint considerations of spectrum/time sharing, power control, and even relay selection under the condition of PU avoidance. Regarding the graph-based routing, it generally represents the network topology by a graph structure $G = (N, E)$, where N is the node set and E is the edge set. Layered graphs are exercised to model the multiple available channels [54] while colored graphs are used to model the spectrum sharing and allocation issues [59]. The Laplacian Matrix in graph theory is crucial for routing, since it can measure the connectivity of the graph (i.e., the network). Moreover, the algebraic connectivity in Laplacian Matrix indicates the hop count of a path and the stability and robustness of the network. The routing protocol can make use of the algebraic connectivity information to guarantee stable and high connectivity [1].

Though it is not a unique issue in CR-MANETs, QoS requirements for routing need to be considered similar to MANETs. In addition to propagation delay, transmission delay, queueing delay, and accessing delay in traditional MANETs, channel switching delay accounts for a large amount in the end-to-end delay. The nodal delay defined in [52] includes the sensing and channel switching, where the switching time is increased with spectrum distance to tune with.

8.2.3 Discussions

The dynamic spectrum availability and the importance of limiting the interference to PUs differentiate CR-MANETs from classical MANETs. Two factors affect spectrum availability. The first one is PU activities. Since CUs are considered as low priority and secondary users of the spectrum allocated to PUs, CUs should sense the spectrum to detect PU activities. Such spectrum sensing can be conducted either non-cooperatively (individually), in which each CU conducts radio detection and makes decision by itself, or cooperatively, in which a group of CUs perform spectrum sensing by collaboration. Due to the capability limitation of some CUs and a large number of possible spectrum bands, spectrum sensing may take a long time [25, 33, 34]. The second factor that affects spectrum availability is CU mobility. In classical MANETs, routes formed by multiple hops may experience frequent disconnections caused by node mobility. This problem may be detected when the next hop node in the path does not reply to messages and the retry limit is exceeded. However, this problem is difficult to solve in CR-MANETs since a node may not be able to transmit if the spectrum becomes unavailable when a node moves into the interference boundary of an active PU. Therefore, correctly inferring mobility conditions in designing effective topology control and routing schemes is critical in CR-MANETs [2].

There are a large number of routing protocols proposed in classical MANETs, such as DSDV [37], DSR [28], AODV [36]. However, it is difficult to apply them directly to CR-MANETs due to the distinct characteristics described above.

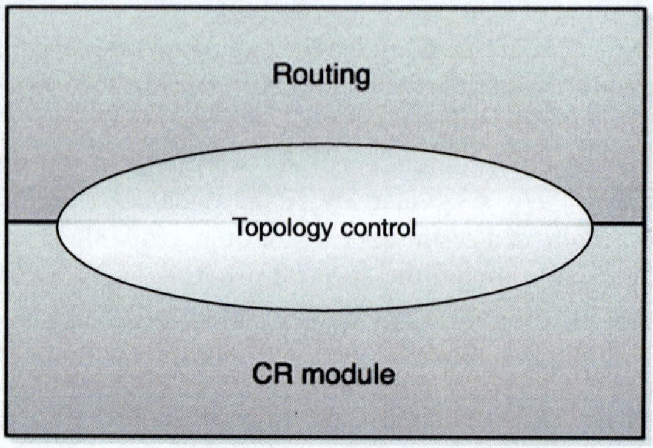

Fig. 8.1 Topology control in the protocol stack: a middleware-like cross-layer module

Although some proposals attempt to modify them for CR-MANETs [7], it may not be desirable to design a new routing protocol dedicated to CR-MANETs due to the maturity and availability (e.g., source code for different operating systems) of existing routing protocols. From this point of view and the principle of *layering as optimisation decomposition* [9], it is better to provision the cognition capability to routing via a middleware-like mechanism, e.g., topology control. An intrinsic feature, also a key challenge of cognitive topology control, is to be intelligent enough to make routing protocols forward looking to the dynamic network condition.

The interaction between topology control and routing refers to a cross-layer design concept with the channel decision at the MAC layer and end-to-end path finding at the routing layer (see Fig. 8.1). Topology control resides between routing and CR module. The dynamic changing environment and the cognition requirement to PU activity necessitate the cross-layer interations in CR-MANETs. Routing strategy is required to change with the dynamic environment accordingly, e.g., if the PUs are highly active, only opportunistic forwarding without pre-established routing is available [30]. In addition to the lower layer changing, network topology and routing need to be traffic aware to improve the route efficiency [55]. In this sense, topology control and routing demand for the abilities of perceiving, learning, forecasting, and proactive response to the network conditions. Specifically, we introduce a prediction-based cognitive topology control (PCTC) [20] in the next section, which enables routing protocols the cognition capability in CR-MANETs in a distributed manner.

8.3 A Prediction-Based Cognitive Topology Control in CR-MANETs

It is desirable that cognitive routing should favor links with long durations to prolong the path survival time and improve the path stability. Existing routing protocols in classical MANETs lacks this cognition capability. Nevertheless, these routing protocols are relatively mature, and the source codes are widely available in different operating systems. Since routing is conducted on a given topology, topology control is essential to provision cognition capability.

PCTC considers a CR-MANET under a underlaying environment. It can capture the topology dynamically based on link prediction to provide efficient and opportunistic link management and routing. The cognitive link prediction model considers both user mobility and interference to PUs, which are factors of paramount importance in CR-MANETs. It predicts both link duration and the probability that this duration may really last to the end of this period. Based on the cognitive prediction, it further describes how to construct a more reliable topology and reduce re-routings. It requires only local neighbor information and network connectivity is preserved in a distributed manner.

In PCTC, topology control constructs a reliable topology based on PU-aware link availability prediction.

8.3.1 Cognitive Link Availability Prediction

In PCTC, topology control constructs a reliable topology based on PU-aware link availability prediction. A large amount of prediction models are available for link availability prediction [3, 26, 56]. A reasonable link availability prediction model proposed in our previous work [26] is recommended. The basic principle of the prediction framework is to provide a predicted time period T_p that the link between two nodes will stay available. In addition, another important parameter, denoted by $L(T_p)$, can be obtained from this framework to estimate the probability that this link may really last to the end of T_p by considering possible changes in velocities. This framework considers if the node does not move at a constant velocity to get $L(T_p)$, which stands for the probability that the velocity will remain constant.

In CR-MANETs, $[T_p, L(T_p)]$ is not sufficient to predict link duration, since the links between CUs are affected not only by CU mobility but also by the interference to PUs. In this situation, a link with a high quality may be regarded unavailable, if a node in that link is moving close to a PU. As a result, in order to avoid interference, the distance between a CU and a PU should be monitored. We propose another pair of $[\hat{T}_p, L(\hat{T}_p)]$ to predict the link availability before nodes moves into the interference boundary of PUs. Similar to T_p, $\hat{T}_p$ is the predicted time till a CU moves into the interference area of a PU, and $L(\hat{T}_p)$ is its corresponding probability. The final link availability is revealed by the combination of $[T_p, L(T_p)]$ and $[\hat{T}_p, L(\hat{T}_p)]$ to enable cognitive link prediction. The link available duration T_a should be set to

$$T_a = \min_{i=1,2;\, j \in \{PUs\}} \left\{ T_p \times L(T_p),\, \hat{T}_{p_i}^{j} \times L\left(\hat{T}_{p_i}^{j}\right) \right\} \tag{8.1}$$

where i is associated with the two ends of a link and $\{PUs\}$ includes all the PUs present in the network. The subscript i and superscript j indicate that a link will be unavailable if any of its ends moves into the interference region of any PU. T_a is enabled the cognitive feature due to its consideration of both node mobility and the interference to PUs.

8.3.2 Cognitive Topology Control and Routing

Equipped with cognitive link estimation, routing protocols can avert using the links within interference regions of PUs since these links are estimated to have $T_a = 0$. However, they cannot avert using the links with low duration to avoid frequent re-routing occurrence. To this end, we propose a topology control and routing scheme in CR-MANETs based on the link prediction presented above.

8.3.2.1 Distributed Topology Construction

It is argued in the literature that a topology with short radio links and small node degree saves energy consumption in MANETs or WSNs. Further, such a topology has less access competition and achieves higher network capacity. However, things are different in CR-MANETs. The dynamic changes due to CU mobility or the interference to PUs, which will result in frequent re-routing, waste large amount of scarce network resource, and achieve low end-to-end performance. The proposed PCTC algorithm aims to solve this problem by constructing a more reliable topology for routing protocols.

In PCTC, a new link reliability metric is defined for topology construction. We first introduce a re-routing penalty denoted by δ. This penalty is a time period that is incurred by re-routing and reduces link availability to $(T_a - \delta)$ in the sequel. Actually, the path duration is not the main concern of an end-to-end transmission. Instead, in nature, it is expected to deliver as many packets as possible before a path failure happens. From this point of view, the only consideration of link available duration T_a is not enough since a link with long duration may have a poor quality resulting in low data rate. In the long run, it needs less re-routings if the selected path consists of links with longer T_a and higher quality. To quantify this measurement, we set the link weights to

$$w = r \times (T_a - \delta) \tag{8.2}$$

where the link data rate r captures its link quality. Herein, the re-routing penalty δ is converted into a capacity loss $r \times \delta$ during the available period. The link weight then presents the traffic carrying ability of this link. We define a path weight as

$$\mathcal{W} = \min w_i, i \in \mathbb{L} \tag{8.3}$$

where $\mathbb{L}$ is the set including all the links along the path. We can see that the definitions of link and path weights reflect its true data transmission ability. A link with long duration but bad link quality or a link with short duration though good quality will result in weak data transmission ability. It is desirable to transmit more data traffic before link failure.

The principle of PCTC is to preserve the reliable path with maximum path weight in (8.3) for any pair of nodes under connectivity guarantee. Similar to other topology control algorithms, the topology construction process consists of three steps: neighbor collection, path search, and neighbor selection. The distributed localized Dijkstra topology control (LDTC) algorithm in our previous work [18], which aims at constructing an energy-efficient topology has some beneficial properties, particularly the 1-spanner property, which preserves the global minimum energy paths in a distributed manner. LDTC runs Dijkstra algorithm over a neighborhood graph. Therefore it requires only local information exchanges. The distributed cognitive PCTC algorithm enhances LDTC to preserve the end-to-end reliable paths for CR-MANETs. As a distributed algorithm, each node runs the following procedures as an initial node:

- *Neighbor collection:*

  1. Collect all of its neighbors, and calculate the edge weights according to (8.2).

- *Path search:*

  2. Set the path weights to infinity for initial node and to zero for neighbors. Mark all the neighbors as unvisited and the initial node as the current one.
  3. Calculate the path weights according to (8.3) from the initial node to unvisited neighbors via the current node. If they are larger than the previously recorded ones, update the path weights to them.
  4. Mark the current node as visited and set the unvisited node with largest path weight as the current node. Then repeat from (3) until all the neighbors are visited.

- *Neighbor selection:*

  5. All of the most reliable paths from the initial node to its neighbors are now found. The resulting topology is the output by preserving all the first hop neighbors of the initial node along these paths.

From the PCTC running procedures, we know that the reliable links are preserved and links with low duration may be removed in the resulting topology, which results in a more reliable and stable topology.

8.3.2.2 Topology Reconfiguration

Wireless links are changing dynamically due to node mobility, fading, barrier, interference, etc. Topology reconfiguration is needed due to the dynamically changing wireless links in CR-MANETs. Topology reconfiguration can also be used to deal with inaccurate link availability predictions since prediction errors are inevitable. Specifically, when links become unavailable unexpectedly, topology reconfiguration process has to be triggered promptly. Accordingly, a perfect prediction technique with 100% accuracy is not necessary in PCTC.

In wireless cellular networks, such as UMTS and LTE [42], wireless nodes measure the wireless links and update the link status frequently (more than 1000 times per second in some situations) to make radio resource management decisions, such as handover and opportunistic scheduling. In CR-MANETs, therefore, an intuitive method for topology reconfiguration is to run the topology control algorithm frequently to keep pace with the dynamical links. However, this frequent topology control requires enormous amount of computational loads and is quite power-consuming. The prediction based link available duration can be used to reduce the topology reconfiguration frequency. The update period of cognitive topology control can be set to

$$\Gamma = \sum_{i=1}^{n} \frac{T_a^i}{n} \tag{8.4}$$

For a distributed topology control algorithm, Γ may be different in each node and is adaptive to the mobile environment. Therefore, it does not require synchronization all over the entire network due to distributed execution. Different from link level update, Γ here is much larger than link update period, which is used by CR module to keep pace with link dynamic.

In order to respond promptly to topology changes, it is desirable to be able to deal with some asynchronization occurrences, such as link appearance, disappearance, and link weight change.

- When a new node wants to joint the network, it first predicts all links to its neighbors based on the cognitive link duration prediction scheme presented in Section 8.3.1. These predictions are then broadcast to neighbors in its range. Any node that has been in the network is not necessary to re-run the topology algorithm to reconfigure the topology. What reconfiguration needs to do is to find a reliable path to the new node from the existing topology. After running PCTC, the initial has obtained the reliable paths to all its original neighbors. Then topology reconfiguration extends these paths by adding the newly appearing links to the newcomer. The one with maximum weight is selected and the first hop neighbor in the path is preserved. The new node may compose more reliable paths to the existing nodes. Therefore, reconfiguration extends the reliable path of the newcomer to other neighbors by added the direct links. The path obtained by PCTC will be replaced if the corresponding extended path has a larger weight. These reconfiguration operations have a complexity of $O(n)$ comparing to $O(n \log n)$ of PCTC.
- If a node or a link disappears occasionally, the topology has to be reconfigured since it may lose the connections to some neighbors. As mentioned above, the links remaining in the final topology are with higher weights than those that are removed. Instead of re-running the algorithm, it is only required to remedy the reliable paths to the next hop nodes of the lost node.

- The cognitive topology control algorithm neglects link changes during topology update period for that link is predicted to be available during this period. It simply updates link weights without executing of the algorithm.

8.3.2.3 Cognitive Routing on the Resulting Topology

Routing is defined to select paths in a network to send data traffic. A challenge in CR-MANETs is to provision cognition capability to routing protocols. With PCTC, existing routing protocols can be easily adopted in CR-MANETs with cognition capability. Based on the cognitive link prediction scheme presented in Section 8.3.1, routing on the PCTC resulting topology is aware of PUs and is forward-looking to link duration. Further, PCTC makes routing adaptive to mobile environment, in which routing favors reliable paths in the network. Take the popular routing protocols, such as DSR and AODV, for examples. They usually send routing request packets (RREQs), to find a path for a source and a destination. When a RREQ reaches an intermediate node, it may be dropped if the transmitter does not exist in the neighborhood relationship generated by the PCTC algorithm. Otherwise, this RREQ is disposed by the intermediate node and re-broadcast. As a result, the links in the vicinity of PUs or with poor quality and short available duration are avoided. The first RREQ is replied in terms of a routing metric of first found path to confirm the found path.

Actually, PCTC optimizes neighborhood relationship among nodes and assigns each path a weight with regard to reliability. With PCTC, we can improve the performance of other routing metrics such as shortest path or QoS routings.

8.3.3 Results and Discussions

Simulation results presented in [20] also necessitates the adoption of prediction techniques in CR-MANETs. We use the simulation results to study the topology control and routing performance with and without predictions.

The properties of the resultant topology is shown in Fig. 8.2. From Fig. 8.2a, we can see that the resulting topology with prediction (i.e., the topology after PCTC) has smaller average node degree and smaller maximum node degree than those without topology control. Small node degree may mitigate contention in the shared wireless medium. Specially for CR-MANETs, smaller node degree means less spectrum bands is required for the medium accessing, which provides flexibility to spectrum allocation schemes and improves spectrum reuse. We note that the average node degree retains a small value as the number of nodes in the network increases, which also makes network scalable. In addition to node degree, the topology control algorithm also results in longer link durations in Fig. 8.2b. This indicates that the resulting topology is more stable and it is possible to reduce re-routings in the network. This result coincides with that in [50], which indicates that prediction techniques improve the stability of clustering schemes.

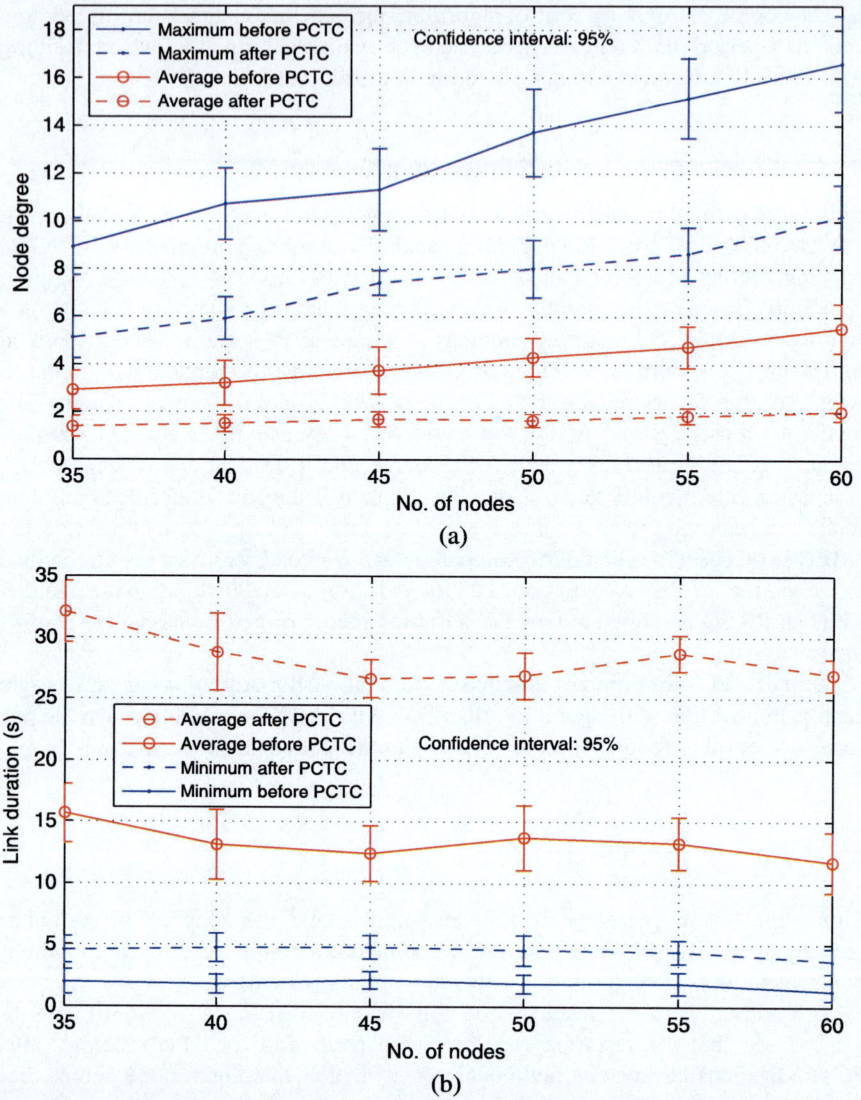

Fig. 8.2 Properties of PCTC resulting topologies. (**a**) Node degree; (**b**) Link duration

Prediction techniques enable topology control to construct a forward looking topology for routing. Another result is also presented to support the adoption of topology control for routing. The following two routing strategies are adopted to evaluate the performance gain by PCTC.

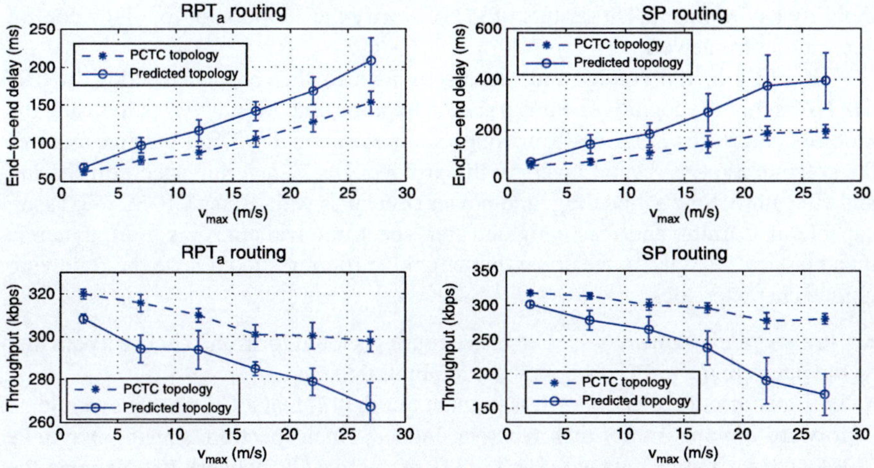

Fig. 8.3 PCTC improves the routing performance. The confidence interval is 95%

- Shortest path (SP): Each flow is transmitted along the path with the minimum number of hops.
- Reliable path with regard to T_a (RPT$_a$): It selects the path with maximum path duration according to (8.3).

Figure 8.3 demonstrates the two main end-to-end performance metrics, i.e., end-to-end throughput and delay, of routing on the predicted topology and cognitive routing on the PCTC resulting topology. The predicted topology has only the knowledge of link prediction but without topology reconstruction. It is clear that the end-to-end routing performance regarding both throughput and delay are improved by PCTC. The reason exists in that the routing protocols perform in a forward-looking way to select reliable paths for data flows.

The simulation results verify the framework of enabling cognitive routing via topology control. Particularly, prediction techniques play a critical role in this framework. In addition, topology control also results in an improvement for the entire network performance.

8.4 Conclusions

Cognitive radio technology will have significant impact on upper layer performance in cognitive radio mobile ad hoc networks (CR-MANETs). In this chapter, we addressed the topology control and routing issues, which work closely with each other. Specially, a prediction-based cognitive topology control (PCTC) provisioning cognition capability to the routing protocols is presented in detail, which is aware of the interference to PUs, forward looking to link available duration, and adaptive to mobility environment. The results of PCTC verify the adoption of topology control and routing framework.

According to the survey and the discussions in this chapter, routing in CR-MANETs (named cognitive routing) should exploits both PU activity pattern and the mobility pattern to use the spectrum opportunities efficiently. This objective requires the collaboration of several layers in the protocol stack, covering spectrum sharing and allocation, time scheduling, and power control as well. Since CR-MANETs are capable of learning and reasoning and link spectrum availability is more dynamic than classical MANETs, we argue that cognitive routing should have the following unique characteristics:

- *PU interference awareness:* Cognitive routing should form and choose a path that the interference to PUs is below the required threshold.
- *Link availability prediction:* It is pointed out in [48] that a CR network should be forward-looking, rather than reactive. Indeed, since spectrum sensing may take a long time and be delayed [25, 33, 34], a reactive CR network will degrade the performance. Therefore, cognitive routing should not only be aware of PUs but also foresee link available periods in terms of its interference to PUs.
- *Adaptive acting:* Cognitive routing should be adaptive to form and choose a path based on the prediction to increase end-to-end throughput and decrease end-to-end delay.

The above cognitive characteristics of cognitive routing demand for the knowledge from the MAC/PHY layers. Accordingly, we suggest topology control as a middleware provisioning cognitive features to routing without major modifications to the existing routing protocols. To achieve cognitive routing, prediction techniques with regard to PU activity pattern, mobility pattern, as well as traffic pattern are critical to topology control and routing. Indeed, reactive response to the network conditions is far from enough in highly CR-MANETs. Learning and forward-looking ability can make the network more intelligent, which is consistent with the concept of *cognitive networks* [4, 15, 16]. To sum up, the topology control and routing framework in this chapter has the following features:

- Topology control serves as a middleware-like mechanism residing between CR module and routing layer.
- Cognitive routing is enabled by topology control, which is in charge of the spectrum availability issues caused by PU activities and mobility.
- Currently existing routing protocols can be applied to CR-MANETs with the cognition capability provisioned by topology control.
- It is flexible to achieve somewhat network-wide objective, e.g., spectrum utilization minimization, network capacity improvement, energy saving.

References

1. Abbagnale, A., Cuomo, F.: Gymkhana: a connectivity-based routing scheme for cognitive radio ad hoc networks. In: INFOCOM IEEE Conference on Computer Communications Workshops (2010), pp. 1–5. San Diego, CA (2010)

2. Akyildiz, I.F., Lee, W., Chowdhury, K.R.: CRAHNs: cognitive radio ad hoc networks. Ad Hoc Netw. **7**(5), 810–836 (2009)
3. Alavi, B., Pahlavan, K.: Modeling of the TOA-based distance measurement error using UWB indoor radio measurements. IEEE Commun. Lett. **10**(4), 275–277 (2006)
4. Benedetto, M.D., Nardis, L.D.: Cognitive routing models in UWB networks. In: 3rd International Conference on Cognitive Radio Oriented Wireless Networks and Communications (CrownCom 2008), pp. 1–6. Singapore (2008)
5. Burkhart, M., von Rickenbach, P., Wattenhofer, R., Zollinger, A.: Does topology control reduce interference? In: Proc. 5th ACM Int. Symposium on Mobile Ad Hoc Networking and Computing. Roppongi Hills, Tokyo, Japan (2004)
6. Butun, I., Talay, A.C., Altilar, D.T., Khalid, M., Sankar, R.: Impact of mobility prediction on the performance of cognitive radio networks. In: 2010 Wireless Telecommunications Symposium (WTS), pp. 1–5. Tampa, FL (2010)
7. Cacciapuoti, A.S., Calcagno, C., Caleffi, M., Paura, L.: CAODV: routing in mobile ad-hoc cognitive radio networks. In: 2010 IFIP Wireless Days (WD), pp. 1–5. Venice, Italy (2010)
8. Cheng, C., Jain, R., van den Berg, E.: Location prediction algorithms for mobile wireless systems. in Wireless internet handbook: technologies, standards, and systems, B. Furht and M. Ilyas, Eds. CRC Press, Inc. Boca Raton, FL (2003). Ch. 11, 245–263
9. Chiang, M., Low, S., Calderbank, A., Doyle, J.: Layering as optimization decomposition: a mathematical theory of network architectures. IEEE Commun. Lett. **95**(1), 255–312 (2007)
10. Chou, C.T., Sai Shankar, N., Kim, H., Shin, K.G.: What and how much to gain by spectrum agility? IEEE J. Sel. Areas Commun. **25**(3), 576 (2007)
11. Chowdhury, K., Felice, M.D.: SEARCH: a routing protocol for mobile cognitive radio ad-Hoc networks. In: IEEE Sarnoff Symposium (SARNOFF'09), pp. 1–6. Princeton, NJ (2009)
12. Dai, F., Wu, J.: Mobility-sensitive topology control in mobile ad hoc networks. IEEE Trans. Parallel Distrib. Syst. **17**(6), 522–535 (2006)
13. De Nardis, L., Guirao, M.D.: Mobility-aware design of cognitive radio networks: challenges and opportunities. In: Proc. 5th Int. Conf. Cognitive Radio Oriented Wireless Networks & Communications (CROWNCOM), pp. 169–177. Cannes, Italy (2010)
14. Ding, L., Melodia, T., Batalama, S., Matyjas, J., Medley, M.: Cross-Layer routing and dynamic spectrum allocation in cognitive radio ad hoc networks. IEEE Trans. Veh. Tech. **59**(4), 1969–1979 (2010)
15. Gelenbe, E.: Steps toward self-aware networks. Commun. ACM **52**(7), 66–75 (2009)
16. Gelenbe, E., Lent, R., Nunez, A.: Self-aware networks and QoS. IEEE Commun. Lett. **92**(9), 1478–1489 (2004)
17. Guan, Q., Ding, Q., Jiang, S.: A minimum energy path topology control algorithm for wireless multihop networks. In: Proc. IWCMC. Leipzig, Germany (2009)
18. Guan, Q., Ding, Q., Jiang, S.: A minimum energy path topology control algorithm for wireless multihop networks. In: Proc. Int. Conf. on Wireless Comm. and Mobile Computing (ICWCMC). Leipzig, Germany (2009)
19. Guan, Q., Jiang, S., Ding, Q.L., Wei, G.: Impact of topology control on capacity of wireless ad hoc networks. In: Proc. IEEE ICCS'08, pp. 588–592. Guangzhou, China (2008)
20. Guan, Q., Yu, F., Jiang, S., Wei, G.: Prediction-based topology control and routing in cognitive radio mobile ad hoc networks. IEEE Trans. Veh. Tech. **59**(9), 4443–4452 (2010)
21. Haykin, S.: Cognitive radio: brain-empowered wireless communications. IEEE J. Sel. Areas Commun. **23**(2), 201–220 (2005)
22. Hou, Y., Shi, Y., Sherali, H.: Spectrum sharing for multi-hop networking with cognitive radios. IEEE J. Sel. Areas Commun. **26**(1), 146–155 (2008)
23. Irwin, R., DaSilva, L.: Channel assignment based on routing decisions (CARD): Traffic-dependent topology control for multi-channel networks. In: IEEE ICC'09 Workshops, pp. 1–5. Dresden, Germany (2009)
24. Jain, K., Padhye, J., Padmanabhan, V.N., Qiu, L.: Impact of interference on multi-hop wireless network performance. In: Proc. 9th Annual Int. Conf. Mobile Computing and Networking. San Diego (2003)

25. Jiang, H., Lai, L., Fan, R., Poor, V.: Optimal selection of channel sensing order in cognitive radio. IEEE Trans. Wireless Commun. **8**(1), 297–307 (2009)
26. Jiang, S., He, D., Rao, J.: A prediction-based link availability estimation for routing metrics in MANETs. IEEE/ACM Trans. Netw. **13**(6), 1302–1312 (2005)
27. Johansson, T., Carr-Motyăková, L.: Reducing interference in ad hoc networks through topology control. In: Proc. Joint Workshop on Foundations of Mobile Computing. Cologne, Germany (2005)
28. Johnson, D.B., Maltz, D.A., Hu, Y.C.: The dynamic source routing protocol for mobile ad hoc networks (DSR). IETF Draft, draft-ietf-manet-dsr-09.txt (2003)
29. Khalife, H., Ahuja, S., Malouch, N., Krunz, M.: Probabilistic path selection in opportunistic cognitive radio networks. In: Proc. GLOBECOM 2008. New Orleans, LA (2008)
30. Khalife, H., Malouch, N., Fdida, S.: Multihop cognitive radio networks: to route or not to route. IEEE Netw. **23**(4), 20–25 (2009)
31. Komali, R., Thomas, R., Dasilva, L., Mackenzie, A.: The price of ignorance: distributed topology control in cognitive networks. IEEE Trans. Wireless Commun. **9**(4), 1434–1445 (2010)
32. Li, L., Halpern, J., Bahl, P., Wang, Y.M., Wattenhofer, R.: A cone-based distributed topology-control algorithm for wireless multi-hop networks. IEEE/ACM Trans. Netw. **13**(1), 147–159 (2005)
33. Li, Z., Yu, F.R., Huang, M.: A distributed consensus-based cooperative spectrum sensing in cognitive radios. IEEE Trans. Veh. Tech. **9**(4), 1370–1379 (2010)
34. Liang, Y.C., Zeng, Y., Peh, E.C.Y., Hoang, A.T.: Sensing-throughput tradeoff for cognitive radio networks. IEEE Trans. Wireless Commun. **7**(4), 1326–1337 (2008)
35. Mitola, J., Maguire, G.: Cognitive radio: making software radios more personal. IEEE Personal Comm. **6**(4), 13–18 (1999)
36. Perkins, C.E., Belding-Royer, E.M., Das, S.R.: Ad hoc on-demand distance vector (AODV) routing. IETF Draft, draft-ietf-manet-aodv-13.txt (2003)
37. Perkins, C.E., Bhagwat, P.: Highly dynamic destination-sequenced distance-vector routing (DSDV) for mobile computers. In: Proc. ACM SIGCOMM'94. London (1994)
38. Rodoplu, V., Meng, T.H.: Minimum energy mobile wireless networks. IEEE J. Sel. Areas Commun. **17**(8), 1333–1344 (1999)
39. Salamch, H., Krunz, M., O.Younis: Throughput-oriented MAC protocol for opportunistic cognitive radio networks. Tech. Rep. UA-ECE-2007-2, University of Arizona (2007)
40. Santi, P.: Topology control in wireless ad hoc and sensor networks. ACM Comput Survey **37**(2), 164–194 (2005)
41. Santi, P.: Topology control in wireless ad hoc and sensor networks. ACM Comput Surveys (CSUR) **37**(2), 164–194 (2005)
42. Sesia, S., Toufik, I., Baker, M.: LTE, The UMTS Long Term Evolution: From Theory to Practice. Wiley, NY (2009)
43. Shih, C., Liao, W.: Exploiting route robustness in joint routing and spectrum allocation in Multi-Hop cognitive radio networks. In: IEEE Wireless Communications and Networking Conference (WCNC), pp. 1–5. Sydney, Australia (2010)
44. Si, P., Yu, F., Ji, H., Leung, V.: Optimal Cooperative Internetwork Spectrum Sharing for Cognitive Radio Systems with Spectrum Pooling. IEEE Trans. Veh. Tech. **59**(4), 1760–1768 (2010)
45. Su, H., Zhang, X.: Cross-Layer Based Opportunistic MAC Protocols for QoS Provisionings Over Cognitive Radio Mobile Wireless Networks. IEEE J. Sel. Areas Commun. **26**(1), 118–129 (2008)
46. Tang, J., Xue, G., Zhang, W.: Interference-aware topology control and QoS routing in multi-channel wireless mesh networks. In: Proc. ACM MobiHoc'05. Urbana-Champaign, IL (2005)
47. Thomas, R., Komali, R., MacKenzie, A., DaSilva, L.: Joint power and channel minimization in topology control: A cognitive network approach. In: IEEE ICC '07, pp. 6538–6543. Glasgow, Scotland (2007)
48. Thomas, R.W., DaSilva, L.A., MacKenzie A.B.: Cognitive networks. In: Proc. IEEE DySPAN'05. Baltimore, M.D (2005)

49. Tseng, Y.C., Ni, S.Y., Chen, Y.S., Sheu, J.P.: The broadcast storm problem in a mobile ad hoc network. Wireless Netw. **8**(2/3), 153–167 (2002)
50. Venkateswaran, A., Sarangan, V., Gautam, N., Acharya, R.: Impact of mobility prediction on the temporal stability of MANET clustering algorithms. In: Proc. 2nd ACM int. workshop on Performance evaluation of wireless ad hoc, sensor, and ubiquitous networks. Montreal, QC (2005)
51. Wattenhofer, R., Zollinger, A.: XTC: A practical topology control algorithm for ad-hoc networks. In: Proc. 18th Int. Parallel and Disttributed Processing Symp.(IPDPS'04). Santa Fe, NM (2004)
52. Wen, Y., Liao, W.: On QoS routing in wireless Ad-Hoc cognitive radio networks. In: IEEE Vehicular Technology Conference (VTC 2010-Spring), pp. 1–5. Taipei, Taiwan (2010)
53. Wysocki, T., Jamalipour, A.: MAC framework for intermittently connected cognitive radio networks. In: Proc. IEEE Personal, Indoor, and Mobile Radio Communications (PIMRC). Tokyo, Japan (2009)
54. Xin, C., Xie, B., Shen, C.: A novel layered graph model for topology formation and routing in dynamic spectrum access networks. In: Proc. IEEE DySPAN'05, pp. 308–317. Baltimore, MD (2005)
55. Xu, Y., Sheng, M., Zhang, Y.: Traffic-Aware routing protocol for cognitive network. In: IEEE Vehicular Technology Conference Fall (VTC 2010-Fall), pp. 1–5. Ottawa, CA (2010)
56. Yang, K., Tsai, Y.: Link stability prediction for mobile ad hoc networks in shadowed environments. In: Proc. IEEE GLOBECOM. San Francisco, CA (2006)
57. Yu, F.R., Huang, M., Tang, H.: Biologically inspired consensus-based spectrum sensing in mobile ad hoc networks with cognitive radios. IEEE Netw. **24**(3), 26–30 (2010)
58. Yuan, Z., Song, J.B., Han, Z.: Interference minimization routing and scheduling in cognitive radio wireless mesh networks. In: IEEE Wireless Communications and Networking Conference (WCNC), pp. 1–6. Sydney, Australia (2010)
59. Zhou, X., Lin, L., Wang, J., Zhang, X.: Cross-layer routing design in cognitive radio networks by colored multigraph model. Wireless Pers Commun. **9**(1), 123–131 (2009)

Chapter 9
Routing Schemes for Cognitive Radio Mobile Ad Hoc Networks

Jun Li, Yifeng Zhou, and Louise Lamont

Abstract In this chapter, we propose a classification of existing routing schemes for cognitive radio mobile ad hoc networks (CR-MANETs) and review these representative CR-MANET routing schemes. Then, we describe a CR-MANET model and present a novel adaptive routing design for the CR-MANET, referred to as ARDC, algorithmically and through examples. ARDC is based on the graph modeling approach, and its most significant contribution is that ARDC adapts to dynamic changes in the network topology much more computationally efficient than other CR-MANET routing schemes. At last, some further research directions on CR-MANET routing are identified.

9.1 Introduction

Wireless services have been witnessing a phenomenal growth since mid-1990s. Under the current static spectrum assignment policy, each wireless device occupies a fixed portion of the spectrum for temporally and spatially exclusive usage. Consequently, the spectrum scarcity problem will be encountered in the near future as more and more wireless services are launched. That is, the frequency spectrum available for those new wireless services will be completely drained off. Following a report by the Federal Communications Commission (FCC) that many statically allocated frequency spectrum bands are under-utilized geographically and/or temporally [9], the emerging cognitive radio technology has been proposed as a solution to the spectrum scarcity problem. In contrast to the fixed and inflexible spectrum occupancy, a cognitive radio technology enabled device, referred to as a secondary user or a cognitive user, is not assigned to any fixed block of the spectrum and can operate on any vacant portion of the spectrum that has previously been allocated to a licensed user, referred to as a primary user. As indicated from their names, a primary user (PU) has high priority in the usage of its pre-assigned portion of the spectrum over a cognitive user (CU), and the activity of a PU will not be interfered

J. Li (✉)
Communications Research Centre Canada, Ottawa, ON K2H 8S2, Canada
e-mail: jun.li@crc.gc.ca

F.R. Yu (ed.), *Cognitive Radio Mobile Ad Hoc Networks*,
DOI 10.1007/978-1-4419-6172-3_9,

by the existence of CUs. To do so, a CU must periodically sense vacant portions of the spectrum, adaptively choose transmission parameters, and dynamically access these (under-utilized) frequency channels.

Studies on cognitive radio communication and networking have mainly been focused on lower layer (i.e., physical and medium access control layer) issues, such as spectrum sensing and opportunistic spectrum access [7, 16, 17]. Due to some fundamental differences between traditional wireless networks and cognitive radio wireless networks, the networking issues also need to be addressed [2, 3], which appear more important in a multi-hop infrastructure-less cognitive radio network, for example, a mobile ad hoc network composed of geographically colocated PUs and CUs (CR-MANET). (A cognitive radio network can also be deployed as an infrastructure-based network [2].) Due to a variety of applications of ad hoc networks in the civilian and military domain [15], in this chapter we focus on the network layer issues in self-organizing cognitive radio networks or CR-MANETs. Specifically, we concentrate on existing and new routing schemes for CR-MANETs that perform computation of end-to-end routing paths joint with channel assignment and maintenance of these routes.

In a CR-MANET, each CU is equipped with one or more pre-defined radios (or transceivers) that can be tuned to a radio frequency band (or a channel) among a range of the spectrum. In addition, a CU has the functionality of scanning available channels at the present moment to avoid inference with the activity of a PU. Through the periodic exchange of beacon information, a CU discovers its neighboring CUs, each of which connects to the CU via one or more scanned channels. A scanned channel is assumed to be symmetric in this chapter. Moreover, the maximum number of channels that can be sensed by a CU is limited. The CR-MANET is heterogenous; i.e., the set of available channels and the number of tunable transceivers may vary from one node to another. In general, the number of pre-defined transceiver is smaller than the maximum number of channels that a node can sense. Links in the CR-MANET may change over time. That is, as the network evolves over time, a new channel may become available to connect a pair of CUs, while an existing channel may disappear in the network, e.g., due to node mobility or start of occupancy of PUs. Based on the network settings discussed above, a variety of routing schemes have been proposed in the literature. In this chapter, we classify those routing schemes for CR-MANETs and review one or more representative ones for each routing class. Then, we present a novel framework of adaptive routing design for CR-MANETs, and give some concrete examples for demonstration of the adaptiveness and performance of the adaptive routing design framework. Some challenges and trends in CR-MANET routing design are also discussed.

The rest of this chapter is organized as follows: Section 9.2 classifies existing CR-MANET routing schemes and reviews these representative ones in each routing class. A novel adaptive routing design for CR-MANETs is presented and concrete examples are given in Section 9.3. Concluding remarks are given in Section 9.4.

9.2 CR-MANET Routing Schemes

In this section, we present a classification of existing routing schemes for CR-MANETs, and in each routing class, we review one or more representative schemes.

9.2.1 Classification of CR-MANET Routing Schemes

Several classification methods have been proposed to categorize existing routing schemes for CR-MANETs. In [2], CR-MANET routing schemes are categorized based on whether a scheme considers support for single or joint functionality among spectrum decision, PU awareness, and reconfigurability. In [4], the authors classify existing CR-MANET routing schemes based on whether or not the spectrum knowledge is fully captured by a CU. In this chapter, we shall present a new classification of CR-MANET routing schemes based on the approaches used for establishing end-to-end routes. As illustrated in Fig. 9.1, at the top level are two general classes. One class includes CR-MANET routing schemes designed by modifying classical MANET routing protocols, while the other contains CR-MANET routing schemes proposed by using various modeling methods. The latter has three subclasses at the second level.

9.2.2 MANET Protocol-Based CR-MANET Routing

In the class of MANET protocol-based CR-MANET routing, a routing scheme for CR-MANETs has been designed by modifying a classical MANET routing scheme. In the following, we mainly focus on SEARCH in [6] and the routing and spectrum assignment protocol in [28], and briefly discuss some other CR-MANET routing schemes, such as these in [5, 10, 18, 21, 26, 30], belonging to this routing class.

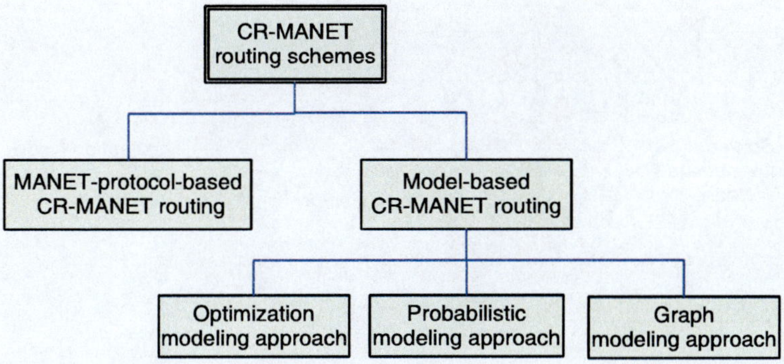

Fig. 9.1 Classification of CR-MANET routing schemes

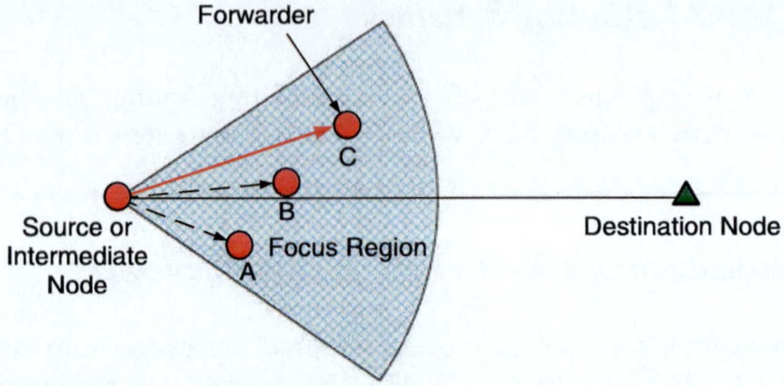

Fig. 9.2 Greeding geographic forwarding mechanism

In [6], Chowdhury and Felice propose a routing protocol for CR-MANETs, referred to as SEARCH, based on GPSR [12], a geographic routing algorithm for MANETs. SEARCH helps a CU find the route of minimum end-to-end latency to another CU with considerations of spectrum selection and avoidance of PU activities. It is composed of two phases: route setup and route maintenance. In the route setup phase, SEARCH uses the greedy geographic forwarding mechanism, as shown in Fig. 9.2, to forward the route request (RREQ), if one or more candidate forwarders are found (i.e., some nodes are within the focus region but not covered by the area of the PU activity). Otherwise, the PU avoidance mechanism, illustrated in Fig. 9.3, is applied to find a node outside the focus region to forward RREQ. Once RREQ reaches the destination, the destination node selects a routing path together with the channels along the path using the joint channel-path optimization mechanism, which aims at minimizing the end-to-end latency. As illustrated in Fig. 9.4, a switch from one channel to another could happen on a selected route, and some extra delay induced by the channel switch has to be considered. After a routing path between the source and the destination node is set up, it might be unusable due

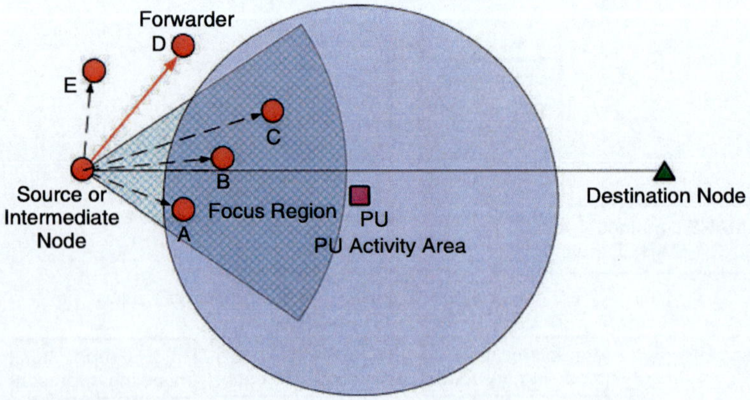

Fig. 9.3 PU avoidance mechanism

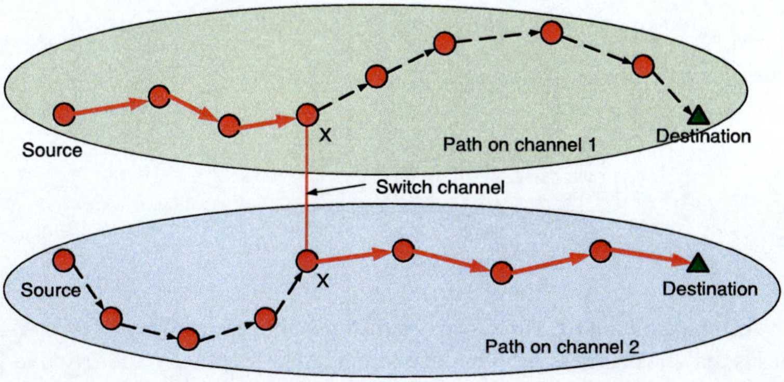

Fig. 9.4 Joint channel-path optimization mechanism

to interference with the PU activity or be disconnected because of node mobility. In either case, the route needs to be maintained in the route maintenance phase, where the node at breaking point of the route sends RREQ to the destination for a path as a replacement of the broken segment of the route. Due to the property of on-demand route setup, SEARCH runs in a decentralized manner. However, it has to be assumed that each node has the location information about its neighboring nodes and the source node knows the location of itself and the destination node. This requires some location service beneath SEARCH. Moreover, RREQ has to be transmitted along each cognitive channel in the route setup phase. All these could result in a significant amount of protocol overhead.

In [28], the authors propose an AODV-based routing and spectrum assignment protocol with local coordination of traffic flows. The protocol operates on a common control channel shared by all CUs. When an end-to-end path between a source–destination pair needs to be established, the source node broadcasts RREQ, which contains the information about its current available cognitive channels, over the common control channel. The forwarding rule for RREQ is discussed as follows. Assume that CU A broadcasts RREQ after updating the channel information in RREQ with the set S_A of available cognitive channels in A and CU B is a neighbor of A. After receiving RREQ, B determines the routing path and sends RREP to the source if B is the destination. Otherwise, the intersection of S_A and S_B is calculated. If the intersection of the two sets is an empty set, it is implied that a route passing through B cannot be established. Otherwise, B broadcasts RREQ after updating its channel information with S_B (i.e., the set of available cognitive channels in B). The procedure of processing RREQ discussed above is illustrated in Fig. 9.5. Under the same network setup as that in [28], e.g., a common control channel is shared by CUs, the authors in [21] study independently the routing problem in CR-MANETs and propose SPEAR, an AODV-based routing scheme as well.

In contrast to the CR-MANET routing schemes in [21, 28], the schemes proposed in [10, 18, 26, 30] do not count on a dedicated common control channel. Instead, RREQ is required to be broadcasted over each cognitive channel such that multiple routes, one per cognitive channel, between a source–destination pair

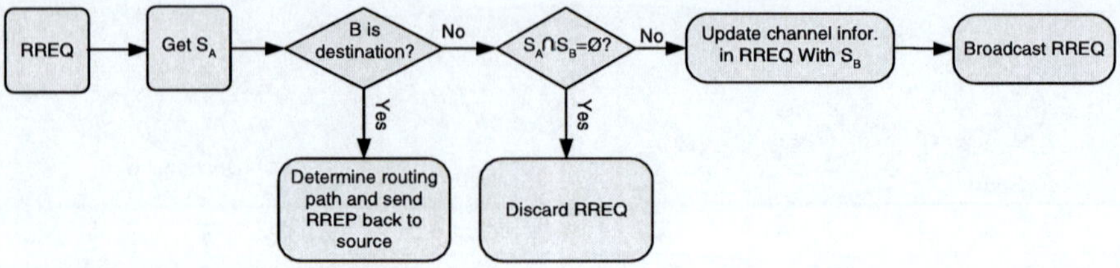

Fig. 9.5 Procedure of processing RREQ

could be established in [10, 18]. (A major difference between the two AODV-based schemes is that the deafness problem defined in [22] is explicitly addressed in [18].) The spectrum-tree-based on-demand routing protocol (STOP-RP) presented in [30] is another AODV-based routing scheme for CR-MANETs. In STOP-RP, a tree, referred to as a spectrum-tree, is constructed for each cognitive channel. Then a route discovery process, which is considered as an extended version of AODV, is conducted on the formed spectrum-trees to find the route between a pair of CUs. Another similar AODV-based routing scheme is reported in [5], where whether RREQ needs to be broadcasted via a dedicated common channel channel is not explicitly described. When a CU is equipped with multiple cognitive transceivers, a DSR-based CR-MANET routing scheme is proposed in [26] to find multiple routes with minimum contention and interference among cognitive channels.

In summary, most MANET protocol-based routing schemes for CR-MANETs are based on AODV and thus are reactive in nature. A fixed common control channel is assumed in design of some MANET protocol-based routing schemes, while it is not in others. While the feasibility of allocating a dedicated control channel in cognitive radio networks is yet to be investigated, performance comparisons of these CR-MANET routing schemes with or without the support of the common control channel need to be carried out.

9.2.3 Model-Based CR-MANET Routing

In the class of model-based CR-MANET routing, a routing scheme is proposed by solving a classic mathematical model that characterizes the settings of a real network. Therefore, a model-based CR-MANET routing scheme does not directly work upon a real network, which becomes a major difference from a MANET protocol-based routing scheme. Based on the specific modeling approaches used for establishing end-to-end routes, model-based CR-MANET routing schemes are further classified into three subclasses: optimization modeling, probabilistic modeling, and graph modeling (see Fig. 9.1).

9.2.3.1 Optimization Modeling Approach

Below we review a representative routing scheme reported in [11] and briefly discuss other CR-MANET routing schemes based on optimization modeling, such as those in [8, 19, 20].

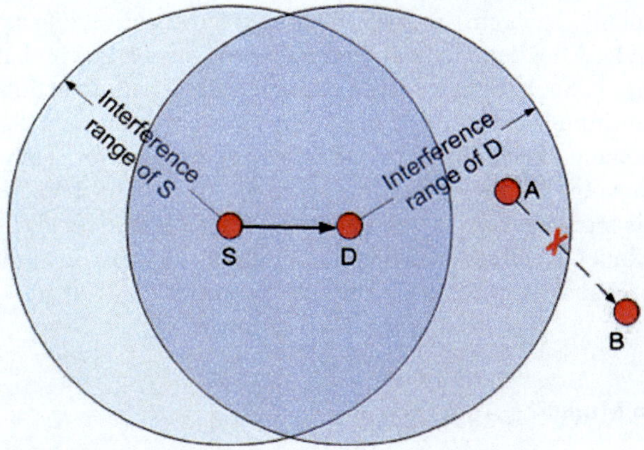

Fig. 9.6 Signal interference constraint

In [11], the authors consider a number of CUs each having a set of spectrum bands available for transmission. Each available spectrum band can be divided into a number of sub-bands of different bandwidths. The spectrum selection mechanism takes into account a signal interference model, which allows at most one communication session to use a specific sub-band in the area covered by the interference ranges of the transmitter and the receiver (see Fig. 9.6). Multiple routes between a source–destination pair can be obtained for a communication session, even though the total rate of the multiple routes needs to be equal to the required data rate of the communication session. Give a set of communication sessions (i.e., a set of source–destination pairs) and their required data rates, a non-linear optimization problem is formulated to find the routes and spectrum sub-bands passed by the routes for each communication session. The object function of the optimization problem is the total spectrum resource in the network fulfilling the communication sessions and their required rates. In contrast, Ma and Tasng [19, 20] consider a fairness factor of a communication session as the object function in their non-linear optimization problem formulation. In addition, the frequency bands of multiple channels are not divided into sub-channels. It is worthwhile noting that, even though the modeling techniques in both studies are mathematically sound, the solutions are so difficult to obtain that only approximate or heuristic results have been developed. In [8], the authors propose a CR-MANET routing scheme by formulating an optimization problem with the objective of maximizing the network throughput.

9.2.3.2 Probabilistic Modeling Approach

The probabilistic modeling approach is used in [13, 23, 24] for CR-MANET routing design. In [13], the authors consider N CUs operating in an area over a maximum number M of frequency channels. PUs are distributed in the area according to a Poisson point process, which leads to a known result that the interference of a CU with a PU approximately follows a lognormal distribution. Based on the standard fact (e.g., Shannon's Theorem), the capacity of a channel connecting a pair of CUs is a random variable. After defining the weight of the channel as the probability that the channel capacity is not less than a fixed required rate, a path selection and channel assignment algorithm is proposed to determine the most probable path between two CUs. The implementation of the proposed routing scheme needs a global view of the network topology. In [23, 24], the authors argue that, if a channel was reliable before the present, it is more likely to be reliable in the future. Based on this argument, the weight of a channel is defined as a function of the history of temporal usage of the channel, and probabilistic routing schemes are proposed.

9.2.3.3 Graph Modeling Approach

The studies in [1, 14, 25, 27, 29] propose CR-MANET routing schemes based on the graph modeling approach. In the following, we mainly discuss the schemes reported in [1, 25, 27, 29].

In [1], Gymkhana, which is a connectivity-based routing scheme for CR-MANETs, is proposed using the graph modeling approach. In Gymkhana, the destination CU first collects the information about all possible paths through RREQs initiated by the source CU. A graph is constructed for each path at the destination, from which the Laplacian matrix of the graph is obtained. (The Laplacian matrix of a graph is defined as the difference between the graph's degree matrix and its adjacent matrix.) By using a known result that the connectivity of a graph can be measured by the second smallest eigenvalue of the Laplacian matrix of the graph, the destination CU selects the path with the highest connectivity. That is, a selected path is guaranteed to have the least interference with primary users. It is worthwhile noting that Gymkhana operates in a distributed manner, while some theoretical aspects of this scheme need be further investigated.

In [27], the authors propose a CR-MANET routing scheme based on graph modeling. The proposed scheme contains two algorithmic components: topology formation algorithm, which constructs a layered graph, and path-centric channel assignment algorithm, which conducts path computation joint with channel assignment. The number of layers in the layered graph corresponds to the maximum number of channels that can be sensed by a node. Each layer contains a set of vertices that is twice the network size (i.e., the number of CUs in the network). With four types of weighted edges connecting vertices, an optimal routing path, which has the smallest number of hops among the paths with minimum adjacent hop interference, is obtained for a pair of nodes. Since the resulting layered graph model contains a larger number of vertices even for a relatively small network size, the overall computational cost of the routing scheme is high. In [29], the authors form a colored multigraph model by using different colors for distinct channels in the network. By applying a novel shortest path algorithm to the colored multigraph model, a locally optimal routing path can be obtained for a pair of nodes. Using graph modeling algorithms, the authors in [25] propose a route selection mechanism that maximizes throughput and minimizes channel interference.

In CR-MANETs, the number of channels that are available to a CU is time varying, i.e., the available channels in the CU change as time evolves. This requires that the routing should explicitly consider the time-varying nature of the channels in an adaptive manner. A majority of above reviewed studies on CR-MANET routing design have assumed static single- or multi-channel networks without consideration of the time-varying availability of network links. In particular, almost all studies on CR-MANET routing using the graph modeling approach only address networks with static links and do not explicitly consider the time-varying nature of link availability. Another interesting aspect in CR-MANET routing design is the exploitation of the channel diversity in a CR-MANET to improve routing efficiency, since two CUs in the CR-MANET are typically connected by multiple paths through different intermediate CUs and channels. In the next section, we propose a framework of adaptive routing design for CR-MANETs using the graph modeling approach. The proposed CR-MANET routing design includes a novel topology formation component and a routing scheme. The topology formation component forms a weighted directional graph model for a given CR-MANET and adapts it to time-varying changes of network links. Based on the graph model, the routing scheme computes optimal routing paths for a pair of CUs.

9.3 ARDC: A Graph Model-Based Routing Scheme

In this section, we describe a cognitive radio mobile ad hoc network model and detail a graph model-based routing design, denoted by ARDC, for the CR-MANET.

9.3.1 CR-MANET Model and Routing Design Framework

A cognitive radio ad hoc network consists of M nodes identified by node $1, 2, \cdots, M$. Each node has one or more pre-defined transceivers that can be tuned to a radio frequency band (or a channel) among a spectrum range using its spectrum mobility and sharing functions. The number of transceivers in node m is denoted by r_m, for $m = 1, 2, \cdots, M$, and the row vector $\mathbf{r} = (r_1, r_2, \cdots, r_M)$ is referred to as the network interface vector. A node periodically scans available channels using its spectrum sensing function and discovers its neighboring nodes, each of which connects to the node via one or more scanned channels. A scanned channel is assumed to be symmetric. The maximum number of channels that can be sensed by a node is assumed to be N, and the N channels are identified by channel $1, 2, \cdots, N$. The network is heterogenous; i.e., the set of available channels and the number of tunable transceivers may vary from one node to another. In general, the number of pre-defined transceivers is smaller than the maximum number of channels that a node can sense. Links in the cognitive radio ad hoc network change over time. That is, as the network evolves over time, a new channel may become available to connect a pair of nodes, while an existing channel may disappear in the network.

We assume that one update in the cognitive radio ad hoc network takes place at a time. Specifically, we use an increasing sequence $\{t_0, t_1, t_2, \cdots\}$ of non-negative real numbers to represent a set of time instants. t_0 denotes an initial time point at which the network starts operation. At each time t_i, for $i \geq 1$, one update in the network occurs, which will be either one of the following four events.

1. A new channel becomes available between a pair of nodes;
2. An existing channel disappears in the network;
3. A communication session is initiated;
4. A communication session is finished.

In the rest of this chapter, a channel update refers to item 1 or 2, and a communication update refers to item 3 or 4. We further assume that the availability of the channels along a chosen optimal routing path for a communication session doesn't change during the period of the session. Based on the above network model, we define that an optimal routing path between a pair of nodes has the smallest number of hops after minimizing the adjacent hop interference. As defined in [29], adjacent hop interference of a route is the number of hops along the route for which each hop uses the same channel as does its previous hop.

The design framework of ARDC, as described in Algorithm 1 using an algorithmic format, consists of two major components. One component is topology formation (i.e., initialization and adaptation steps in Algorithm 1) and the other is routing scheme. The topology formation component is responsible for channel updates as well as initial graph construction, while the routing scheme deals with communication updates. The proposed routing design framework addresses routing issues that occur in the network layer. Below the network layer is the medium access control (MAC) layer, where some MAC protocol is assumed. Since a MAC protocol can regulate the access of the network nodes to the transmission channels, in practice one channel could simultaneously be used by more than one routing path. Therefore, we allow a channel to be concurrently selected by multiple routes in ARDC.

Algorithm 1: Framework of ARDC

```
begin
    Initialization according to Algorithm 2 (see Section 9.3.2.1);
    i ← 1;
    while (algorithm running) do
        if (a channel update occurs at t_i) then
            Adaptation according to Algorithm 3 (see Section 9.3.2.2);
        end
        if (a communication update occurs at t_i) then
            Routing scheme according to Algorithm 4 (see Section 9.3.3);
        end
        i ← i + 1;
    end
end
```

In the remainder of this section, we discuss these two components, algorithmically and through examples.

9.3.2 Topology Formation

In this section, the topology formation component, which includes both the initialization and the adaptation steps of Algorithm 1, is explained in detail. The key concept used in topology formation algorithms is graph modeling, through which a simple directed graph is formed as a representation of the up-to-date physical network. Based on the simple directed graph model, the routing scheme, which will be detailed in Section 9.3.3, computes optimal routing paths for a requested communication session.

We denote a simple directed graph by

$$\mathcal{G}_i = (\mathcal{V}_i, \mathcal{E}_i) \tag{9.1}$$

which will be a representation of the physical network at time t_i, for $i = 0, 1, 2, \cdots$. In (1), $\mathcal{V}_i$ and $\mathcal{E}_i$ represent the vertex set and the edge set, respectively. A directional edge $\mu\nu$ represents a directional connection from vertex μ to vertex ν, while a bidirectional edge $\mu\nu$ represents two directional edges $\mu\nu$ and $\nu\mu$. In the initialization step of Algorithm 1, an initial graph $\mathcal{G}_0$ is constructed with the network connection information at the initial time t_0 and the network interface vector $\mathbf{r}$ being inputs. If a new channel becomes available or an existing channel disappears at time t_i, for $i \geq 1$, an adaptive graph $\mathcal{G}_i$ is created by adapting $\mathcal{G}_{i-1}$ to the channel update at t_i in the adaptation step.

In the constructed graph $\mathcal{G}_i$, for each $i = 0, 1, 2, \cdots$, a vertex $m_n \in \mathcal{V}_i$ implies that node m has channel n available for communicating with a neighbor of the node at time t_i. Directional edges from/to vertex m_n are added between vertices associated with neighbors of node m. A directional edge $m_n\hat{m}_{\hat{n}}$ implies that data packets could enter node m over channel n and be routed from node m to node $\hat{m}$ over channel $\hat{n}$.

We take into account the adjacent hop interference metric and the effect of transceivers in optimal routing path computation by assigning a weight associated with each directional edge. First of all, on a selected routing path, it is preferable that two adjacent hops use different channels to minimize adjacent hop interference. This is accomplished by differentiating directional edges of $\mathcal{G}_i$. That is, the weight associated with a directional edge $m_n\hat{m}_{\hat{n}}$ for $n \neq \hat{n}$ must be smaller than that associated with edge $m_n\hat{m}_{\hat{n}}$ for $n = \hat{n}$. Moreover, a node with more than one transceivers can relay packets simultaneously using two different channels (one for reception and the other for transmission), and thus should be considered a preferable intermediate node on an optimal routing path. Therefore, the assigned weight of a directional edge $m_n\hat{m}_{\hat{n}}$ for $n \neq \hat{n}$ depends on the number r_m of transceivers in node m. In summary, if we let w be the weight associated with a directional edge $m_n\hat{m}_{\hat{n}}$ for $n = \hat{n}$, w_1 the weight associated with a directional edge $m_n\hat{m}_{\hat{n}}$ for $n \neq \hat{n}$ and $r_m = 1$, and w_2 the weight associated with a directional edge $m_n\hat{m}_{\hat{n}}$ for $n \neq \hat{n}$ and $r_m \geq 2$, then we have

$$w > w_1 > w_2 \geq 0 \tag{9.2}$$

In examples and simulation studies conducted in this Chapter, we use $w = 3$, $w_1 = 2$, and $w_2 = 1$ for illustrating the algorithm. A study on optimized settings of these weights could be carried out as a future research topic.

9.3.2.1 Initialization Step

The algorithm for constructing the initial graph $\mathcal{G}_0$ is given by Algorithm 2. In this algorithm, we create vertices associated with each node in the CR-MANET. A vertex m_n is added to the vertex set $\mathcal{V}_0$ if node m can potentially communicate with a neighbor of node m over channel n. A directional edge $m_n\hat{m}_{\hat{n}}$ indicates that data packets can be received by node m via channel n and routed from node m to node $\hat{m}$ over channel $\hat{n}$. This directional edge can result in minimum (or zero) adjacent hop interference if $n \neq \hat{n}$, and thus should be favorably chosen on an optimal routing path. Therefore, the assigned weight of directional edge $m_n\hat{m}_{\hat{n}}$ for $n \neq \hat{n}$ and $r_m > 1$ (w_2 in Algorithm 2) is smaller than that for $n \neq \hat{n}$ and $r_m = 1$ (w_1 in Algorithm 2), which is smaller than that for $n = \hat{n}$ (w in Algorithm 2).

Algorithm 2: Initialization Step

```
input : The physical network at t_0 and the interface vector r = [r_1, r_2, ··· , r_M].
output: Graph G_0 with all edges assigned to weights.
begin
    V_0 ← ∅;
    E_0 ← ∅;
    for m ← 1 to M do
        temp ← V_0;
        for n ← 1 to N do
            if channel n is available at node m then
                add a new vertex m_n into V_0;
                foreach vertex m̂_n̂ in temp do
                    if (n ≠ n̂) & (m and m̂ is connected via channel n̂) then
                        add directional edge m_n m̂_n̂ to E_0;
                        assign weight of directional edge m_n m̂_n̂ to w_1 if r_m = 1;
                        assign weight of directional edge m_n m̂_n̂ to w_2 if r_m ≥ 2;
                    end
                    if (n ≠ n̂) & (m and m̂ is connected via channel n) then
                        add directional edge m̂_n̂ m_n to E_0;
                        assign weight of directional edge m̂_n̂ m_n to w_1 if r_m̂ = 1;
                        assign weight of directional edge m̂_n̂ m_n to w_2 if r_m̂ ≥ 2;
                    end
                    if (n = n̂) & (m and m̂ is connected via channel n) then
                        add bidirectional edge m_n m̂_n to E_0;
                        assign weight of directional edge m_n m̂_n to w;
                    end
                end
            end
        end
    end
end
```

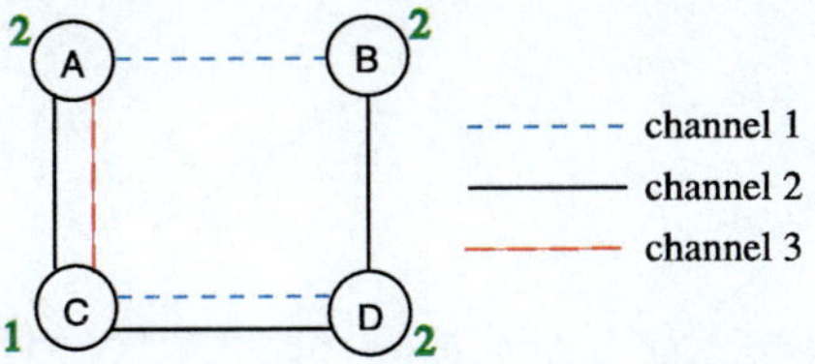

Fig. 9.7 Cognitive radio ad hoc network

In the following, Example 1 illustrates Algorithm 2, which creates the initial graph $\mathcal{G}_0$.

Example 1 Assume that a cognitive radio ad hoc network consists of nodes identified by A, B, C, D ($M = 4$). The maximum number of channels that can be sensed by a node is 3 ($N = 3$), and these channels are identified by channel 1, 2, and 3. The physical network as shown in Fig. 9.7, where the integer near a node represents the number of transceivers of the node, is given at some initial time t_0.

Initially, we set two empty sets $\mathcal{V}_0$ and $\mathcal{E}_0$. Since $M = 4$, we need four iterations, each of which corresponds to one of these four nodes, to complete construction of graph $\mathcal{G}_0$. Without loss of generality, we use the alphabetic order of the nodes for the construction process. In the first iteration, we consider node A, whose available channels are channels 1, 2, and 3. After this iteration, $\mathcal{V}_0 = \{A_1, A_2, A_3\}$ and $\mathcal{E}_0$ is empty. $\mathcal{G}_0$ after the first iteration is shown in Fig. 9.8a. The second iteration involves node B, which has channels 1 and 2 available and is connected with node A over channel 1. Then two vertices B_1 and B_2 are added to the set $\mathcal{V}_0$. Since $r_A = r_B = 2$, these added directional edges have weight 1 except bidirectional edge A_1B_1 with weight 3. Graph $\mathcal{G}_0$ after this iteration is shown in Fig. 9.8b. Node C is dealt with in the third iteration. It has three available channels, and thus three new vertices C_1, C_2, and C_3, are added into the set $\mathcal{V}_0$. Since $r_C = 1$, these directional edges $C_nA_{\hat{n}}$, for which $n \neq \hat{n}$, have weight 2. $\mathcal{G}_0$ after the third iteration is shown in Fig. 9.9.

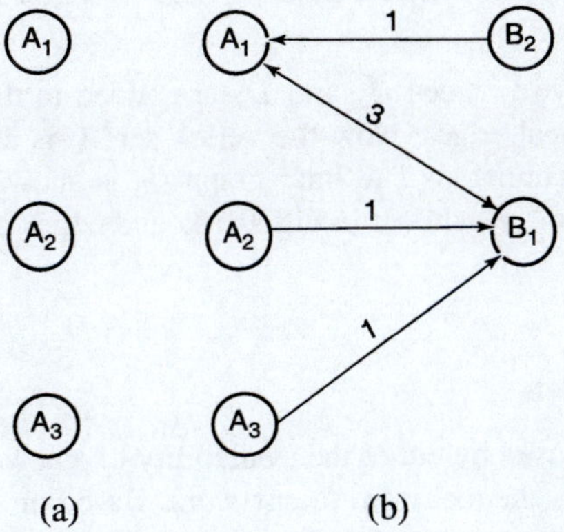

Fig. 9.8 First and second iterations

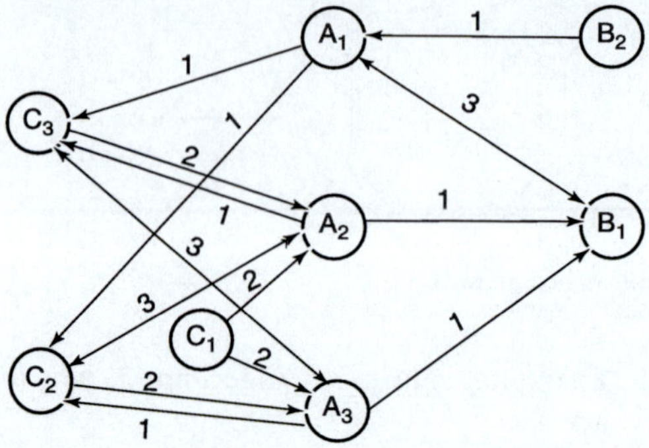

Fig. 9.9 Third iteration

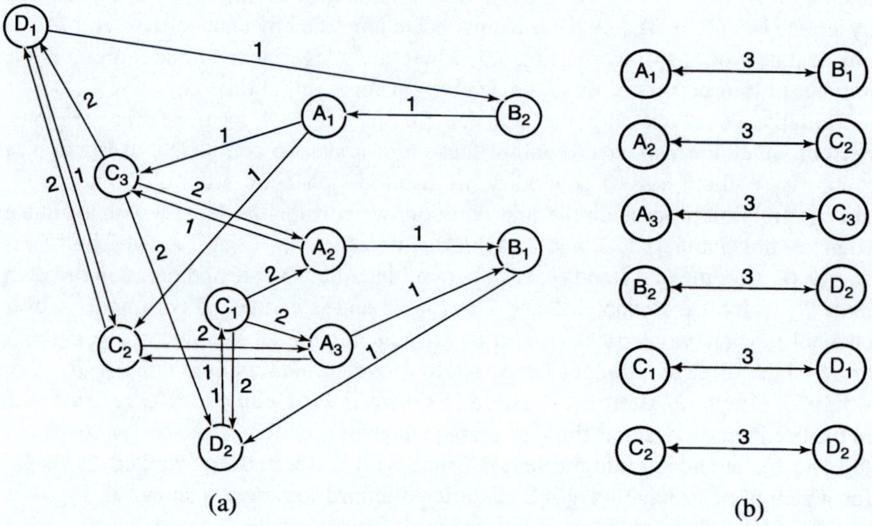

Fig. 9.10 Graph $\mathcal{G}_0$. (**a**) Edges of weights 1 and 2; (**b**) Edges of weight 3

In the last iteration, two vertices D_1 and D_2 are added to the set $\mathcal{V}_0$ followed by corresponding directional edges. Now the vertex set $\mathcal{V}_0$ is complete and the construction process is terminated. The final graph $\mathcal{G}_0$ is shown in Fig. 9.10, where edges of weights 1 and 2 are shown in Fig. 9.10a and edges of weight 3 are shown in Fig. 9.10b.

9.3.2.2 Adaptation Step

A channel update is caused by either the availability of a new channel or the disappearance of an existing channel from the network. Based on $\mathcal{G}_{i-1}$ and the channel update at time t_i, an adaptive algorithm for constructing $\mathcal{G}_i$ is given in Algorithm 3. As we can see from the algorithm, a channel update only affects the topology

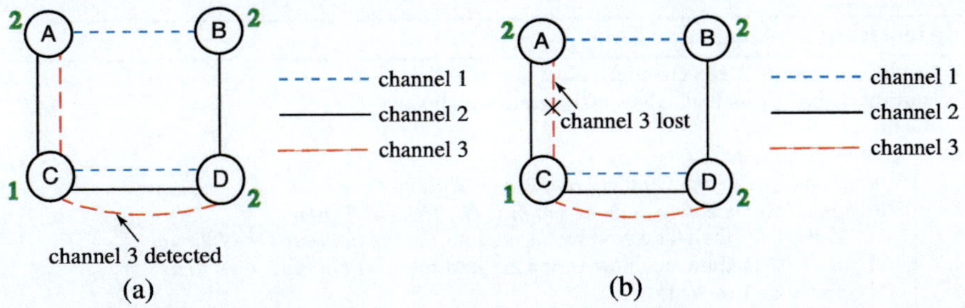

Fig. 9.11 Channel updates. (**a**) Gain of a channel; (**b**) Loss of a channel

formation locally. Thus we can locally adapt $\mathcal{G}_{i-1}$ with the channel update. This results in the adaptive graph $\mathcal{G}_i$, which represents the topology formation of the physical network at t_i. Algorithm 3 in the adaptation step features a very low computational cost, which is promising for practical implementations in CR-MANETs. The following example illustrates Algorithm 3 with $\mathcal{G}_{i-1}$ and a channel update at t_i being inputs.

Example 2 At time t_0, the physical network as shown in Fig. 9.7 with corresponding initial graph $\mathcal{G}_0$ formed in Fig. 9.10 is given. At some time t_1 after t_0, a channel update (i.e., gain of channel 3 connecting C and D) takes place, as shown in Fig. 9.11a. This results in $\mathcal{G}_1$. At some time t_2 after t_1, another channel update (i.e., loss of channel 3 connecting A and C before t_2) occurs, as shown in Fig. 9.11b. This results in $\mathcal{G}_2$. We construct adaptive graphs $\mathcal{G}_1$ and $\mathcal{G}_2$ below based on Algorithm 3.

To obtain adaptive graph $\mathcal{G}_1$, graph $\mathcal{G}_0$, and the channel update at time t_1 (i.e., channel 3 between C and D) are used as inputs to Algorithm 3. Initially, a new vertex D_3 is added to the vertex set $\mathcal{V}_1$. (Vertex C_3 is already in $\mathcal{V}_1$.) Then, directional edges are added to the edge set $\mathcal{E}_1$. The resulting graph $\mathcal{G}_1$ after adapting $\mathcal{G}_0$ to the acquisition of channel 3 at t_1 is shown in Fig. 9.12, where the newly added vertex and edges are shown in dotted lines.

At time t_2, channel 3, which connected A and C before t_2, becomes unavailable. Graph $\mathcal{G}_1$ and loss of channel 3 are used as inputs to Algorithm 3. In order for $\mathcal{G}_1$ to adapt to the update, directional edges A_1C_3, A_2C_3, A_3C_3, C_1A_3, C_2A_3, and C_3A_3 are removed from graph $\mathcal{G}_1$. Since channel 3 becomes unavailable in A at time t_2, vertex A_3 and all corresponding edges involving A_3 are removed from $\mathcal{G}_1$. The resulting graph $\mathcal{G}_{t_2}$ after adapting to the loss of channel 3 at t_2 is shown in Fig. 9.13, where the removed vertex and edges are in dashed lines.

9.3.3 Routing Scheme

The routing scheme computes optimal routing paths for a requested communication session and updates the topology formation for a completed communication session.

Algorithm 3: Adaptation Step

input : Graph $\mathcal{G}_{i-1}$ and channel update at t_i.
output: Graph $\mathcal{G}_i$ with all edges assigned to weights.
begin
  $\mathcal{V}_i \longleftarrow \mathcal{V}_{i-1}$, and $\mathcal{E}_i \longleftarrow \mathcal{E}_{i-1}$;
  $boolean1 \longleftarrow False$, and $boolean2 \longleftarrow False$;
  if *channel k connecting node m and $\hat{m}$ is GAINED at t_i* **then**
    if $m_k \notin \mathcal{V}_i$ **then** add new vertex m_k into $\mathcal{V}_i$, and $boolean1 \longleftarrow True$;
    if $\hat{m}_k \notin \mathcal{V}_i$ **then** add new vertex $\hat{m}_k$ into $\mathcal{V}_i$, and $boolean2 \longleftarrow True$;
    for $n \leftarrow 1$ **to** N **do**
      if *$(n \neq k)$ & (m_n is in $\mathcal{V}_i$)* **then**
        add directional edge $m_n\hat{m}_k$ to $\mathcal{E}_i$;
        assign weight of directional edge $m_n\hat{m}_k$ to w_1 **if** $r_m = 1$ OR to w_2 **if** $r_m \geq 2$;
      end
      if *$(n \neq k)$ & ($\hat{m}_n$ is in $\mathcal{V}_i$)* **then**
        add directional edge $\hat{m}_n m_k$ to $\mathcal{E}_i$;
        assign weight of directional edge $\hat{m}_n m_k$ to w_1 **if** $r_{\hat{m}} = 1$ OR to w_2 **if** $r_{\hat{m}} \geq 2$;
      end
      if *$(n \neq k)$ & (m and $\hat{m}$ is connected via channel n)* **then**
        if *(boolean1 is True)* **then**
          add directional edge $m_k\hat{m}_n$ $\mathcal{E}_i$;
          assign the weight of directional edge $m_k\hat{m}_n$ to w_1 **if** $r_m = 1$ OR to w_2 **if** $r_m \geq 2$;
        end
        if *(boolean2 is True)* **then**
          add directional edges $\hat{m}_k m_n$ to $\mathcal{E}_i$;
          assign the weight of directional edge $\hat{m}_k m_n$ to w_1 **if** $r_{\hat{m}} = 1$ OR to w_2 **if** $r_{\hat{m}} \geq 2$;
        end
      end
      if *(n is equal to k)* **then**
        add bidirectional edge $m_n\hat{m}_k$ into $\mathcal{E}_i$;
        assign the weight of bidirectional edge $m_k\hat{m}_k$ to w;
      end
    end
  end
  if *channel k connecting node m and $\hat{m}$ is LOST at t_i* **then**
    for $n \leftarrow 1$ **to** N **do**
      if *m_n (respectively, $\hat{m}_n$) is in $\mathcal{V}_i$* **then**
        remove directional edge $m_n\hat{m}_k$ (respectively, $\hat{m}_n m_k$) from $\mathcal{E}_i$;
      end
    end
    if *channel k is not available in node m (respectively, node $\hat{m}$) at t_i* **then**
      remove all edges involving m_k (respectively, those involving $\hat{m}_k$) from $\mathcal{E}_i$;
      remove vertex m_k (respectively, vertex $\hat{m}_k$) from $\mathcal{V}_i$;
    end
  end
end

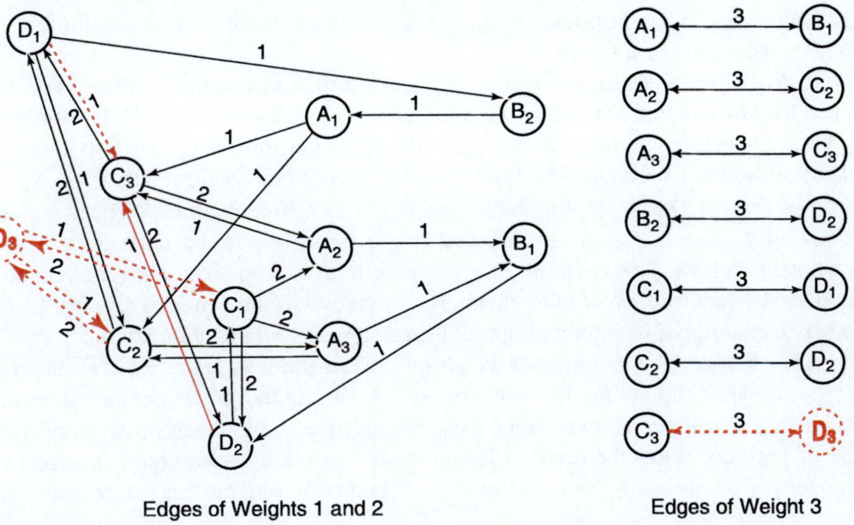

Fig. 9.12 Adaptive graph $\mathcal{G}_1$

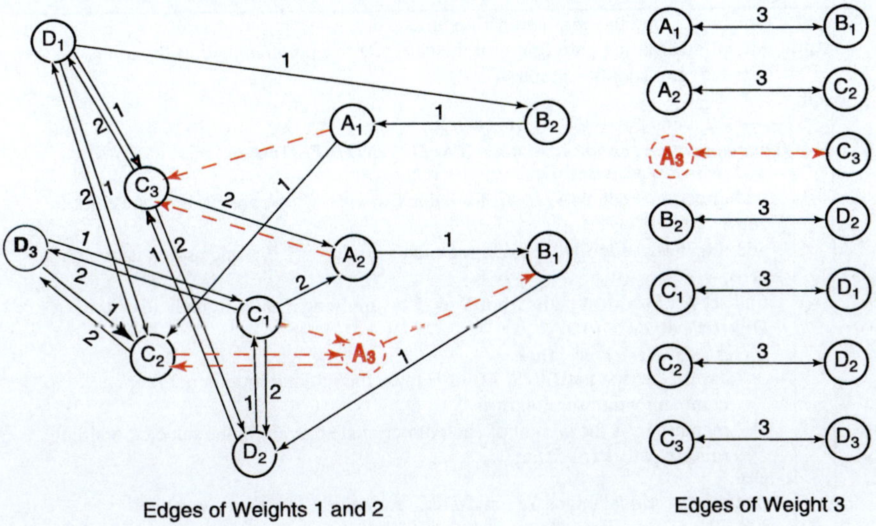

Fig. 9.13 Adaptive graph $\mathcal{G}_2$

We assume that, at time t_i for some $i \geq 1$, a communication update from source node S to destination node D is made. When the communication session from node S to D is initiated at t_i, the set of available channels in S is assumed to be $\{s_1, \cdots, s_P\}$ and the set of available channels in D is $\{d_1, \cdots, d_Q\}$, for some $P, Q = 1, \cdots, N$. In this case, the routing algorithm, as described in Algorithm 4, is used to identify an optimal routing path for the communication session and to update the topology formation.

In Algorithm 4, vertices S and D, and some directional edges involving S and D are added to assist in computing the optimal routing paths from S to D. By applying a known shortest path algorithm (e.g., Dijkstra's algorithm) to $\mathcal{G}_i$, we obtain the set $\mathcal{P}$ of all optimal routing paths from S to D. An optimal routing path $S \rightarrow S_x \rightarrow m_n \rightarrow \hat{m}_{\hat{n}} \rightarrow D_y \rightarrow D$, for some $x = s_1, \cdots, s_P$, intermediate vertices m_n and $\hat{m}_{\hat{n}}$, and $y = d_1, \cdots, d_Q$, implies that packets will be sent by node S to m over channel n, relayed from m to $\hat{m}$ via channel $\hat{n}$, relayed from $\hat{m}$ to D over channel y, and finally received by D over channel y. A study will be conducted in our future work to examine if a computed optimal routing path could contain a loop.

After the set $\mathcal{P}$ is computed, we simply select the first path $P(S, D)$ in set $\mathcal{P}$ to be used for the communication session. As discussed at the beginning of this section, a channel is allowed to be used by multiple communication sessions via a MAC protocol below the network layer. In such a case, a packet could experience a longer period of waiting time before transmission. To mitigate this effect, we adjust

Algorithm 4: Routing Scheme

```
input : Graph G_{i-1} and the communication update at t_i.
output: An optimal routing path (only for the case where a communication session is
        initiated) and adaptive graph G_i.
begin
    V_i <-- V_{i-1}, and E_i <-- E_{i-1} to get G_i;
    if (communication session from node S to D is INITIATED) then
        add two vertices S and D to V_i;
        add directional edge SS_x to E_i, for each x = s_1, ..., s_P, and assign its weight to
        zero;
        add directional edge D_yD to E_i, for each y = d_1, ..., d_Q, and assign its weight to
        zero;
        find set P of shortest paths from S to D by applying a shortest path algorithm (e.g.,
        Dijkstra's algorithm) to G_i;
        if (set P is NOT empty) then
            assign the first path P(S, D) in P to be the optimal routing path for
            communication session from S to D;
            increase by 1 the weight of each directional edge along the selected optimal
            routing path P(S, D) in G_i;
        end
        remove directional edges SS_x and D_yD, for each x = s_1, ..., s_P and
        y = d_1, ..., d_Q, from E_i;
        remove vertices S and D from V_i;
        return P(S, D) and graph G_i;
    end
    if (communication session from node S to D is COMPLETED) then
        in G_i, decrease by 1 the weight of each directional edge along the optimal routing
        path used for the communication session from node S to D;
        return graph G_i;
    end
end
```

the edge weights in graph $\mathcal{G}_i$ that correspond to channels that are being used for the communication session. This reduces the possibility of a channel being used by two simultaneous communication sessions. The adjustment process should reflect these explicit MAC requirements; this adjustment process in Algorithm 4 is implemented by increasing the weight of each directional edge along the selected optimal routing path $P(S, D)$ by 1. Once the communication session is complete, a reverse operation is performed to update the graph. That is, the weight of each directional edge along the selected optimal routing path $P(S, D)$ is decreased by 1.

The routing path computation in Algorithm 4 provides at least one optimal routing path to be used for packet transmission from S to D, as long as they are connected to each other. Since the computed set $\mathcal{P}$ may consist of multiple elements, multiple routing paths from S to D can be obtained, each of which has the same number of hops from S to D with the same minimized adjacent hop interference. In the following, an example is given to illustrate how to compute optimal routing paths and obtain graph $\mathcal{G}_i$ based on Algorithm 4.

Example 3 At time t_i, a physical network as shown in Fig. 9.14 is given, and a request for transmission from source node A to destination node D is made. From Example 2, the corresponding topology formation before t_i, which is denoted by

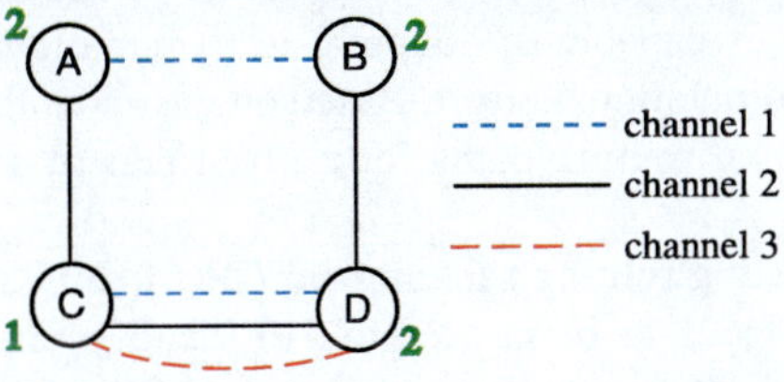

Fig. 9.14 Example 3

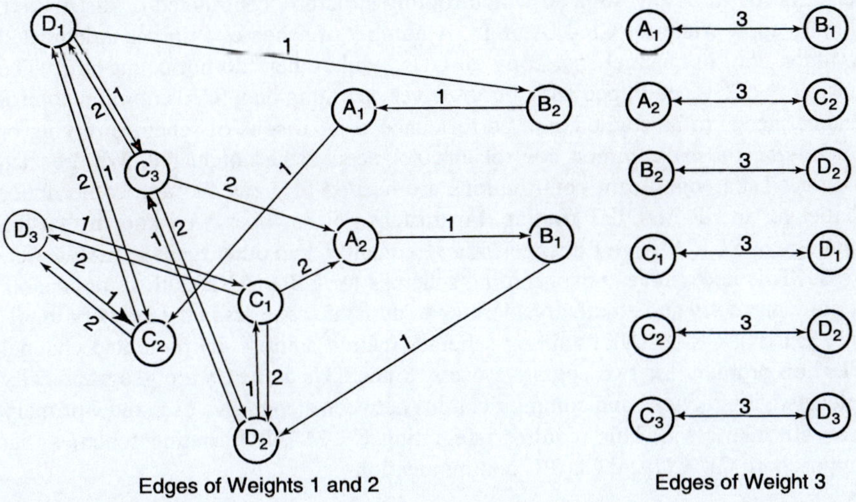

Fig. 9.15 Graph model

$\mathcal{G}_{i-1}$, is given in Fig. 9.15. We need to identify the optimal routing path(s) for transmission from A to D and the updated graph $\mathcal{G}_i$ based on Algorithm 4.

Starting from graph $\mathcal{G}_{i-1}$, two vertices A and D are first added to the graph. Then directional edges AA_1, AA_2, D_1D, D_2D, and D_3D are added with weight 0. After running Dijkstra's algorithm on the graph from A to D, $\mathcal{P} = \{A \rightarrow A_2 \rightarrow B_1 \rightarrow D_2 \rightarrow D\}$ gives a unique optimal routing path $P(A, D)$. The unique transmission route means that packets shall be transmitted from A (the source node) to B through channel 1 and from B to D (the destination node) via channel 2. During this transmission process, the weights of both directional edge A_2B_1 and directional edge B_1D_2 are increased by 1, respectively, which results in graph $\mathcal{G}_i$.

9.4 Conclusion and Discussion

In this chapter, we categorized existing routing schemes for CR-MANETs based on the approaches used for solving the routing problem. For each routing class, we reviewed one or more representative routing schemes. At last, we presented ARDC, a graph model-based routing design for CR-MANETs. In contrast to most relevant studies that do not address the time-varying nature of the availability of links in CR-MANETs, ARDC is capable of adapting to dynamic changes in the network topology with little computational effort. Based on the up-to-date network topology, ARDC computes optimal routing paths for a given pair of source and destination nodes.

From review of existing routing schemes for CR-MANETs, we strongly believe that further research needs to be carried out to identify pros and cons for these routing schemes. For instance, a number of proposed CR-MANET routing schemes have to be operated in a centralized unit, while others operate in a distributed manner. It needs to be investigated which routing structure, centralized or distributed, is more appropriate for CR-MANETs. A number of schemes assumes a dedicated common control channel shared by all CUs, while others do not count on it. The feasibility of CUs equipped with a transceiver operating on a fixed common control channel needs to be studied, and performance comparisons of schemes relying or not relying on the common control channel need to be conducted. We are also convinced that significant contributions are needed to break through some major challenges in CR-MANET routing. For instance, it has been well known that the topology of a CR-MANET changes more frequently than other types of mobile networks. This necessitates novel routing schemes for CR-MANETs that are capable of more promptly and efficiently adapting to dynamic changes in the network topology. Existing CR-MANET routing schemes mainly address the route and channel selection problem for two cognitive users. Since CUs and PUs are geographically colocated, interactions and communications between a cognitive user and a primary user will take place. This requires integrating CR-MANET routing schemes that support both CU-CU and CU-PU communications.

Acknowledgments This work was supported by Defence Research and Development Canada (DRDC).

References

1. A. Abbagnale and F. Cuomo, "Gymkhana: a connectivity-based routing scheme for cognitive radio ad hoc networks," *Proc. 29th IEEE International Conference on Computer Communications (INFOCOM 2010)*, pp. 1–5, San Diego, CA, Mar. 2010.
2. I.F. Akyildiz, W.Y. Lee, M.C. Vuran, and S. Mohanty, "NeXt generation/dynamic spectrum access/cognitive radio wireless networks: A survey," *Elsevier Computer Networks*, vol. 50, no. 13, pp. 2127–2159, Sept. 2006.
3. I.F. Akyildiz, W.Y. Lee, M.C. Vuran, and K.R. Chowdhury, "CRAHNs: Cognitive radio *ad hoc* networks," *Elsevier Ad Hoc Networks*, vol. 7, no. 5, pp. 810–836, Jul. 2009.
4. M. Cesana, C. Francesca, and E. Eylem, "Routing in cognitive radio networks: Challenges and solutions," *Elsevier Ad Hoc Networks*, vol. 9, no. 3, pp. 228–248, May 2011.
5. G. Cheng, W. Liu, Y. Li, and W. Cheng, "Joint on-demand routing and spectrum assignment in cognitive radio networks," *Proc. IEEE International Conference on Communications (ICC 2007)*, pp. 6499–6503, Scotland, Jun. 2007.
6. K.R. Chowdhury and M.D. Felice, "SEARCH: A routing protocol for mobile cognitive radio ad-hoc networks," *Elsevier Computer Communications*, vol. 32, no. 18, pp. 1983–1997, Dec. 2009.
7. C. Cormio and K.R. Chowdhury, "A survey on MAC protocols for cognitive radio networks," *Elsevier Ad Hoc Networks*, vol. 7, no. 7, pp. 1315–1329, Sept. 2009.
8. L. Ding, T. Melodia, S.N. Batalama, and M.J. Medley, "ROSA: distributed joint routing and dynamic spectrum allocation in cognitive radio *ad hoc* networks," *Proc. the 12th ACM International Conference on Modeling, Analysis and Simulation of Wireless and Mobile Systems (MSWiM 2009)*, pp. 13–20, Spain, Oct. 2009.
9. FCC, "Spectrum policy task force report," *ET Docket No. 02 - 135*, 2002.
10. T. Fujii and Y. Yamao, "Multi-band routing for ad-hoc cognitive radio networks," *Proc. SDR'06*, Orlando, FL, Nov. 2006.
11. Y.T. Hou, Y. Shi, and H.D. Sherali, "Spectrum sharing for multi-hop networking with cognitive radios," *IEEE Journal on Selected Areas in Communications*, vol. 26, no. 1, pp. 146–154, Jan. 2008.
12. B. Karp and H.T. Kung, "GPSR: greedy perimeter stateless routing for wireless networks," *Proc. of 6th Annual ACM/IEEE International Conference on Mobile Computing and Networking (MobiCom 2000)*, pp. 243–254, Boston, MA, Aug. 2000.
13. H. Khalife, S. Ahuja, N. Malouch, and M. Krunz, "Probabilistic path selection in opportunistic cognitive radio networks," *Proc. IEEE Global Communications Conference (GLOBECOM 2008)*, pp. 4861–4865, New Orleans, LA, Nov. 2008.
14. Y.R. Kondareddy and P. Agrawal, "A graph based routing algorithm for multi-hop cognitive radio networks," *Proc. the 4th Annual International Conference on Wireless Internet (WICON'08)*, Maui, Hawaii, Nov. 2008.
15. C. Kopp, *NCW101: An introduction to network centric warfare, Part 4 - Ad Hoc Networking*, Air Power Australia, 2008.
16. K.B. Letaief and W. Zhang, "Cooperative communications for cognitive radio networks," *Proc. of the IEEE*, vol. 97, no. 5, pp. 878–893, May 2009.
17. Z. Li, F.R. Yu, and M. Huang, "A distributed consensus-based cooperative spectrum sensing in cognitive radios," *IEEE Transactions on Vehicular Technolgy*, vol. 59, no. 1, pp. 383–393, Jan. 2010.
18. H. Ma, L. Zheng, X. Ma, and Y. Luo, "Spectrum aware routing for multi-hop cognitive radio networks with a single transceiver," *Proc. 3rd International ICST Conference on Cognitive Radio Oriented Wireless Networks and Communications (CrownCom 2008)*, Singapore, May 2008.

19. M. Ma and D.H.K. Tsanga, "Joint spectrum sharing and fair routing in cognitive radio networks," *Proc. 5th IEEE Consumer Communications and Networking Conference (CCNC 2008)*, pp. 978–982, Las Vegas, NV, Jan. 2008.
20. M. Ma and D.H.K. Tsanga, "Joint design of spectrum sharing and routing with channel heterogeneity in cognitive radio networks," *Elsevier Physical Communication*, vol. 2, no. 1–2, pp. 127–137, Mar.–Jun. 2009.
21. A. Sampath, L. Yang, L. Cao, H. Zheng, and B.Y. Zhao, "High throughput spectrum-aware routing for cognitive radio networks," *Proc. 3rd International ICST Conference on Cognitive Radio Oriented Wireless Networks and Communications (CrownCom 2008)*, Singapore, May 2008.
22. J. So and N.H. Vaidya, "A routing protocol for utilizing multiple channels in multi-hop wireless networks with a single transceiver," Technical Report, University of Illinois at Urbana-Champaign, Oct. 2004.
23. A.C. Talay and D.T. Altilar, "RACON: A routing protocol for mobile cognitive radio networks," *Proc. the 2009 ACM Workshop on Cognitive Radio Networks*, pp. 73–78, Beijing, China, Sept. 2009.
24. A.C. Talay and D.T. Altilar, "ROPCORN: Routing protocol for cognitive radio ad hoc networks," *Proc. International Conference on Ultra Modern Telecommunications & Workshops (ICUMT'09)*, pp. 1–6, St. Petersburg, Oct. 2009.
25. Q. Wang and H. Zheng, "Route and spectrum selection in dynamic spectrum networks," *Proc. IEEE Consumer Communications and Networking Conference (CNCC 2006)*, vol. 1, pp. 625–629, Jan. 2006.
26. X. Wang, T. Kwon, and Y. Choi, "A multipath routing and spectrum access (MRSA) framework for cognitive radio systems in multi-radio mesh networks," *Proc. 2009 ACM Workshop on Cognitive Radio Networks*, pp. 55–60, Beijing, China, Sept. 2009.
27. C. Xin, L. Ma, and C.C. Shen, "A path-centric channel assignment framework for cognitive radio wireless networks," *ACM/Springer Mobile Networks and Applications*, vol. 13, no. 5, pp. 463–476, Oct. 2008.
28. Z. Yang, G. Cheng, W. Liu, W. Yuan, and W. Cheng, "Local coordination based routing and spectrum assignment in multi-hop cognitive radio networks," *Springer Mobile Networks and Applications*, vol. 13, no. 1–1, pp. 67–81, Apr. 2008.
29. X. Zhou, L. Lin, J. Wang, and X. Zhang, "Cross-layer routing design in cognitive radio networks by colored multigraph model," *Springer Wireless Personal Communications*, vol. 49, no. 1, pp. 123–131, Jul. 2009.
30. G.M. Zhu, I.F. Akyildiz, and G.S. Kuo, "STOD-RP: A spectrum-tree based on-demand routing protocol for multi-hop cognitive radio networks," *Proc. IEEE Global Communications Conference (GLOBECOM 2008)*, pp. 1–5, New Orleans, LA, Nov. 2008.

Chapter 10
Delay in Cognitive Radio Networks

Yaling Yang, Chuan Han, and Bo Gao

Abstract This chapter presents analysis for delays for both multihop cognitive radio networks and single-hop cognitive radio networks. For multihop cognitive radio networks, we analyze the amount of time that a packet spends to travel over the intermittent relaying links over multiple relaying hops and characterize it with the metric called *information propagation speed*. Optimal relaying node placement strategies are derived to maximize information propagation speed. For single-hop cognitive radio networks, we will analyze how delay is affected by multiple cognitive radio design options, including the number of channels to be aggregated, the duration of transmission, the channel separation constraint on channel aggregation, and the time needed for spectrum sensing and protocol handshake. How these different options may affect the delay under different secondary and primary user traffic loads is revealed. Methods for computing optimal cognitive radio design and operation strategy are derived.

10.1 Introduction

Cognitive radio networks (CRN) have a lot of potentials to improve wireless spectrum efficiency. Understanding the fundamental performance characteristics of this new type of networks is important for the optimal planning of CRN and the design of CRN applications. Hence, in this chapter, one important aspect of CRN performance, the delay, is analyzed.

For a multihop CRN, information propagation speed (IPS) is used as a metric for understanding the multihop end-to-end packet delays. Models of IPS and methods to maximize IPS in two cases are introduced. The first case, named the maximum network IPS, maximizes IPS across a network topology over an infinite plane. The second case, named the maximum flow IPS, maximize the IPS between a given pair of source and destination nodes separated by a fixed distance. The analysis shows that both maximum IPS are determined by the activity level of primary users (PU) and the placement of secondary user (SU) nodes. Optimal relay placement strategies will be identified to maximize these two IPS under different primary users' activity levels.

Y. Yang (✉)
Virginia Polytechnic Institute and State University, Blacksburg, VA, USA
e-mail: yyang8@vt.edu

F.R. Yu (ed.), *Cognitive Radio Mobile Ad Hoc Networks*,
DOI 10.1007/978-1-4419-6172-3_10,

For a single-hop CRN, we will analyze how delay is affected by multiple SU design options, including the number of channels to be aggregated, the duration of transmission, the channel separation constraint on channel aggregation, and the time needed for spectrum sensing and protocol handshake. How these different options may affect the delay under different SU and PU traffic loads is revealed. Methods for computing optimal CRN design and operation strategy are derived.

10.2 Optimal Information Propagation Speed Analysis in Multihop Cognitive Radio Networks

Similar to the delay in any networks, delay in CRN is the combination of two components [1]: the information propagation delay and the queuing delay. The information propagation delay is the amount of time that a packet spends to travel over the intermittent relaying links in a CRN and is determined by the underlying communication capabilities of the network. The queuing delay is the amount of time that a packet spends in waiting for other packets to finish their transmission. Queuing delay is determined by the specific traffic load, traffic pattern, and the scheduling algorithms adopted at all hops of CRN.

In this section, we will study the information propagation delay and understand how to plan node placement to minimize this important delay component in multihop CRN. The analysis of information propagation delay in CRN is different from it in other types of networks due to the unique two-tier architecture of CRN, where the information propagation delay in a CRN is related to not only the settings of the CRN network itself but also the traffic activities of primary users. Hence, existing works about multihop delay analysis for other types of networks [1–4] cannot be applied to CRN.

As a means to interpret the information propagation delay independent of propagation distance, we use *Information Propagation Speed (IPS)* as its measurement metric. Information propagation speed is defined as the speed that a piece of information (e.g., a packet of very small size) can be transmitted over a multi-hop CRN. In the remainder of this section, we will establish a model of IPS in CRN and categorize IPS maximization problem into two cases. The first case, named the maximum network IPS problem, maximizes IPS across a network topology over an infinite plane. The second case, named the maximum flow IPS problem, maximizes the IPS between a given pair of source and destination nodes separated by a fixed distance. We will reveal that both maximum network and flow IPS are determined by the primary user (PU) activity level and the placement of SU relay nodes. We will introduce numerical methods to compute the two maximum IPS and built optimal relay placement strategies to realize these two maximum IPS under different PU activity levels.

The results of the IPS study can be used as a benchmark for network design. It can be used to check whether a certain delay-sensitive traffic is supportable in a particular network setting. The optimal node placement settings that can achieve the maximum IPS can also be used as useful design guidelines in CRN planning.

The rest of this section is organized as follows. The network model and IPS model are presented in Sections 10.2.1 and 10.2.2. The analyses for the network and flow IPS are in Sections 10.2.3 and 10.2.4, respectively. The analytical results are validated by simulations and numerical results in Section 10.2.5.

10.2.1 Network Model

A cognitive radio network is modeled as a network formed by secondary users (SU) and an overlaid primary user (PU) network in an infinite two-dimensional region. The location distribution of PU nodes is assumed to be a two-dimensional Poisson point processes. Assume that there are K channels and an active PU or SU only uses one channel. Based on the measurement study of realistic wireless PU activities in cellular networks [5], the PU traffic can be accurately modeled as a Poisson arrival process with the mean arrival rate λ_P per unit area, while the service time of PU is usually not Poisson.

We assume duplex communications between the SU transmitter and the receiver (e.g., the SU receiver sends ACK back for received data). Hence, both SU transmitter and SU receiver need to avoid interfering with PU receivers. Denote d_s as the sensing radius of SU for PU receivers[1] and let $U(d)$ be the union of the sensing regions of a pair of SU transmitter and receiver that are d distance apart. The shape of $U(d)$ is shown by the gray region in Fig. 10.1 (II). When there are no active PU receivers within $U(d)$, the SU transmitter and receiver can communicate and we call that the SU link between the SU transmitter and the receiver is feasible. Note that here we implicitly assume that SUs can cancel the interference from PU transmitters to the SU receiver through interference cancelation schemes [6].

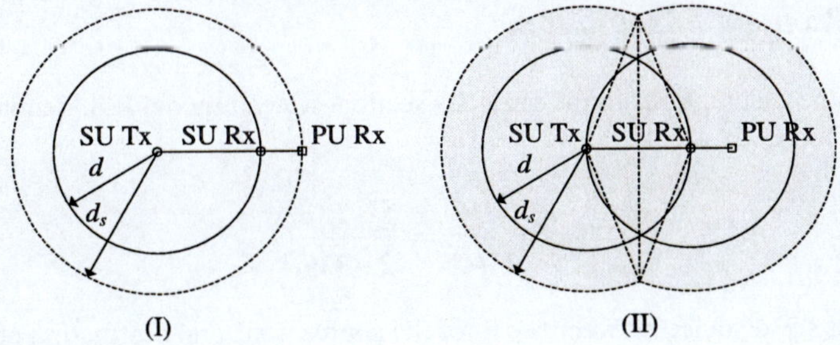

Fig. 10.1 One hop sensing region

[1] PU receiver detection can be realized by exploiting the feedback mechanisms in two-way PU communications as shown by [7, 8].

It is important to note that given cognitive radio's capability to adapt transmission power and hence its potential interference to PUs, the sensing radius of SU should not be treated as a fixed value. In our model, we assume that a SU controls its transmit power so that its communication range equals its distance d to its receiver. In this way, the interference to PUs is minimized. Under this optimal power control policy, we get:

$$\frac{P_t}{d^\alpha} = T_r \tag{10.1}$$

where $\alpha \geq 2$ is the attenuation exponent, P_t is the transmit power of the SU transmitter and T_r is the receiver's sensitivity level. Assume that when the signal of a SU exceeds the interference threshold T_s at a PU receiver, the PU receiver is interfered by the SU transmission. Therefore, SU's sensing radius d_s can be expressed as:

$$\frac{P_t}{d_s^\alpha} = T_s \tag{10.2}$$

Combining (10.1) and (10.2), we have

$$d_s = \left(\frac{T_r}{T_s}\right)^{\frac{1}{\alpha}} d = C_1 d \tag{10.3}$$

where $C_1 = \left(\frac{T_r}{T_s}\right)^{\frac{1}{\alpha}}$. The above equation shows that the sensing radius d_s is proportional to the SU communication range that equals the distance d between the SU transmitter and SU receiver under the optimal power control policy.

10.2.2 Problem Formulation

Based on the assumptions in the previous section, we can model IPS, denoted as w, in a multihop CRN as:

$$w = \frac{D}{\tau(D)} = \frac{\sum_i P(d_i)}{\sum_i \tau(d_i)} \tag{10.4}$$

where D is the distance between a pair of SU source node and destination node, and $\tau(D)$ is the expected information propagation delay over this distance D. d_i is the transmission distance of the ith hop, and $P(d_i)$ is the projection of d_i on the straight line from the source to the destination as shown in Fig. 10.2. $\tau(d_i)$ is the expected information propagation delay over the ith hop.

By observing (10.4), we could see a trade-off exists in the setting of d_i and w. While a large d_i increases the numerator in (10.4), it also increase $\tau(d_i)$ since it has

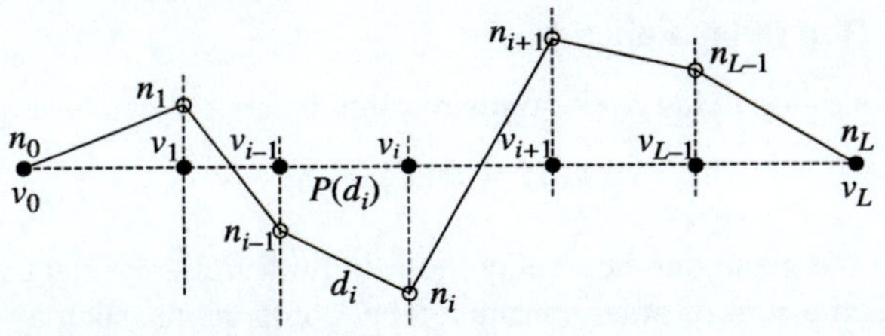

Fig. 10.2 One-hop progress distance

a large $U(d_i)$ and a SU link cannot be used if there is any active PU receiver in the area of $U(d_i)$. The objective of our research, hence, can be formulated as

$$\max_{\mathbf{d}} \frac{\sum_i P(d_i)}{\sum_i \tau(d_i)} \tag{10.5}$$

$$\text{subject to } d_i \le r_c, \forall d_i \in \mathbf{d} \tag{10.6}$$

where r_c is the SU communication range constrained by the SU radio's maximum transmission power and $\mathbf{d}$ is the set of all hops.

In the following two sections, we study two different cases for solving the above optimization problem by optimally setting $\mathbf{d}$. In Section 10.2.3, we maximize IPS while assuming the network is over an infinite plane and all the SU nodes can be flexibly placed. We call it maximum network IPS. In Section 10.2.4, we maximize IPS for a specific flow with fixed source and destination locations. We call it maximum flow IPS.

10.2.3 Network IPS

Based on (10.6), when a CRN is over an infinite plane and all nodes' locations are flexible, the maximum network IPS, denoted as W_u, is as follows:

$$W_u = \max_{\mathbf{d}} \frac{\sum_i P(d_i)}{\sum_i \tau(d_i)} = \frac{P(d^*)}{\tau(d^*)} \tag{10.7}$$

where d^* is the optimal one-hop distance that results in the maximum IPS in that hop. Equation (10.7) means that the maximum network IPS is achieved when every hop has the same optimal one-hop distance d^*. Therefore, the problem for deriving the IPS upper bound in a multi-hop cognitive radio network has been transformed to the problem of finding the maximum one-hop IPS. In the reminder of this section, we derive the expected one-hop delay function, and then study its monotonicity and convexity, based on which we derive the optimal one-hop distance d^* that achieves the maximum network IPS.

10.2.3.1 One-Hop Delay Function

The expected one-hop delay over a transmission distance d can be expressed as

$$\tau(d) = E\{T\} + \tau_0 \tag{10.8}$$

where $E\{T\}$ is the mean one-hop delay induced by waiting for PU traffic to vacant a channel, τ_0 is the sum of other constant delay components, such as channel sensing delay for determining the channel availability, transmission delay determined by the channel capacity, and packet processing delay determined by the hardware processing capability.

While τ_0 is fixed given the cognitive radio design, the value of $E\{T\}$ is equivalent to the time that it takes for a channel to be vacanted by PU when a SU is ready to transmit a packet. Note that in Section 10.2.1, we have pointed out that measurement results [5] show that realistic PU traffic follows a Poisson arrival process and has complex service time distribution. Hence, we can treat PU as a high priority flow, SU as a low priority flow, and the K channels as K servers in a two-priority preemptive M/G/K queuing model. Under this model, $E\{T\}$ is equivalent to the queueing delay of the low priority flow when the packet service time of the low priority flow approaches 0.

The anlaytical work of two-priority preemptive M/G/K queue in [9] shows that the queuing delay of the low priority flow can be approximated as follows:

$$\text{delay}_1 = \begin{cases} \dfrac{1}{2K\rho_1}\left[\dfrac{s_1\rho_1 + s_2\rho_2}{1-\rho_1-\rho_2}\dfrac{(\rho_1+\rho_2)^K + (\rho_1+\rho_2)}{2} - \dfrac{s_2\rho_2}{1-\rho_2}\dfrac{\rho_2^K+\rho_2}{2}\right] \\ \rho_2 > 0.7, \rho_1+\rho_2 > 0.7; \\ \dfrac{1}{2K\rho_1}\left[\dfrac{s_1\rho_1 + s_2\rho_2}{1-\rho_1-\rho_2}\dfrac{(\rho_1+\rho_2)^K + (\rho_1+\rho_2)}{2} - \dfrac{s_2\rho_2}{1-\rho_2}\rho_2^{\frac{K+1}{2}}\right] \\ \rho_2 < 0.7, \rho_1+\rho_2 > 0.7; \\ \dfrac{1}{2K\rho_1}\left[\dfrac{s_1\rho_1 + s_2\rho_2}{1-\rho_1-\rho_2}(\rho_1+\rho_2)^{\frac{K+1}{2}} - \frac{s_2\rho_2}{1-\rho_2}\rho_2^{\frac{K+1}{2}}\right] \\ \rho_2 < 0.7, \rho_1+\rho_2 < 0.7. \end{cases} \tag{10.9}$$

where subscript 1 is for low priority flow and 2 is for high priority flow and $s_i := \frac{1+C_{B_i}^2}{\mu_i}$, $C_{B_i}^2 := \frac{\sigma_i^2}{\bar{X_i}^2}$, $\rho_i := \lambda_i \bar{X}_i / K$ Here, X_i is the service time for priority i traffic, σ_i^2 is the variance of X_i, and $\bar{X}_i$ is the mean value of X_i.

Therefore, assuming that the traffic load of PU is reasonable (a.k.a. $\rho_2 < 0.7$), we can get $E\{T\}$ as

$$E\{T\} = \lim_{\rho_1 \to 0, s_1 \to 0} \text{delay}_1 \tag{10.10}$$

$$= \frac{\frac{K+1}{2} s_2 \rho_2^{\frac{K+1}{2}}}{2K(1-\rho_2)} + \frac{s_2 \rho_2^{\frac{K+3}{2}}}{2K(1-\rho_2)^2} \tag{10.11}$$

The reason that we assume $\rho_2 < 0.7$ is because CRN is usually used in scenarios where PUs have low utilization of licensed spectrum bands. Hence, the assumption that PU traffic load is light to medium is valid in our application scenario. The ρ_2 in (10.11) can be computed as follows. Note that within an unit area, the PU traffic is a Poisson arrival process with parameter λ_P. Hence, the aggregate PU traffic arrival rate within region $U(d)$ is Poisson with parameter $\lambda_a = \lambda_P A(d)$, where $A(d)$ is the area of region $U(d)$ as shown in Fig. 10.1. By (10.3) and Fig. 10.1 (II), it can be shown that $A(d) = Cd^2$, where $C = 2\pi C_1^2 - 2C_1^2 \cos^{-1} \frac{1}{2C_1} + \sqrt{C_1^2 - \frac{1}{4}}$ and $C_1 = \left(\frac{T_r}{T_s}\right)^{\frac{1}{\alpha}}$. Therefore, we have

$$\rho_2 = Cd^2 \lambda_P \bar{X}_p / K = Cd^2 \rho \tag{10.12}$$

where $\rho := \lambda_P \bar{X}_p / K$ and $\bar{X}_p$ is the mean active duration of a primary user.

Hence, combining (10.11) and (10.8), we get:

$$\tau(d) = \frac{(K+1) s_2 \rho_2^{\frac{K+1}{2}}}{4K(1-\rho_2)} + \frac{s_2 \rho_2^{\frac{K+3}{2}}}{2K(1-\rho_2)^2} + \tau_0 \tag{10.13}$$

10.2.3.2 Properties of One-Hop Delay Function

Next, we study two important properties of $\tau(d)$: monotonicity and convexity. We show that the one-hop delay function $\tau(d)$ is monotonically increasing and strictly convex. To simplify the mathematical derivation, we denote

$$h_1(\rho_2) = \frac{(K+1) s_2 \rho_2^{\frac{K+1}{2}}}{4K(1-\rho_2)} \tag{10.14}$$

and

$$h_2(\rho_2) = \frac{s_2 \rho_2^{\frac{K+3}{2}}}{2K(1-\rho_2)^2} \tag{10.15}$$

We can prove the following lemmas.

Lemma 1 *For (10.14) and (10.15), it follows*

$$h_1'(\rho_2) > 0, h_2'(\rho_2) > 0 \tag{10.16}$$

and

$$h_1''(\rho_2) > 0, h_2''(\rho_2) > 0 \tag{10.17}$$

Proof See our technical report [10].

Lemma 2 *The function* $\tau(d), 0 < d \le r_c$ *is monotonically increasing.*

Proof See our technical report [10].

Lemma 3 *The function* $\tau(d), 0 \le d \le r_c$ *is strictly convex.*

Proof See our technical report [10].

10.2.3.3 Speed Upper Bound Analysis

Clearly, the IPS can only be maximized when SU nodes are aligned on the straight line between the source and the destination. When this happens, the one-hop progress distance along the straight line from the source to the destination equals the one hop distance, i.e., $P(d) = d$. Therefore, the optimal one-hop distance d^* in (10.7) can be computed as

$$d^* = \arg \min_{0<d\le r_c} \left\{ \frac{\tau(d)}{d} \right\} \tag{10.18}$$

Since the physical meaning of $\frac{\tau(d)}{d}$ is the slope of the line passing through the origin and a point on the function $\tau(d)$ curve, it follows that d^* is the d value that minimizes the line slope.

- *Optimal One-Hop Distance Analysis*
  By Lemma 2 and 3, $\tau(d), 0 < d \le r_c$ is monotonically increasing and strictly convex. As shown in Fig. 10.3, there are two possibilities when determining d^*. Consider the tangent line of the curve $\tau(d)$ that passes the origin and touches $\tau(d)$ at a point $(d_0, \tau(d_0))$. When there is such a tangent line as shown in Fig. 10.3 (II), we have $d^* = d_0$. When there is no such a tangent line as shown in Fig. 10.3 (I), we have $d^* = r_c$. Mathematically, we have

$$d^* = \begin{cases} d_0, \text{ if } \exists 0 < d_0 \le r_c, \; s.t. \; \tau'(d_0) = \frac{\tau(d_0)}{d_0} \\ r_c, \text{ if } \nexists 0 < d_0 \le r_c, \; s.t. \; \tau'(d_0) = \frac{\tau(d_0)}{d_0} \end{cases} \tag{10.19}$$

To determine d^*, we need to determine if there is a real root to equation $\tau'(d) = \frac{\tau(d)}{d}, 0 < d \le r_c$. To solve this root existence problem, we define

$$\begin{aligned} f(\rho, d) &= \tau'(d) - \frac{\tau(d)}{d} \\ &= 2Cd\rho h_1'(Cd^2\rho) + 2Cd\rho h_2'(Cd^2\rho) \\ &\quad - \frac{h_1(Cd^2\rho) + h_2(Cd^2\rho)}{d} - \frac{\tau_0}{d} \end{aligned} \tag{10.20}$$

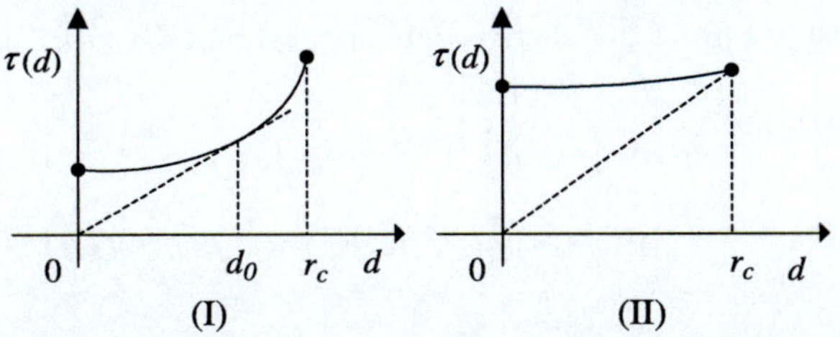

Fig. 10.3 Two examples of $\tau(d)$: (I) $\exists 0 < d_0 \leq r_c,\ s.t.\ \tau'(d_0) = \frac{\tau(d_0)}{d_0}$; (II) $\nexists 0 < d_0 \leq r_c,\ s.t.\ \tau'(d_0) = \frac{\tau(d_0)}{d_0}$

and study the root existence problem for the following equation

$$f(\rho, d) = 0, 0 < d \leq r_c \tag{10.21}$$

- *Threshold Property of d^**
  Next, let us solve the root existence problem in (10.21) and determine d^*. The analytical results are summarized in the following proposition.

Proposition 1 *There exists a threshold* $0 < \rho_u(r_c) < \frac{1}{Cr_c^2}$, $f(\rho_u(r_c), r_c) = 0$ *such that* $d^* = r_c$ *when* $\rho < \rho_u(r_c)$ *and* $d^* < r_c$ *when* $\rho > \rho_u(r_c)$.

Proof By Lemma 2 and 3, $\tau(d), 0 < d \leq r_c$ is monotonically increasing and strictly convex. By (10.20), it follows that $f(\rho, d)$ is an increasing function of d. Since $\lim_{d \to 0+} f(\rho, d) = -\infty$, it follows that $f(\rho, d)$ increases from $-\infty$ to $f(\rho, r_c)$ when d increases from 0 to r_c. When $f(\rho, r_c) > 0$, there must exist a real root to the problem in (10.21). By (10.19), we have $d^* < r_c$. When $f(\rho, r_c) < 0$, we have $f(\rho, d) < 0, \forall 0 < d \leq r_c$, i.e., there does not exist a real root to the problem in (10.21). By (10.19), we have $d^* = r_c$.

Next, we study the positivity of $f(\rho, r_c)$. Since r_c is fixed, the positivity of $f(\rho, r_c)$ depends on ρ. We study the positivity of $f(\rho, r_c)$ when ρ changes. By (10.20), we have

$$\begin{aligned} f(\rho, r_c) &= 2Cr_c\rho h_1'\left(Cr_c^2\rho\right) + 2Cr_c\rho h_2'\left(Cr_c^2\rho\right) \\ &\quad - \frac{h_1\left(Cr_c^2\rho\right) + h_2\left(Cr_c^2\rho\right)}{r_c} - \frac{\tau_0}{r_c} \end{aligned} \tag{10.22}$$

It can be shown that $\lim_{\rho \to 0+} f(\rho, r_c) < 0$ and $\lim_{\rho \to \left(\frac{1}{Cr_c^2}\right)^-} f(\rho, r_c) = \infty$. Therefore, there is at least one real root to equation

$$f(\rho, r_c) = 0, 0 < \rho < \frac{1}{Cr_c^2} \tag{10.23}$$

In the following, we prove that there is only one real root. By (10.22), we have

$$\frac{\partial f(\rho, r_c)}{\partial \rho} = Cr_c h_1'\left(Cr_c^2\rho\right) + Cr_c h_2'\left(Cr_c^2\rho\right) + 2\rho C^2 r_c^3 h_1''\left(Cr_c^2\rho\right) + 2\rho C^2 r_c^3 h_2''\left(Cr_c^2\rho\right) \tag{10.24}$$

By (10.16) and (10.17) in Lemma 1, we have $\frac{\partial f(\rho, r_c)}{\partial \rho} > 0$, i.e., $f(\rho, r_c)$ is an increasing function with respect to ρ. Therefore, there exists only one real root to equation $f(\rho, r_c) = 0, 0 < \rho < \frac{1}{Cr_c^2}$. Denote the root as $\rho_u(r_c)$. Recall that $\lim_{\rho \to 0^+} f(\rho, r_c) < 0$ and $\lim_{\rho \to \left(\frac{1}{Cr_c^2}\right)^-} f(\rho, r_c) = \infty$. It follows that $f(\rho, r_c) > 0$ when $\rho > \rho_u(r_c)$ and $f(\rho, r_c) < 0$ when $\rho < \rho_u(r_c)$. Therefore, when $\rho > \rho_u(r_c)$ we have $d^* < r_c$, and when $\rho < \rho_u(r_c)$ we have $d^* = r_c$.

Although it is difficult to derive a closed form formula for the threshold $\rho_u(r_c)$ and d^*, we can numerically derive it from (10.23). The physical intuition behind Proposition 1 is as follows. When the actual ρ is small, the delay of SU traffic caused by yielding to PU transmissions in the region $U(d)$ is negligible. The SU's IPS is only constrained by the maximum transmission power. Therefore, the optimal one-hop distance $d^* = r_c$. When ρ is large, the delay caused by PU transmissions in the region $U(d)$ dominates other delay components. Hence, a shorter one-hop distance d incurs a smaller $U(d)$ size, resulting a smaller delay. Therefore, the optimal one-hop distance $d^* < r_c$.

10.2.4 Flow IPS

Beyond the network IPS upper bound for all possible flows, we are also interested in the IPS upper bound for a particular given flow, called the flow IPS upper bound. Since in this case the source–destination distance is fixed, the problem of maximizing the IPS is equivalent to minimizing the total propagation delay from the source to the destination. Therefore, for the flow IPS case, by (10.8), the IPS upper bound W_f can be modeled as

$$W_f = \sup\left\{\frac{D}{\sum_i \tau(d_i)}\right\} = \frac{D}{\inf\left\{\sum_i \tau(d_i)\right\}} \tag{10.25}$$

where $\tau(\cdot)$ is the expected one-hop propagation delay. It is clear that the total delay is minimized when all the SU nodes are placed on the straight line between the source and the destination. Mathematically, the problem of optimal node placement is transformed to

$$\min \sum_i \tau(d_i),\ s.t. \sum_i d_i = D. \tag{10.26}$$

We decompose the above minimization problem to two subproblems: how to place SU nodes given a fixed number of relay nodes and how many of them should be added between the source and the destination. We show that the IPS is maximized when an optimal number of relay nodes are evenly spaced along a straight line between the source and the destination.

10.2.4.1 Optimal Node Placement

We first study the problem of how to place relay nodes to minimize the total delay when the total number of relay nodes is fixed. We decompose the problem of multi-hop path delay to a series of two-hop path problems. Our analysis shows that the total delay is minimized when the inter-node distances are equal.

Lemma 4 *Consider a K-hop ($K \geq 2$) SU path between a pair of given source and destination nodes. The total expected delay from the source to the destination is minimized when all the $K - 1$ relay nodes on the path are evenly placed on the straight line from the source to the destination.*

Proof Consider a two-hop SU path, whose source node and destination node is $0 < y < 2r_c$ distance apart. Then, the total delay of the two-hop path is

$$\tau_2(x) = \tau(x) + \tau(y - x) \tag{10.27}$$

where $y - r_c < x \leq r_c, r_c \leq y < 2r_c$ or $0 < x < r_c, 0 < y < r_c$. By Lemma 2 and 3, we have $\tau_2''(x) = \tau''(x) + \tau''(y - x) > 0$, and $\tau_2(x)$ is strictly convex. Therefore, $\tau_2(x)$ is minimized when $\tau_2'(x) = \tau'(x) - \tau'(y - x) = 0$, i.e., $x = \frac{y}{2}$. Physically, the total delay of a two-hop path is minimized when the relay node is placed in the middle point between the source node and the destination node.

Next, we prove the lemma by contradiction. Given a fixed number of relay nodes, suppose that the minimum total expected delay from the source to the destination is achieved when nodes are not evenly spaced along the straight line between the source and the destination. Denote such a path as P_u. It follows that there exists a two-hop subpath on the path P_u such that the middle SU node of the two-hop subpath is not in the middle position between the source and the destination of the subpath. By placing the middle SU node to the middle position, the total expected delay of path P_u decreases. This contradicts that path P_u minimizes the total expected delay from the source to the destination. Therefore, the lemma follows.

10.2.4.2 Optimal Number of Relay Nodes

Next, we determine the optimal number of relay nodes to minimize the total delay between the source and the destination. Note that to guarantee connectivity between the source and the destination, there are at least $\left\lfloor \frac{D}{r_c} \right\rfloor$ SU nodes placed on the straight line between the source and the destination. Denote n as the number of SU nodes to add, and $m = n + 1$ as the number of hops between the source and the destination.

It follows that $n \geq \left\lfloor \frac{D}{r_c} \right\rfloor$ and $m \geq \left\lfloor \frac{D}{r_c} \right\rfloor + 1$. By (10.26) and Lemma 4, given a m hop SU path ($n = m - 1$ relay nodes), the minimum total delay is

$$t(m) = m\tau\left(\frac{D}{m}\right) \tag{10.28}$$

Therefore, the optimization problem in (10.26) can be transformed to

$$\min t(m) = m\tau\left(\frac{D}{m}\right), s.t.\ m \geq \left\lfloor \frac{D}{r_c} \right\rfloor + 1, m \in Z^+ \tag{10.29}$$

Consider its continuous counterpart problem

$$\min t(x) = x\tau\left(\frac{D}{x}\right), s.t.\ x \geq \left\lfloor \frac{D}{r_c} \right\rfloor + 1, x \in R^+ \tag{10.30}$$

It can be shown that

$$t'(x) = \tau\left(\frac{D}{x}\right) - \frac{D}{x}\tau'\left(\frac{D}{x}\right) \tag{10.31}$$

and

$$t''(x) = \frac{D^2}{x^3}\tau''\left(\frac{D}{x}\right) > 0 \tag{10.32}$$

By (10.32), $t(x)$ is a strictly convex function over $x \geq \left\lfloor \frac{D}{r_c} \right\rfloor + 1$. There are two possibilities when solving the problem in (10.30). When $t'\left(\left\lfloor \frac{D}{r_c} \right\rfloor + 1\right) > 0$, the optimal solution $x^* = \left\lfloor \frac{D}{r_c} \right\rfloor + 1$. When $t'\left(\left\lfloor \frac{D}{r_c} \right\rfloor + 1\right) \leq 0$, the optimal solution $x^* = x_0$, where $t'(x_0) = 0$, $x_0 \geq \left\lfloor \frac{D}{r_c} \right\rfloor + 1$. Therefore, the optimal solution to the problem in (10.29) is as follows:

$$m^* = \begin{cases} \left\lfloor \frac{D}{r_c} \right\rfloor + 1, \text{ if } t'\left(\left\lfloor \frac{D}{r_c} \right\rfloor + 1\right) > 0, \\ \arg \min\limits_{m \in \{m_1, m_2\}} \{t(m)\}, \text{ if } t'\left(\left\lfloor \frac{D}{r_c} \right\rfloor + 1\right) \leq 0 \end{cases} \tag{10.33}$$

where $m_1 = \lfloor x^* \rfloor$, $m_2 = \lceil x^* \rceil$, and $t'(x^*) = 0$.

10.2.4.3 An Iterative Method of Calculating m^*

Note that directly computing m^* from (10.33) involves solving the equation $t'(x_0) = 0$, $x_0 \geq \left\lfloor \frac{D}{r_c} \right\rfloor + 1$, which may be computationally intensive. This motivates us to find alternative methods to determine m^*. Since we have proved that $t(x)$, $x \geq \left\lfloor \frac{D}{r_c} \right\rfloor + 1$ is convex, $t(m)$, $m \geq \left\lfloor \frac{D}{r_c} \right\rfloor + 1$ can be either monotonically increasing or first monotonically decreasing and then monotonically increasing. Therefore, m^* is the smallest m such that $t(m+1) > t(m)$. Mathematically,

$$m^* = \min\{m | (m+1)\tau\left(\frac{D}{m+1}\right) > m\tau\left(\frac{D}{m}\right)$$
$$m \geq \left\lfloor \frac{D}{r_c} \right\rfloor + 1, m \in Z^+\Big\} \tag{10.34}$$

Based on (10.34), it is straight forward to develop an iterative algorithm to compute m^*.

10.2.4.4 A Table Look-Up Method Based on Threshold Property of m^*

While it is possible to determine m^* by (10.34), the iterative algorithm may incur heavy computation overheads. It may take many steps before finding m^*, when m^* is much larger than $\left\lfloor \frac{D}{r_c} \right\rfloor + 1$. This motivates us to find another easy method of determining m^*.

Our basic idea is to determine m^* by considering whether adding a relay node decreases the total delay. Our analysis shows that there exists a threshold PU activity level when deciding whether to add a relay SU node. By (10.34), adding a relay decreases the total delay when $(m+1)\tau\left(\frac{D}{m+1}\right) < m\tau\left(\frac{D}{m}\right)$. To determine m^*, we define

$$\begin{aligned} g(\rho, m) &= (m+1)\tau\left(\frac{D}{m+1}\right) - m\tau\left(\frac{D}{m}\right) \\ &= (m+1)\left[h_1\left(\frac{\rho CD^2}{(m+1)^2}\right) + h_2\left(\frac{\rho CD^2}{(m+1)^2}\right)\right] \\ &\quad - m\left[h_1\left(\frac{\rho CD^2}{m^2}\right) + h_2\left(\frac{\rho CD^2}{m^2}\right)\right] + \tau_0 \end{aligned} \tag{10.35}$$

When $g(\rho, m) < 0$, adding a relay node decreases the total relay. When $g(\rho, m) > 0$, adding a relay node increases the total relay. Given m, the positivity of $g(\rho, m)$ depends on ρ. Next, we study the positivity of $g(\rho, m)$ when ρ changes.

Lemma 5 *Consider a m-hop SU path, whose source–destination distance is D. There exists* $0 < \rho_f(m) < \frac{m^2}{CD^2}$, $g(\rho_f(m), m) = 0$ *such that the following properties hold.*

- *When* $\rho > \rho_f(m)$, *adding a relay node and evenly spacing all relay nodes decreases the total delay.*

- *When $\rho < \rho_f(m)$, adding extra relay nodes increases the total delay.*
- *The function $\rho_f(m)$ is monotonically increasing.*

Proof Note that $\lim_{\rho\to 0^+} g(\rho, m) = \tau_0 > 0$, and $\lim_{\rho\to\left(\frac{m^2}{CD^2}\right)^-} g(\rho, m) = -\infty < 0$.

Therefore, there is at least one real root to equation $g(\rho, m) = 0, 0 < \rho < \frac{m^2}{CD^2}$. Next, we show that there is only one real root. We prove this by showing that $g(\rho, m)$ is monotonically decreasing with respect to ρ. By (10.35), we have

$$\frac{\partial g(\rho, m)}{\partial \rho} = \frac{CD^2}{m+1}\left[h_1'\left(\frac{\rho CD^2}{(m+1)^2}\right) + h_2'\left(\frac{\rho CD^2}{(m+1)^2}\right)\right] - \frac{CD^2}{m}\left[h_1'\left(\frac{\rho CD^2}{m^2}\right) + h_2'\left(\frac{\rho CD^2}{m^2}\right)\right] \tag{10.36}$$

By (10.16), we have $h_1'\left(\frac{\rho CD^2}{(m+1)^2}\right) + h_2'\left(\frac{\rho CD^2}{(m+1)^2}\right) > 0$, and

$$\frac{\partial g(\rho, m)}{\partial \rho} < \frac{CD^2}{m}\left[h_1'\left(\frac{\rho CD^2}{(m+1)^2}\right) + h_2'\left(\frac{\rho CD^2}{(m+1)^2}\right)\right] - \frac{CD^2}{m}\left[h_1'\left(\frac{\rho CD^2}{m^2}\right) + h_2'\left(\frac{\rho CD^2}{m^2}\right)\right]. \tag{10.37}$$

By (10.17), it follows that $h_1'(\rho_2)$ and $h_2'(\rho_2)$ are monotonically increasing. Therefore, by (10.37) we have $\frac{\partial g(\rho,m)}{\partial \rho} < 0$. Hence, there is only one real root to equation $g(\rho, m) = 0, 0 < \rho < \frac{m^2}{CD^2}$. Denote the root as $\rho_f(m)$. Recall that $\lim_{\rho\to 0^+} g(\rho, m) > 0$, and $\lim_{\rho\to\left(\frac{m^2}{CD^2}\right)^-} g(\rho, m) < 0$. We have the following conclusions.

There exists a threshold value $0 < \rho_f(m) < \frac{m^2}{CD^2}$ such that $g(\rho_f(m), m) = 0$. When $\rho < \rho_f(m)$, it follows that $g(\rho, m) > 0$. By (10.35), adding a relay node increases the total delay. When $\rho > \rho_f(m)$, it follows that $g(\rho, y) < 0$. By (10.35), adding a relay node and placing all the nodes equal distance apart decreases the total delay.

Next, we prove that $\rho_f(m)$ is a monotonically increasing function of m, i.e., $\rho_f(m+1) > \rho_f(m)$. Recall that $\lim_{\rho\to 0^+} g(\rho, m) = \tau_0 > 0$, which is not dependent on m, and $\frac{\partial g(\rho,m)}{\partial \rho} < 0$. To show $\rho_f(m+1) > \rho_f(m)$, it is equivalent to show that $g(\rho_f(m), m+1) > 0$. By (10.35) and the definition of $\rho_f(m)$, we have $g(\rho_f(m), m) = (m+1)\tau\left(\frac{D}{m+1}\right) - m\tau\left(\frac{D}{m}\right) = 0$. Since we have proved that $t(x) = x\tau\left(\frac{D}{x}\right)$ is strictly convex, we have $t(m+2) - t(m+1) > t(m+1) - t(m)$, i.e., $(m+2)\tau\left(\frac{D}{m+2}\right) - (m+1)\tau\left(\frac{D}{m+1}\right) > (m+1)\tau\left(\frac{D}{m+1}\right) - m\tau\left(\frac{D}{m}\right)$. Therefore, given $\rho = \rho_f(m)$, we have $g(\rho_f(m), m+1) = (m+2)\tau\left(\frac{D}{m+2}\right) -$

$(m+1)\tau\left(\frac{D}{m+1}\right) > (m+1)\tau\left(\frac{D}{m+1}\right) - m\tau\left(\frac{D}{m}\right) = g(\rho_f(m), m) = 0$. Hence, $\rho_f(m)$ is a monotonically increasing function of m.

The significance of Lemma 5 is that it can be used to determine the value interval of ρ corresponding to a given optimal m^* value. Given $m \geq \left\lfloor \frac{D}{r_c} \right\rfloor + 1$, by equation $g(\rho, m) = 0$, we can numerically derive the threshold $\rho_f(m)$. The optimal hop count m^* can be determined as follows.

Proposition 2 *Given an actual* ρ, *we have* $m^* = m$, *when*

- $\rho \in (0, \rho_f(m)], m = \left\lfloor \frac{D}{r_c} \right\rfloor + 1$;
- *or* $\rho \in (\rho_f(m-1), \rho_f(m)], m > \left\lfloor \frac{D}{r_c} \right\rfloor + 1$.

Proof When $\rho > \rho_f(m)$, adding a relay node decreases the total delay and increments m, which in turn increases the threshold value $\rho_f(m)$ by Lemma 5. Keep incrementing m, until ρ is less than the new threshold value $\rho_f(m)$. At this stage, adding extra relay nodes increases the total delay. Therefore, the optimal hop count $m^* = m$, for $\rho \in (0, \rho_f(m)], m = \left\lfloor \frac{D}{r_c} \right\rfloor + 1$ or $\rho \in (\rho_f(m-1), \rho_f(m)], m > \left\lfloor \frac{D}{r_c} \right\rfloor + 1$.

Note that each interval of ρ corresponds to an optimal hop count m^*. Since $\rho_f(m)$ can be numerically computed, they can be computed off-line and stored in a table. When there are different ρ values, the optimal hop count m^* can be derived simply by looking up the table. This saves a lot of online computation overhead. With m^* computed, the optimal number of relay nodes $n^* = m^* - 1$ can be easily determined. Therefore, we conclude the following proposition.

Proposition 3 *In the flow IPS case, the IPS upper bound is achieved when* $n^* = m^* - 1$ *relay nodes are evenly spaced along the straight line between the source and the destination, where* m^* *is given by* (10.33), (10.34), *or the table lookup method.*

10.2.5 Simulation and Numerical Validation

In this section, we validate the correctness of our upper bounds by simulations and show the correctness of the analytical results in Proposition 1 and 2 by numerical experiments.

10.2.5.1 Validation of the Theoretical Upper Bound

- *Network IPS Case*
  To validate the correctness of our theoretical results, we next compare our theoretical IPS upper bound with the actual IPS computed from simulations. The

simulation region is a square with edge length 10,000 m. The PU transmitters are uniformly distributed within the simulation region. We simulate one-hop, two-hop, and three-hop SU paths. For each path length, we generate 50 paths and for each path, we generate 50 PU transmitter distribution. For each of the setting, we measure the delays for 20,000 packet deliveries between the source and the destination. In the simulation, $K = 20$, $r_c = 110$ m, $\tau_0 = 0.1$ ms, $\frac{T_r}{T_s} = 2$, and $\mu_P^{-1} = 1$ ms. Three possible PU service time distributions are simulated: exponential distribution, uniform distribution, and constant. Their simulation results are shown in Fig. 10.4a–c, respectively.

We perform two sets of simulations for each distribution. In the first set of simulations, we randomly position SU nodes. The maximum IPSs are shown in Fig. 10.4a (I), b (I), and c (I). The mean IPSs and the standard derivations are shown in Fig. 10.4a (II), b (II), and c (II). The maximum IPSs from the simulation are below the theoretical IPS upper bound, validating the correctness of the IPS upper bound. When the path hop count increases, the simulated IPS decreases. This is because a longer SU path has a higher probability that SU nodes may not be aligned on the straight line between the source and the destination, causing an excessive delay. When the ρ value increases, the simulated IPS decreases. This is because that the PU traffic becomes heavier when ρ increases, causing a larger delay. Also, we observe that our theoretical upper bound is tight compared with the maximum IPS, validating the correctness of our approximation.

In the second set of simulations, SU nodes are evenly spaced along the straight line between the source and the destination. We focus on examining the delay of a 3-hop SU path. The one-hop distance d is set to d^*, $0.8d^*$, and $1.2d^*$. When the one-hop distance $d > r_c$, it is rounded to r_c. The simulated mean IPSs and their standard deviations are shown in Fig. 10.4a (III), b (III), and c (III). When $d = d^*$, the simulated mean IPS curves almost match the theoretical IPS upper bound curve. When $d = 0.8d^*$, $1.2d^*$, the simulated mean IPS curves are below the theoretical upper bound curves. This proves that our IPS upper bound can be achieved when SU nodes are optimally deployed.

- *Flow IPS Case*

  Next, we compare our theoretical IPS upper bound with the actual IPS computed from simulations in the flow IPS case. The simulation region is a square with edge length 10, 000 m. The PU transmitters are uniformly distributed within the simulation region. The source–destination distance is 500 m. We simulate m^*-hop, $m^* + 1$-hop, and $m^* + 2$-hop SU paths, where m^* is the optimal number of relay nodes. For each path length, we generate 50 paths and for each path, we generate 50 PU transmitter distribution. For each of the setting, we measure the delays for 20, 000 packet deliveries between the source and the destination. In the simulation, $K = 20$, $r_c = 110$ m, $\tau_0 = 0.1$ ms, $\frac{T_r}{T_s} = 2$, and $\mu_P^{-1} = 1$ ms. Three possible PU service time distributions are simulated: exponential distribution, uniform distribution, and constant. Their simulation results are shown in Fig. 10.5(a–c), respectively.

We perform two sets of simulations for each distribution. In the first set of simulations, we randomly position SU nodes as long as they maintain connectivity between

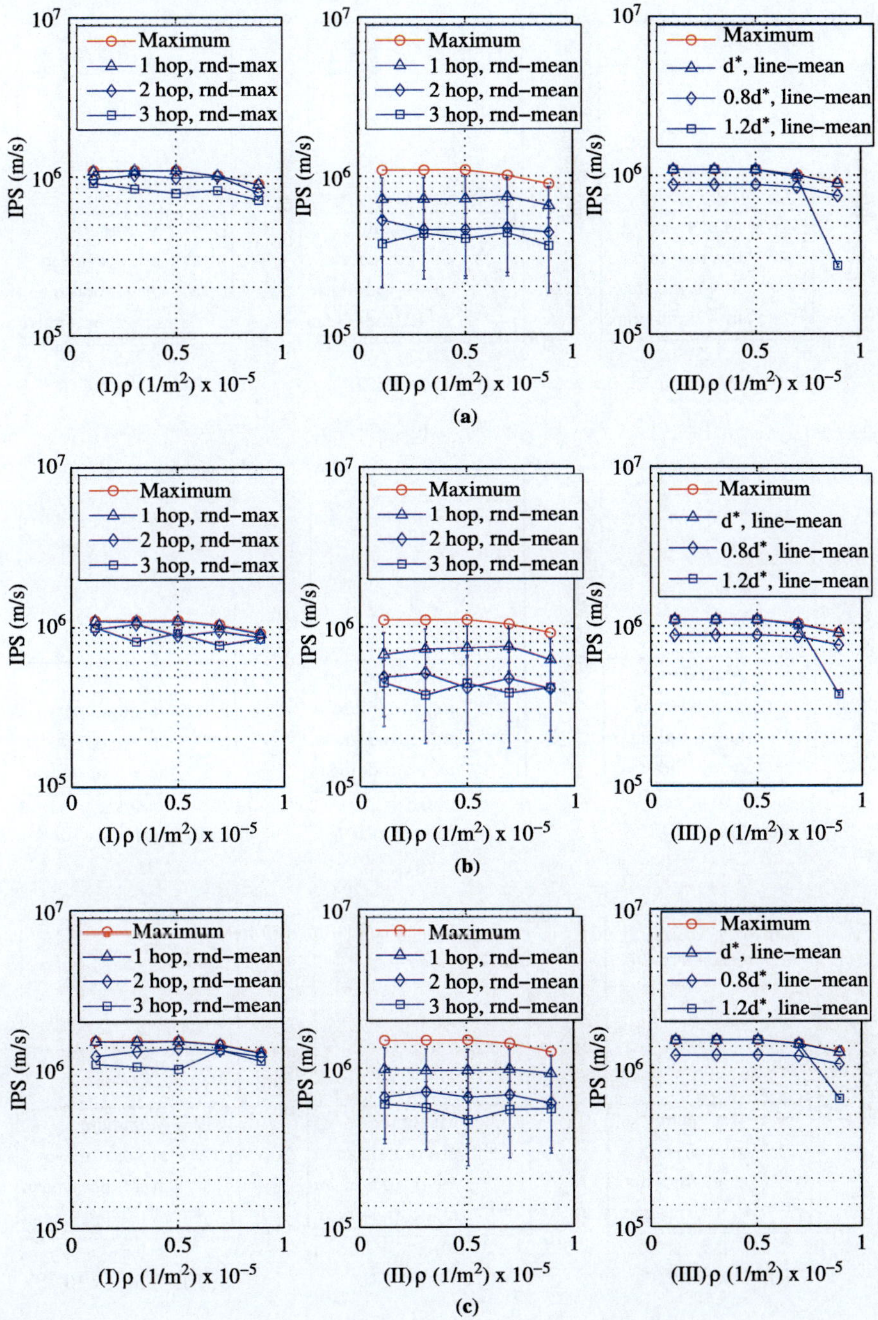

Fig. 10.4 Theoretical IPS upper bound and the simulated IPS for the network IPS case. **(a)** Exponential distribution service time; **(b)** Uniform distribution service time; **(c)** Constant service time

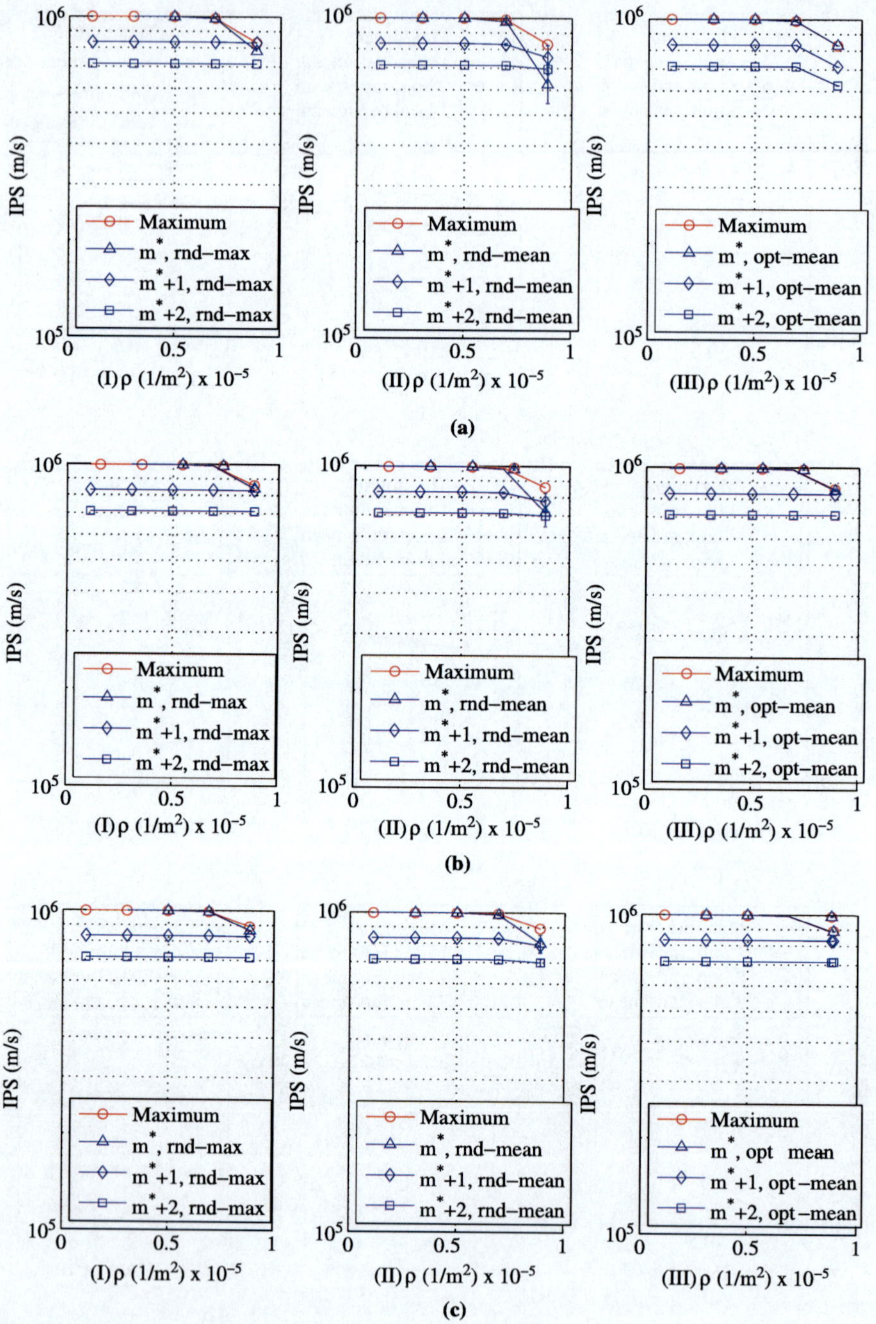

Fig. 10.5 Theoretical IPS upper bound and the simulated IPS for the flow IPS case. **(a)** Exponential distribution service time; **(b)** Uniform distribution service time; **(c)** Constant service time

the source and the destination. The maximum IPSs are shown in Fig. 10.5a (I), b (I), and c (I). The mean IPSs and the standard derivations are shown in Fig. 10.5a (II), b (II), and c (II). The maximum IPSs from the simulation are below the theoretical IPS upper bound, validating the correctness of the IPS upper bound. Also, we observe that our theoretical upper bound is tight compared with the maximum IPS, validating the correctness of our approximation.

In the second set of simulations, SU nodes are evenly spaced along the straight line between the source and the destination. The simulated mean IPSs and their standard deviations are shown in Fig. 10.5a (III), b (III), and c (III). When $m = m^*$, the simulated mean IPS curves match the theoretical IPS upper bound curve. When $m = m^* + 1, m^* + 2$, the simulated mean IPS curves are below the theoretical upper bound curves. This proves the correctness of Lemma 4 and Proposition 3.

10.2.5.2 Optimal One-Hop Distance, Optimal Number of SU Relay Nodes, and Theoretical Upper Bounds

- *Network IPS Case*
  To demonstrate the correctness of Proposition 1, we numerically compute the optimal one-hop distance d^* and the corresponding IPS upper bound for different network settings, e.g., communication range r_c, and PU service time distribution. The d^* is derived from (10.19) and (10.21). We perform numerical experiments for multiple cases with different r_c values and PU service time distributions. In the experiments, we have $K = 20$, $\frac{T_r}{T_s} = 2$, $\alpha = 3$, $\tau_0 = 0.1$ ms, and $\mu_P^{-1} = 1$ ms. We consider three PU service time distributions: exponential distribution, uniform distribution and constant as shown in Fig. 10.6(a–c), respectively.

  For each case, the optimal one-hop distances d^* and the corresponding theoretical IPS upper bounds with respect to ρ values are shown. As shown in these figures, for all different r_c values there is a threshold ρ value. Below this threshold, the optimal one-hop distance $d^* = r_c$. Above this threshold, the optimal one-hop distance $d^* < r_c$. We also observe that the optimal one-hop distance d^* decreases when the ρ value increases. This is because a higher ρ value indicates heavier PU traffic, which causes more excessive delay. Therefore, for a higher ρ value, the optimal one-hop distance d^* is shorter. These results validate the correctness of Proposition 1. For all different r_c values, the bound decreases when the ρ value increases. This is because a higher ρ value indicates heavier PU traffic, which slows down the IPS.
- *Flow IPS Case*
  To demonstrate the correctness of Proposition 2, we numerically compute the optimal number of relay nodes n^* and the corresponding theoretical IPS upper bound for different network settings, e.g., communication range r_c and PU service time. The n^* is iteratively derived from (10.34) and the fact that $n^* = m^* - 1$. We perform numerical experiments for multiple cases with different r_c values and PU service time distributions. In the experiments, we have $K = 20$, $\frac{T_r}{T_s} = 2$,

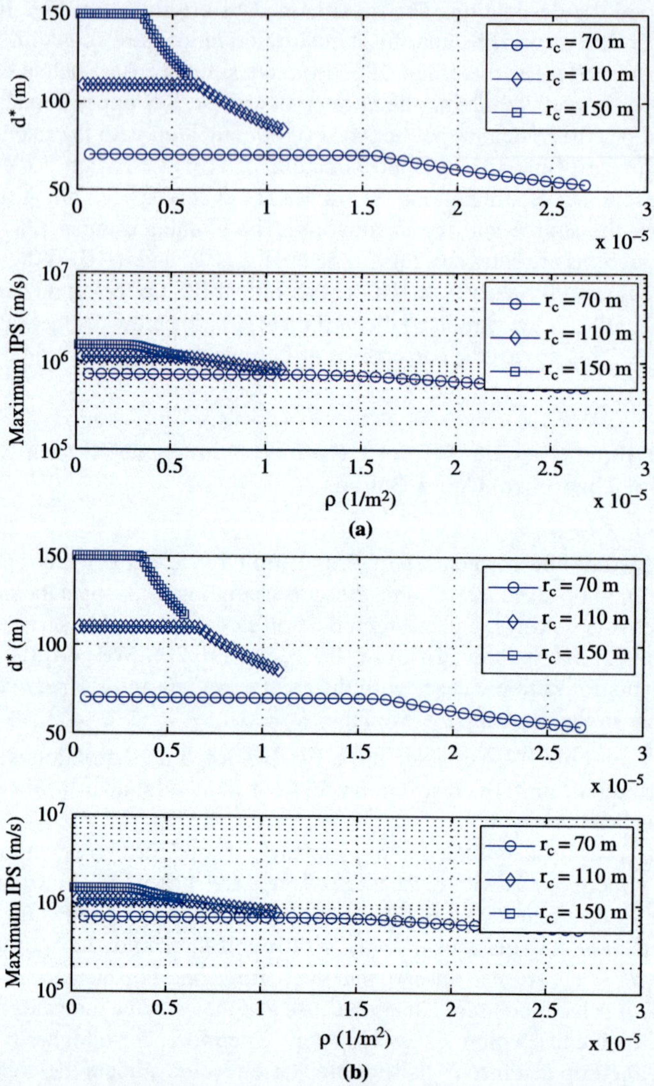

Fig. 10.6 Optimal one-hop distance d^* and IPS upper bound for the network IPS case. **(a)** Exponential distribution service time; **(b)** Uniform distribution service time; **(c)** Constant service time

$\alpha = 3$, $\tau_0 = 0.1$ ms, and $\mu_P = 1$ ms. We consider three PU service time distributions: exponential distribution, uniform distribution, and constant as shown in Fig. 10.7(a–c), respectively.

For each case, the optimal number of relay nodes n^* and the corresponding theoretical IPS upper bounds with respect to ρ values are shown. As shown in these figures, for each r_c value there are threshold ρ values. Above these thresholds,

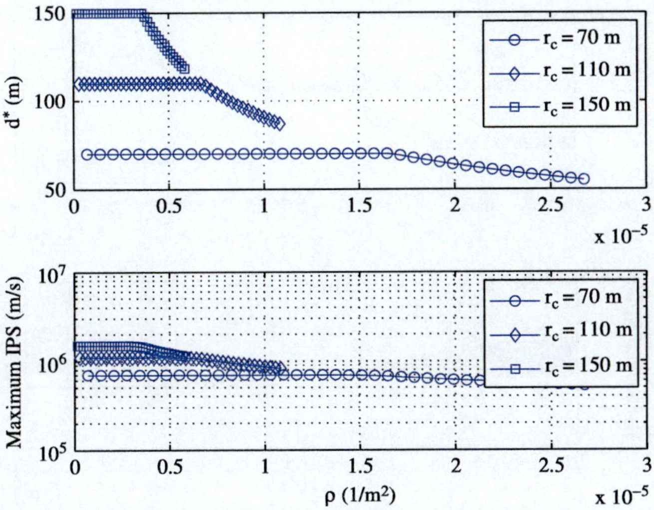

(c)

Fig. 10.6 (continued)

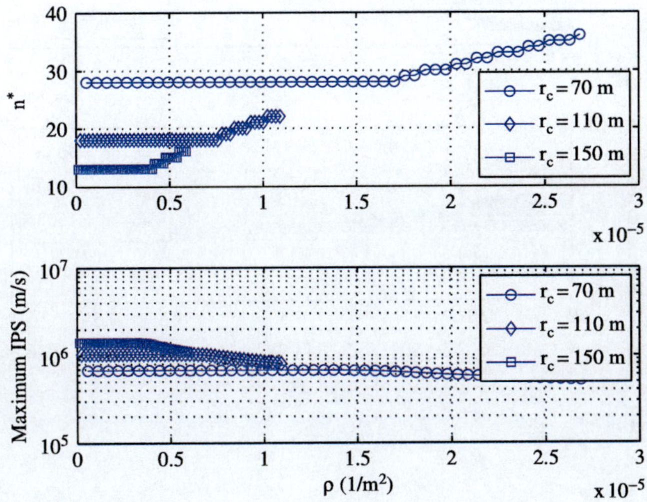

(a)

Fig. 10.7 Optimal number of relay SU nodes n^* and IPS upper bound for the flow IPS case. **(a)** Exponential distribution service time; **(b)** Uniform distribution service time; **(c)** Constant service time

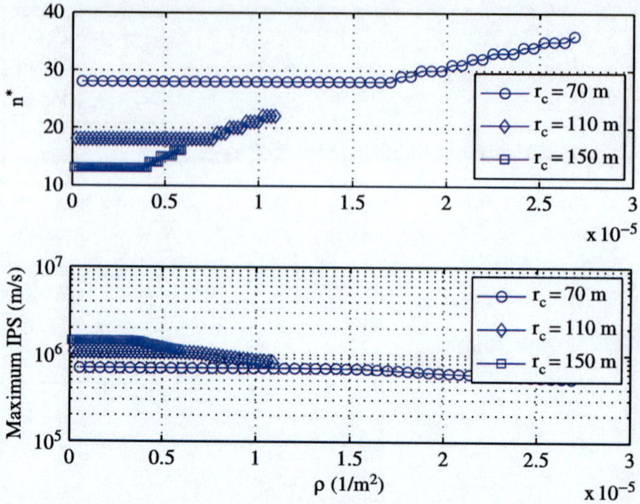

(b)

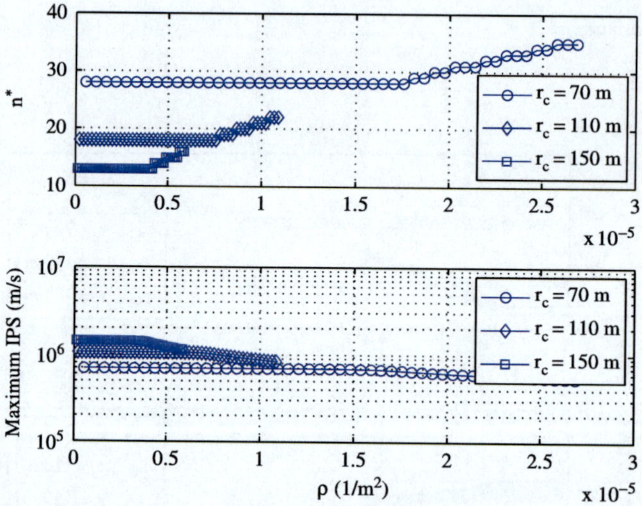

(c)

Fig. 10.7 (continued)

n^* increments. We also observe that n^* is a non-decreasing function of ρ. This is because when ρ is large, the IPS is mainly constrained by the interference from PU traffic. A larger number of relay nodes results in a shorter one-hop distance and a shorter sensing range, rendering less interference from PU traffic. Therefore, n^* is a non-decreasing function of ρ. These results validate the correctness of Proposition 2. The general trends and underlying rationales are the same as that of the network IPS case. The bounds are slightly below the bounds of the network IPS case. This is because the source–destination distance is not necessarily a multiple of d^* in the network IPS case, which decreases the IPS.

10.3 Delay Analysis in Single-Hop Cognitive Radios Networks

As discussed in Section 10.2, the total delay of a flow is a combination of both information propagation delay and queueing delay. While Section 10.2 gives an analysis for information propagation delay in multihop CRN, accurate analysis of queueing delays in such networks is still an open problem due to the complex correlations between packet loss, queue length, scheduling algorithms, and interference among all the hops of a flow. However, in a single-hop CRN network, it is possible to provide accurate analysis of the total delay that includes both information propagation delay and queueing delay with detailed consideration of many CRN design characteristics. In this section, we will provide such an analysis.

One design characteristic that we consider in CRN is *channel aggregation*. Usually, licensed spectrum is divided into a number of discrete channels. As in the Shannon's theorem, channel capacity is proportional to channel width (bandwidth). Hence, efficient utilization of *white spaces* can be achieved by properly enabling each SU to access multiple channels at a time [11]. This assembling of noncontiguous channels for communication is called *channel aggregation* as defined in IEEE 802.22 draft [12]. Technically, channel aggregation can be implemented based on orthogonal frequency division multiplexing (OFDM) [13–15] or multiple radios [16]. Other design characteristics that are considered in our analysis include the duration of each transmission attempt, the delay in channel sensing and channel switching, and the handshake delay for channel negotiation among communication peers.

This section is organized as follows. First, we propose a new channel usage model to investigate the impact of both PU and SU behaviors on the availability of white spaces for channel aggregation. Unlike the ON-OFF process, this general model can capture a wider range of user behaviors. Next, we derive the delay costs for performing channel aggregation under this model. User demands in both frequency and time domains are considered to evaluate the costs for making negotiation and renewing transmission. Further, an optimal channel aggregation strategy is defined in order to minimize the cumulative delay for transmitting data. Finally, numerical analysis and discrete-event simulation are used to illustrate and validate our model and the optimal channel aggregation strategy.

10.3.1 System Model

10.3.1.1 Basic Assumptions

A SU is assumed to be equipped with a dedicated radio for operating on data channels in vacant licensed spectrum bands and another *scanner radio* for sending/receiving control messages on a dedicated control channel in unlicensed spectrum band [17, 18]. In addition, the scanner radio is responsible for sensing the licensed spectrum bands to discover spectrum white spaces, denoted as $\mathcal{W}_n$, in its sensing region V_n. Denoting $\mathcal{F}=\{f_1, \cdots, f_K\}$ as the set of K channels in the licensed spectrum bands, we have $\mathcal{W}_n \subseteq \mathcal{F}$. For these data channels, each SU with a b-channel *bandwidth demand* can assemble b channels at a time so as to form an *aggregated channel* $\mathcal{A}^{(b)}=\{f_{1'}, \cdots, f_{b'}\} \subset \mathcal{F}$ for data communication. Due to limitations on radio design complexity, there exists a limit B on b such that $b \leq B$. Usually, B is a small positive number. For example, in the IEEE 802.22 standard, $B = 3$. In addition, if any two channels, say f_l and $f_{l+\delta}$, are too far apart in $\mathcal{F}$, they cannot be aggregated. This constraint for the channel separation is denoted as Δ (a.k.a. $\delta \leq \Delta$). In other words, an $\mathcal{A}^{(b)}$ can only be selected from $\mathcal{C}_{l,\Delta}=\{f_l, \cdots, f_{l+\Delta}\} \subset \mathcal{F}$, which is a set of *candidate channels* satisfying the Δ constraint.

Whenever a sender n tries to send data packets to its receiver n', a *negotiation* between n and n' via control channel is necessary for an agreement on the forming of the aggregated channel $\mathcal{A}^{(b)}$. It is assumed that each SU does not have full knowledge of spectrum usage in its vicinity. Thus, multiple negotiation attempts between n and n' may be needed to finalize $\mathcal{A}^{(b)}$ that satisfy the Δ constraint.

Specifically, n should first sense a set of channels to find common white spaces in both V_n and $V_{n'}$. As shown in the example in Fig. 10.8, first, n picks a spectrum range $\mathcal{C}_{l,\Delta}$ (e.g., $\mathcal{C}_{7,4} = \{f_7, f_8, f_9, f_{10}\}$ in the 1st attempt) that satisfies the Δ constraint to sense. This leads to the discovery of the white space $\mathcal{T}_{l,\Delta} = \mathcal{C}_{l,\Delta} \cap \mathcal{W}_n$ (e.g., $\mathcal{T}_{7,4} = \{9\}$ in 1st attempt). Then, n checks if $|\mathcal{T}_{l,\Delta}| \geq b = 2$. If not the case (e.g., 1st attempt), we call it a blocking incident at the sender and n will pick another spectrum range $\mathcal{C}_{l,\Delta}$ to sense for white spaces until finally it finds a white space $\mathcal{T}_{l,\Delta}$ that is larger than b. Then, n initiates a *handshake* with n' to see if enough channels in $\mathcal{T}_{l,\Delta}$ is also available in n' such that $|\mathcal{P}_{l,\Delta}| = |\mathcal{T}_{l,\Delta} \cap \mathcal{W}_{n'}| \geq b$. If not true (e.g., attempt 2nd), a blocking incident at the receiver happens and n is informed to go

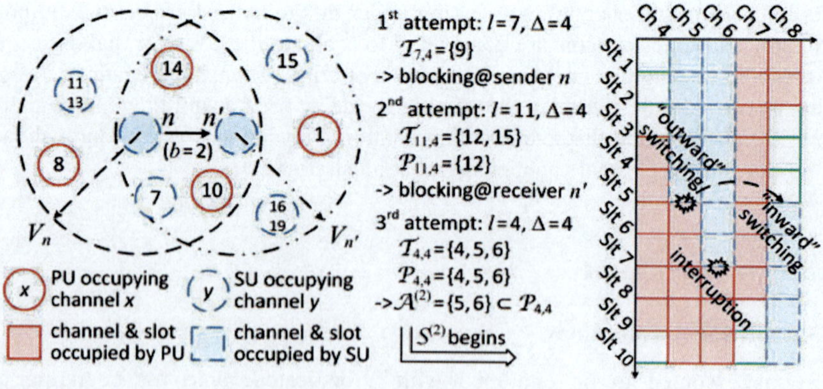

Fig. 10.8 An example of negotiation and transmission between n and n'

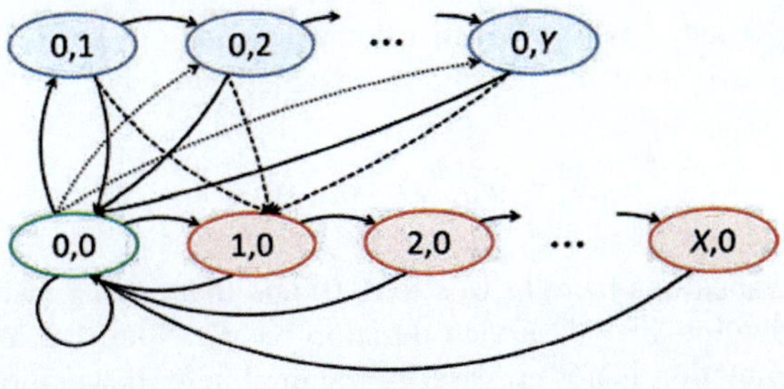

Fig. 10.9 Channel usage model

over all the spectrum sensing and handshake steps again until finally a viable $\mathcal{P}_{l,\Delta}$ is found (e.g., attempt 3rd). Then, n' selects an $\mathcal{A}^{(b)} \subseteq \mathcal{P}_{l,\Delta}$ and replies to n. Then, a *transmission* $\mathcal{S}^{(b)} = \{s_{1'}, \cdots, s_{b'}\}$ is initiated, which includes b parallel subflows with a d-slot *duration demand*.

However, a successful negotiation does not mean a reliable transmission, attributed to the low priority of SU service. To overcome this, *spectrum switching* is employed. Specifically, whenever a PU arrival to any $f_k \in \mathcal{A}^{(b)}$ is detected, the pair of n and n' needs to vacate the preempted f_k immediately and then tries to renew the corresponding s_k on a *backup channel* $f_{k'} \in \mathcal{B}_{l,\Delta}$, where $\mathcal{B}_{l,\Delta} = \mathcal{P}_{l,\Delta} \backslash \mathcal{A}^{(b)}$. For ease of presentation, such spectrum switching is divided into two steps: "outward" switching from f_k and "inward" switching to $f_{k'}$. If $|\mathcal{B}_{l,\Delta}| = 0$, an *interruption* incident occurs to the expelled s_k. But all the other ongoing ones in $\mathcal{S}^{(b)}$ may not be affected as long as the independence of these parallel subflows is guaranteed [19].

10.3.1.2 Channel Usage Model

In a certain SU n's vicinity, any channel $f_c \in \mathcal{F}$ may be occupied by an active PU or SU service for a period of time. As in Fig. 10.9, the average channel occupancy on such f_c is modeled as a Markov chain, in which channel state transits on a slot basis with a τ-second slot duration. Three groups of channel states are defined as follows: (i) *idle state* (0,0), in which f_c is a white space; (ii) *PU service states* $(x,0)$'s, $x \in \{1, \cdots, X\}$, in which f_c has been occupied by a PU service for x slots; (iii) *SU service states* $(0,y)$'s, $y \in \{1, \ldots, Y\}$, in which f_c has been occupied by a SU subflow for y slots. Both X and Y are large enough such that $\Pr[x > X]$ and $\Pr[y > Y]$ are negligible. If there are multiple PU or SU services sharing f_c, the statistical data of the service with maximum duration can be applied.

The availability of white spaces is characterized by the steady-state probabilities of channel states, denoted by $\pi_{(\text{current state})}$'s, especially $\pi_{(0,0)}$ for idle state. To derive them, the transition probabilities, denoted by $\omega_{(\text{current state})}^{(\text{next state})}$'s, are obtained as follows.

(1) First, state transitions from (0,0) and $(0,y)$'s to (1,0) represent a PU arrival to f_c. Each transition from $(0,y)$ to (1,0) also indicates an "outward" switching of the SU from f_c. The transition probabilities are actually equal to the PU arrival

probability, denoted by λ_α, which further depends on the PU arrival process with average arrival rate α_n learnt in V_n. Namely, we have

$$\omega_{(0,0)}^{(1,0)} = \omega_{(0,y)}^{(1,0)} = \lambda_\alpha,\ y \in \{1,\ldots,Y\} \tag{10.38}$$

(2) Next, state transitions from $(x,0)$'s to (0,0) and that among $(x,0)$'s are defined by the distribution of PU service duration on f_c. Note that any closed-form distribution function is not necessarily required here. Instead, one can directly input the statistical distribution of service duration collected from a real network to determine the following transition probabilities

$$\begin{cases} \omega_{(x,0)}^{(0,0)} = \Pr\left[(x-1)\tau < S_{\text{pu}} \le x\tau\right],\ x \in \{1,\cdots,X\}; \\ \omega_{(x,0)}^{(x+1,0)} = 1-\omega_{(x,0)}^{(0,0)},\ x \in \{1,\cdots,X-1\} \end{cases} \tag{10.39}$$

where S_{pu} denotes the random variable of PU service duration.

(3) In a similar way, state transitions from $(0,y)$'s to (0,0) and that among $(0,y)$'s are defined by the distribution of SU service duration on f_c, which can also be general. Hence, we have

$$\begin{cases} \omega_{(0,y)}^{(0,0)} = \left(1-\omega_{(0,y)}^{(1,0)}\right)\Pr\left[(y-1)\tau < S_{\text{su}} \le y\tau\right],\ y \in \{1,\cdots,Y\}; \\ \omega_{(0,y)}^{(0,y+1)} = 1-\omega_{(0,y)}^{(1,0)}-\omega_{(0,y)}^{(0,0)},\ y \in \{1,\cdots,Y-1\} \end{cases} \tag{10.40}$$

where S_{su} denotes the random variable of SU service duration.

(4) State transition from (0,0) to (0,1) is triggered by a SU arrival. The SU arrival probability, denoted by λ_β, is determined by the SU arrival process with average arrival rate β_n learnt in V_n.

State transition from (0,0) to $(0,y)$ where $y > 1$ is triggered by an "inward" switching to f_c. Each transition from (0,0) to $(0,y)$ indicates the case that a subflow $s_{c'}$ switches to f_c when it has last y-1 slots on another channel $f_{c'}$ and has been forced to leave $f_{c'}$ due to the arrival of PU activities on $f_{c'}$. The subflow is renewed on f_c starting from the yth slot.

To derive the state transition probability in the above two cases, we first derive the probability that a certain subflow has successfully switched into f_c, denoted by γ. Due to Δ constraints, only the $2\cdot\Delta$ channels excluding f_c in $\mathcal{C}_{c-\Delta,2\Delta}$ can perform an "inward" switching to f_c. If there are u preempted channels and v idle channels out of such $2\cdot\Delta$ channels, γ is equivalent to the probability that one of the u expelled subflows successfully chooses f_c out of the total v+1 idle channels for an "inward" switching. Here the worst case is analyzed, in which any incoming subflow neglects the idle channels that are not included in $\mathcal{C}_{c-\Delta,2\Delta}$. Then, we have

$$\gamma = \sum_{u=1}^{2\Delta} \sum_{v=0}^{2\Delta-u} \frac{(2\Delta)!}{u!v!(2\Delta-u-v)!} \left[\lambda_\alpha \left(\sum_{y=1}^{Y} \pi_{(0,y)}\right)\right]^u \left(\pi_{(0,0)}\right)^v$$
$$\cdot \left[1-\lambda_\alpha \left(\sum_{y=1}^{Y} \pi_{(0,y)}\right) - \pi_{(0,0)}\right]^{2\Delta-u-v} \min\left\{\frac{u}{v+1}, 1\right\} \tag{10.41}$$

Given that a subflow, say $s_{c'}$, has switched into f_c, we further derive $\xi^{(0,y)}$, the probability that $s_{c'}$ has finished (y–1) slots on $f_{c'}$. Assuming that the primary user arrival is independent of the secondary user activities, we have

$$\xi^{(0,y)} = \sum_{z=y}^{Y} \Pr\left[(z-1)\tau < S_{su} \le z\tau\right], \ y \in \{2, \ldots, Y\} \tag{10.42}$$

Using (10.4) and (10.5), we have the following transition probabilities

$$\begin{cases} \omega_{(0,0)}^{(0,1)} = \left(1-\omega_{(0,0)}^{(1,0)}\right) \lambda_\beta; \\ \omega_{(0,0)}^{(0,y)} = \left(1-\omega_{(0,0)}^{(1,0)}\right) \gamma \xi^{(0,y)}, \ y \in \{2, \ldots, Y\} \end{cases} \tag{10.43}$$

At last, we have the transition probability from (0,0) to itself

$$\omega_{(0,0)}^{(0,0)} = 1-\omega_{(0,0)}^{(1,0)} - \sum_{y=1}^{Y} \omega_{(0,0)}^{(0,y)} \tag{10.44}$$

10.3.2 Delay Analysis Under Channel Aggregation

Both the negotiation and the transmission between a sender n and its receiver n' can involve service failures and delay costs. In this section, we investigate the corresponding service failure probabilities and model the delay costs for performing channel aggregation under the influence of PU activity.

10.3.2.1 Delays in Negotiation Process

The efficiency of negotiation between n and n' is restricted by the availability of white spaces in both V_n and $V_{n'}$. In general, the delays for making a successful negotiation include: (i) *sensing delay* $T_{\mathrm{ss}}^{(b)}$, which is the time required for sensing channels at both n and n'; (ii) *handshake delay* $T_{\mathrm{hs}}^{(b)}$, which is the time required for accessing control channel and making handshakes on it back and forth. In the following, these delays in negotiation processes are analyzed.

Note that after each blocking incident caused by $|\mathcal{P}_{l,\Delta}| < b$, a new round of handshake needs to be started until the maximum limit of blocking incidents, denoted as $\hat{N}_{\mathrm{bl}}$, is reached. Hence, negotiation delays are related to the number of blocking incidents during a negotiation process.

To analyze the number of blocking incidents, note that the blocking probability, denoted as $\theta^{(b)}$, can be computed as:

$$\theta^{(b)} = \sum_{v=0}^{b-1} \binom{\Delta+1}{v} \left(\pi_{(0,0)}\right)^{v} \left(1 - \pi_{(0,0)}\right)^{(\Delta+1)-v} \tag{10.45}$$

where $\pi_{0,0}$ is the probability that a channel is idle in the joint sensing range of a pair of SUs n and n' (a.k.a. $V_n \cup V_{n'}$). The PU and SU's arrival rates in $V_n \cup V_{n'}$ can be derived as $\alpha_{n,n'} = c_{n,n'} \cdot \alpha_n$ and $\beta_{n,n'} = c_{n,n'} \cdot \beta_n$ respectively, where $c_{n,n'}$ denotes the correlation factor between the arrival rate in V_n and the arrival rate in $V_n \cup V_{n'}$. Replacing α_n and β_n with $\alpha_{n,n'}$ and $\beta_{n,n'}$ in the channel model in Section 10.3.1.2, $\pi_{(0,0)}$ in (10.45) can be computed.

Assume that an entire negotiation process is considered failed when the number of blocking incident reaches the threshold $\hat{N}_{\text{bl}}$ with no success. Then, the negotiation failure probability, denoted as $\varepsilon^{(b)}$, can be expressed as:

$$\varepsilon^{(b)} = 1 - \sum_{r=0}^{\hat{N}_{\text{bl}}} \left(\theta^{(b)}\right)^{r} \left(1 - \theta^{(b)}\right) \tag{10.46}$$

With (10.45) and (10.46), the expected number of blocking incidents in a successful negotiation process becomes:

$$N_{\text{bl}}^{(b)} = \frac{1}{1-\varepsilon^{(b)}} \sum_{r=0}^{\hat{N}_{\text{bl}}} \left(\theta^{(b)}\right)^{r} \left(1 - \theta^{(b)}\right) r \tag{10.47}$$

Note that not every blocking incident costs the delay of a handshake since n initiates a handshake only when $|\mathcal{T}_{l,\Delta}| \geq b$. Therefore, we also need to obtain the blocking probability due to $|\mathcal{T}_{l,\Delta}| < b$, which is denoted as $\tilde{\theta}^{(b)}$. The same formula in (10.45) can be used to compute $\tilde{\theta}^{(b)}$ in a similar way as $\theta^{(b)}$. The only difference is that the $\pi_{(0,0)}$ in the expression of $\tilde{\theta}^{(b)}$ is computed using α_n and β_n, which are the PU and SU arrival rates in V_n. With $\tilde{\theta}^{(b)}$ computed, the expected number of handshake attempts for a successful negotiation can be computed as:

$$N_{\text{hs}}^{(b)} = N_{\text{bl}}^{(b)} \left(1 - \frac{\tilde{\theta}^{(b)}}{\theta^{(b)}}\right) + 1 \tag{10.48}$$

where $1 - \tilde{\theta}^{(b)}/\theta^{(b)}$ denotes the probability that $|\mathcal{T}_{l,\Delta}| \geq b$ but $|\mathcal{P}_{l,\Delta}| < b$.

Assuming that sequential sensing is used [20, 21], based on (10.48) and (10.47), we can compute the sensing delay $T_{\text{ss}}^{(b)}$ as follows. In sequential sensing, whenever a new $\mathcal{C}_{l,\Delta}$ is chosen, n needs to sense the channels in it to keep $\mathcal{T}_{l,\Delta}$ fresh, while n' senses the channels in $\mathcal{T}_{l,\Delta}$ to complete the entire negotiation. Hence, we can get:

$$T_{\text{ss}}^{(b)} = N_{\text{hs}}^{(b)} \left[\sum_{v=b}^{\Delta+1} \binom{\Delta+1}{v} \left(\pi_{(0,0)}\right)^{v} \left(1 - \pi_{(0,0)}\right)^{(\Delta+1)-v} v\right] \tau_{\text{ss}} + \left(N_{\text{bl}}^{(b)} + 1\right) (\Delta+1)\, \tau_{\text{ss}}, \tag{10.49}$$

where τ_{ss} denotes the average time for sensing one channel, and $\pi_{(0,0)}$ is computed using α_n and β_n. We can also compute the handshake delay as:

$$T_{\text{hs}}^{(b)} = N_{\text{hs}}^{(b)} \left(\tau_{\text{ma}} + \tau_{\text{rt}}\right), \tag{10.50}$$

where τ_{ma} denotes the average time for accessing control channel, which would be given by classic analytical models [22]; and τ_{rt} denotes the round-trip time for one handshake.

10.3.2.2 Delays in Transmission

The success of transmission cannot be guaranteed due to the occurrence of interruption incidents. In general, the delay costs for completing a successful transmission include: (i) *switching delay* $T_{\text{sw}}^{(b,d)}$, which is the time required for vacating the preempted channels and renewing the corresponding subflows; (ii) *transmission delay* T_{tx}, which is the amount of time it takes to transmit a given size of data. Typically, such transmission costs depend not only on b but also on d, i.e., the number of slots demanded for service duration after each successful negotiation. Sometimes, dividing a large size of data into smaller segments and transmitting them separately in several shorter periods can be a better choice due to reduced switching needs in each transmission periods. In the following, the switching delay $T_{\text{sw}}^{(b,d)}$ and transmission delay T_{tx} will be analyzed.

Transmission delay T_{tx} depends on the data transmission rate R_{tx}. To get the data transmission rate, note that when a PU arrival to a subchannel in an aggregated channel $\mathcal{A}^{(b)}$ is detected, the switching subflow in $\mathcal{A}^{(b)}$ is interrupted if the SU finds no available white space in the spectrum range that it is sensing (a.k.a. $|\mathcal{B}_{l,\Delta}|=0$). Then, the transmission on that subflow fails and the overall capacity of the aggregated channel reduces. Hence, to study such bandwidth reduction of $\mathcal{A}^{(b)}$ in each slot, we define a one-step interruption probability matrix

$$\boldsymbol{\Phi} = \begin{pmatrix} \varphi_{0,0} & & & & & \\ \varphi_{1,0} & \varphi_{1,1} & & & & \\ \varphi_{2,0} & \varphi_{2,1} & \varphi_{2,2} & & & \\ \varphi_{3,0} & \varphi_{3,1} & \varphi_{3,2} & \varphi_{3,3} & & \\ \vdots & \vdots & \vdots & & \ddots & \\ \varphi_{B,0} & \varphi_{B,1} & \varphi_{B,2} & \cdots & \cdots & \varphi_{B,B} \end{pmatrix} \tag{10.51}$$

in which each $\varphi_{i,j}$ denotes the probability that $(i-j)$ subflows in $\mathcal{A}^{(i)}$ are interrupted in one slot. Note that here we assume that more than one subflow in $\mathcal{A}^{(b)}$ can be interrupted in the same slot. $\varphi_{i,j}$ can be expressed as

$$\begin{cases} \varphi_{i,j} = \sum_{u=i-j}^{i} \binom{i}{u} (\lambda_\alpha)^u (1-\lambda_\alpha)^{i-u} \binom{(\Delta+1)-i}{u-i+j} \left(\pi_{(0,0)}\right)^{u-i+j} \\ \qquad \cdot \left(1-\pi_{(0,0)}\right)^{(\Delta+1)-u-j}, \; i\in\{1,\cdots,B\},\; j\in\{0,\cdots,i-1\}; \\ \varphi_{i,i} = 1-\sum_{j=0}^{i-1} \varphi_{i,j},\; i\in\{1,\cdots,B\} \end{cases} \tag{10.52}$$

Further for a d-slot transmission, the d-step interruption probability matrix $\boldsymbol{\Phi}^d = \boldsymbol{\Phi}^{d-1}\boldsymbol{\Phi}$ is used instead, in which each $\varphi_{i,j}^{(d)}$ defines the corresponding bandwidth reduction of $\mathcal{A}^{(i)}$ within d slots. Note that $\varphi_{b,0}^{(d)}$ is the complete transmission failure probability, where all the subflows of $\mathcal{A}^{(b)}$ are interrupted. With $\boldsymbol{\Phi}^d$, the average transmission rate for a transmission attempt can be expressed as

$$R_{tx} = \left(\sum_{j=0}^{b} \varphi_{b,j}^{(d)} R j d\tau \right) \tag{10.53}$$

where R denotes the bit rate on one channel and τ is the duration of a time slot.

Next, the switching delay $T_{\text{sw}}^{(b,d)}$ for a successful d-slot subflow is analyzed. Given that there is no interruption, let $\chi^{(b)}$ be the probability that a switching operation succeeds in one slot. We have

$$\chi^{(b)} = \frac{\lambda_\alpha \left[1 - \left(1 - \pi_{(0,0)}\right)^{(\Delta+1)-b} \right]}{1 - \lambda_\alpha \left(1 - \pi_{(0,0)}\right)^{(\Delta+1)-b}} \tag{10.54}$$

in which $\pi_{(0,0)}$ is computed using $\alpha_{n,n'}$ and $\beta_{n,n'}$. Within d slots, the expected number of switching operations is

$$N_{\text{sw}}^{(b,d)} = \sum_{z=0}^{d} \binom{d}{z} \left(\chi^{(b)}\right)^z \left(1 - \chi^{(b)}\right)^{d-z} z \tag{10.55}$$

Accordingly, for a successful d-slot subflow, we have

$$T_{\text{sw}}^{(b,d)} = N_{\text{sw}}^{(b,d)} \tau_{\text{sw}} \tag{10.56}$$

where τ_{sw} denotes the time required for one switching operation. The sensing time for locating backup channels can be negligible due to the simultaneous operations of the cognitive radio and scanning radio.

10.3.3 Optimal Bandwidth Duration Decision

Based on the derived negotiation and transmission costs for performing channel aggregation, in this section, we further define an optimal channel aggregation strategy that minimizes the cumulative delay costs.

10.3.3.1 Cumulative Delay

As shown above, the delay costs for channel aggregation are closely related to the values of b and d, i.e., the user demands on aggregated bandwidth and service duration. On one hand, the choice of b should consider the trade-off between channel capacity and blocking (interruption) probability during a negotiation (transmission). More white spaces are needed to meet a higher requirement of b. On the other hand, the choice of d should consider the trade-off between negotiation overhead and interruption probability during a transmission. With certain data to transmit, one can choose to divide the entire data into d-slot segments. A larger value of d results in fewer data segments and thus fewer negotiation operations. However, there may be more spectrum resources wasted due to higher interruption probability. Therefore, there exists an optimal combination of b and d to achieve optimal efficiency of channel aggregation.

The optimal efficiency is represented by a metric named *cumulative delay*, which is the total amount of time needed for transmitting a M-bit size data. Note that the average amount of data that can be successfully transmitted by each attempt is

$$\tilde{M}^{(b,d)} = \left(1-\varepsilon^{(b)}\right) R_{\text{tx}} \tag{10.57}$$

In addition, an attempt can meet three cases: (i) fails at negotiation stage, (ii) succeeds in negotiation but fails at transmission; and (iii) succeeds in both negotiation and transmission. The average cumulative delay, denoted as $T_{\text{cm}}^{(b,d)}$, must account for all the cases. Hence,

$$\begin{aligned} T_{\text{cm}}^{(b,d)} = \frac{M}{\tilde{M}^{(b,d)}} \Bigg\{ & \varepsilon^{(b)} \left(\hat{T}_{\text{ss}}^{(b)} + \hat{T}_{\text{hs}}^{(b)}\right) + \left(1-\varepsilon^{(b)}\right) \Bigg[T_{\text{ss}}^{(b)} + T_{\text{hs}}^{(b)} \\ & + \varphi_{b,0}^{(d)} \left(T_{\text{sw}}^{\left(b,\frac{d}{2}\right)} + \frac{d}{2}\tau \right) + \left(1-\varphi_{b,0}^{(d)}\right) \left(T_{\text{sw}}^{(b,d)} + d\tau\right) \Bigg] \Bigg\} \end{aligned} \tag{10.58}$$

in which $\hat{T}_{\text{ss}}^{(b)}$ and $\hat{T}_{\text{hs}}^{(b)}$ are computed by replacing the expected negotiation times $N_{\text{bl}}^{(b)}$ with negotiation failure threshold $\hat{N}_{\text{bl}}$ in (10.49) and (10.50), respectively, and the expected duration of failed service related to $\varphi_{b,0}^{(d)}$ is assumed to be $d/2$.

10.3.3.2 Optimal Channel Aggregation Strategy

A *channel aggregation strategy* (b, d) is defined as the combination of both bandwidth and duration demands. The objective is to find the optimal (b^*, d^*) that minimizes $T_{\text{cm}}^{(b,d)}$:

$$(b^*, d^*) = \underset{(b,d)\in\mathcal{G}}{\arg\min}\, T_{\text{cm}}^{(b,d)} \tag{10.59}$$

It is not hard to find (b^*, d^*) by searching the finite set of all possible (b, d)'s. Note that in CR networks, both PU and SU behaviors that affect the availability of white spaces are stochastic. Hence, the optimal channel aggregation strategy defined in (10.59) is actually optimal in the sense of average performance.

10.3.4 Numerical Analysis and Simulation Results

To illustrate and validate the analytical results, figures that are derived from numerical analysis and discrete-event simulation of a few of our analytical results are shown in this section, including the negotiation failure probability $\varepsilon^{(b)}$ in (10.46), the transmission failure probability $\varphi_{b,0}^{(d)}$ defined in Section 10.3.2.2, and the cumulative delay $T_{\text{cm}}^{(b,d)}$ in (10.58). These figures will show the impact of PU activity and channel aggregation strategy on the efficiency of negotiation and transmission between secondary users.

In the numerical analysis, a pair of PU sender and receiver, n and n', in a distributed CR network is considered. Poisson PU arrival process is assumed in both V_n and $V_{n'}$ areas. The distribution of PU service duration is set according to the statistical distribution of call duration collected from a real cellular network [5]. As for SU behavior, Poisson SU arrival process and random SU service duration are assumed. Note that the choice of service duration for the pair of n and n' is a part of their optimal decision, but we fix the patterns of SU activity in the background. The constant parameters are set as follows: $K = 50$; $B = 3$; $\beta_n = 0.02$ user/s; $c_{n,n'} = 1.5$; $E[Y] = 3$ s; $\tau = 10$ ms; $\tau_{\text{ss}} = 10$ ms [18, 21]; $\tau_{rt} = 200$ ms; $\tau_{\text{sw}} = 600$ ms [20]; $\hat{N}_{\text{bl}} = 5$; $M = 50$ Mb; $R = 5$ Mb/s. The others are viewed as variables.

10.3.4.1 Illustration of Negotiation Failure Probability

As in (10.46), $\varepsilon^{(b)}$ defined for negotiation failure which characterizes the repeated blocking incidents caused by lack of enough common white space at n and n' (a.k.a. $|\mathcal{P}_{l,\Delta}| < b$). The impact of α_n and Δ on $\varepsilon^{(b)}$ with fixed transmission duration of $d \cdot \tau = 3$ s is plotted in Fig. 10.10. Generally, the numerical results (marked as "ana") and simulation results (marked as "sim") match well with each other under the same settings. It can be seen that with a higher demand on b or a drop in the availability of white spaces, $\varepsilon^{(b)}$ increases. In addition, a relaxation of the hardware limitation on Δ offers more candidate channels and thus lowers $\varepsilon^{(b)}$.

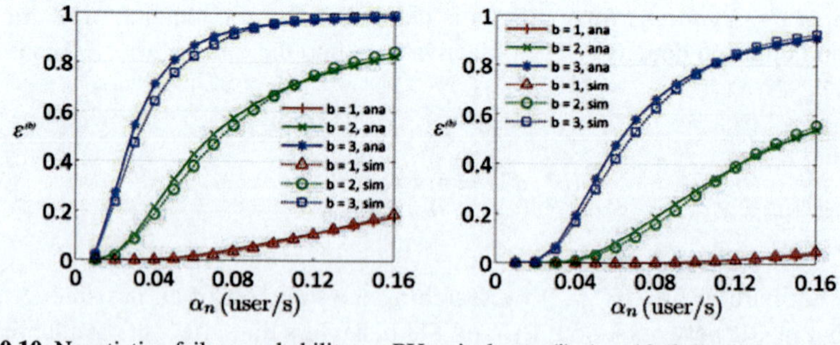

Fig. 10.10 Negotiation failure probability vs. PU arrival rate: (i) $\Delta = 10$ (*left*); (ii) $\Delta = 20$ (*right*)

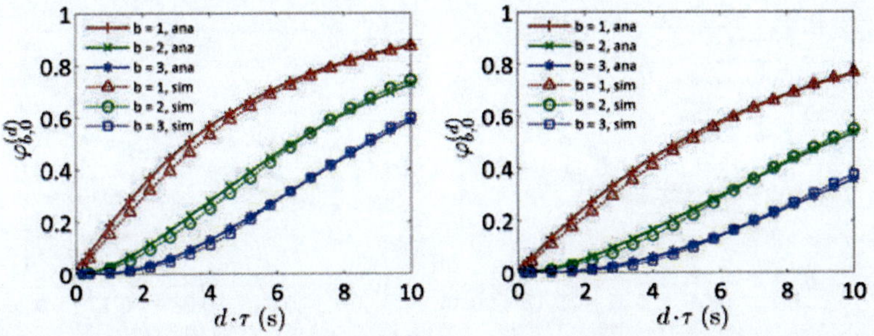

Fig. 10.11 Transmission failure probability vs. duration demand: (i) $\Delta = 10$ (*left*); (ii) $\Delta = 20$ (*right*)

10.3.4.2 Illustration of Transmission Failure Probability

The impact of transmission duration d on transmission failure probability $\varphi_{b,0}^{(d)}$ is shown at Fig. 10.11, where the PU traffic arrival rate is fixed at $\alpha_n = 0.1$ user/s. Defined in Section 10.3.2.2, $\varphi_{b,0}^{(d)}$ is the probability that all subflows of an aggregated channel are interrupted halfway during a transmission by PUs. Clearly, with longer transmission duration $d\tau$ or a smaller channel separation constraint Δ, the $\varphi_{b,0}^{(d)}$ increases. Interestingly, unlike negotiation failure probability $\varepsilon^{(b)}$, a larger transmission duration b actually reduces transmission failure probability $\varphi_{b,0}^{(d)}$. Intuitively, this is because the transmission consisting of more subflows would tolerant more interruption incidents. Hence, a trade-off obviously exists among the negotiation failure probability and transmission failure probability to achieve the overall optimal transmission strategy.

10.3.4.3 Illustration of Cumulative Delay

For the transmission of M-bit data, the related cumulative delay $T_{\mathrm{cm}}^{(b,d)}$ has been chosen as our objective function as in (10.58). In Figs. 10.12 and 10.13, the impact of α_n and (b, d) on $T_{\mathrm{cm}}^{(b,d)}$ is plotted, respectively. It can be seen that $T_{\mathrm{cm}}^{(b,d)}$ rises rapidly with the increase of α_n due to the increase of $\varepsilon^{(b)}$ and $\varphi_{b,0}^{(d)}$ in such cases. A larger channel separation constraint Δ lowers $T_{\mathrm{cm}}^{(b,d)}$. To achieve the lowest $T_{\mathrm{cm}}^{(b,d)}$, the optimal channel aggregation strategy (b^*, d^*) is evaluated under different settings in Fig. 10.13. The marked point that represents the optimal decision varies significantly with the availability of white spaces. Obviously, when there are plenty of white spaces as in Fig. 10.13-i, larger b and d are the optimal solution to achieve the highest utilization of licensed spectrum. Note that the range of d differs for different values of b for transmitting the same size of data. However, if there are few white spaces as in Fig. 10.13-iii, both b and d should be low to avoid the huge costs for repeated negotiation and transmission attempt.

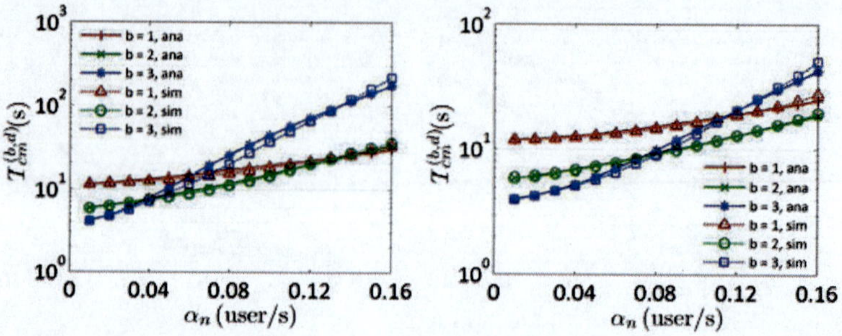

Fig. 10.12 Cumulative delay (in log scale) vs. PU arrival rate: (i) $\Delta = 10$ (*left*); (ii) $\Delta = 20$ (*right*)

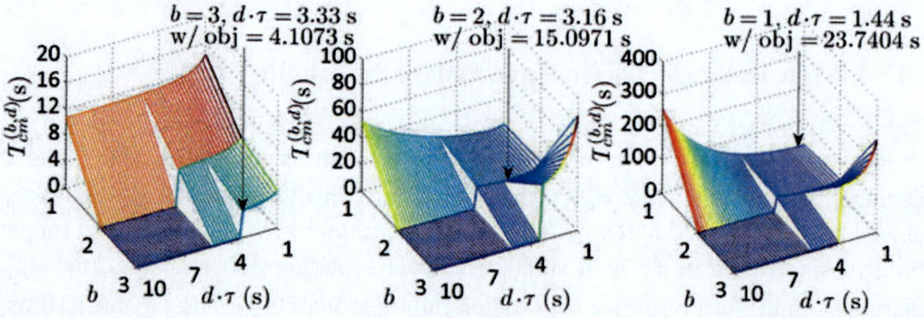

Fig. 10.13 Cumulative delay vs. channel aggregation strategy: (i) $\alpha_n = 0.01$ user/s (*left*); (ii) $\alpha_n = 0.1$ user/s (*middle*); (iii) $\alpha_n = 0.16$ user/s (*right*)

10.4 Summary

In this section, we analyzed the delay for both multihop and single-hop CRN.

For multihop CRN, we derive the IPS upper bound under two cases: the network IPS and the flow IPS. In the network IPS case, we discover that the network IPS upper bound is related to a threshold value of the PU activity level. Below the threshold, the IPS upper bound is achieved when the one-hop distance equals the communication range of cognitive radios. Above the threshold, the upper bound speed is achieved when using an optimal one-hop distance which is less than the communication range. We design efficient numerical methods to compute the optimal one-hop distance and the corresponding IPS upper bound. In the flow IPS case, we discover that the IPS upper bound is achieved when an optimal number of SU relay nodes are evenly spaced on the straight line between the source node and the destination node. The optimal number of relay nodes shows a stair-like incremental trend, when the PU activity level increases. We design multiple numerical methods to compute the optimal number of SU relay nodes. The simulation and numerical results prove the correctness of our analysis.

For single-hop CRN, we have studied the delay under considerations of various practical constraints and costs. A new channel usage model based on general assumptions is introduced to investigate the negotiation and transmission costs for utilizing channel aggregation under the influence of PU activity. We have found that user demands on both aggregated bandwidth and service duration affect the delay performance. Hence, an optimal channel aggregation strategy has been defined and validated to achieve the lowest delay for data transmission.

Acknowledgments This work was supported in part by the US National Science Foundation under grant CNS-0831865 and the Institute for Critical Technology and Applied Science (ICTAS) of Virginia Tech.

References

1. Y. Xu and W. Wang, "The speed of information propagation in large wireless networks," in *Proc. of IEEE INFOCOM*, Phoenix, AZ, 2008.
2. R. Zheng, "Information dissemination in power-constrained wireless networks," in *Proc. of IEEE INFOCOM*, Barcelona, Catalunya, Spain, 2006.
3. P. Jacquet, B. Mans, P. Muhlethaler, and G. Rodolakis, "Opportunistic routing in wireless ad hoc networks: Upper bounds for the packet propagation speed," in *Proc. of IEEE MASS*, Atlanta, Georgia, USA, 2008.
4. P. Jacquet, B. Mans, and G. Rodolakis, "Information propagation speed in mobile and delay tolerant networks," in *Proc. of IEEE INFOCOM*, Rio de Janeiro, Brazil, 2009.
5. D. Willkomm, S. Machiraju, J. Bolot, and A. Wolisz, "Primary user behavior in cellular networks and implications for dynamic spectrum access," *IEEE Communications Magazine*, vol. 47, pp. 88–95, 2009.
6. P. Popovski, H. Yomo, K. Nishimori, R. D. Taranto, and R. Prasad, "Opportunistic interference cancellation in cognitive radio systems," in *Proc. of IEEE DySPAN*, Dublin, Ireland, 2007.
7. R. Zhang and Y.-C. Liang, "Exploiting hidden power-feedback loops for cognitive radio," in *Proc. of IEEE DySPAN*, Chicago, Illinois, 2008.
8. G. Zhao, G. Y. Li, and C. Yang, "Proactive detection of spectrum opportunities in primary systems with power control," *IEEE Transactions on Wireless Communications*, vol. 8, pp. 4815–4823, 2009.
9. G. Bolch, S. Greiner, H. de Meer, K. S. Trivedi, H. de Meer, and K. S. Trivedi, *Queueing Networks and Markov Chains: Modeling and Performance Evaluation With Computer Science Applications*. Wiley-Interscience, Wiley, 2006.
10. C. Han and Y. Yang, "Information propagation speed in cognitive radio networks: Network and flow analysis," Virginia Tech, Tech. Rep., 2010.
11. R. Chandra, R. Mahajan, T. Moscibroda, R. Raghavendra, and P. Bahl, "A Case for Adapting Channel Width in Wireless Networks," in *Proc. ACM SIGCOMM'08*, Seattle, WA, Aug. 2008.
12. IEEE 802.22 WG, "IEEE P802.22/D0.1 Draft Standard for Wireless Regional Area Networks Part 22: Cognitive Wireless RAN Medium Access Control (MAC) and Physical Layer (PHY) Specifications: Policies and Procedures for Operation in TV Bands," *IEEE Standard*, May 2006.
13. S. Haykin, "Cognitive Radio: Brain-Empowered Wireless Communications," *IEEE Journal on Selected Areas in Communications*, Vol. 23, No. 2, pp. 201–220, Feb. 2005.
14. H. Kim, and K. G. Shin, "Efficient Discovery of Spectrum Opportunities with MAC-Layer Sensing in Cognitive Radio Networks," *IEEE Transactions on Mobile Computing*, Vol. 7, No. 5, pp. 533–545, May 2008.
15. F. Huang, W. Wang, H. Luo, G. Yu, and Z. Zhang, "Prediction-Based Spectrum Aggregation with Hardware Limitation in Cognitive Radio Networks," in *Proc. IEEE VTC'10-Spring*, Taipei, Taiwan, May 2010.

16. P. Bahl, A. Adya, J. Padhye, and A. Wolman, "Reconsidering Wireless Systems with Multiple Radios," *ACM SIGCOMM Comp. Comm. Rev.*, Vol. 34, No. 5, pp. 39–46, Oct. 2004.
17. Y. Yuan, P. Bahl, R. Chandra, P. Chou, J. Ferrell, T. Moscibroda, S. Narlanka, and Y. Wu, "KNOWS: Kognitiv Networking Over White Spaces," in *Proc. IEEE DySPAN'07*, Dublin, Ireland, Apr. 2007.
18. Y. Yuan, P. Bahl, R. Chandra, T. Moscibroda, and Y. Wu, "Allocating Dynamic Time-Spectrum Blocks in Cognitive Radio Networks," in *Proc. ACM MobiHoc'07*, Montreal, QC, Canada, Sep. 2007.
19. J. Lee, and J. So, "Analysis of Cognitive Radio Networks with Channel Aggregation," in *Proc. IEEE WCNC'10*, Sydney, Australia, Apr. 2010.
20. D. Xu, E. Jung, and X. Liu, "Optimal Bandwidth Selection in Multi-Channel Cognitive Radio Networks: How Much Is Too Much?," in *Proc. IEEE DySPAN'08*, Chicago, Illinois, Oct. 2008.
21. T. Shu, and M. Krunz, "Throughput-Efficient Sequential Channel Sensing and Probing in Cognitive Radio Networks under Sensing Errors," in *Proc. ACM MobiCom'09*, Beijing, China, Sep. 2009.
22. G. Bianchi, "Performance Analysis of IEEE 802.11 Distributed Coordination Function," *IEEE Journal on Selected Areas in Communications*, Vol. 18, No. 3, pp. 535–547, Mar. 2000.

Part IV
Multimedia Transmissions

Chapter 11
Real-Time Multimedia Transmission over Cognitive Radio Networks

Haiyan Luo, Song Ci, Dalei Wu, Zhiyong Feng, and Hui Tang

Abstract Cognitive radio (CR) has been proposed as a promising solution to improve connectivity, self-adaptability, and efficiency of spectrum usage. When used in video applications, user-perceived video quality experienced by secondary users is a very important performance metric to evaluate the effectiveness of CR technologies. However, most current research only considers spectrum utilization and effectiveness at MAC and PHY layers, ignoring the system performance of upper layers. Therefore, in this chapter we aim to improve the user experience of secondary users for wireless video services over cognitive radio networks. We propose a quality-driven cross-layer optimized system to maximize the expected user-perceived video quality at the receiver end, under the constraint of packet delay bound. By formulating network functions such as encoder behavior, cognitive MAC scheduling, transmission, as well as modulation and coding into a distortion-delay optimization framework, important system parameters residing in different network layers are jointly optimized in a systematic way to achieve the best user-perceived video quality for secondary users in cognitive radio networks. Furthermore, the proposed problem is formulated into a MIN-MAX problem and solved by using dynamic programming. The performance enhancement of the proposed system is evaluated through extensive experiments based on H.264/AVC.

11.1 Introduction

With the fast development of wireless communication technologies, the limited unlicensed spectrum bands can no longer meet the increasing requirement. Wireless multimedia applications require significant bandwidth with relatively tight delay constraints. With limited radio spectrum, bandwidth is considered to be one of the major bottlenecks for high-quality multimedia wireless applications. Frequency spectrum has become the scarcest resource in the next-generation wireless multimedia networks. However, statistics indicate that a large amount of the licensed spectrum bands are under-utilized, due to the spectrum strategy of static allocation and centralized management. Research also shows that the utilization of both licensed

S. Ci (✉)
College of Engineering, University of Nebraska-Lincoln, Omaha, NE, USA
e-mail: sci@engr.unl.edu

F.R. Yu (ed.), *Cognitive Radio Mobile Ad Hoc Networks*,
DOI 10.1007/978-1-4419-6172-3_11,

and unlicensed frequency bands changes with time and location. Thus, efficient methods are necessary for spectrum sharing among different systems, applications, and services in a dynamic wireless environment [1]. Therefore, "cognitive radio" has emerged as a new design paradigm for next-generation wireless networks, aiming to increase the utilization of radio spectrum for both licensed and unlicensed bands.

Generally speaking, cognitive radio is an intelligent wireless communication system that is aware of its surrounding environment and uses the methodology of understanding-by-building to learn from the environment. Then, its internal states are adapted to respond to statistical variations of the incoming RF stimuli by changing certain operating parameters in real time. Here, there are two primary objectives: (1) to provide highly reliable communications whenever and wherever needed and (2) to achieve efficient utilization of the radio spectrum [2]. Figure 11.1 illustrates a typical scenario of wireless video transmission over cognitive radio networks, where primary users and secondary users share a block of frequency spectrum. In cognitive radio networks, only primary users are authorized to use the radio spectrum. Thus, secondary users have to search the idle channels to use at the beginning of every slot by performing channel sensing. Based on the sensing outcomes, secondary users will decide whether or not to access the sensed channels. It has been reported that some frequency bands in the radio spectrum are largely unused, while some are

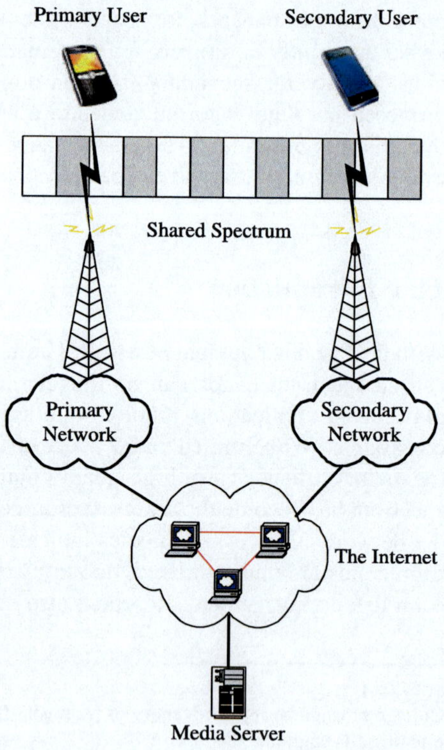

Fig. 11.1 Wireless video transmission over cognitive radio networks, where primary users and secondary users access the network through shared radio frequency band

heavily used. In particular, while a frequency band is assigned to a primary wireless system/service at a particular time and location, the same frequency band is unused by this wireless system/service in other times and locations. This results in spectrum holes (a.k.a spectrum opportunities) [3]. Therefore, by allowing secondary users to utilize these spectrum holes, spectrum utilization can be improved substantially.

Indeed, improving system performance of secondary users is a vital factor that hinges on the success of cognitive radio technologies for wireless video applications. Actually, if upper layer system performance is not well considered, the video quality degradation perceived by secondary users can largely impede the successful deployment of cognitive radio technologies. Also, user experience is by far the most important performance metric for end users in wireless video transmission. Therefore, it is critical to improve the user-perceived video quality at the receiver side for secondary users in cognitive radio networks. Although a lot of research activities have been conducted in cognitive radio networks, most of them only consider sensing effectiveness and spectrum utilization at the system design level. Other design metrics of the upper layers, such as the user-perceived video quality, have been mostly ignored.

In this chapter, we first briefly introduce two major new design technologies that can be used to improve the overall network performance of video applications over cognitive radio networks. Then, to achieve the best user-perceived video quality at the receiver side for secondary users of real-time wireless video transmission over cognitive radio networks, we present a quality-driven cross-layer system for joint optimization of system parameters residing in the entire network protocol stack. In the presented system, time variations of primary network usage and wireless channels are modeled, based on which, the encoder behavior, cognitive MAC scheduling, transmission, and modulation, and coding are jointly optimized for secondary users in a systematic way under a distortion-delay framework for the best video quality perceived by secondary users. The presented problem is formulated as a MIN-MAX problem and solved by using dynamic programming [4, 5]. Furthermore, the experiments prove that the proposed joint optimization system can greatly improve user experience at the receiver side for real-time wireless video services over cognitive radio networks.

11.2 Design Background

To increase resource utilization and improve the overall performance of cognitive radio networks, different design ideas have been proposed in recent years. Generally, two schools of approaches were presented: game theoretic approach and cross-layer-based optimization. In game theoretic approach, the behaviors of different competing users in the same frequency are modeled and a maximized utility function is formulated to dynamically adapt the channel selection strategy. In cross-layer optimization strategy, the interactions of different layers are jointly considered to adapt to the varying wireless channel quality to make better usage of the wireless channel resources.

11.2.1 Game Theory

Game theory is a discipline aimed at modeling situations in which decision makers have to make specific actions that have mutual, possibly conflicting, consequences. It has been used primarily in economics to model competition between companies. For example, should a given company enter a new market, considering that its competitors could make similar (or different) moves [6, 7]?

Game theory has also been applied to other areas, including politics and biology. Not surprisingly, it has also been applied to networking, in most cases to solve routing and resource allocation problems in a competitive environment. In recent years, it has been applied by an increasing number of researchers to resolve traditional issues in wireless networks. The common scenario is that the decision makers in the game are rational users or networks operators who control their communication devices. These devices have to cope with a limited transmission resource (i.e., the radio spectrum) that imposes a conflict of interests. In an attempt to resolve this conflict, they can make certain moves such as transmitting now or later, changing their transmission channel, or adapting their transmission rate. Felegyhazi et al. in [8] summarized how game theory can be used to model the radio communication channel. By leveraging on four simple running examples, the authors in this chapter have introduced the most fundamental concepts of non-cooperative game theory.

In [9], the authors proposed a dynamic channel-selection solution for autonomous wireless users transmitting delay-sensitive multimedia applications over cognitive radio networks, in which various rate requirements and delay deadlines of heterogeneous multimedia users were also considered. An information exchange mechanism was proposed to manage available spectrum resources in a decentralized manner. Based on this, a priority virtual queue interface was proposed that determines the required information exchanges and evaluates the expected delays experienced by various priority traffic. Then, a dynamic strategy learning (DSL) algorithm is deployed at each user that exploits the expected delay and dynamically adapts the channel selection strategies to maximize the user's utility function. In their simulation, almost 2 dB of video quality improvement can be achieved by reducing packet loss rate.

In another paper [10], the scalable and delay-sensitive characteristics of multimedia data and the resulting impact on users' viewing experiences of multimedia content are explicitly involved in the proposed utility function. The spectrum allocation problem was then formulated as an auction game and a distributively auction-based spectrum allocation scheme.

Generally, using game theoretic approach in multimedia applications over cognitive radio networks is a new research direction that was emerging a few years ago. The coming years is expected to see more research findings in this regard.

11.2.2 Cross-layer Optimization

Cross-layer design over cognitive radio networks was briefly discussed in [11, 12]. In the proposed frameworks, each node in the network can sense and learn from the wireless environment and then respond to environment changes by adapting system parameters. However, how to integrate the different layers was not discussed. Also, there were no experimental results to verify the proposed frameworks.

In [13, 14], a rake optimized power-aware scheduling (ROPAS) architecture was proposed for mobile ad hoc networks (MANs) to deal with the utilization of cognitive radio (CR) for dynamic channel allocation among the requesting applications while limiting the average power transmitted in each sub-band. In that work, the cross-layer interaction between medium access control (MAC) and physical (PHY) layers of cognitive radio networks was considered. The joint power control and link scheduling can reduce adjacent channel and multi-access interference.

Su et al. in [15] have proposed an opportunistic multi-channel MAC protocol, in which the spectrum sensing at the physical layer and the packet scheduling at the MAC layer are integrated for wireless ad hoc networks. The proposed MAC protocol enables secondary users to identify and utilize the leftover frequency spectrum to reduce the interference level. Two different policies on channel sensing, which includes random sensing policy and negotiation-based sensing policy, have been proposed to detect the availability of unused licensed channels.

In [16], a framework called cognitive resource manager (CMR) was proposed to optimize the network protocol stack as a whole. The exchange of network information between CMRs can avoid harmful interactions arising from local optimization methods. In addition, the proposed framework can adapt MAC and link parameters to choose the best possible settings for the applications running on top. However, no implementation details and experimental analysis were discussed.

Overall, most current research efforts on cross-layer design over cognitive radio networks only focus on the joint consideration of MAC layer and physical layer. The design objectives are limited due to the fact that only sensing effectiveness and spectrum utilization are used as the design criteria, while the performance at the upper layers has been largely ignored. In this chapter, we will focus on addressing this issue.

11.3 System Model for Video Transmission

In this section, we present a quality-driven cross-layer optimized system that includes different modules, such as video encoder module, cognitive MAC module, modulation and coding module, cross-layer optimization module, as well as wireless video transmission module. These system modules actually represent different network functions residing in different network layers. For example, the video encoder resides in the application layer. The cognitive MAC module resides in the MAC layer, while the modulation and coding module is in the physical layer. As shown in Fig. 11.2, the cross-layer optimization module is able to communicate with other system modules to adjust the network functions, by selecting the optimal system parameters within a distortion-delay optimization framework. In this way, the major network functions are jointly optimized to achieve the best user-perceived video quality over cognitive radio networks under the current network conditions.

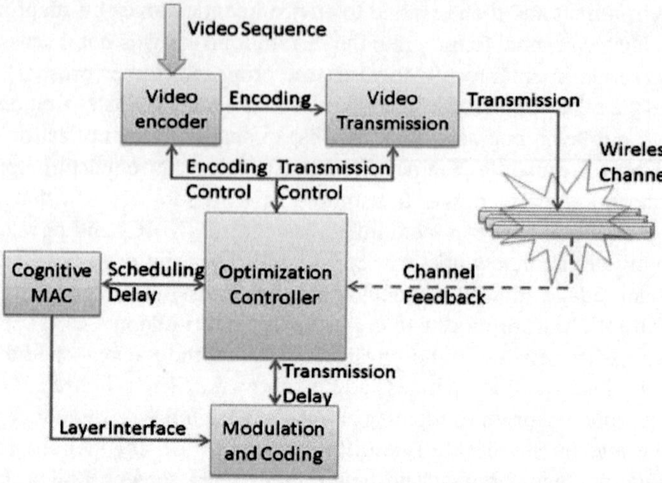

Fig. 11.2 The quality-driven cross-layer optimized system model for real-time wireless video transmission over cognitive radio networks

In the presented system, the expected video distortion calculated at the video encoder of the application layer is adopted as the objective function, which can be represented by encoding parameters (such as quantization step size q) and packet loss rate ρ. Furthermore, the expected packet delay, which is jointly represented by system parameters that affect encoder behavior, MAC scheduling, transmission, and modulation and channel coding, is used as the design constraint. Therefore, with the feedback information from the network such as RTT, queue length, and packet loss rate, the distortion-delay optimization module can choose the optimal set of parameters through the proposed cross-layer optimized system to achieve the best user-perceived video quality.

11.4 Video Quality Performance Metric

For video applications, the estimated video distortion is popularly used as the most important design metric. In this section, we present a formulation of end-to-end video distortion at the video encoder that can be used as the objective function in the proposed design.

Video encoder situates at the application layer. Without losing generality, we consider H.264 video codec in this chapter. In H.264 codec, each video frame is represented by block-shaped units of the associated luminance and chrominance samples (16×16 pixel region) called macroblocks (MBs). Furthermore, macroblocks can be both intra-coded and inter-coded from samples of previous video frames [17].

Intra-coding is performed in the spatial domain by referring to neighboring samples of previously coded blocks which are on the left and/or above the block to be predicted. Meanwhile, inter-coding is performed with temporal prediction from samples of previous video frames.

To estimate the end-to-end video distortion accurately, we need to consider all possible factors, which include source coding, error propagation, channel coding. Many research efforts can be found in literature on distortion estimation for hybrid motion-compensated video coding and transmission over lossy channels [18–21]. For real-time source coding, the estimated distortion caused by quantization, packet loss, and error concealment at the encoder can be calculated by using the "Recursive Optimal Per-pixel Estimate" (ROPE) method [18], providing an accurate optimization metric to the proposed system based on video quality. For the source coding parameter, we consider quantization step size (QP) q in this chapter.

According to the H.264 standard, one packet is set to be one row of macroblocks, which is also called one slice [17]. Therefore, slice and packet are two interchangeable concepts in this chapter. Given the dependencies introduced by error concealment scheme, the expected distortion of packet x of video frame n can be calculated at the encoder by using ROPE method as

$$E[D_{n,x}] = (1-\rho_x)E\left[D^r_{n,x}\right] + \rho_x(1-\rho_{x-1})E\left[D^{lr}_{n,x}\right] + \rho_x\rho_{x-1}E\left[D^{ll}_{n,x}\right] \tag{11.1}$$

where ρ_x is the loss probability of packet x with consideration of packet delay bound $T_n^{\max}$. $E\left[D^r_{n,x}\right]$ is the expected distortion of packet x when it is successfully received. Furthermore, depending on whether packet $(x-1)$ is received or lost, $E\left[D^{lr}_{n,x}\right]$ and $E\left[D^{ll}_{n,x}\right]$ are the corresponding expected distortion after concealment when packet x is lost. Therefore, the expected distortion of the whole video frame n can be represented as

$$E[D_n] = \sum_{x=1}^{X_n} E[D_{n,x}] \tag{11.2}$$

where X_n is the total number of packets in the video frame n. Thus, the expected end-to-end video distortion is accurately calculated by ROPE under instantaneous network conditions, which becomes the objective function in the proposed optimized system. For a given video packet x, the expected packet distortion only depends on packet error rate ρ_x and QP q. Considering the fact that the individual contribution of each path is continuously updated, this parameter is updated after each packet is encoded. In addition, the prediction and calculation of packet loss rate ρ_x will be discussed in Section 11.7. Readers can also refer to [4, 18] for detailed information regarding the calculation of the expected video distortion.

11.5 Channel Model

In this chapter, we assume that wireless channels are frequency flat, remaining time-invariant during a packet, but may vary from packet to packet. The channel quality is captured by the received signal-to-noise ratio (SNR) ξ. We adopt the Rayleigh channel model to describe ξ statistically. Therefore, the received SNR per packet is a random variable with a probability density function (pdf):

$$p(\xi) = \frac{1}{\bar{\xi}}\exp\left(-\frac{\xi}{\bar{\xi}}\right), \qquad \xi \geq 0 \tag{11.3}$$

where $\bar{\xi} := E\{\xi\}$ is the average received SNR. We also assume that the receiver has perfect channel side information and hence knows the instantaneous values of channel state information (CSI), while the transmitter has no such knowledge [22].

Moreover, we employ a cognitive channel model in which secondary users will try to transmit data when primary users are in presence. Secondary users will first perform channel sensing to detect the activity of primary users and then decide whether to transmit the data immediately or wait for the next available time slot depending on the detection result.

The different possible channel states for the primary usage are defined in Table 11.1.

Based on this definition, we depict the state transition model for cognitive radio transmission as shown in Fig. 11.3. In this figure, the state transition model is completely described by its stationary distribution of each channel state i and the state transition probability from state s_i to state s_j at the beginning of each time slot, which is denoted as p_{ij} $(1 \leq i, j \leq 4)$. Given the knowledge of channel fading, state transition probabilities, and primary user usage, the channel transition matrix $\mathbf{R}$ can be written as

$$\mathbf{R} = \{p_{ij}\} \quad (1 \leq i \leq 4, 1 \leq j \leq 4) \tag{11.4}$$

where

$$p_{ij} \geq 0, \forall i, j$$
$$\sum_{j=1}^{4} p_{ij} = 1, \forall i$$

Next, we will derive the transition matrix $\mathbf{R}$. In this chapter, we adopt noisy observations y_i under the Neyman–Pearson formulation to detect correlated random signals [23–25]. Therefore, channel sensing can be formulated as a hypothesis testing problem between the noise w_i and the signal s_i in noise.

$$\begin{aligned} H_0 &: y_i = w_i, \quad i = 1, \ldots, S_{\mathrm{tot}} \\ H_1 &: y_i = s_i + w_i \quad i = 1, \ldots, S_{\mathrm{tot}} \end{aligned} \tag{11.5}$$

Table 11.1 Possible channel states for the primary channel

State ID	State name	State description
s_1	correct detection	channel is idle, detected as idle
s_2	false alarm	channel is idle, detected as busy
s_3	missed detection	channel is busy, detected as idle
s_4	correct detection	channel is busy, detected as busy

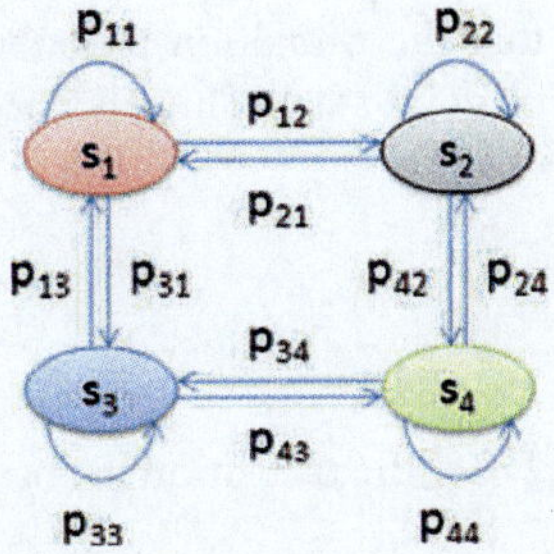

Fig. 11.3 The proposed state transition model for cognitive radio channel

where S_{tot} is the total symbols within a duration that is allocated to sense the channel. Received noise $\{w_i\}$ is a zero-mean, complex Gaussian random variable with variance σ_w^2 for all i, denoted as $\{w_i\} \backsim S_{\text{tot}}\left(0, \sigma_w^2\right)$. $\{s_i\}$ is the sum of the active primary users' faded signals arriving at the secondary receiver, which has a circularly symmetric complex Gaussian distribution with zero-mean and variance σ_s. Both $\{s_i\}$ and $\{w_i\}$ are assumed to be independent and identically distributed (i.i.d) [25].

Therefore, the optimal Neyman–Pearson detector of the above detection problem is given by

$$Y = \frac{1}{S_{\text{tot}}} \sum_{i=1}^{S_{\text{tot}}} \left| y_i^2 \right| \gtrless_{H1}^{H_0} t_{\text{d}} \tag{11.6}$$

where t_{d} is the detection threshold. Thus, test statistic Y is chi-square distributed with $2S_{\text{tot}}$ degrees of freedom. Therefore, the probabilities of detection p_{d} and false alarm p_{f} can be represented as follows, respectively.

$$p_{\text{d}} = P_r(Y > t_{\text{d}}|H_1) = 1 - P\left(\frac{S_{\text{tot}} t_{\text{d}}}{\sigma_w^2 + \sigma_s^2}, S_{\text{tot}}\right) \tag{11.7}$$

$$p_{\text{f}} = P_r(Y > t_{\text{d}}|H_0) = 1 - P\left(\frac{S_{\text{tot}} t_{\text{d}}}{\sigma_w^2}, S_{\text{tot}}\right) \tag{11.8}$$

where $P(x, z)$ denotes the regularized lower gamma function. Denote $\Gamma(z)$ and $\gamma(x, z)$ the Gamma function and the lower incomplete gamma function, respectively, then $P(x, z)$ can be represented as

$$P(x, z) = \frac{\gamma(x, z)}{\Gamma(z)} \tag{11.9}$$

Therefore, the probability of channel being idle and detected as idle can be calculated from

$$p_{11} = (1 - p_f)(1 - \zeta_b) \tag{11.10}$$

where ζ_b is the prior probability of channel being busy. Equation (11.10) indicates that the transition probability only depends on the current state, regardless of the original state. This also holds true for the other state transition probabilities. Denote

$$\begin{aligned} &p_{i1} = p_1; \quad p_{i2} = p_2 \\ &p_{i3} = p_3; \quad p_{i4} = p_4 \\ &\text{where} \quad 1 \le i \le 4 \end{aligned} \tag{11.11}$$

Then, the transition probabilities can be represented as

$$\begin{aligned} p_1 &= (1 - p_f)(1 - \zeta_b) \\ p_2 &= p_f(1 - \zeta_b) \\ p_3 &= (1 - p_d)\zeta_b \\ p_4 &= p_d\zeta_b \end{aligned} \tag{11.12}$$

Hence, the 4×4 transition matrix $\mathbf{R}$ can be derived as

$$\mathbf{R} = \begin{pmatrix} p_1, p_2, p_3, p_4 \\ p_1, p_2, p_3, p_4 \\ p_1, p_2, p_3, p_4 \\ p_1, p_2, p_3, p_4 \end{pmatrix} \tag{11.13}$$

11.6 MAC Scheduling Delay

To formulate the MAC frame scheduling delay for secondary users in cognitive radio networks, we first denote T_0 the duration of a time slot, and t_s the corresponding channel sensing time allocated for each time slot, as shown in Fig. 11.4.

In the proposed channel model shown in Fig. 11.3, if the current channel is in state s_2 or s_4, the MAC frame has to wait for the next time slot. When new time slot arrives, it has to wait again if the channel is still in state s_2 or s_4. This process repeats until a time slot becomes available or until a maximum waiting threshold in terms of the number of time slots is reached, denoted as $N_s^{\max}$. If this maximum threshold is

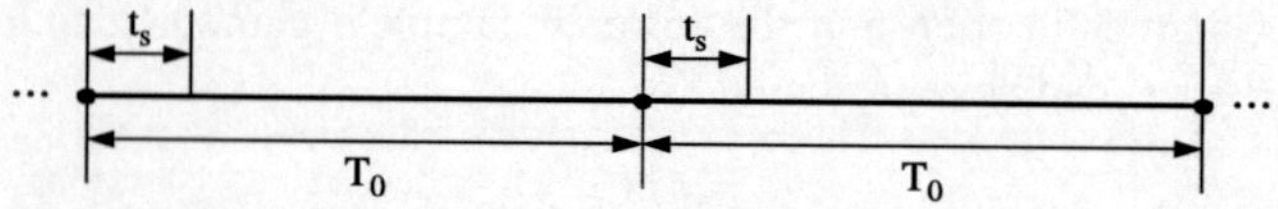

Fig. 11.4 Illustration of time slot duration and channel sensing time over cognitive radio networks

reached, the waiting packet has to be dropped from the sending queue. We call this truncated MAC scheduling. Furthermore, to ensure real-time video transmission, every video packet has to meet a delay bound. Therefore, all the video packets of the same video frame have the same delay bound. We denote $T_n^{\max}$ the delay bound of the video frame n. Then, $N_s^{\max}$ can be calculated as

$$N_s^{\max} = \left\lfloor \frac{T_n^{\max}}{T_0} \right\rfloor \tag{11.14}$$

Note that forward error control coding (FEC) and automatic repeat request (ARQ) are the two major error resilient approaches used by video encoder. However, ARQ is not always feasible for real-time video transmission, due to the excessive delay caused by retransmissions [26, 27]. Therefore, in this chapter, we do not consider ARQ. Instead, we will optimize system parameters in a holistic way to improve the overall system performance for secondary users.

As shown in Fig. 11.3, only when channel is in state s_1 can it be effectively used to transmit data for secondary users. When the channel is in state s_2, s_3, or s_4, it is not available for secondary users or it is not detected by the secondary users as available. Therefore, the probability p_w that a given MAC frame has to wait for the next time slot can be expressed as

$$p_w = 1 - (1 - p_{\mathrm{f}})(1 - \zeta_{\mathrm{b}}) \tag{11.15}$$

Thus, assuming the availability of time slots are independent, the average scheduling time at hop h can be represented as

$$\begin{aligned} t_{\text{sched}}^h &= \left(p_w + p_w^2 + \cdots + p_w^{N_s^{\max}}\right) * T_0 \\ &= \left(\frac{p_w - p_w^{N_s^{\max}}}{1 - p_w}\right) * T_0 \end{aligned} \tag{11.16}$$

11.7 Transmission Delay

In this chapter, adaptive modulation and coding (AMC) technique at the physical layer is adopted on a packet-by-packet basis to enhance the throughput. With AMC, the optimal combination of different modulation constellations and different rates of error-control codes is selected based on the time-varying channel quality. For example, in good channel conditions, AMC schemes with larger constellation sizes and higher channel coding rate will guarantee the required packet error rate for QoS provisioning [28, 29]. Usually, bit error rate (BER) $\varepsilon(\xi)$ can be calculated from the following approximated expression:

$$\varepsilon(\xi) = a_m \mathrm{e}^{-b_m \xi} \tag{11.17}$$

Table 11.2 AMC schemes at the physical layer

AMC mode (m)	$m = 1$	$m = 2$	$m = 3$	$m = 4$	$m = 5$	$m = 6$
Modulation Scheme	BPSK	QPSK	QPSK	16-QAM	16-QAM	64-QAM
Coding Rate (c_m)	1/2	1/2	3/4	9/16	3/4	3/4
r_m (bits/sym.)	0.50	1.00	1.50	2.25	3.00	4.50
a_m	1.1369	0.3351	0.2197	0.2081	0.1936	0.1887
b_m	7.5556	3.2543	1.5244	0.6250	0.3484	0.0871

where coefficients a_m and b_m can be obtained by fitting (11.17) to the exact BER, as shown in Table 11.2. Therefore, the frame error rate $\rho(\xi)$ can be expressed as

$$\rho(\xi) = 1 - (1 - \varepsilon(\xi))^{\ell} \tag{11.18}$$

where ℓ is the MAC frame size and ξ is the received signal-to-noise ratio (SNR).

Assume that the xth packet of video frame n is fragmented into Z MAC frames, and these MAC frames are transmitted along a path with H hops. Thus, the packet loss rate for a given packet x can be expressed as

$$\rho_x = 1 - \prod_{z=1}^{Z} \left(1 - \rho_x^z(\xi)\right) \tag{11.19}$$

where $\rho_x^z(\xi)$ is the frame error rate of MAC frame z of packet x.

Denote BW_h the bandwidth (symbols/second) of hop h, $\ell_x^z(m,h)$ the bit number of the zth physical frame of packet x at hop h. Therefore, the transmission delay can be represented as

$$t_{\text{trans}}^{h} = \left\lceil \frac{\frac{\ell_x^z(m,h)}{r_{m,h} \times \mathrm{BW}_h}}{T_0 - t_s} \right\rceil * T_0 \tag{11.20}$$

where $r_{m,h}$ is the rate (bits/symbol) of AMC mode m at hop h as shown in Table 11.2.

11.8 Problem Formulation and Optimal Solution

To achieve the best video quality at the receiver side, the expected end-to-end video distortion under the constraint of video packet delay should be minimized. With the proposed distortion-delay framework, the source coding, MAC scheduling, transmission, and modulation and coding are jointly optimized in a cross-layer fashion.

In wireless video transmission, all packets of a given frame f_n are constrained by a frame delay bound $T_n^{\max}$. Therefore, all packets of the video frame f_n have the same delay constraint $T_n^{\max}$. Denote $\mathbf{Q}$ as all possible operating points of source coding parameter (such as quantization step size $q_{n,x}$) of packet x of frame n, $\mathbf{M}$

as all possible modulation, and channel coding schemes $m_{n,x}$. Thus, the proposed problem can be formulated as

$$\min_{\{q_{n,x}\in \mathbf{Q}, m_{n,x}\in \mathbf{M}\}} \sum_{n=1}^{N}\sum_{x=1}^{X_n} E[D_{n,x}]$$
$$s.t.:\ \max_{1\le x\le X_n} t_{n,x} \le T_n^{\max},\ \forall x \tag{11.21}$$

where N is the total number of video frames of the given video sequence, and X_n is the total number of packets generated from the nth video frame. In addition,

$$t_{n,x} = \sum_{z=1}^{Z}\sum_{h=1}^{H}\left(t_{\text{sched}}^{z,h} + t_{\text{trans}}^{z,h}\right) \tag{11.22}$$

where $t_{\text{sched}}^{z,h}$ and $t_{\text{trans}}^{z,h}$ have already been derived by using (11.16) and (11.20), respectively. Z and H are the MAC frame number and the hop number, respectively. In other words, the xth packet of video frame n is fragmented into Z MAC frames, and these MAC frames are transmitted on a path with H hops. Therefore, the proposed problem has been formulated into a MIN-MAX problem [30]. Denote α as the index of video packet x over the entire video clip. Thus, the parameter vector of packet x of video frame n can be represented as

$$\mathcal{V}_\alpha := [q_{n,x}, m_{n,x}] \tag{11.23}$$
$$\text{where } q_{n,x} \in \mathbf{Q},\ \ m_{n,x} \in \mathbf{M}$$
$$\text{and } \ 1 \le \alpha \le N \times X_n$$

where $q_{n,x}$ and $m_{n,x}$ are the quantization step size, the AMC mode of packet x of the nth video frame, respectively. In addition, $\mathbf{Q}$ and $\mathbf{M}$ are the sets of all the possible values of $q_{n,x}$ and $m_{n,x}$, respectively.

To solve the MIN-MAX problem as shown in (11.21), we first convert it into an unconstrained optimization problem. According to the formulation (11.21), any parameter vector $\mathcal{V}_\alpha$ resulting in the expected packet delay greater than the constraint $T_n^{\max}$ cannot be the optimal parameter vector $\mathcal{V}_\alpha^*$, defined as $\mathcal{V}_\alpha^* := \left[q_{n,x}^*, m_{n,x}^*\right]$. Therefore, the objective function can be re-written as

$$E[D_{n,x}] = \begin{cases} \infty : & t_{n,x} > T_n^{\max} \\ E[D_{n,x}] : & t_{n,x} \le T_n^{\max} \end{cases} \tag{11.24}$$

where the average distortion of a packet with expected delay greater than the delay bound $T_n^{\max}$ is set to infinity, meaning that the corresponding parameter vector of the possible solution will not satisfy the packet delay bound $T_n^{\max}$. In this way, the minimum distortion problem with delay constraint is transformed into an unconstrained optimization problem. Note that most modern source codecs such as H.264 [17] adopt error concealment strategies to improve the visual quality, which also introduces dependencies among slices/packets. Therefore, if the error concealment algorithm uses the motion vector of the previous slice to recover the lost slice, it will cause the calculation of the expected distortion of the current slice to depend on its previous slice. As mentioned earlier, a packet is generated from one slice. Therefore, without losing generality, we assume that the current packet depends on its previous θ packets ($\theta \geq 0$) in error concealment. To solve the formulated optimization problem in (11.21), we define a cost function $f_{\Bbbk}(\mathcal{V}_{\Bbbk-\theta}, ..., \mathcal{V}_{\Bbbk})$ to represent the minimum average distortion up to and including the $\Bbbk$th packet, where $\mathcal{V}_{\Bbbk-\theta}, ..., \mathcal{V}_{\Bbbk}$ are decision vectors of the $(\Bbbk - \theta)$th to $\Bbbk$th packets. Denote τ the total packet number of the video sequence, where $\tau := N \times X_n$. Therefore, $f_{\tau}(\mathcal{V}_{\tau-\theta}, ..., \mathcal{V}_{\tau})$ represents the minimum distortion incurred by all packets of the given video sequence. Thus, solving (11.21) is essentially to solve the following equation

$$\min_{\{\mathcal{V}_{\tau-\theta}, ..., \mathcal{V}_{\tau}\}} f_{\tau}(\mathcal{V}_{\tau-\theta}, ..., \mathcal{V}_{\tau}) \tag{11.25}$$

Therefore, given the cost function $f_{\Bbbk-1}(\mathcal{V}_{\Bbbk-\theta-1}, ..., \mathcal{V}_{\Bbbk-1})$ and the $\theta+1$ decision vectors $\mathcal{V}_{\Bbbk-\theta-1}, ...\mathcal{V}_{\Bbbk-1}$ for the $(\Bbbk - \theta - 1)$th to the $(\Bbbk - 1)$th packets, the selection of the next decision vector $\mathcal{V}_{\Bbbk}$ is independent of the selection of the previous decision vectors $\mathcal{V}_1, \mathcal{V}_2, ..., \mathcal{V}_{\Bbbk-\theta-2}$. This means that the cost function can be expressed recursively as

$$f_{\Bbbk}(\mathcal{V}_{\Bbbk-\theta}, .., \mathcal{V}_{\Bbbk}) = \min_{\{\mathcal{V}_{\Bbbk-\theta-1}, ..., \mathcal{V}_{\Bbbk-1}\}} \{f_{\Bbbk-1}(\mathcal{V}_{\Bbbk-\theta-1}, ..., \mathcal{V}_{\Bbbk-1}) + E[D_{\Bbbk}]\} \tag{11.26}$$

which implies that the next step of the optimization process for the cost function is independent of its past steps, forming the foundation of dynamic programming.

Essentially, (11.26) can be further converted into and solved as a graph theory problem of finding the shortest path in a directed acyclic graph (DAG) [31]. By using dynamic programming to solve this shortest path problem, the computational complexity of the algorithm is decreased to $\mathbf{O}(\tau \times |\mathcal{V}|^{\theta+1})$ (where $|\mathcal{V}|$ is the cardinality of $\mathcal{V}$), depending directly on the value of θ. For most cases, θ is a small number, so the computational complexity of the algorithm is effectively decreased, compared with the exponential computational complexity of exhaustive search algorithm [32, 33].

11.9 Experimental Analysis

11.9.1 Experimental Environment

In this chapter, video coding is performed by H.264/AVC JM 15.1 codec. The video sequence "Foreman" is adopted for performance analysis. The first 100 frames of the QCIF (176 × 144) video clip are coded at frame rate of 30 frames/s, and each I frame is followed by 9 P frames. Assume the whole packet/slice is lost if one of the MAC frames of the packet is lost, which is reasonable since usually intra-prediction is derived from the decoded samples of the same decoded slice. To avoid prediction error propagation, a 10% macroblock level intra-refreshment is used during the experiments. When a packet is lost during transmission, the temporal-replacement error concealment strategy will be used. The motion vector of a missing MB can be estimated as the median of motion vectors of the nearest three MBs in the preceding row. If that row is also lost, the estimated motion vector is set to zero. The pixels in the previous frame, pointed by the estimated motion vector, are used to replace the missing pixels in the current frame.

In the experiments, quantization step size (QP) q and AMC mode m of each packet are considered as the parameters to be optimized. The possible values of QP are chosen from 1 to 50, while the available AMC schemes are 1–6 as shown in Table 11.2. According to [25], the channel is assumed to be busy with an average probability of $\zeta_b = 0.1$. Also, since performance enhancement of secondary users in cognitive radio networks is the main focus of this chapter, we set the detection threshold $t_d = 1.35$, so that false alarm probabilities are effectively decreased. Therefore, under this setting, the channel sensing is reliable, and the interference to primary users is minimal.

Given an average SNR $\bar{\xi}$, the instantaneous link quality ξ can be randomly produced from (11.3). In this chapter, the link bandwidth is set to 100k symbols/s. Moreover, without losing generality, a single hop scenario is considered in the experiments to verify the performance of the proposed framework. Similar conclusions derived from the single hop scenarios may straightly apply to the multi-hop scenarios when the channel state information (CSI) of each hop is available.

Also, the delay bound is set in accordance with the frame rate, as adopted in literature [34–36]. As the most important performance metric for video applications [37, 38], peak signal-to-noise ratio (PSNR) of the received video frames of secondary users is used as the performance metric to compare the proposed system with the existing system, which has fixed AMC schemes.

11.9.2 Performance Evaluation

The performance enhancement of the proposed system for secondary users under various packet delay bounds is verified in Fig. 11.5, where time slot duration T_0 is set to 5 ms, average SNR ξ is set to 15 dB, and channel sensing time t_s is set to 0.5 ms and 1 ms, respectively. From the figure, we can observe that by jointly optimizing the system parameters residing in different network layers under the proposed system, significant performance improvement can be achieved. Another observation is that as the delay bound becomes more and more stringent, the

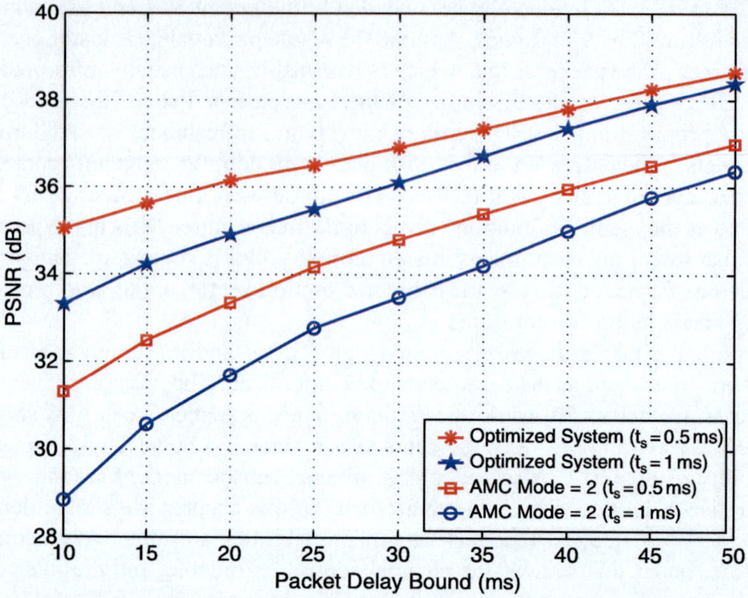

Fig. 11.5 Average PSNR comparison between the proposed quality-driven cross-layer optimized system and the existing system for real-time wireless video transmission over cognitive radio networks with various packet delay bounds $T_n^{\max}$

performance gain becomes higher and higher. This implies that the proposed system is especially suitable for real-time video transmission over cognitive radio networks with stringent delay bound. Furthermore, we can also observe that under the same network conditions, if the channel sensing time becomes longer, the performance gain actually becomes higher. This indicates that the proposed system might be useful for cognitive networks, due to the fact that when more time is allocated to perform channel sensing, MAC scheduling, and ARQ management, the negative impact on the overall system performance is minimized.

In Fig. 11.6, the visual comparison of one video frame (the 40th frame of the Foreman video clip) is presented with the packet delay bound being set to 30 ms and t_s 0.5 ms. Other environment settings remain the same with those of the above figure. Thus, it is observable that the user-perceived video quality has been greatly improved.

We also evaluate the relationship between the channel quality SNR ξ and the perceived video quality under the proposed system as shown in Fig. 11.7. In this experiment, time slot duration T_0 is set to 5 ms and the channel sensing time t_s is 0.5 ms. The packet delay bound is set to 20 ms and 30 ms, respectively. From the figure, we can observe that the proposed system can significantly improve PSNR performance, especially when the channel quality is not good or/and the delay bound is more stringent.

(a) (b) (c)

Fig. 11.6 The visual comparison between the proposed quality-driven cross-layer optimized system and the existing system for real-time wireless video transmission over cognitive radio networks. (**a**) Original frame; (**b**) Without optimization; (**c**) Optimized

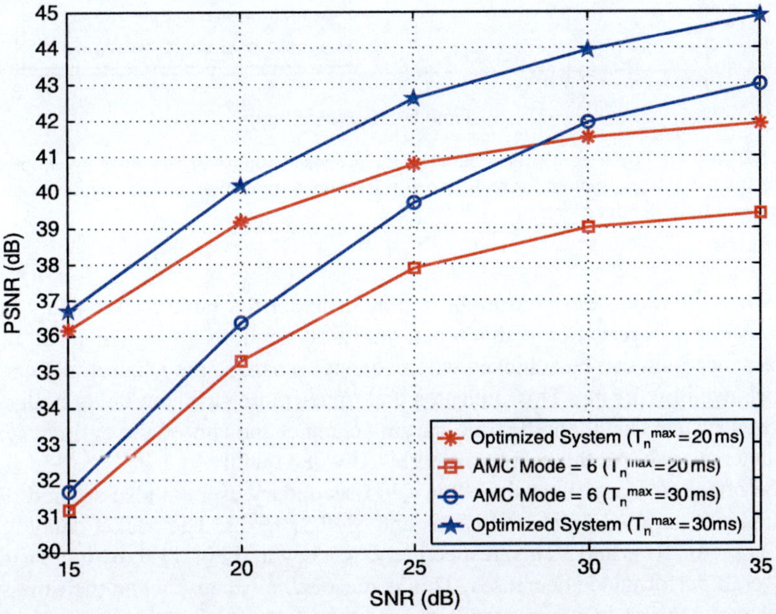

Fig. 11.7 Average PSNR comparison between the proposed quality-driven cross-layer optimized system and the existing system for real-time wireless video transmission over cognitive radio networks with various SNRs

Furthermore, the impact of time slot duration T_0 on secondary users is shown in Fig. 11.8. In this case, the channel sensing time t_s is set to 0.5 ms and average SNR ξ is 15 dB. The packet delay bound $T_n^{\max}$ is set to 20 ms and 30 ms, respectively. From the figure, we can observe that the proposed system achieves higher performance

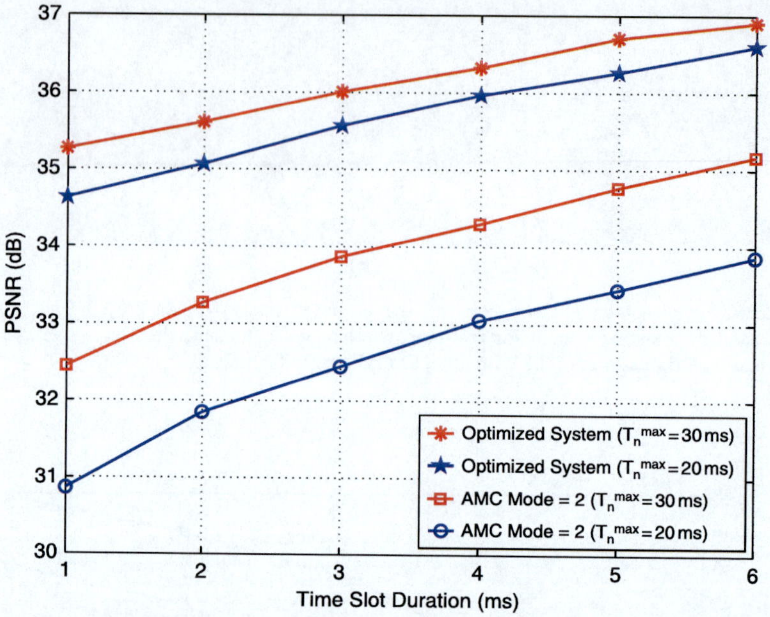

Fig. 11.8 Average PSNR comparison between the proposed quality-driven cross-layer optimized system and the existing system for real-time wireless video transmission over cognitive networks with various time slot durations T_0

gain when the time slot duration T_0 is smaller under the same packet delay bound T_n^{max}. Another observation is that when time slot durations T_0 are the same, higher performance gain can be achieved when channel sensing time t_s is set to 1 ms than that achieved by 0.5 ms. This indicates that under more stringent delay bound, the proposed system can adapt the time-varying channel and choose the optimal system parameters to achieve the best user-perceived video quality.

The effect of channel sensing time t_s on secondary users is also studied. Here, the time slot duration T_0 is set to 5 ms and SNR is 15 dB. The packet delay bound T_n^{max} is set to 20 ms and 30 ms, respectively. As shown in Fig. 11.9, by increasing t_s, the overall performance decreases. This is reasonable because when the time spent on channel sensing increases, the time spent on transmission $(T_0 - t_s)$ decreases accordingly. However, the proposed system is able to achieve higher performance gain when more time is spent on channel sensing, thus minimizing the negative impact on the overall system performance.

In summary, all the experimental results demonstrate the significant performance enhancement of the proposed system for secondary users in cognitive radio networks. The experimental results also indicate that the performance gain is usually higher when the wireless channel experiences bad quality in a more stringent delay-bounded video application.

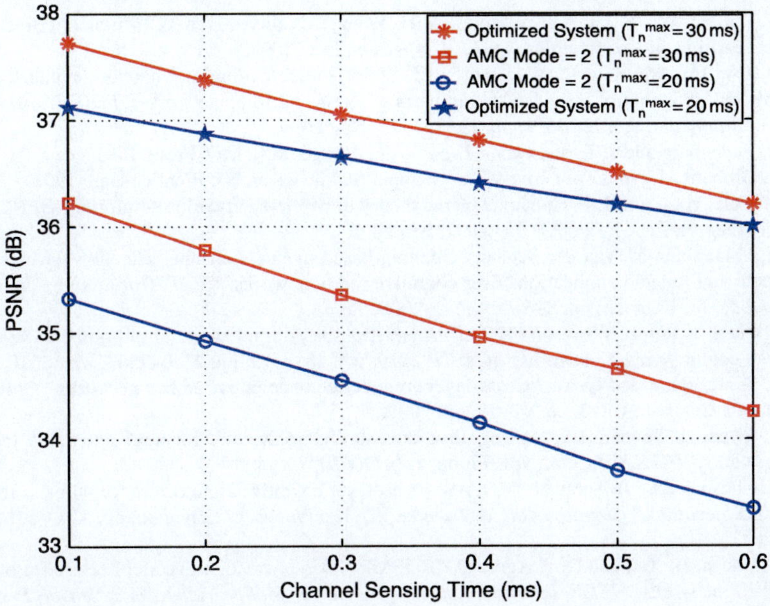

Fig. 11.9 Average PSNR comparison between the proposed quality-driven cross-layer optimized system and the existing system for real-time wireless video transmission over cognitive networks for various channel sensing times t_s

11.10 Conclusions

In this chapter, a cross-layer optimized system for real-time video transmission over cognitive radio networks has been studied. The proposed system can achieve the best possible video quality for secondary users, significantly improving the user experience of secondary users in cognitive radio networks, leading to the possibility of wide deployment of CR technologies in video applications. The design problem is presented to minimize the expected video distortion under the constraint of packet delay bound, which has been formulated as a MIN-MAX problem and solved by using dynamic programming. Experimental results have validated the effectiveness of the proposed system.

References

1. D. Niyato and E. Hossain, "Cognitive radio for next-generation wireless networks: An approach to opportunistic channel selection in IEEE 802.11-based wireless mesh," *IEEE Wireless Commun.*, vol. 16, pp. 46–54, Feb 2009.
2. S. Haykin, "Cognitive radio: Brain-empowered wireless communications," *IEEE J. Sel. Areas Commun.*, vol. 23, pp. 201–220, Feb 2005.
3. R. Tandra, S. Mishra, and A. Sahai, "What is a spectrum hole and what does it take to recognize one?" *Proceedings of the IEEE*, Apr 2009.

4. H. Luo, D. Wu, S. Ci, A. Argyriou, and H. Wang, "Quality-driven TCP friendly rate control for real-time video streaming," *IEEE GlobeCom*, Dec. 2008.
5. H. Luo, A. Argyriou, D. Wu, and S. Ci, "Joint source coding and network-supported distributed error control for video streaming in wireless multi-hop networks," *IEEE Transactions on Multimedia*, vol. 11, no. 7, pp. 1362–1372, Nov 2009.
6. D. Fudenberg and J. Tirole, *Game Theory*. Cambridge, MA: MIT Press, 1991.
7. R. Gibbons, *A Primer in Game Theory*. Upper Saddle River, NJ: Prentice Hall, 1992.
8. M. Felegyhazi and J.-P. Hubaux, "Game theory in wireless networks: A tutorial," *EPFL technical report, LCA-REPORT-2006-002*, Feb 2006.
9. H. Shiang and M. van der Schaar, "Queuing-based dynamic channel selection for heterogeneous multimedia applications over cognitive radio networks," *IEEE Transactions on Multimedia*, vol. 10, no. 5, pp. 896–909, Aug 2008.
10. Y. Chen, Y. Wu, B. Wang, and K. Liu, "An auction-based framework for multimedia streaming over cognitive radio networks," *ICASSP 2010*, vol. 48, no. 2, pp. 2350–2353, Mar 2010.
11. Y. Yu, L. Wang, and Q. Yu, "Cross-layer architecture in cognitive ad hoc networks," *Commun. Mobile Comput.*, vol. 2, pp. 47–51, Jan 2009.
12. S. Ci and J. Sonnenberg, "A cognitive cross-layer architecture for next-generation tactical networks," *IEEE MILCOM*, vol. 77, pp. 1–6, Oct 2007.
13. C. Ghosh and D. P. Agrawal, "ROPAS: Cross-layer cognitive architecture for wireless mobile adhoc networks," *Cognitive Radio Oriented Wireless Networks Commun.*, pp. 514 – 518, Aug 2007.
14. C. Ghosh, B. Xie, and D. P. Agrawal, "ROPAS: Cross-layer cognitive architecture for mobile UWB networks," *IEEE International Conference on Mobile Adhoc and Sensor Systems,* pp. 1–7, Pisa, Italy, Oct. 2007.
15. H. Su and X. Zhang, "Cross-layer based opportunistic MAC protocols for QoS provisionings over cognitive radio wireless networks," *IEEE J. Sel. Areas Commun.*, vol. 26, pp. 118–129, Jan. 2008.
16. P. Mahonen, M. Petrova, J. Riihijarvi, and M. Wellens, "Cognitive wireless networks: Your network just became a teenager," *In Proceedings of IEEE INFOCOM*, Barcelona, 2006.
17. "Draft ITU-T Recommendation and Final Draft International Standard of Joint Video Specification (ITU-T Rec. H.264ąłISO/IEC 14496-10 AVC)," *ftp://ftp.imtc-files.org/jvt-experts/2003-03-Pattaya/JVT-G50r1.zip*, May 2003.
18. R. Zhang, S. L. Regunathan, and K. Rose, "Video coding with optimal inter/intra-mode switching for packet loss resilience," *IEEE J. Select. Areas Commun.*, vol. 18, no. 6, pp. 966–976, Jun 2000.
19. D. Wu, T. Hou, W. Zhu, H.-J. Lee, T. Chiang, Y.-Q. Zhang, and H. J. Chao, "On end-to-end architecture for transporting MPEG-4 video over the Internet," *IEEE Trans. Circuits Syst. Video Technol.*, vol. 10, pp. 923–941, Sep. 2000.
20. G. Cote, S. Shirani, and F. Kossentini, "Optimal mode selection and synchronization for robust video communications over error prone networks," *IEEE J. Sel. Areas Commun.*, vol. 18, pp. 952–965, Jun 2000.
21. A. Argyriou, "Real-time and rate-distortion optimized video streaming with TCP," *Elsevier Signal Process. Image Commun.*, vol. 22, pp. 374–388, Apr 2007.
22. Q. Liu, S. Zhou, and G. Giannakis, "Cross-layer combining of adaptive modulation and coding with truncated arq over wireless links," *IEEE Trans. Wireless Commun.*, vol. 3, no. 5, pp. 1746–1755, Sept 2004.
23. H. Poor, *An Introduction to Signal Detection and Estimation*, 2nd ed. Springer, New York, NY, 1994.
24. Y. Sung, L. Tong, and H. V. Poor, "Neyman-Pearson detection of Gauss-Markov signals in noise: Closed-form error exponent and properties," *IEEE Trans. Inf. Theory*, vol. 52, pp. 1354–1365, Apr 2006.
25. S. Akin and M. C. Gursoy, "Effective capacity analysis of cognitive radio channels for quality of service provisioning," *CoRR*, vol. abs/0906.3888, 2009.

26. H. Wang, S. Tsaftaris, and A. K. Katsaggelos, "Joint source-channel coding for wireless object-based video communications utilizing data hiding," *IEEE Trans. Image Process.*, vol. 15, no. 8, pp. 2158–2169, Sept 2008.
27. T. Stockhammer, M. M. Hannuksela, and T. Wiegand, "H.264/AVC in wireless environments," *IEEE Trans. Circuits Syst. Video Technol.*, vol. 13, no. 7, pp. 657–673, Jul 2003.
28. M. S. Alouini and A. J. Goldsmith, "Adaptive modulation over Nakagami fading channels," *Kluwer J. Wireless Commun.*, vol. 13, pp. 119–143, May 2000.
29. A. Doufexi, S. Armour, M. Butler, A. Nix, D. Bull, J. McGeehan, and P. Karlsson, "A Comparison of the HIPERLAN/2 and IEEE 802.11a Wireless LAN Standards," *IEEE Commun. Magazine*, vol. 40, pp. 172–180, May 2002.
30. Z. Li, G. Schuster, and A. Katsaggelos, "MINMAX optimal video summarization," *IEEE Trans. Circuits Syst. Video Technol.*, vol. 15, pp. 1245–1256, Oct 2005.
31. G. M. Schuster and A. K. Katsaggelos, *Rate-Distortion Based Video Compression: Optimal Video Frame Compression and Object Boundary Encoding*. Norwell, MA: Kluwer, 1997.
32. A. Ortega and K. Ramchandran, "Rate-distortion methods for image and video compression," *Signal Process.*, vol. 15, pp. 23–50, Nov 1998.
33. D. Wu, S. Ci, and H. Wang, "Cross-layer optimization for video summary transmission over wireless networks," *IEEE J. Sel. Areas Commun.*, vol. 25, no. 4, pp. 841–850, May 2007.
34. Y. Andreopoulos, N. Mastronade, and M. van der Schaar, "Cross-layer optimized video streaming over wireless multi-hop mesh networks," *IEEE J. Sel. Areas Commun.*, vol. 24, no. 11, pp. 2104–2115, Nov 2006.
35. P. Pahalawatta, R. Berry, T. Pappas, and A. Katsaggelos, "Content-aware resource allocation and packet scheduling for video transmission over wireless networks," *IEEE J. Sel. Areas Commun.*, vol. 25, no. 4, pp. 749–759, May 2007.
36. E. Maani, P. Pahalawatta, R. Berry, T. Pappas, and A. Katsaggelos, "Resource allocation for downlink multiuser video transmission over wireless lossy networks," *IEEE Trans. Image Process.*, vol. 17, no. 9, pp. 1663–1671, Sept 2008.
37. H. Luo, S. Ci, D. Wu, and H. Tang, "Cross-layer design for real-time video transmission in cognitive wireless networks," *IEEE INFOCOM 2010 Workshop on Cognitive Wireless Communications and Networking*, San Diego, Mar 2010.
38. H. Luo, S. Ci, D. Wu, and H. Tang "Quality-driven cross-layer optimized video delivery over lte," *IEEE Commun.*, vol. 48, no. 2, pp. 102–109, 2010.

Part V
Applications of Cognitive Radio Mobile Ad Hoc Networks

Chapter 12
An Adaptive WiFi/WiMAX Networking Platform for Cognitive Vehicular Networks

Dusit Niyato, Ekram Hossain, and Teerawat Issariyakul

Abstract This chapter presents an adaptive networking platform using WiFi/WiMAX technologies for cognitive vehicle-to-roadside communications, which can be used to transfer safety messages and provide Internet access for mobile users inside vehicles. The proposed platform is based on a heterogeneous multi-hop cluster-based vehicular network, where a vehicular node can choose to play the role of a gateway or a client. The gateway nodes communicate directly with a roadside base station through a WiMAX link. The client nodes connect to the gateways through WiFi links. Traffic from client nodes are relayed by the gateways to a roadside base station. The vehicular nodes are the self-interest (i.e., rational) and have capability to learn and adapt decision to achieve their objectives independently. A decision-making framework is proposed for this WiFi/WiMAX platform. This distributed decision-making framework, which enables the vehicular nodes with cognitive capability, is modeled and analyzed using game theory. Also, a Q-learning algorithm is used in vehicular nodes to provide the cognitive capability to learn and adapt their decision. Dynamics of Q-learning algorithm can be modeled as an evolutionary game.

12.1 Introduction

Wireless vehicle-to-vehicle (V2V) and vehicle-to-roadside (V2R) communications and networking technologies are the keys to providing Internet connectivity to mobile users in the vehicles. Vehicular networks using wireless access technologies (e.g., WiFi and WiMAX technologies) can support data communications for safety and intelligent transportation systems (ITS) applications (e.g., reporting traffic condition to the driver) and infotainment applications (e.g., providing interactive media and advertisement to the passengers). The vehicular nodes can form a heterogeneous *cognitive* multihop wireless network, where each node is able to dynamically choose among different radio access technologies for V2V and V2R communications. Illustrated in Fig. 12.1, two major components in a cognitive vehicular network are the network model and the decision-making framework. A network model incorporates all the basic functionalities necessary for data communications. A decision-making

D. Niyato (✉)
School of Computer Engineering, Nanyang Technological University, Singapore
e-mail: dniyato@ntu.edu.sg

F.R. Yu (ed.), *Cognitive Radio Mobile Ad Hoc Networks*,
DOI 10.1007/978-1-4419-6172-3_12,

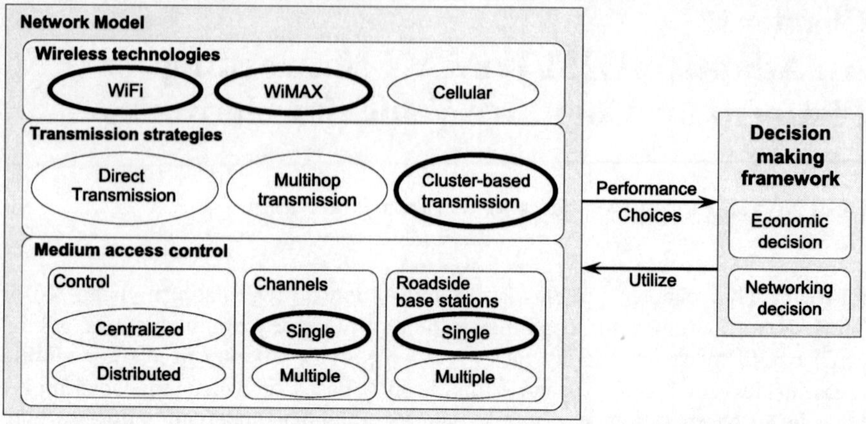

Fig. 12.1 Cognitive vehicular network model: The *rectangles* represent the key components of the model, while the *ellipses* stand for choices for a component. The *ellipses* with *thick edges* correspond to the components considered in this chapter

framework, which is required for packet routing and distributed resource management in a vehicular network, is composed of the economic and the networking decisions to optimize the utility of the vehicular nodes in terms of both performance and cost. In a vehicular network, which can be considered as a distributed dynamic system, the vehicular nodes can be considered as an independent rational agents. That is, each vehicular node is an autonomous computational entity with a flexible dynamic behavior in an unpredictable environment. In such an environment, a vehicular node must be able to learn and adapt to the ambient environment to achieve its goal.

The different components of the cognitive vehicular networking model are described next.

12.1.1 Wireless Technologies

Currently, there are several enabling technologies for V2V and V2R communications. The IEEE 802.11-based WiFi technology supports short-range high-speed data transmission. However, its short transmission range leads to frequent transmission interruption (e.g., while vehicle speed is high) and the high deployment cost (e.g., many access points have to be deployed along the road) [1, 2]. IEEE 802.11-based services would be viable in a congested area where the vehicles move slowly. In this scenario, the users would benefit from high data rate and infrequent transmission interruption, while the service provider needs to install only few roadside access points at the selected hot spots.

The IEEE 802.16-based WiMAX technology provides large coverage area and high-speed connectivity [3]. While WiMAX helps overcome the range limitation of WiFi, its achievable data rate for low mobility may not be as high as that of WiFi.

In addition, the price for WiMAX access is comparatively higher than that for WiFi access.

Besides WiMAX and WiFi, the 3G cellular wireless technology for V2R communications provides a very broad coverage and supports high-mobility vehicles [4]. Due to lower data rate, the services in a wireless cellular network are usually less expensive than that in an IEEE 802.16 network.

12.1.2 Transmission Strategies

A transmission strategy determine how data packets are delivered from a vehicular node to a roadside base station (and vice versa). The strategy can be *direct transmission* where a roadside base station can be reached directly from a vehicular node (e.g., [5, 8]). If a roadside base station is located far away from a vehicular node, a *multihop transmission strategy* can be employed. In this scenario, the data packets from a vehicular node are relayed by other vehicular nodes until these data packets reach the designated roadside base station (e.g., coordinated external peer communications (CEPEC) in [3]). In a multihop vehicular network, traffic from a vehicular client node can be relayed through a vehicular gateway node to a roadside base station. Since the traffic from multiple vehicular nodes are aggregated and transmitted through this gateway, the utilization of the vehicle-to-roadside wireless link can be improved while the cost of a network is reduced due to bandwidth sharing. This *client–gateway* model is similar to the cluster-based vehicular network which was proposed in [6].

In a *cluster-based transmission strategy*, the vehicular nodes form groups (i.e., clusters) of vehicles, delegate a representative (i.e., a cluster head or a gateway) for each group, and transmit data through this selected representative [6]. A cluster-based vehicular network can improve the communication efficiency by not only reducing the signaling overhead but also alleviating congestion of the channel access which is fully controlled by a cluster head. In the *client–gateway* model, the gateway node acts as a cluster head which controls the transmission of traffic from the cluster members to a roadside base station. In this scenario, a vehicular node can use WiFi radio for local communications with the cluster head and a cluster head can use WiMAX radio for broadband communications with a roadside base station.

12.1.3 Medium Access Control Protocols

Medium access control (MAC) protocols refer to how vehicular nodes and roadside base stations share common radio channels. The MAC protocols can be classified based on three following criteria:

- *Centralized or distributed MAC protocols*: With a centralized MAC protocol, the decision of when and how the channels are accessed is determined by a central controller (e.g., scheduling in [6, 8]). With complete node information, a centralized MAC protocol can be optimally designed at the expense of overhead needed to acquire such information. A distributed MAC protocol, on the other hand, determines how the channels are accessed based on local information (e.g., contention in [6, 7]). Despite decreasing overhead, a distributed MAC protocol is usually not optimal due to incomplete node information.

- *Single or multiple channels*: In presence of multiple channels (possibly different technologies), a MAC protocol needs to select the channels to satisfy application requirements. For example, a high data rate high attenuation-sensitive channel (e.g., in 5.8 GHz) should be used for typical data exchange, while a low data rate (usually more robust) channel (e.g., VHF or UHF) should be used for transmitting control and safety messages [5].
- *Single or multiple roadside base stations*: When considering multiple roadside base stations, a MAC protocol needs to control the handover process when a vehicle moves from one roadside base station to another. For example, [7] designed a distributed MAC protocol which quickly associates and disassociates a vehicular node with a roadside base station. The maximum freedom last (MFL) scheme minimizes the handover rate subject to a given delay constraint [8].

12.1.4 Distributed Decision-Making Framework

The decision-making framework, which considers both network pricing and network quality-of-service (QoS) issues, enables the vehicular nodes with cognitive radio capability. A pricing model characterizes the service fee (i.e., price) for using wireless access service. To access the radio resources in a wireless system, a mobile node has to pay to the radio resource owner (i.e., the service provider). Similarly, in most V2R communication scenarios, every vehicular node needs to pay to the service provider. For example, if a vehicular node (e.g., a gateway) uses direct WiMAX link to a roadside base station, the price has to be paid to the corresponding WiMAX service provider. However, if a vehicular node (e.g., client) uses a gateway to relay its traffic to a roadside base station, a price has to be paid to the gateway. In a vehicular network, the decision on price setting has to be optimally made by the wireless service provider.

In a V2R communications scenario, a vehicular node also has to make different networking decisions. In a cluster-based vehicular network, a vehicular node can choose to act as a *client* (i.e., a cluster member) or as a *gateway* (i.e., a cluster head). As a client, a vehicular node forwards its data traffic through the associated gateway. As a gateway, a vehicular node shares the link to the roadside base station with its client. Also, a client has to select the best gateway to relay its traffic to gain the highest benefit.

In general, a vehicular node can be considered as an independent and rational entity in a vehicular network. It will make a decision to maximize its benefits. For example, a vehicular node may decide to become a gateway and use WiMAX interface to provide the relaying functionality for other vehicular nodes if a gateway receives high benefit from bandwidth sharing or "reselling." Alternatively, a vehicular node may decide to become client and use WiFi interface to transmit data to the gateway. Here, the decision-making framework for each vehicular node needs to be implemented in a *distributed* manner, considering both the networking aspects and the economic aspects. To this end, a supporting theory to obtain a stable solution for the above decision-making framework is required.

This chapter presents an adaptive decision-making framework for cognitive vehicle-to-roadside communications in a vehicular network. In this network, WiFi and WiMAX interfaces are used adaptively for client-to-gateway and gateway-to-roadside base station communications, respectively. A vehicular node with this platform forms a cluster-based network. With the assumption that a vehicular node is independent and rational to maximize its benefit (i.e., net utility), a vehicular node has to make a decision according to the vehicular network condition to use different wireless interface for data transmission. The first decision is whether a vehicular node should become a client or a gateway (i.e., role selection). If a vehicular node decides to become a client, it uses WiFi interface for data transmission and selects a gateway to relay its traffic. However, if a vehicular node decides to become a gateway, it uses WiMAX interface and determines the price of bandwidth sharing to be charged to its clients. The decision of a vehicular node affects not only its own benefit but also the benefits of other vehicular nodes. For example, if many clients select the same gateway, the portion of bandwidth given to each sharing node will decrease. Also, if a gateway charges high price, its clients will switch to other gateways which offer lower price of bandwidth sharing.

The rest of this chapter is organized as follows. Section 12.2 reviews the related work on cognitive vehicular networks. Section 12.3 presents an overview of distributed decision making based on game theory and reinforcement learning. The adaptive WiFi/WiMAX framework is described in Section 12.4. Section 12.5 presents the game model for the distributed decision-making framework. Performance evaluation results for the proposed framework are presented in Section 12.6. Section 12.7 summarizes the contribution of the chapter.

12.2 Cognitive Vehicular Networks: Related Work

Research on dynamic spectrum access-based cognitive vehicular networking has become popular recently. The spectrum sensing problem for cognitive-radio-enhanced vehicular ad hoc networks was addressed in [9] and [10]. In [11], communication protocols were proposed for universal wireless access in vehicular networking scenarios. In such a scenario, for V2R and V2V communications, a vehicular node is able to communicate on multiple frequency bands using different medium access control (MAC) and physical (PHY) layer interfaces. Since there are multiple interfaces, a new routing protocol is required to efficiently forward data packets. A cognitive communication for vehicular networking (CCVN) layer over multiple

MAC and PHY interfaces was introduced to support optimal connectivity, seamless mobility management, and QoS guarantee.

In [12], a cognitive MAC protocol, namely, CMV protocol, was proposed for multi-channel access in vehicular ad hoc networks. The protocol can support high mobility and spectrum handover by introducing the concepts of long-term and short-term spectrum access. For a long-term spectrum access, the channel is probed for every CCH period defined in the IEEE 1609.4 standard. Then, the spectrum status table (SST) is updated. For a short-term spectrum access, a spectrum pooling technique is used so that the best channel can be selected and the packet loss probability can be reduced. The performance evaluation results showed that the proposed CMV protocol outperforms existing multi-channel MAC protocols by up to 72%.

In [13], a dynamic spectrum access technique was adopted for inter-vehicle communications. Specifically, dynamic per-hop channel switching schemes for multi-hop VANET were proposed. These schemes were referred to as metric-based dynamic channel selection schemes with/without spatial awareness. These schemes are based on transmission rate, rate and utilization, and rate and idle probability metrics which are used together with the information about spatial movement of vehicular nodes (e.g., transmission range) to adapt the channel access. Simulation results showed the advantages of the proposed schemes in terms of communication duration and amount of transmitted data especially in the multi-hop and highly congested environments with high-speed mobility.

A framework for optimal channel access for vehicular nodes utilizing the exclusive-use and shared-use channels in cognitive radio network was proposed in [14]. The objective is to maximize the utility of data transmission by cluster members under QoS constraints (e.g., packet loss probability due to buffer overflow, average packet delay) and collision probability with primary users. Three major components in this framework are the queue-aware opportunistic access to shared-use channels, the reservation of bandwidth in the exclusive-use channel, and the cluster size control. To optimally design these components, a hierarchical optimization model was developed. With this framework, the cost of channel access to support various ITS applications can be minimized while guaranteeing the QoS requirements for the mobile nodes.

In [15], the vehicular public safety cognitive radio (VPSCR) platform was introduced. VPSCR has the ability to scan the radio spectrum over multiple public safety frequency bands. Then, commonly used public safety waveforms and networks can be identified such that VPSCR can adapt the spectrum access for network interoperation accordingly. This VPSCR platform was designed to communicate with a personal digital assistant (PDA) through existing fixed infrastructure (e.g., IEEE 802.11 or Bluetooth) to remotely control and access services.

In [16], a cognitive security protocol for sensor based VANET (S-VANET) was introduced. This protocol can distribute the security information to support the prevention of data aging, efficient QoS, and robustness against denial-of-service (DoS) attack. The reliability and optimality of the protocol were evaluated in terms of response time, ability to maintain message authentication, integrity, confidentiality, and non-repudiation.

12.3 Distributed Decision Making

Three mechanisms to achieve the distributed decision making are the agent-based computing, intelligent algorithm, and game theory. Since it is impossible for a vehicular node to anticipate and estimate the consequence of all situations to encounter in a dynamic vehicular environment, cognitive or learning capability becomes crucial. With the use of a learning algorithm (specifically reinforcement learning), evolutionary game theory can be used to study the dynamics of a multiagent system. An evolutionary game theory model can be used to obtain the solution of rational agents (i.e., agent with self-interest). The relationship among multiagent systems, evolutionary game theory, and reinforcement learning is shown in Fig. 12.2 [17].

12.3.1 Evolutionary Game Theory

Evolutionary game theory is a branch of game theory developed to provide a basis to understand rational decision making in an uncertain environment. Evolutionary game theory complements traditional noncooperative game theory in following aspects.

- *Refinement of traditional solution concept*: In a traditional noncooperative game, the Nash equilibrium is the most common solution concept. However, in any game, the Nash equilibrium cannot be guaranteed to exist if the player is restricted to use only pure strategy. Also, there could be multiple Nash equilibria in the game. In this case, the solution of evolutionary game theory (i.e., evolutionary stable strategies (ESS) or evolutionary equilibrium) can serve as a refinement to the Nash equilibrium especially when multiple Nash equilibria exist.
- *Bounded rationality*: In a traditional noncooperative game, the agent is assumed to be rational. That is, an agent will always maximize the payoff in which this assumption is derived from the utility theory. This rationality of agent requires complete information and well-defined and consistent set of choices. However, in reality, this assumption is rarely held. Evolutionary game theory has been developed to model the behavior of biological agents (e.g., insects and animals) which does not require the strong rationality assumption. Therefore, evolutionary game theory will be suitable for the problem which involves human being as the agents. These agents may not have hyperrational behavior.

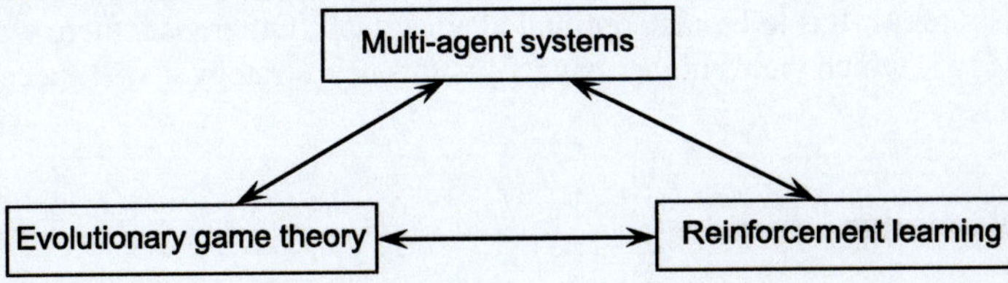

Fig. 12.2 Relationship among multiagent systems, evolutionary game theory, and reinforcement learning [17]

- *Dynamics of game*: Traditional noncooperative game has been developed mostly for the static analysis. It cannot model the adaptation of agents to change their strategies and to reach the equilibrium solution. Evolutionary game theory is based on the evolutionary process which is dynamic in nature. Evolutionary game establishes the dynamics model of interactions among agents in the population (i.e., strategy adaptation over time).

In evolutionary game, a game is played repeatedly by the agents. These agents are randomly selected from a large population. Two major mechanisms of an evolutionary process are mutation and selection. While the mutation mechanism is used to provide diversity in the population, the selection mechanism is used to promote the agents with higher fitness over other agents. In an evolutionary game, the mutation mechanism is described by the evolutionary stable strategies (ESS). The selection mechanism is described by the replicator dynamics. In other word, ESS is used to study a static evolutionary game while replicator dynamics is used for a dynamic evolutionary game.

12.3.1.1 Evolutionary Stable Strategies (ESS)

With a large population, let most of the players adopt the same strategy (e.g., mixed strategy s), and there is a small fraction $\varepsilon \in (0, 1)$ of a population adopting a different strategy (e.g., mixed strategy s'). Then, if the reproductive success of the new strategy s' is smaller than the original strategy s, the entire population will not be overruled by the new strategy s' and this new strategy s' will disappear eventually. In this case, the original strategy s is said to be an ESS which is robust against the evolutionary pressure from any appearing mutant strategy s'. Specifically, the payoff of the player adopting original strategy s is denoted as $\mathscr{U}(s, (1-\varepsilon)s + \varepsilon s')$, where $\mathscr{U}(\cdot, \cdot)$ is the utility function whose first parameter is the current strategy and the second parameter is the strategy of an opponent. Then, the payoff of a player adopting new strategy s' is denoted as $\mathscr{U}(s', (1-\varepsilon)s + \varepsilon s')$. Strategy s is an ESS if $\forall s' \neq s$, there exists $\delta \in (0, 1)$ such that the following condition holds:

$$\mathscr{U}(s, (1-\varepsilon)s + \varepsilon s') > \mathscr{U}(s', (1-\varepsilon)s + \varepsilon s'), \quad \forall \varepsilon : 0 < \varepsilon < \delta. \tag{12.1}$$

In general, ESS is a subset of the Nash equilibrium, since the conditions for an ESS are stricter than those of the Nash equilibrium. That is, the Nash equilibrium of a player is required to be the best response to the strategy of the opponent. To be ESS, this strategy s has to be also optimal against itself. Otherwise, there would be other strategy s' which yields higher payoff and this new strategy s' will successfully invade strategy s.

12.3.1.2 Replicator Dynamics

In an evolutionary game, the dynamic process is related to the evolution of population adopting different strategies. As has been mentioned before, the evolution is based on selection and mutation. While selection is used to select the fraction of a population with the higher payoff, a mutation provides a variety of strategies in the population. The replicator dynamics is a system of differential equations used to describe the selection process of evolution, i.e., how the population choosing different strategies changes over time. Each replicator represents a pure strategy s. The new offspring is reproduced for different strategy which can be modeled by

$$\frac{\mathrm{d}x_s}{\mathrm{d}t} = \dot{x}_s = x_s \left(\mathcal{U}(s) - \overline{\mathcal{U}}(\mathbf{x}) \right) \tag{12.2}$$

where x_s represents the fraction of population adopting strategy s, $\mathbf{x}$ is a vector x_s, and it is referred to as the state of population. $\mathcal{U}(s)$ is a payoff of player adopting strategy s, and $\overline{\mathcal{U}}(\mathbf{x})$ is the average payoff of the population. At the steady state (i.e., for time $t \to \infty$), the replicator dynamics will be $\dot{x}_s = 0$ if the strategy is stable. The fraction of the population adopting different strategy at the stable steady state is referred to as the evolutionary equilibrium. It is known that every Nash equilibrium is an evolutionary equilibrium of replicator dynamics. However, an evolutionary equilibrium may not be a Nash equilibrium.

The theory of evolutionary game has been adopted to solve various problems in wireless networks (e.g., [18–22]). In [18], an evolutionary game theory was used to model the network selection behavior of the mobile users in a heterogeneous wireless network, which is composed of multiple wireless access technologies (e.g., cellular, broadband wireless access, and WLAN). The mobile users can adapt their network selection strategy based on the perceived performance and the cost of a wireless connectivity. In [19], the traffic routing problem was modeled by an evolutionary game. The players can choose the routing path to avoid any congestion so that the performance is maximized. In [20], a similar game model was developed for the IEEE 802.16 multihop wireless backhaul. In this case, the traffic routing has to also take the wireless channel quality into account. In [21], an evolutionary game theory was applied to study the problem of power allocation in the cooperative relay networks. In such a network, a relay node can select the different power levels for relaying traffic from the source nodes. The power level to be used can evolve based on the benefit of relaying (e.g., higher transmission rate). In [22], the cooperative spectrum sensing problem for cognitive radios was modeled by an evolutionary game. The cooperation behavior of the secondary users can evolve due to benefit of performing cooperative spectrum sensing with other secondary users to detect the primary user.

12.3.2 Reinforcement Learning

Distributed decision making using reinforcement learning algorithm is based on the optimization model of Markov decision process (MDP). An MDP is defined by a set of states, a set of actions, and a set of rewards. At each time t, the agent observes the state x_t. Then, the agent chooses an action s_t given state x_t. Then, the system transits to the new state and the agent receives the reward u_t (or experiences the cost). The agent with a learning algorithm (e.g., reinforcement learning) will develop a policy, which is a mapping from state to action, to maximize the long-term reward. This long-term reward can be the sum of a immediate reward in the finite time horizon case with limit T (i.e., $U = \sum_{t=0}^{T} u_t$) or the sum of a discounted immediate reward (i.e., $U = \sum \lim_{t=0}^{\infty} \gamma^t u_t$, where γ for $0 < \gamma < 1$ is a discounting factor). The most popular reinforcement learning algorithm for MDP is the Q-learning. A simple example of this algorithm is shown in Algorithm 1 where $rand()$ is a random number generator, α is the learning rate, and γ is the discounting factor. This algorithm is divided into two steps, i.e., exploration and exploitation. The algorithm performs exploration step randomly with a certain probability (line 4 of Algorithm 1). In this exploration step, the algorithm tries different action randomly so that the knowledge (i.e., Q-value) of the action can be obtained. In the exploitation step, this knowledge is used to make the optimal decision (line 6 of Algorithm 1). This Q-value is updated according to the equation in line 8 of Algorithm 1.

Algorithm 1 Q-learning algorithm

1: Initialize q-value $Q(x_t, u_t)$ where x_t is state and s_t is action at time $t \leftarrow 0$
2: **loop**
3:     **if** $rand()$ < Exploration probability **then**
4:         Select action s_t randomly given state x_t at time t
5:     **else**
6:         Select the best action $s_t = \arg\max_s Q(x_t, s)$
7:     **end if**
8:     $Q(x_t, s_t) \leftarrow Q(x_t, s_t)(1 - \alpha) + \alpha\,(u_t + \gamma \max_s Q(x_{t+1}, s))$
9: **end loop**

The Q-learning algorithm has been applied to solve distributed decision-making problems in wireless networks (e.g., [23–28]). In [23], Q-learning was used to obtain the distributed handoff decisions for the mobiles in a heterogeneous wireless network. The objective is to maximize the expected total utility of a connection subject to the constraint on the total access cost. The utility is defined as the quality of wireless connection, with a penalty on the signal and call dropping. In [24], Q-learning was used to obtain a distributed buffer management policy for mobiles transmitting biosignal data from patients to different wireless access networks in a heterogeneous wireless telemedicine system. The objective is to minimize the cost while the QoS requirements (i.e., delay and loss) are met. In [25], a Q-learning algorithm was adopted in a cognitive radio network where the secondary base station chooses a wireless channel to access given the states of the channels. The reward was defined in terms of the signal-to-interference and noise ratio (SINR). In [27], a Q-learning algorithm was used for solving a routing problem in the multihop cognitive radio networks. The number of available channels is estimated and the optimal route is selected based on this information. In [28], a Q-learning algorithm was used to optimize the spectrum sensing in the cognitive radio networks. The reward is defined in terms of the accuracy of channel sensing result.

12.3.3 Reinforcement Learning and Evolutionary Game Theory

Reinforcement learning (i.e., Q-learing) of agents can be modeled as an evolutionary game [17]. For a system with two players, let $\mathbf{U}_1$ and $\mathbf{U}_2$ denote payoff matrices of players 1 and 2, respectively. The dynamics of player 1 can be expressed as follows:

$$\frac{dx_{s,1}}{dt} = \dot{x}_{s,1} = x_{s,1}\alpha((\mathbf{U}_1\mathbf{x}_2)_s - \mathbf{x}_1\mathbf{U}_1\mathbf{x}_2) + x_{s,1}\alpha \sum_{s'} x_{s',1} \ln\left(\frac{x_{s',1}}{x_{s,1}}\right) \quad (12.3)$$

and dynamics of player 2 can be expressed as follows:

$$\frac{dx_{s,2}}{dt} = \dot{x}_{s,2} = x_{s,2}\alpha((\mathbf{U}_2\mathbf{x}_1)_s - \mathbf{x}_2\mathbf{U}_2\mathbf{x}_1) + x_{s,2}\alpha \sum_{s'} x_{s',2} \ln\left(\frac{x_{s',2}}{x_{s,2}}\right). \quad (12.4)$$

These dynamics represent the evolution of both players using a Q-learning algorithm in terms of a probability for selecting strategy (i.e., $x_{s,j}$ is a probability of selecting strategy s of player j). It can be observed that the first terms of (12.3) and (12.4), which account for the strategy selection process of players, are the same as those of replicator dynamics. The second terms account for the mutation process. Specifically, the mutation and selection processes can be considered as the exploration and exploitation steps in the reinforcement learning. Alternatively, the evolutionary game formulation for a Q-learning algorithm can also be modeled as a Markov chain since the population can make decision randomly due to bounded rationality [29]. In this case, the fractions of population selecting different strategies are modeled as the states of the Markov chain. The transition rates or transition probabilities are determined by the payoffs corresponding to different strategies.

12.4 Adaptive WiFi/WiMAX Networking Platform

In this section, we present an adaptive multihop and clustered WiFi/WiMAX-based cognitive vehicular networking platform.

12.4.1 Network Model

Consider a cluster-based vehicular network with N vehicular nodes moving in the same direction (Fig. 12.3). Each of these N nodes is equipped with a dual-mode WiFi/WiMAX transceiver. A WiFi/WiMAX transceiver conforms to the IEEE 802.11 and the IEEE 802.16 MAC protocols.

All vehicular nodes need to communicate with a roadside base station. These vehicular nodes may establish a direct wireless link to the roadside base station using a WiMAX transceiver unit. These nodes are referred to as *gateways*. Others, referred to as *clients*, communicate with the roadside base station through one of the gateways. These clients connect to a gateway using a WiFi transceiver and share the WiMAX link with the gateway. n_g and n_c denote the numbers of gateways and clients, respectively, where $n_g + n_c = N$. Also, the number of clients associated with gateway i is denoted by $n_{c,i}$. It is assumed that the bandwidth on a WiMAX

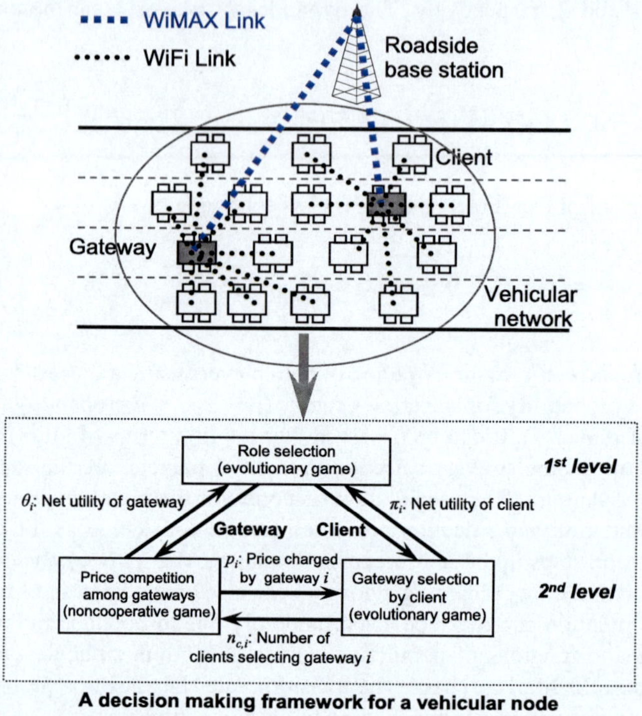

Fig. 12.3 A cluster-based vehicular network: A gateway is represented in *gray* and has a direct WiMAX link to the roadside base station. A client is represented in *white* and connects to the roadside base station via a gateway. Both the gateway and the client use a two-level decision making framework

link is B_b bps. This link is shared among $1 + n_{c,i}$ nodes (a gateway and its clients). Correspondingly, each of the $1 + n_{c,i}$ vehicular nodes is allocated with a logical WiMAX link to a roadside base station with bandwidth B_b, $(1 + n_{c,i})$ bps. Let B_c denote the aggregated bandwidth on all WiFi links associated with a gateway. The bandwidth between gateway i and each of its clients would be $B_c, n_{c,i}$, bps in which the IEEE 802.11 MAC protocol is based on point coordination function (PCF). Under this model, the bandwidth of a link between a vehicular node and a roadside base station is $b = \min(B_b/(1 + n_{c,i}), B_c, n_{c,i})$ bps.

12.4.2 Decision-Making Framework

Communication services are offered to the vehicular nodes by the service provider and the gateways. The service provider offers WiMAX services with bandwidth B_b. The price for bandwidth B_b bps is P_b monetary units (MUs). A gateway which purchases a WiMAX link may share the link with its clients. Gateway i offers the traffic relaying service to its clients and charges price $p_i < P_b$ MUs to each client.

Under the above network model, each vehicular node needs to make a two-level decision (Fig. 12.3). In the first level, a vehicular node decides whether to be a client or to be a gateway. An evolutionary game model based on a Markov chain is used to analyze this decision of vehicular nodes implementing the Q-learning algorithm. After deciding its role, a vehicular node defines its parameters based on the selected role. As a client, a vehicular node selects a gateway for relaying its traffic. The decision process for gateway selection is modeled by an evolutionary game. As a gateway, a vehicular node determines the price to charge its clients for bandwidth sharing. The decision on price setting can be modeled as a noncooperative game. Given the decisions of other gateway nodes, a vehicular gateway node makes its pricing decision to maximize its net utility. At an equilibrium point, none of the vehicular nodes would be willing to change its decision. That is, in the second level, the clients and gateways determine their gateway and competitive price, respectively.

It is assumed that every vehicular node is interested in maximizing its own satisfaction, which is modeled by the so-called *net utility*. The net utility depends on the bandwidth (b) from itself to the roadside base station, the price (p), and the revenue (r) gained from other nodes. Mathematically, net utility is defined as the rate utility $\mathscr{U}(b)$ minus cost p plus revenue r, i.e., $\mathscr{N}(b, p, r) = \mathscr{U}(b) - p + r$. The cost ($p$) of a gateway and a client denote, respectively, the price for a WiMAX link charged by the service provider, and the price charged by a gateway to a client node to relay traffic over the WiMAX link. Attributed to a gateway, the revenue r is earned by sharing the purchased WiMAX link with its clients. Finally, the rate utility is characterized by a concave logarithm utility function $\mathscr{U}(b) = u_1 \log(1 + u_2 b)$, where u_1 and u_2 are the parameters of the function.

Note that similar pricing models for traffic relaying can be found in [30, 31]. In these models, pricing was used as an incentive for one node to relay traffic for other nodes [30]. The optimal price can be determined from the bandwidth demand of neighboring nodes based on an auction mechanism [31]. However, most of the work ignored the issues of gateway selection and price competition which are common in a cluster-based network (e.g., a vehicle platoon on the highway). Also, similar networking decision model can be found in [32]. However, there are many major differences. First, in [32], the number of gateways is fixed. Second, the users are assumed to possess linear bandwidth demand function. Third, the price competition among the gateways was not considered in [32]. In short, the decision-making framework presented in this chapter is more general which can capture the independent and rational decision-making behavior of a mobile node which is common in a cognitive vehicular network.

12.5 Hierarchical Game Formulation for Distributed Decision Making Framework

The decision-making framework for independent and rational vehicular nodes is developed using a hierarchical game structure. This framework consists of two levels and three game formulations (see Fig. 12.3). In the first level, each vehicular node applies an *evolutionary game* to determine its role (as a client or as a gateway). In the second level, a gateway applies a *noncooperative game* to obtain a competitive price, while a client uses an *evolutionary game* for the gateway selection. The following two-step backward induction procedure is applied to solve the entire hierarchical game: (1) Obtain the solution in the second level for both clients and gateways and (2) Use the solution of the second level to select the role (i.e., gateway or client) of a vehicular node in the first level.

12.5.1 Gateway Selection by Client Nodes

We use an evolutionary game model [33] for the gateway selection problem. Here, each client observes the price periodically (e.g., for every 10 s) broadcast by all the gateways, computes the expected net utility, and selects the gateway which gives rise to the highest net utility. The utility depends on both price and bandwidth. While a gateway specifies the price, its offered bandwidth depends on the number of associated clients. Therefore, the value of net utility can change after the clients make a decision (e.g., change the gateway). In this iterative algorithm, at each iteration (e.g., every 10 s) a client chooses a gateway. After reaching the equilibrium, the net utility will remain unchanged over the rest of the adaptation interval, and the clients will stick to one gateway which maximizes their net utility.

An evolutionary game for gateway selection is formulated as follows. A *player* is a client. A *population* is a group of n_c clients. The *strategy* of a client is the selection of a gateway. The set of strategies correspond to the set of gateways. The *payoff* is given by the net utility of a client. Each client decides to join one of n_g groups (i.e., gateways) which maximizes its net utility. In an evolutionary game, the proportion, x_i, of the clients selecting a gateway i can be determined, where $\sum_{i=1}^{n_g} x_i = 1$. An evolutionary equilibrium is defined as a point where no strategy can lead to a change in the proportion of clients $x_i, \forall i$. In particular, it can be expressed as

$$\dot{x}_i = \frac{dx_i}{dt} = x_i\,(\pi_i - \overline{\pi}) = 0 \qquad (12.5)$$

where $\pi_i = \mathscr{N}(b_i, p_i, 0)$ denotes the payoff of each client selecting gateway i, and $\overline{\pi} = \sum_{i=1}^{n_g} x_i \pi_i$ denotes the average payoff of the entire population. Since the proportion x_i ceases to vary at the equilibrium, the number of clients associated with gateway i, $n_{c,i} = x_i(N - n_g)$ ceases to change. Here, the net utility of each client remains unchanged, and each client sticks to a gateway which maximizes its utility.

12.5.2 Price Competition Among Gateway Nodes

Since a client can select and switch to the gateway which provides a higher net utility, the price offered by each gateway has to be carefully chosen. For example, if the price is high, only a few clients will select this gateway to relay their traffic, and only small revenue can be generated. However, if the price is low, a number of clients will select this gateway, and the end-to-end bandwidth can be degraded due to congestion. Since each gateway makes its decision independently and noncooperatively, the desirable solution in terms of price has to maximize the net utility of the gateway. Therefore, a noncooperative game [34] is formulated to obtain this competitive price. This game can be described as follows. A *player* of this game is a gateway. The *strategy* of a player is the offered price. The *payoff* of a gateway is defined as

$$\theta_i = \mathcal{N}(b_i, P_b, n_{c,i} p_i) = \mathcal{U}(b_i) - P_b + n_{c,i} p_i \tag{12.6}$$

where b_i is the end-to-end bandwidth, and the revenue from the clients is a chosen price p_i multiplied by the total number of clients $n_{c,i}$ selecting this gateway i. Since the net utility of the gateway is a function of $n_{c,i}$ which is again a function of price offered by other gateways, the net utility can be written as $\theta_i(p_i, \mathbf{p}_{-i})$, where $\mathbf{p}_{-i}$ is a vector of prices from other gateways except gateway i.

A noncooperative game is used since each gateway wants to achieve the highest payoff in terms of net utility by increasing the price. However, if one gateway increases its offered price, it is likely that other gateways will reduce the price to attract more clients, and the gateway with high price loses revenue. In this competitive situation, the Nash equilibrium is considered as a solution of this noncooperative game. The Nash equilibrium has the property that the payoff of one gateway is maximized, given the price chosen by other gateways. At the Nash equilibrium, this property applies to all gateways. Therefore, none of the gateways would unilaterally change the strategy to improve its payoff.

The Nash equilibrium can be obtained by using the best response function which is the best price from one gateway given the prices from other gateways. In particular, the best response function of a gateway is obtained by formulating a payoff maximization problem. The best response function of gateway i can be defined as follows:

$$p_i^* = \mathcal{B}_i(\mathbf{p}_{-i}) = \arg\max_{p_i} \theta_i(p_i, \mathbf{p}_{-i}). \tag{12.7}$$

This best response can be obtained by a numerical method. Then, the Nash equilibrium can be obtained from $p_i^* = \mathcal{B}_i\left(\mathbf{p}_{-i}^*\right)$ for all i, where $\mathbf{p}_{-i}^*$ is a vector of best response of all gateways except gateway i.

The distributed algorithm that achieves the Nash equilibrium for the gateways works as follows. Gateway i observes the price broadcast by other gateways. Then gateway i chooses the price to maximize its payoff. This can be done by observing the responses of the clients to a small variation in the current price (i.e., more or fewer number of clients will select gateway i due to lower or higher prices, respectively). From these responses, the gateway can estimate marginal payoff due to variation in price. This marginal payoff is then used to obtain the best response for this gateway. This procedure is repeated for all gateways until there is no change in price offered by all gateways. Since there is no change in the strategy adopted by all players, the solution of this algorithm is the Nash equilibrium.

12.5.3 Role Selection by Vehicular Nodes

After the solutions of the gateway selection and the price competition are obtained where the corresponding net utilities are given by π_i and θ_i, respectively, we backtrack to the first-level decision on whether the vehicular node decides to become a gateway node or a client node. Since the net utilities for being a gateway and a client are not known by the node a priori (i.e., other gateways and other clients do not reveal their net utility information), the vehicular node must learn by trials. In this case, a vehicular node can randomly become a client or a gateway and observe the net utility. For instance, if becoming a client yields a higher net utility, this node will choose to become a client in the future.

The Q-learning-based algorithm of a vehicular node to decide whether to become a gateway or a client works as follows:

A vehicular node randomly chooses to become a gateway or a client.
Q-value $Q(s_t)$ for strategy $s_t \in$ {gateway, client} is initialized at time $t = 0$.
loop
  if Node is gateway **then**
    if Q(gateway) $<$ Q(client) **then**
      Gateway switches back to become a client. {Becoming client yields higher net utility (exploitation)}
    else
      Gateway randomly becomes a client with exploration rate ρ (e.g., $\alpha =$ 0.1) {Learning by trial (exploration) }
    end if
    A vehicular node observes its net utility (i.e., π_i for client).
    $Q(s_t) \leftarrow Q(s_t)(1-\alpha) + \alpha\left(\pi_i + \gamma \max_{s_{t+1}} Q_{s_{t+1}}\right)$
  else
    if Q(client) $<$ Q(gateway) **then**
      Client switches to gateway {Becoming a gateway yields higher net utility (exploitation)}
    else
      Client randomly becomes a gateway with exploration rate ρ. {Learning by trial (exploration) }
    end if
    A vehicular node observes its net utility (i.e., θ_i for gateway).
    $Q(s_t) \leftarrow Q(s_t)(1-\alpha) + \alpha\left(\theta_i + \gamma \max_{s_{t+1}} Q_{s_{t+1}}\right)$
  end if
end loop

This algorithm is performed periodically until a vehicular node finishes the data transfer with a roadside base station or the node leaves the network. Note that the random actions in lines 8 and 16 of the above algorithm are used to try an alternative strategy periodically. This trial is required to avoid a vehicular node being locked up in the sub-optimal decision due to an obsolete information about net utility when the network condition changes.

This learning algorithm can be modeled as a stochastic evolutionary game where a vehicular node gradually learns by randomly trying the different available strategies. The game can be described as follows. A *player* is a vehicular node. A *strategy* is to become either a gateway or a client. *Payoff* is the net utility of a node. The solution of this evolutionary game is an equilibrium which can be analytically obtained by formulating a finite discrete-state and continuous-time Markov chain. The state space of this Markov chain is a random integer between 1 and N representing the current number of gateways. The transition rate between each state is a function of the net utility received by a vehicular node. In particular, if the net utility of a gateway is higher than that of a client, the transition rate from state n_g to $n_g + 1$ (i.e., the number of gateways increases by one) is $(N - n_g)(\theta_i - \pi_i)$. While $(\theta_i - \pi_i)$ indicates the "incentive" for each client to become a gateway, $(N - n_g)$ indicates that all clients can observe the higher net utility of a gateway. Therefore, every client has an equal chance to become a gateway. However, there is a small chance (i.e., ρ) of trying different strategy. As a result, a gateway can switch to become a client, although becoming a gateway can yield a higher net utility than becoming a client. The transition rate from state n_g to $n_g - 1$ is denoted by $n_g\rho$, which is non-zero. In particular, every gateway has an equal chance to become a client. The transition rates for the case that the net utility of a client is higher than that of a gateway can be obtained in a similar way. Finally, the steady-state probability can be computed from this continuous-time Markov chain which can be used to calculate the average number of active gateways in the network.

12.6 Performance Evaluation

We consider a highway with 4 lanes. The average speed of a vehicular node is 64 km/h. The roadside base station allocates 1 Mbps of bandwidth to each connection from gateway. The price of this roadside connection is fixed with $P_b = 10$ MUs. The transmission range of a WiMAX base station is 10 km, while that of WiFi is 100 m. The constants in the utility function are as follows $u_1 = 1$ and $u_2 = 1$. The simulator is developed by using MATLAB with the mobility model for the vehicular nodes similar to that in [35].

12.6.1 Gateway Selection

We first investigate an effect of the price on the number of clients selecting the gateways (Fig. 12.4). In this scenario, the number of gateways is fixed to 3. The

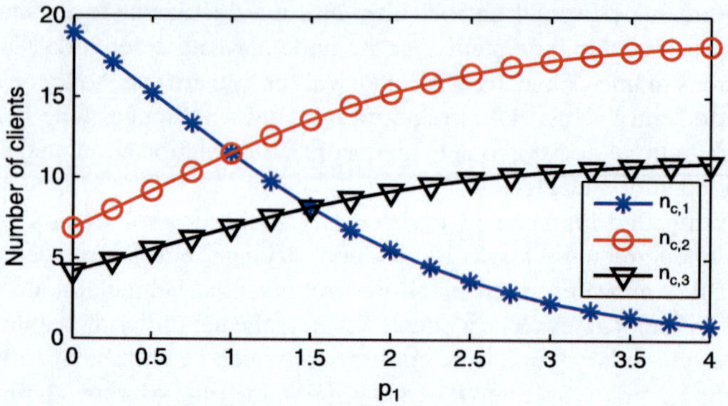

Fig. 12.4 Number of clients associated with each gateway

price of gateway 1 is varied from 0 to 4, while that of gateway 2 and gateway 3 is fixed at $p_2 = 1.0$ and $p_3 = 1.5$. When the price of the gateway 1 increases, the number of clients selecting this gateway decreases. Since the net utility decreases due to higher price, the client deviates to gateway 2 and gateway 3 with lower price to yield the higher net utility. Since the available bandwidth assigned by the roadside base station to each gateway is identical, when the prices offered by two gateways are the same (e.g., $p_1 = p_2 = 1.0$ and $p_1 = p_3 = 1.5$), the number of clients selecting all gateways is the same.

12.6.2 Gateway Selection and Price Competition

Since a client can select the gateway which yields the highest payoff, the number of clients at a particular gateway increases as the price offered by this gateway decreases. From this behavior of the clients, a gateway can optimize its price to the Nash equilibrium such that the optimal revenues can be achieved given the strategies of other gateways. In this case, the variation of the Nash equilibrium for the prices from gateway under different average speeds of the vehicular nodes is shown in Fig. 12.5. When the speed increases, the distance between each vehicle decreases [35]. Also, the density of vehicles decreases, and fewer nodes are in the same network. As the number of nodes in a network decreases, the gateway can increase its price to achieve a higher revenue and subsequently a higher net utility. The number of gateways is also varied in this scenario. The more the number of gateways, the higher the level of competition among the gateways. In such a scenario, to attract more clients, a gateway node decreases the price. Note that as the number of gateways increases, the total end-to-end bandwidth for all clients in the network increases.

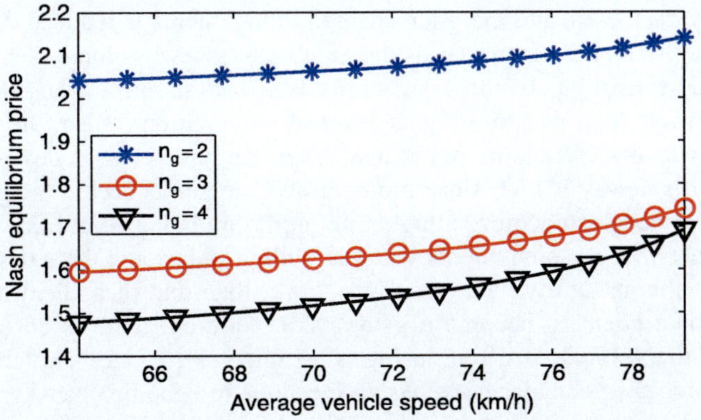

Fig. 12.5 Nash equilibrium under different speeds of vehicular nodes. Increasing mobility of the vehicular nodes increases the price charged by the gateway

12.6.3 Individual Net Utility of Gateway and Client

The individual net utility is a major factor for a vehicular node to decide whether a node should become a gateway or a client. To investigate the incentive of a vehicular node to become a gateway or client, the number of gateways in a network is varied. The net utilities of gateway and client are shown in Fig. 12.6. As the number of gateways increases, the net utility of gateway decreases since the number of clients

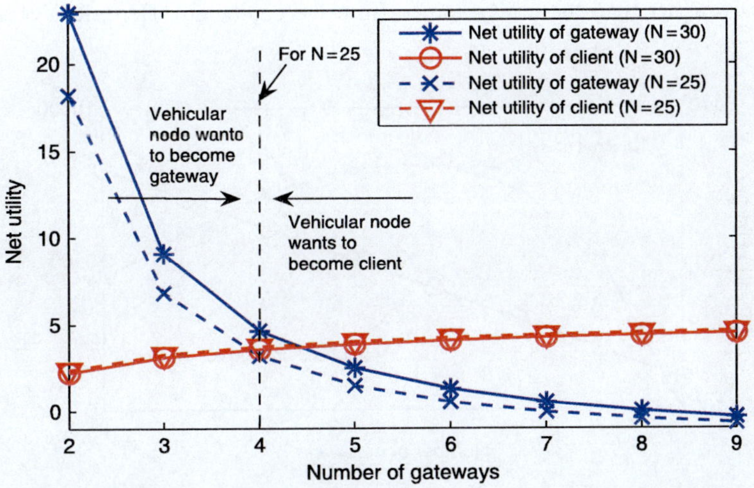

Fig. 12.6 Utility of gateway and client. As the number of gateways increases, the net utility of the clients increases, while that of the gateways decreases

per gateway decreases, and the price charged to the clients is reduced. Conversely, the net utility of a client increases as the number of gateways increases, this is due to larger end-to-end bandwidth assigned by base station and lower price charged by the gateways (i.e., due to a higher level of competition). Also, the number of nodes in a network affects the net utility. When the number of nodes is large, the net utility of gateway is high since more number of clients access the gateway. In this case, if a node can achieve a higher net utility by being a gateway rather than a client, there will be some clients who are willing to become the gateway nodes. However, if the net utility of a gateway is lower than that of a client, there is no incentive for a client to become a gateway. In addition, there is an equilibrium point (e.g., the vertical dash line in Fig. 12.6 for $N = 25$) for the left and right sides of which the vehicular node has an incentive to become gateway and client, respectively.

12.6.4 Total Net Utility and Total End-to-End Bandwidth

The total net utility (i.e., sum of net utility from all nodes in a network) versus the total end-to-end bandwidth is shown in Fig. 12.7 for $N = 30$. Evidently, when the number of gateways increases, the end-to-end bandwidth for all nodes increases linearly. On the other hand, due to decision of a node to become either a gateway or a client, and the price competition among gateways, the total net utility of the network first increases as the number of gateways increases. This increase in the total net utility is due to larger end-to-end bandwidth. However, at a certain point, this net utility decreases, since the price charged by the base station becomes higher than the utility gained from the bandwidth. Here, it is observed

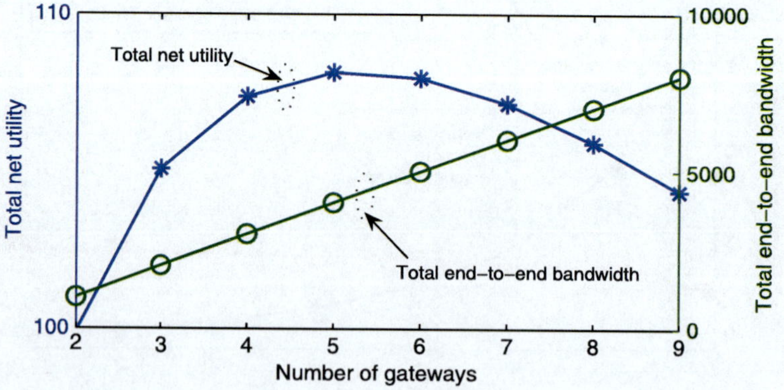

Fig. 12.7 Total utility of the network. As the number of gateways increases, the end-to-end bandwidth linearly increases, while the total utility of an entire network first increases and then decreases. There is an optimal number of gateways the total net utility is maximized

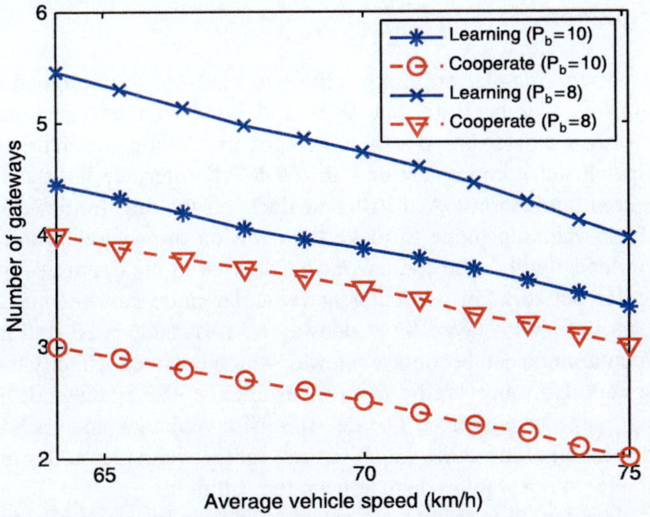

Fig. 12.8 Variation in the number of gateways when the vehicular nodes make their decision independently (i.e., by learning) and cooperatively

that there is an optimal number of gateways for which the highest total net utility is achieved.

12.6.5 Number of Gateways Under Different Vehicle Speeds

Next, the number of gateways in the case that all nodes make decision independently and in the case that all nodes make a cooperation to achieve the highest total net utility are compared. In the former case, an evolutionary game is applied to obtain the average number of gateways. In the latter case, where all nodes fully cooperate, the number of gateways is determined from the location where the total utility is maximized (e.g., 5 gateways in Fig. 12.7). The number of gateways in both cases are shown in Fig. 12.8.

As the speed of vehicles increases, the aggregated bandwidth demand (from fewer vehicular nodes) decreases, and the number of gateways decreases due to change in this demand. In particular, the vehicular node observes that becoming a client yields a higher net utility since the revenue decreases as the number of nodes in the network decreases. Also, the price charged by the base station has a significant effect to the maximum number of gateways. That is, when the base station charges a high price, the net utility of the gateway decreases. As a result, the node is reluctant to become a gateway. In the case that all nodes cooperate, the price paid to the base stations can be reduced. Therefore, the number of gateways is smaller than that in the case when all nodes make their decisions independently.

12.7 Conclusion

An adaptive networking platform for a vehicle-to-roadside communication has been introduced in this chapter. With dual WiFi and WiMAX interfaces, the vehicular nodes can form a cluster-based vehicular network. While the WiFi interface is used for intra-cluster communications, the WiMAX interface is used for cluster-to-roadside communications. A distributed decision-making framework has been developed for a vehicular node to make the decision on a wireless access intelligently and independently. The framework is adaptive to the dynamics of a cluster-based vehicular network. In particular, a vehicular node can become a client in which case its traffic is relayed by a gateway to a roadside base station. Alternatively, a vehicular node can become a gateway which connects directly to a roadside base station and also relays traffic from other clients. The framework is based on a hierarchical game formulation. The decision of a vehicular node is based on the equilibrium solution of the game which ensures that all vehicular nodes in a network are satisfied and do not want to deviate from the solution.

Based on the simulation results, the observations can be summarized as follows:

- The number of clients selecting a particular gateway depends on the price charged by that gateway.
- Vehicle mobility impacts the price at the Nash equilibrium of the gateways and the number of gateways in the network.
- As the number of gateways in the network increases, net utility of gateways decreases, while net utility of clients increases.
- There is an equilibrium number of gateways such that the net utilities of client and gateway are identical. This equilibrium number of gateways can be reached if all vehicular nodes in a network make decision independently.
- There is an optimal number of gateways in the network for which the total net utility is maximized.

Experimental evaluation of the performance and behavior of the adaptive decision framework in a practical vehicular network needs further investigation. The impact of variations in the channel state and node mobility VANET environment need to be investigated.

Acknowledgment This work was supported by the AUTO21 NCE research grant for the project F303-FVT.

References

1. Susan R. Dickey, C.-L. Huang, and X. Guan, "Field Measurements of Vehicle to Roadside Communication performance," in *Proceedings of IEEE Vehicular Technology Conference (VTC) Fall*, pp. 2179–2183, September-October 2007.
2. H. Cai and Y. Lin, "Design of a Roadside Seamless Wireless Communication System for Intelligent Highway," in *Proceeding of IEEE Networking, Sensing and Control*, pp. 342–347, March 2005.

3. K. Yang, S. Ou, H.-H. Chen, and J. He, "A Multihop Peer-Communication Protocol with Fairness Guarantee for IEEE 802.16-Based Vehicular Networks," *IEEE Transactions on Vehicular Technology*, vol. 56, no. 6, Part 1, pp. 3358–3370, November 2007.
4. L. Wischhof, A. Ebner, H. Rohling, M. Lott, and R. Halfmann, "Adaptive Broadcast for Travel and Traffic Information Distribution Based on Inter-Vehicle Communication," in *Proceedings of IEEE Intelligent Vehicles Symposium*, pp. 6–11, June 2003.
5. O. Maeshima, S. Cai, T. Honda, and H. Urayama, "A Roadside-to-Vehicle Communication System for Vehicle Safety using Dual Frequency Channels," in *Proceedings of IEEE Intelligent Transportation Systems Conference (ITSC) 2007*, pp. 349–354, September–October 2007.
6. H. Su and X. Zhang, "Clustering-Based Multichannel MAC Protocols for QoS Provisionings Over Vehicular Ad Hoc Networks," *IEEE Transaction Vehicular Technology*, vol. 56, no. 6, pp. 3309–3323, November 2007.
7. B. Sikdar, "Design and Analysis of a MAC Protocol for Vehicle to Roadside Networks," in *Proceedings of IEEE Wireless Communications and Networking Conference (WCNC)*, pp. 1691–1696, March–April 2008.
8. C.-J. Chang, R.-G. Cheng, H.-T. Shih, and Y.-S. Chen, "Maximum Freedom Last Scheduling Algorithm for Downlinks of DSRC Networks," *IEEE Transactions on Intelligent Transportation Systems*, vol. 8, no. 2, pp. 223–232, June 2007.
9. X. Y. Wang and P.-H. Ho, "A Novel Sensing Coordination Framework for CR-VANETs," *IEEE Transactions on Vehicular Technology*, vol. 59, no. 4, pp. 1936–1948, May 2010.
10. H. Li and D. K. Irick, "Collaborative Spectrum Sensing in Cognitive Radio Vehicular Ad hoc Networks: Belief Propagation on Highway," in *Proceedings of IEEE Vehicular Technology Conference (VTC)-Spring*, pp. 1–5, May 2010.
11. Z. Ahmed, H. Jamal, S. Khan, R. Mehboob, and A. Ashraf, "Cognitive Communication Device for Vehicular Networking," *IEEE Transactions on Consumer Electronics*, vol. 55, no. 2, pp. 371–375, May 2009.
12. S. Chung, J. Yoo, and C. Kim, "A Cognitive MAC for VANET Based on the WAVE Systems," in *Proceedings of International Conference on Advanced Communication Technology (ICACT)*, vol. 1, pp. 41–46, February 2009.
13. K. Tsukamoto, Y. Omori, O. Altintas, M. Tsuru, and Y. Oie, "On Spatially-Aware Channel Selection in Dynamic Spectrum Access Multi-hop Inter-Vehicle Communications," in *Proceedings of IEEE Vehicular Technology Conference (VTC)-Fall*, pp. 1–7, September 2009.
14. D. Niyato, E. Hossain, and P. Wang, "Optimal Channel Access Management with QoS Support for Cognitive Vehicular Networks," *IEEE Transactions on Mobile Computing*, vol. 10, no. 4, pp. 573–591, February 2011.
15. R. Rangnekar, F. Ge, A. Young, M. D. Silvius, A. Fayez, and C. W. Bostian, "A Remote Control and Service Access Scheme for a Vehicular Public Safety Cognitive Radio," in *Proceedings of IEEE Vehicular Technology Conference Fall (VTC)-Fall*, pp. 1–5, September 2009.
16. R. Muraleedharan and L. A. Osadciw, "Cognitive Security Protocol for Sensor Based VANET Using Swarm Intelligence," in *Proceedings of Asilomar Conference on Signals, Systems and Computers*, pp. 288–290, November 2009.
17. K. Tuyls, A. Nowe, T. Lenaerts, and B. Manderick, "An Evolutionary Game Theoretic Perspective on Learning in Multi-agent Systems," in *Synthese Knowledge, Rationality & Action*, vol. 139, no. 2, pp. 297–330, March 2004.
18. D. Niyato and E. Hossain, "Dynamics of Network Selection in Heterogeneous Wireless Networks: An Evolutionary Game Approach," *IEEE Transactions on Vehicular Technology*, vol. 58, no. 4, pp. 2008–2017, May 2009.
19. E. Altman, R. ElAzouzi, Y. Hayel, and H. Tembine, "An Evolutionary Game Approach for the Design of Congestion Control Protocols in Wireless Networks," in *Proceedings of International Symposium on Modeling and Optimization in Mobile, Ad Hoc, and Wireless Networks and Workshops (WiOPT)*, pp. 547–552, April 2008.
20. M. P. Anastasopoulos, P.-D. M. Arapoglou, R. Kannan, and P. G. Cottis, "Adaptive Routing Strategies in IEEE 802.16 Multi-hop Wireless Backhaul Networks Based on Evolutionary Game Theory," *IEEE Journal on Selected Areas in Communications*, vol. 26, no. 7, pp. 1218–1225, September 2008.

21. J. Hu, B. Wang, J. Wei, and S. Huang, "Evolutionary Game for Distributed Power Allocation over Cooperative Relay Networks," in *Proceedings of International Conference on Wireless Communications, Networking and Mobile Computing (WiCom)*, September 2009.
22. B. Wang, K. J. Ray Liu, and T. C. Clancy, "Evolutionary Cooperative Spectrum Sensing Game: How to Collaborate?," *IEEE Transactions on Communications*, vol. 58, no. 3, pp. 890–900, March 2010.
23. C. Sun, E. Stevens-Navarro, and V. W. S. Wong, "A Constrained MDP-based Vertical Handoff Decision Algorithm for 4G Wireless Networks," in *Proceedings of IEEE International Conference on Communications (ICC)*, pp. 2169–2174, May 2008.
24. D. Niyato, E. Hossain, and S. Camorlinga, "Remote Patient Monitoring Service Using Heterogeneous Wireless Access Networks: Architecture and Optimization," *IEEE Journal on Selected Areas in Communications*, vol. 27, no. 4, pp. 412–423, May 2009.
25. A. Galindo-Serrano and L. Giupponi, "Distributed Q-learning for Aggregated Interference Control in Cognitive Radio Networks," *IEEE Transactions on Vehicular Technology*, vol. 59, no. 4, pp. 1823–1834, May 2010.
26. J. Gummeson, D. Ganesan, M. D. Corner, and P. Shenoy, "An Adaptive Link Layer for Range Diversity in Multi-radio Mobile Sensor Networks," in *Proceedings of IEEE INFOCOM*, pp. 154–162, April 2009.
27. B. Xia, M. H. Wahab, Y. Yang, Z. Fan, and M. Sooriyabandara, "Reinforcement Learning Based Spectrum-aware Routing in Multi-hop Cognitive Radio Networks," in *Proceedings of International Conference on Cognitive Radio Oriented Wireless Networks and Communications (CROWNCOM)*, June 2009.
28. M. Li, Y. Xu, and J. Hu, "A Q-Learning Based Sensing Task Selection Scheme for Cognitive Radio Networks," in *Proceedings of International Conference on Wireless Communications & Signal Processing (WCSP)*, November 2009.
29. W. H. Sandholm, *Population Games and Evolutionary Dynamics*. The MIT Press, January 2011.
30. M. Lindstrom and P. Lungaro, "Resource Delegation and Rewards to Stimulate Forwarding in Multihop Cellular Networks," in *Proceedings of IEEE Vehicular Technology Conference (VTC) Spring*, vol. 4, pp. 2152–2156, May–June 2005.
31. K. Chen, Z. Yang, C. Wagener, and K. Nahrstedt, "Market Models and Pricing Mechanisms in a Multihop Wireless Hotspot Network," in *Proceedings of International Conference on Mobile and Ubiquitous Systems: Networking and Services (MobiQuitous)*. pp. 73–82, July 2005.
32. D. Niyato and E. Hossain, "Integration of WiMAX and WiFi: Optimal Pricing for Bandwidth Sharing," *IEEE Communications*, vol. 45, no. 5, pp. 140–146, May 2007.
33. T. L. Vincent and J. S. Brown, *Evolutionary Game theory, Natural selection, and Darwinian Dynamics*. Cambridge University Press, 2005.
34. M. J. Osborne, *An Introduction to Game Theory*. Oxford University Press, 2004.
35. M. J. Lighthill and G. B. Whitham, "On Kinematic Waves. II. A Theory of Traffic Flow on Long Crowded Roads," in *Proceedings of the Royal Society of London. Series A, Mathematical and Physical Sciences*, vol. 229, no. 1178, pp. 317–345, May 1955.

Chapter 13
Cognitive Radio Mobile Ad Hoc Networks in Healthcare

Ziqian Dong, Shamik Sengupta, S. Anand, Kai Hong, Rajarathnam Chandramouli, and K.P. Subbalakshmi

Abstract Low-cost automated health monitoring system sees a high demand with the Presidents' proposal on health care reform. Legacy health care monitoring systems demand a great amount of resources such as health care personnel and medical equipments. This increases the cost of health care making it unaffordable to the majority of our society. This chapter introduces an architecture and design of a health care automation network. The health care automation network uses a cognitive radio-based infrastructure to monitor real-time patients' vital signs, collect, and document medical information. The health care automation network can be implemented in hospitals or in senior communities. This network can leverage the existing infrastructure and reduce the cost of implementation. Research challenges in development of cognitive radio health care automation network are also discussed.

13.1 Introduction

Cognition is the scientific term for the process involved in knowing, which includes perception and judgement. Such process intelligently detects surrounding environment and makes decision based on what is learnt. In this chapter, we apply the concept of cognition to health care network design using a cognitive radio-based infrastructure for automated 24×7 medical surveillance. The objective of our study is to provide a low-cost automation network for health care information collection and documentation.

The cost of health care expenditures increases rapidly every year. Expenditures in the United States on health care surpassed \$2.2 trillion in 2007, more than three times the \$714 billion spent in 1990 and over eight times the \$253 billion spent in 1980 [1]. As population aging is unprecedented, the growth of health care expenditure will continue to rise in the future [2, 3]. United Nations (UN) report has shown that twenty-first century has witnessed more rapid aging than the past century. Treatment of chronic illnesses and long-term care account for an estimated 75% of the health care expenditure [4]. Legacy health care monitoring systems demand a great

Z. Dong (✉)
Department of Electrical and Computer Engineering, New York Institute of Technology, New York, NY 10023, USA
e-mail: ziqian.dong@nyit.edu

F.R. Yu (ed.), *Cognitive Radio Mobile Ad Hoc Networks*,
DOI 10.1007/978-1-4419-6172-3_13,

amount of resources such as health care personnel and medical equipments resulting in expensive health care. With the economy slowdown, workers' wages grow at a much slower pace than the health care costs. Many care receivers face difficulty in affording the health care spending. As the health care reform is taking place in the United States, innovations of low-cost health care system are being promoted. As focused by the health care reform, preventative care is given more attention. Low-cost health care monitoring system will be an important factor in preventative care. *How can we provide a low-cost automated health monitoring system?* This chapter addresses this problem with a simple system architecture design and discusses research challenges in development of such systems.

13.1.1 Technology Advancement Has Made Health Care Automation Network Possible

Recent advancements in sensor technologies coupled with miniaturization and its huge potential to function without much human intervention makes the deployment of an automated health care monitoring system feasible. Body sensors (wearable, implantable, or portable) are fast emerging and are envisioned to be a promising approach to monitor the physiological conditions and disease progression. This enables monitoring physiologies without affecting users' normal day-to-day life. Many such body sensors have recently been proposed and are already in the verge of commercial use.

Wearable devices that measure the vital signs of patients have drawn much research attention. Nonin [5] developed wearable bracelet-sensors that read the oxygen saturation (SpO_2) and blood pressure. Harvard University proposed miniaturized sensor motes in the Code Blue project [6–8]. Sensatex [9] and Vivometrics [10] developed wearable t-shirts with in-built sensor technologies that gather information on human vital signs periodically which can be used for health monitoring. Recently, Medrad developed a portable magnetic resonance imaging (MRI) device to enable wireless MRI monitoring [11]. These and other related research can be incorporated into a network of sensor devices, databases, and computing devices that collects, stores, and processes medical data.

13.1.2 Wireless Technologies for Data Transmission

The end of the analog TV broadcasting opens a broad spectrum bands in the 400 MHz and 700 MHz range for wireless transmission for the public safety operations and other commercial services. A wide range of applications will be trialed and competing for this band. The excellent propagation characteristics of this band through buildings and other obstructions makes it an ideal band for Multimedia Home Networks [12]. It can be considered a good medium for at-home medical information collection and transmission. The transmission of medical information such as vital signs can be performed using any or a combination of the existing wireless technologies, e.g., cellular, WiFi, Zigbee, Bluetooth. Each of these technologies provides different advantages in terms of coverage (e.g., cellular) or bandwidth (e.g., WiFi). With the advent of heterogeneous wireless access networks, wireless service providers (WSPs) can combine the complementary advantages of different wireless access networks operating on both licensed and unlicensed bands to serve an increasing amount of automated health monitoring demands. This, however, poses a unique challenge of switching between different wireless networks to provide ubiquitous connectivity. Also, devices that are close to each other and transmitting using the same frequency band may cause interference to each other. This may result in faulty data or service interruption. It is thus essential to design a system that can switch between different access networks and different frequency bands effectively in an intelligent manner. Cognitive radio (CR) based on dynamic spectrum access (DSA) [13] has the capability of dynamically accessing different frequency bands depending on the environment and location and thus is anticipated to enable dynamic switching between different wireless networks.

How can the CR know when to switch frequency band and what frequency band to use? In a hospital or senior health care center, many medical devices are present and may be functioning at a certain frequency band. These devices may be physically placed in a room or moved around. The CR need to know the location of itself and the location of neighboring devices and their frequency bands. When a CR senses the location of a device that is transmitting in a frequency band that may cause interference between them, it will switch its frequency band to avoid interference with that device. Here, location information can be used to predict and model interference among the functional medical devices and assist CR to make appropriate decision on frequency band selection upon sensing possible interference. This requires the CR to have location sensing capability of both itself and the other devices. Since the monitoring devices are usually attached to patients who may not show much mobility, the environment is rather static in a sense that the radio environment may not change much. Therefore, spectrum and interference measurement can be used to model spectrum availability and resource allocation.

An important criteria for the CR nodes to decide when to switch access networks or frequency channels is the location of the transmitting devices. In an automated health care monitoring system, it is crucial to provide accurate location information of either health care receivers or medical devices. This not only ensures timely response to emergent situations by providing accurate location but also avoids interference among the functioning medical devices by smartly sensing and switching transmission channels.

Paramedics require real-time accurate patient location information to provide immediate medical attention. E-911 service requires location information provided by service provider. Providing location information of mobile medical devices ensures the choice of the proper transmission frequency to avoid interference with other devices.

Global positioning system (GPS) provides outdoor location information; however, a major challenge in outdoor/wide area network (WAN) environments lies in that the existing wireless network protocols often do not support GPS. Moreover, for indoor locations, the situation is even worse as it is hard for the commercially deployed GPS to provide accurate location information [14] due to Non-Line-of-Sight (NLoS). In recent times, there have also been few research works focusing on IP geolocation using commercial databases. However, the record in the database is often not updated on time, therefore, makes it unreliable when the IP address of the network devices changes constantly.

Why is location information important to the system in an indoor environment? In a hospital or health care center, various medical devices are present which are sensitive to interference. To provide an untethered connection among different mobile health monitoring devices and to the medical data center, the devices need to be able to flexibly use a limited available wireless spectrum. It is mandatory that the devices do not cause interference to each other as well as to other high-priority equipments like MRI or surgical machines. Note that two devices cause interference to each other when they are present within a certain physical proximity and operate using the same frequency bands. The knowledge of the location of such devices helps prevent this interference while making mobile devices smart enough to avoid the interference. Intelligent mechanisms are therefore necessary to avoid interference and provide high throughput at the same time.

When transmitting large amounts of medical data over wireless channels using small CR devices, power consumption is of great concern. It is also crucial to adopt lossless compression methods to reduce storage space. Therefore, it is important to adopt an efficient data compression and channel coding scheme to losslessly compress the transmitted data on the monitoring devices.

In this chapter, the challenges of designing an automation health network using cognitive radio are discussed and an architecture design of a health care automation network is introduced. The health care automation network provides a location aware cognitive radio-based infrastructure to deploy an automation network for real-time vital sign monitoring and information collection and documentation. It is an intelligent network that adapts itself in order to avoid interference between various medical devices/appliances while documenting the monitored data. It can be implemented in hospitals or in senior care centers. It leverages existing networking hardware infrastructure which minimizes the cost of implementation.

13.2 System Architecture

In a CR health care automation network, users could have tiny body sensors attached to them as shown in Fig. 13.1. The body sensors collect necessary vital signs information in a periodic or on-demand manner and transmit them to a nearby CR node, which can be connected to a bed in a hospital or home environment or even be worn to enable mobility. The CR node transfers the information to any base station or

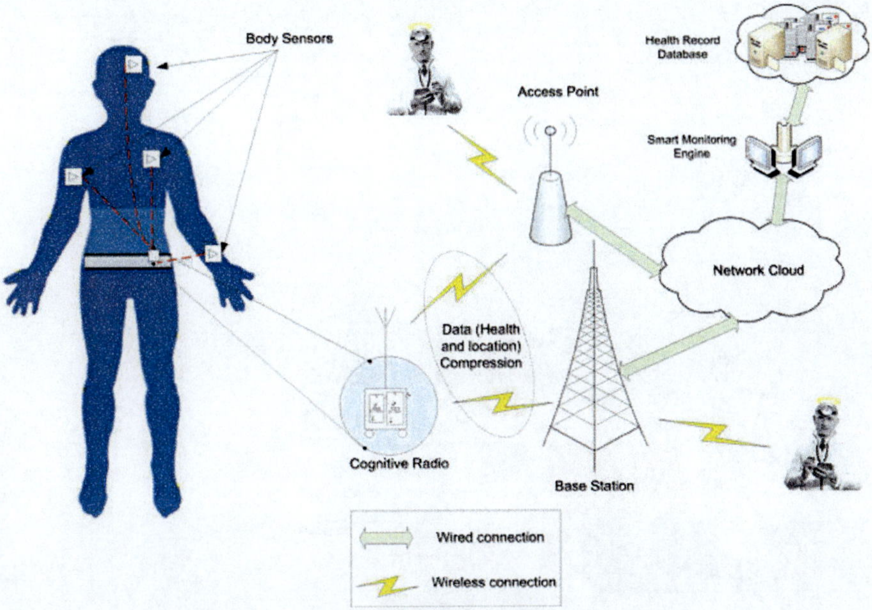

Fig. 13.1 Cognitive radio node collecting vital signs information from body sensors and dynamically transferring information to health database using available access technology

access point (the choice of which depends on several factors like location, wireless channel characteristics, wireless resource availability). The base stations/access points update the information at a health record database through a smart monitoring engine. The smart monitoring engine is capable of performing complex computations if necessary.

A simple illustration of the proposed system is shown in Fig. 13.2 where a geographical region is enabled with heterogeneous overlapping wireless access networks.

The CR node could access a WiFi access point (AP) inside a hospital or a home environment. When the user moves out from an indoor environment (e.g., inside a hospital building or house) to an outdoor environment (e.g., on the street or in a car), the CR node can switch to connect to a cellular or WiMAX access network. The operating frequencies for the indoor and outdoor environments are different. The Federal Communications Commission (FCC) has recently allowed usage of spectrum in the sub-900 MHz bands to unlicensed services. The 402–405 MHz bands have been allocated for the purpose of indoor medical communication services [15]. Other spectrum bands, e.g., 600 MHz, 700 MHz, are also being considered by the FCC. The outdoor cellular services operate in 900 MHz while WiMAX operates at 2–11 GHz. Therefore, the CR node needs to switch spectrum in a dynamic manner to provide ubiquitous connectivity.

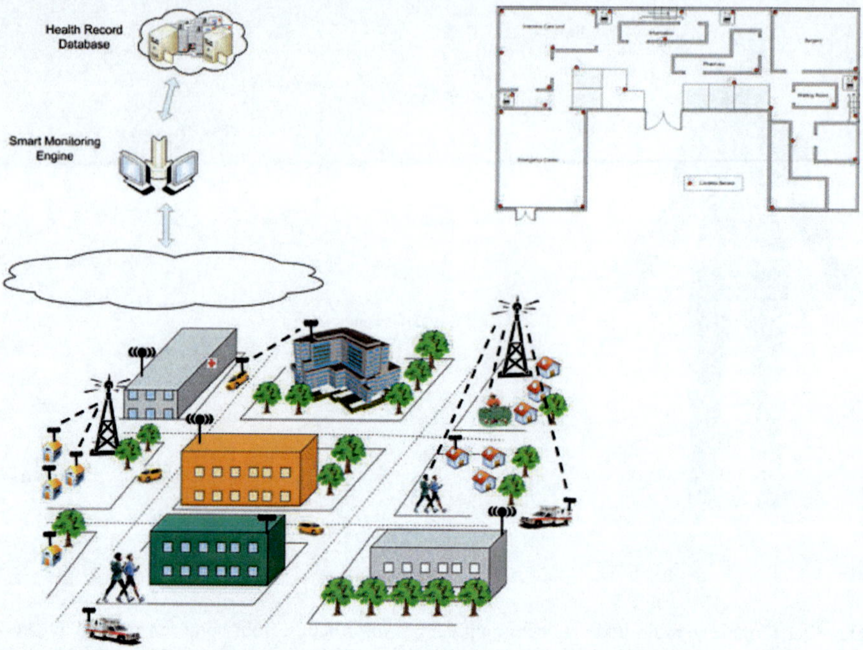

Fig. 13.2 Geolocation in the heterogeneous overlapping wireless access networks

13.3 Cognitive Radio for Health Care Automation Network: Research Challenges

A wireless regional area network architecture that is based on the concept of cognitive radio for supporting automated health monitoring and health record update can be considered in the health care automation network design. In this architecture, a central smart monitoring engine typically manages the heterogeneous access networks by controlling the on-air activity for efficient resource management.

13.3.1 Location-Assisted Dynamic Spectrum Access

To allocate spectrum resource efficiently, knowledge of the location of the functioning devices is critical. Advances in wireless technology enables many medical devices to collect and transmit medical data using wireless connections. To incorporate these devices in a health care automation network, it is important to avoid interference among all the devices. The network needs to know the locations of the transmitting devices that may be attached to or in close vicinity of users and medical devices to choose the appropriate transmitting channel or access network. Furthermore, location information is required by E-911 in an event of medical emergency (e.g., whereabouts of Alzheimer patients and unconscious patients). It also assists the paramedics in accurately locating a patient who may not be able to provide his/her location and providing timely medical attention on site.

Indoor geolocation not only provides location information of patients but also that of medical equipments. The challenge for indoor geolocation is that commercially deployed global positioning system (GPS) cannot provide accurate indoor location information [14]. The challenge in outdoor/wide area network (WAN) environments is that the existing wireless network protocols do not support GPS. Geolocation using commercial databases is available. However, the record in the database is not reliable when the IP address of the network devices changes constantly when DHCP (Dynamic Host Configuration Protocol) is commonly used to obtain configuration information by the devices rendering obsolete information in the database.

For indoor geolocation systems, the key parameters are received signal strength indicator (RSSI), carrier signal phase of arrival (POA), and time of arrival (TOA) of the received signal. These parameters are used in estimating distance between the monitoring device and the location sensors deployed inside a building.

13.3.2 Interference Awareness Among Health Monitoring Devices

Due to the scarcity of the wireless spectrum, to support a large number of devices, frequency bands need to be reused. Different health monitoring devices (CR nodes) need to share the limited available wireless spectrum. It is mandatory that the devices do not cause interference to each other as well as to other high priority equipments like MRI or surgical machines. Two devices cause interference to each other when they are present within a specified physical proximity and simultaneously transmit using the same frequency bands. Intelligent mechanisms are therefore necessary to support as many users as possible, while avoiding interference among the devices.

To avoid interference among the functioning medical devices, CR nodes will have location sensing functions that have knowledge of whereabouts of the transmitting devices and other medical devices. Location sensing of the CR nodes can be achieved using a hybrid database and measurement-based geolocation system [16], where location information of static devices is registered in the database and the location of mobile objects such as users and mobile medical devices is calculated based on network measurement. Localization in wireless network relies on several key parameters such as received signal strength indicator (RSSI), carrier signal phase of arrival (POA), and time of arrival (TOA) of the received signal. These parameters are used in estimating distances between the monitoring device and the location sensors deployed inside a building as shown in Fig. 13.3. Algorithms for location estimation in wireless networks have been studied extensively (e.g., [17, 18]).

Location information of CR nodes and medical devices can not only be used in interference measurement but also serve as an input in modeling the spectrum

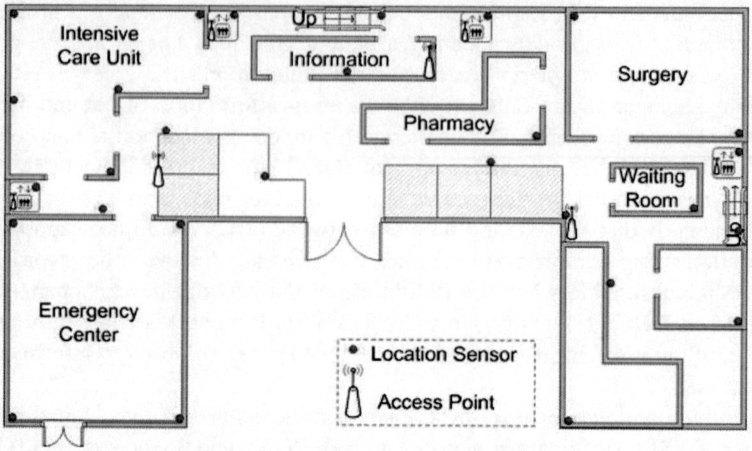

Fig. 13.3 Sample location sensor deployment in one floor of a hospital

availability and assisting spectrum resource allocation among the devices without causing interference to each other. The problem of spectrum sharing among the devices in a non-interfering basis can be modeled as a *graph coloring problem* [19]. *Graph coloring* is a classic problem in *Graph theory*, which involves assigning colors to nodes of an undirected graph with as few colors as possible, such that no adjacent nodes (i.e., nodes with a link between them) share a color. Here, different medical devices represent the nodes in a graph. Two medical devices are adjacent if they are located within a specified proximity of each other. The color used for a node represents the spectrum band that the node can use. As adjacent nodes cannot use the same spectrum bands due to interference, they cannot use the same color.

To illustrate, we consider N overlapping access networks using a graph theoretic model. An undirected graph $G = \{V, E, B\}$ is defined, where V represents the set of health monitoring devices in a region, E represents the set of all undirected edges denoting mutually interfering devices, and B represents the available spectrum band. The spectrum sharing between the devices is then modeled as a **graph coloring** problem [20]. Associated with the spectrum bands is a *spectral efficiency*, which provides a measure of the maximum rate at which data can be transmitted in the bands. This, in turn, provides a means to characterize the maximum number of medical devices that can be simultaneously supported in the system. Traditional graph coloring algorithms only aim to use the minimum number of colors (separate spectrum sets) needed to avoid interference. However, it does not focus on intelligent assignment of spectrum such that spectrum reuse can be maximized resulting in maximization of number of users supported. Admitting maximum number of users incurs minimum cost for each user. Maximizing reuse enables efficient spectrum management among the access networks and admission of larger number of users and thus minimizes cost.

To get rid of the potential drawbacks offered by traditional graph coloring algorithms, an extension of the traditional graph coloring algorithm, called *Utility Graph Coloring* (UGC) [19] based on *cognitive assignment*, can be applied. This new mechanism works in two phases and not only attempts to minimize the number of colors (to create minimum spectrum divisions) but also attempts to maximize the spectrum reuse through cognitive assignment and thus enabling maximum number of users admittance.

In the first phase, the traditional graph coloring algorithm can be applied to determine the number of colors, m, required to color all the vertices. Let $C_1, C_2, \cdots, C_m$ be the m colors required to color all the vertices. This provides the number of divisions into which the available spectrum band needs to be divided. Note that, bandwidth of each division is yet unknown.

In the second phase, the *occurrence* of the colors in the graph is calculated. Let color C_i occur N_i times $(1 \leq i \leq m)$, such that $N_1, +, N_2+, \cdots, +N_m = N$. Then for each of the colors, cognitive assignment parsing is executed to maximize the reuse of the spectrum among the devices subject to various system constraints. The complete UGC algorithm is presented in [19].

To illustrate the UGC algorithm, consider the graph shown in Fig. 13.4, where health monitoring devices (CR1–CR6) operate along with an MRI machine and a surgical machine. Both the traditional graph coloring algorithm and the cognitive assignment allocate three spectrum bands (i.e., three colors), C_1, C_2, and C_3, to the devices. Traditional graph coloring algorithm assigns color C_1 to the surgical machine, CR3, and CR4; color C_2 to CR1, CR2, and the MRI machine and color C_3 to CR2 and CR5. If the devices operate in the 402–406 MHZ band, then the total available spectrum is 4 MHz (i.e., 4000 KHz). When this spectrum is shared by the devices with the concept of proportionality, the traditional graph coloring results in a bandwidth of 1500, 1500, and 1000 KHz corresponding to colors C_1, C_2, and C_3, respectively. Hence, for a typical spectral efficiency of 5 b/s/Hz, traditional graph coloring algorithm provides data rates of 7.5, 7.5, and 5 Mbps corresponding to C_1, C_2, and C_3, respectively. Utility is given by [19]

$$U = 3B(C_1) + 3B(C_2) + 2B(C_3) \tag{13.1}$$

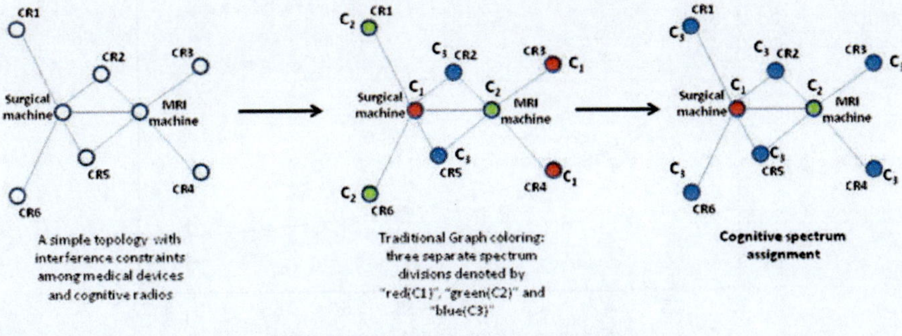

Fig. 13.4 An illustrative example of cognitive spectrum assignment

where $B(C_i)$ is the bandwidth assigned to color C_i. Intuitively, a higher value of utility indicates larger available data rate in the network. In this case, the system utility can be obtained as 11000 units from (13.1). In contrast, when the spectrum bands are allocated in cognitive manner, the surgical machine is assigned color C_1, the MRI machine is assigned color C_2, and CR1–CR6 are assigned color C_3. This corresponds to a bandwidth of 500 KHz corresponding to colors C_1 and C_2 and 3000 KHz corresponding to color C_3. For the same spectral efficiency of 5 b/s/Hz, this cognitive assignment provides data rates of 2.5 Mbps for spectrum bands represented by C_1 and C_2 and 15 Mbps for spectrum band represented by C_3. Note that the maximum data rate provided by the cognitive assignment is 15 Mbps as against 7.5 Mbps provided by the traditional graph coloring algorithm. This is a 100% improvement in the peak data rate which corresponds to 100% improvement in the maximum number of users that can be simultaneously supported. Thus, the UGC algorithm results in reduced cost for each user. The system utility using UGC is $\hat{U} = B(C_1) + B(C_2) + 6B(C_3)$ [19], which yields a system utility of 19000 units, thus resulting in an improvement of 72%. Hence, it is observed that the cognitive assignment maximizes spectrum reuse while avoiding interference among the medical devices and other devices. The UGC algorithm can be further extended to handle multiple system constraints which include fairness constraints like equal distribution of resources among all health care devices or improvement of Jain's fairness index [21].

Another important parameter that measures the data rate in wireless networks is spectral efficiency. This provides the average throughput that can be obtained for each unit of available spectrum. Figure 13.5 presents the improvement in spectral efficiency that can be obtained using the UGC algorithm. It is observed that UGC improves the spectral efficiency by a factor of almost 5. This implies that the UGC

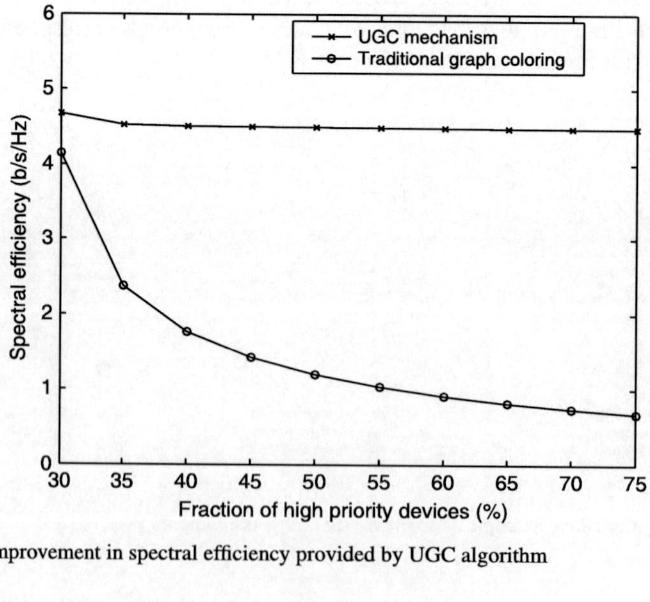

Fig. 13.5 Improvement in spectral efficiency provided by UGC algorithm

algorithm simultaneously supports about five times as many health care devices as that supported by the traditional graph coloring algorithm.

Our approach presents a centralized algorithm for resource management. The large volume of health care devices located in wide geographical regions will require distributed approaches to resource management in addition to the centralized one. In such cases, the graph model of the network depends on the location information of the monitoring devices. As an example, if a patient moves to an MRI room, the location information will be sent to the smart monitoring engine to determine an appropriate spectrum band. Distributed autonomous location aware graph coloring approaches that take into account the location of the monitoring devices in addition to the network characteristics can be considered. The distributed location aware graph coloring is expected to result in increased robustness as there is no single point of failure. The distributed nature of the algorithm is expected to lead to faster convergence and hence better connectivity.

13.3.3 Power Aware Data Compression and Channel Coding

The monitored patient data are transmitted from the CR node through the wireless channel. Some of these vital signs parameters exhibit dependence with each other. As an example, consider a patient whose blood pressure is being monitored remotely. The systolic and diastolic pressures have been shown to exhibit correlation coefficients greater than 0.7 [22]. Similarly, consider a cycle in a typical ECG waveform as shown in Fig. 13.6. The Q, R, and S waveforms of the ECG represent activities of the right ventricle and are shown to exhibit correlations greater than 0.9 [23, 24]. Such dependence between the parameters can be exploited to obtain loss-less compression of data to save the wireless spectrum bandwidth and transmit power of the CR node.

The amount of compression that can be obtained depends on the inherent redundancy in the transmitted data. Redundancy is measured in terms of the entropy or the average information content of a source [25]. Consider two sensors X and Y measuring a patient's systolic and diastolic pressure, respectively. In order to digitally transmit the tuple (x, y) where x is the measured systolic pressure and y is the diastolic pressure, one can separately transmit the two parameters. However, this would be inefficient because it fails to exploit the dependencies between the transmitted parameters. For two sources X and Y with entropies $H(X)$ and $H(Y)$, respectively, and joint entropy $H(X, Y)$, a well known result in information theory is that [25].

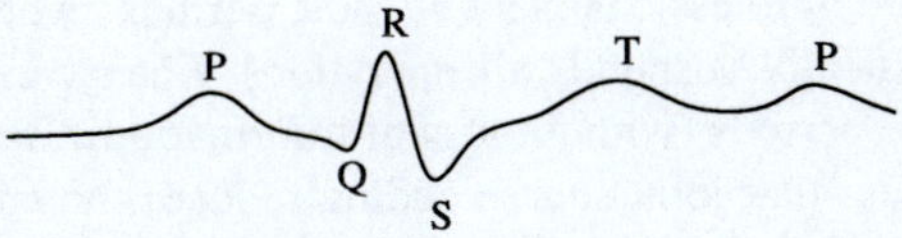

Fig. 13.6 A cycle in a typical ECG wave

$$H(X, Y) \leq H(X) + H(Y) \tag{13.2}$$

with equality if and only if X and Y are independent. Intuitively, the joint entropy $H(X, Y)$ represents the minimum number of bits on an average required to represent the tuple (x, y). $H(X) + H(Y)$ represents the average number of bits required to represent the tuple (x, y) when the two parameters are transmitted individually. Equation (13.2) then indicates that the number of bits required to jointly represent two dependent sources is lesser than that required when representing each source individually. In general, for N sources $X_1, X_2, \cdots, X_N$,

$$H(X_1, X_2, \cdots, X_N) \leq \sum_{i=1}^{N} H(X_i) \tag{13.3}$$

This result is used as a key motivation for joint compression of sources for bandwidth saving. The wireless channel may introduce errors in the transmitted data, which can be corrected by channel coding [25]. Error correction is performed by introducing redundancies in the transmitted data. The complementary requirements of source coding (i.e., minimizing redundancies) and channel coding (i.e., introducing redundancies) can be combined to provide efficient data transmission. In [26], a joint source channel coding mechanism for wireless channels with memory is presented. For sources exhibiting joint correlation greater than 0.7, the quality of the signal can be improved by a factor greater than 50%, which leads to an increase in the bandwidth by similar proportions.

The joint compression of sources is expected to improve utilization of system resources. Consider a patient for whom the ECG and the systolic and diastolic pressures are being monitored. The CR node transmits these parameters to the smart monitoring engine. The CR node thus is equivalent to seven *virtual users* transmitting seven parameters, viz., systolic pressure, diastolic pressure, P, Q, R, S, and T waveforms of the ECG. Using the correlation between the systolic and diastolic pressures and that between the Q, R, and S waveforms, the CR node can then transmit the joint systolic and diastolic pressures, P, QRS, and T waveforms, thus resulting in four virtual sources instead of seven. If the network supported 50 such users without joint source coding, then by exploiting the cross-correlation, the system can support approximately 75 users thus resulting in a 50% increase in the capacity. This is particularly useful during times when epidemics strike because the number of patients is expected to be very large and efficient mechanisms are necessary to reduce the traffic in the wireless medium without losing information.

Every wireless device contains "active periods" while actually transmitting or receiving data over the wireless channel and "idle periods" when it performs channel sensing or periodically suspends all operations. Energy can be conserved by minimizing the active periods (without loss of information). Based on the previous discussion, it is evident that joint source coding reduces the amount of traffic that is to be generated in the wireless medium. This implies that the CR node has lesser volume of data to transmit and hence has a reduced active period. The reduction in the active period automatically results in lesser power consumption and larger energy savings. As an illustration, if a wireless node has an energy capacity of 1 Kb/J (i.e., can transmit 1 Kb for each unit of energy consumed), then, in the absence of joint source coding, a source transmitting at an average rate of 10 Kbps of data with 25% active period consumes 4 J of energy. Using joint source coding with sources having a correlation of greater than 0.8, the energy consumption to 2.5 J thus saving 40% energy or increasing the battery life time by 40%.

Although the health care monitoring devices can save transmit power through compression, they also spend power to perform the computations corresponding to the compression algorithms. Thus, the CR node needs to trade-off between the complexity of the compression mechanism and the power consumption due to wireless transmission. The trade-off depends on the location of the health monitoring device (i.e., the patient) as well as the access network to which the device is connected. The use of the interference avoidance techniques and the information on the patient location can improve the performance of CR and assist in developing location and network aware compression mechanisms.

13.4 Cognitive Radio Testbed for Health Care Automation Network

As a preliminary study, a cognitive radio network testbed is designed and tested transmitting large medical image files over this testbed and traditional WiFi network. This cognitive radio prototype is based on a software abstraction layer implemented over off-the-shelf IEEE 802.11a/b/g stack supported by Atheros hardware chipsets (Orinoco 802.11 a/b/g PCMCIA wireless card). The CR nodes can sense and access frequency bands dynamically in the IEEE 802.11a/b/g network (all 16 channels), thus providing us the ability to study the effectiveness of dynamic spectrum switching. The prototype (consisting three cognitive radio nodes) is presented in Fig. 13.7a. Figure 13.7b displays the spectrum bands chosen by the CR nodes according to the quality of the bands.

(a) (b)

Fig. 13.7 **(a)** Cognitive radio testbed for health care automation network. **(b)** Demonstration of CR node switching frequencies as needed

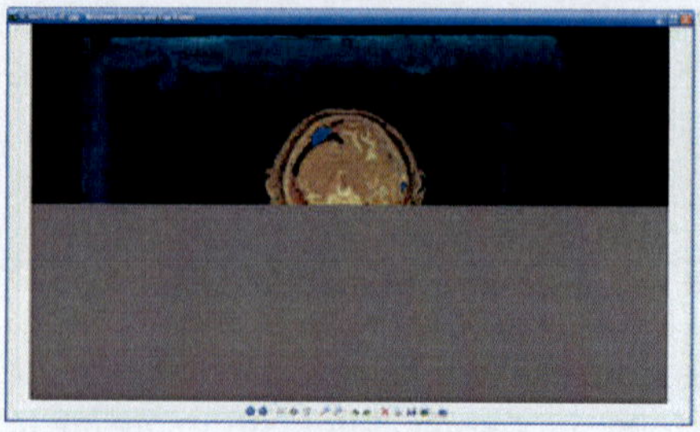

(a)

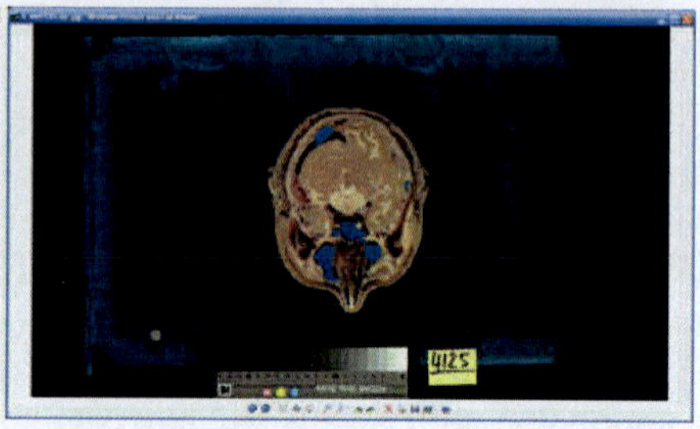

(b)

Fig. 13.8 **(a)** Medical image transfer using standard WiFi under interference; **(b)** Medical image transfer using CR under interference

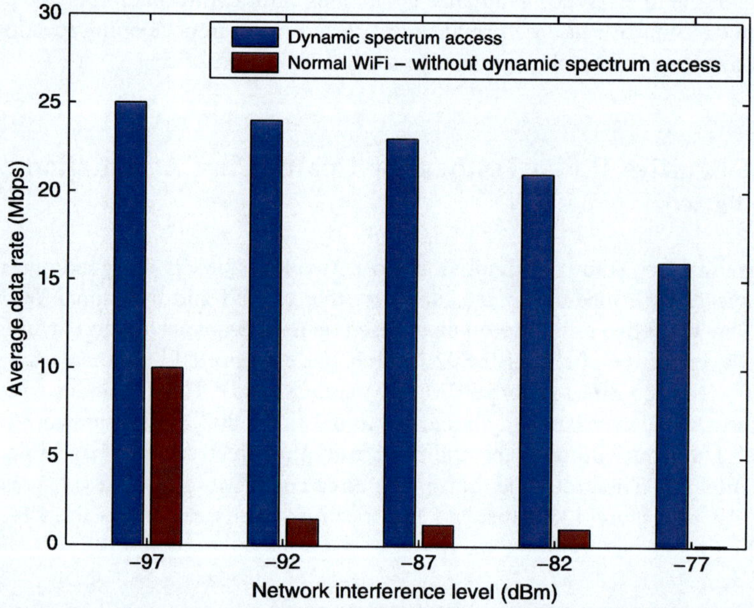

Fig. 13.9 Average data rates achieved by the proposed CR prototype based on dynamic spectrum access and the traditional WiFi network

A large medical image is transmitted over a traditional WiFi network and our CR prototype. The data loss and the transmission rates experienced by the CR prototype and the traditional WiFi system are compared, in the presence of interference. Fig. 13.8a, b show the received image using traditional WiFi and CR, respectively. It is observed that the image transmission over traditional WiFi was terminated due to channel interference (Fig. 13.8a), resulting in an incomplete received image. However, the complete image is received over CR (Fig. 13.8b) because the CR was able to select and transmit over a different transmission channel with a better quality.

Figure 13.9 presents the average data rate achieved by the proposed CR prototype (with DSA capability) and for the traditional WiFi network (without DSA capability) at various levels of network interference. The experiment is conducted to evaluate the performance of the proposed system in comparison to the traditional WiFi access networks where dynamic spectrum access is not available. The results are plotted against different network interference levels (dBm). As evident from the figure, the proposed CR system demonstrates significantly higher data rates.

13.5 Conclusion

This chapter introduced the design of a health care automation network based on cognitive radio (CR) infrastructure. Some of the technical challenges in the design of such systems, such as location sensing, interference avoidance, and spectrum allocation among devices, are discussed. A CR testbed for preliminary research is presented. The effectiveness of CR based dynamic spectrum access was demonstrated, in terms of the quality of the received medical data and increased data rates. The potential impact of this system can be multi-fold. It can reduce health care cost by better utilization of limited health care resources. The system utilizes the existing infrastructure with minimal additional requirements. It provides non-intrusive monitoring for the patients and the elderly due to ubiquitous wireless connectivity. The system can also be designed to efficiently track the location of the patients and elderly to provide timely medical services in an emergency.

Acknowledgment This research was partially funded by NSF # 0917008 and NSF # 0916180 and partially funded by 2009-92667-NJ-IJ.

References

1. Jane An, Romy Saloner, Rebecca Tisdale, and Usha Ranji, U.S. Health Care Costs, Kaiser Family Foundation. Updated: July 2009. [Online]. Available: http://www.kaiseredu.org.
2. K. Kinsella and V. Velkoff, An Aging World: 2001, U.S.C. Bureau, U.S. Government Printing Office, 2001. [Online]. Available: http://www.census.gov/prod/2001pubs/p95-01-1.pdf.
3. Table 094. Midyear population, by age and sex. International database 2004. [Online]. Available: http://www.census.gov/.
4. K. Levit, et al., Trends in U.S. health care spending 2001. *Health Affairs*, Issue 22, No. 1, pp. 154–164, 2003.
5. [Online]. Available: http://www.nonin.com/.
6. [Online]. Available: http://www.eecs.harvard.edu/~mdw/proj/codeblue/.
7. V. Shnayder, B. Chen, K. Lorincz, T.F. Jones, M. Welsh, "Sensor Networks for Medical Care", Technical Report TR-08-05, Division of Engineering and Applied Sciences, Harvard University, 2005.
8. D. Malan, T.F. Jones, M. Welsh, S. Moulton, "CodeBlue: An Ad Hoc Sensor Network Infrastructure for Emergency Medical Care", Workshop on Applications of Mobile Embedded Systems (WAMES 2004), Boston, MA, 2004.
9. [Online]. Available: http://www.sensatex.com/.

10. [Online]. Available: http://www.vivometrics.com/.
11. [Online]. Available: http://news.thomasnet.com/fullstory/801186.
12. P. Henry, Tech View: Cognitive Radio for Multimedia Home Networks [Online]. Available:http://www.research.att.com.
13. J. Mitola, G.Q. Maguire Jr., "Cognitive radio: making software radios more personal", *IEEE Personal Communications*, vol. 6, no 4, pp. 13–18, Aug. 1999.
14. K. Tan and C. Law, "GPS and UWB integration for indoor positioning", *6th International Conference on Information, Communications and Signal Processing*, pp. 1–5, Singapore, Dec. 2007.
15. FCC Rules and Regulations, "MICS Band Plan", Table of Frequency Allocations, Part 95, Jan. 2003.
16. R. Perera, Z. Dong, R. Chandramouli, "Method and Apparatus for Estimating Geographic Location of Internet Hosts", U.S. Provisional Application No. 61/164,578.
17. N. Patwari, A. O. Hero, III, M. Perkins, N. S. Correal, and R. J. O'Dea, "Relative Location Estimation in Wireless Sensor Networks", *IEEE Transactions on Signal Processing*, vol. 51, no. 8, pp. 2137–2148, Aug. 2003.
18. L. Doherty, K. Pister, and L. El Ghaoui, "Convex optimization methods for sensor node position estimation", *IEEE Infocom 2001*, Anchorage, Alaska, April 22–26, 2011.
19. S. Sengupta, S. Brahma, M. Chatterjee and S. Shankar, "Enhancements to cognitive radio based IEEE 802.22 air-interface", *IEEE International Conference on Communications (ICC'2007)*, Glasgow, Scotland, UK, June 24–28, 2007.
20. T.H. Cormen, C.E. Leiserson, R.L. Rivest, and C. Stein, "Introduction to Algorithms", 2nd Edition, The MIT Press and McGraw-Hill, Cambridge, Massachusetts, 2001.
21. R. Jain, "The Art of Computer Systems Performance Analysis: Techniques for Experimental Design, Measurement, Simulation and Modeling", Wiley, New York, NY, ISBN 0-471-50336-3.
22. F. Molinari, M. Eichinger, F. Risse, C. Plathow, M. Puderbach, S. Ley, F. Herth, L. Bonomo, H. Kauczor, and C. Fink, "Navigator trigerred oxygen enhanced MRI with simultaneous cardiac and respiratory synchronization for the assessment of interstitial lung disease", *Journal of Magentic Resonance Imaging*, vol. 26, pp. 1523–1529, 2007.
23. P. Guzik, J. Piskorski, T. Krauze, R. Schneider, K. H. Wessling, A. Wykretowicz, and H. Wysocki, "Correlations between the poincare plot and conventional heart rate variability parameters assessed during paced breathing", *Journal of Physiological Sciences*, vol. 57, no. 1, pp. 63–71, Feb. 2007.
24. T. Last, C. D. Nungent, and F. J. Owens, "Multi-component based cross-correlation beat detection in electrocardiogram analysis", *Biomedical Engineering Online*, Jul. 2004. [Online]. Available: http://www.biomedical-engineering-online.com/content/3/1/26.
25. T. M. Cover and J. A. Thomas, "Elements of Information Theory." Wiley, Hoboken, NJ, 2006.
26. M. A. Haleem and K. P. Subbalakshmi, "Optimal source-channel decoders for correlated Markov sources over additive Markov channels," *Proc., Data Compression Conference (DCC'2007)*, Snowbird, Utah, USA, March 27–29, 2007.

Chapter 14
Interoperability Between IEEE 802.11e and HSDPA: Challenges from Cognitive Radio

Orlando Cabral, João M. Ferro, and Fernando J. Velez

Abstract In this chapter we propose a scenario for interoperability between high-speed downlink packet access (HSDPA) and Wi-Fi. This scenario involves the end-user traveling in a public transportation system and requesting multimedia services to the operator. The interoperability between HSDPA and Wi-Fi (IEEE 802.11e standard) radio access technologies (RATs) is first addressed, a topology in which the user has access to both RATs was considered, together with a common radio resource management (CRRM) to manage the connections. We reached the conclusion that the CRRM enables to increase the system throughput when the load thresholds are set to 0.6 for HSDPA and 0.53 for Wi-Fi. Then, spectrum aggregation is implemented in HSDPA. A resource allocation (RA) algorithm allocates user packets to the available radio resources (in this case Node Bs operating at 2 and 5 GHz are available) in order to satisfy user requirements. Simulation results show that gains up to 22% may be achieved. We have also sought the most efficient way to manage routing packets inside the Wi-Fi network. The proposal which uses links with higher throughputs enables to reach the best results, with gains up to 300% in the packet delivery ratio. Finally, we discuss the challenges that need to be addressed in order to materialise the envisaged cognitive radio scenario in public transportation.

14.1 Introduction

In this chapter we address a network of wireless networks consisting of a backbone infrastructure provided by high-speed downlink packet access (HSDPA) radio towers hierarchically bonded with a flexible wireless ad hoc network running IEEE 802.11e standard. The purpose of this hybrid network is to serve users with high-quality video and audio content, using the already existing infrastructure. We first discuss the interoperability between HSDPA and Wi-Fi and provide a brief presentation of the simulator built for this study. Then, we discuss spectrum aggregation, and

O. Cabral (✉)
Instituto de Telecomunicações, Covilhã, Portugal
e-mail: orlandoc@ubi.pt

F.R. Yu (ed.), *Cognitive Radio Mobile Ad Hoc Networks*,
DOI 10.1007/978-1-4419-6172-3_14, © Springer Science+Business Media, LLC 2011

some results obtained by that simulator in an HSDPA/HSDPA scenario. Two Node Bs (NBs) operating at 2 and 5 GHz are available, and the operator automatically switches the user between them. In the context of the Wi-Fi network, the effects of changing the routing metrics in the service quality for an ad hoc network are studied, mostly the number of packets delivered and latency. This study manages to find the best approach, which is used to compute the paths for the hierarchical HSDPA/Wi-Fi scenario that is being suggested and will be simulated in the future.

14.2 HSDPA/Wi-Fi Interoperability

In recent years, cooperation has been gaining an increased interest in the context of wireless networks. The definition of wireless mesh networks (WMN) and the upcoming standard IEEE 802.11s demonstrates the interest in this type of networks [1]. The cooperation between heterogeneous networks is also an hot topic, like demonstrated by several IST projects such as CAUTION [2] or AROMA [3].

In our scenario, the mesh network involves the use of two different radio access technologies, the HSDPA and the Wi-Fi ones. In this section, we present our study on cooperation and coexistence for these networks.

14.2.1 Interoperability Between HSDPA and Wi-Fi

To study the cooperation between HSDPA and Wi-Fi (IEEE 802.11e) Radio Access Technologies (RATs), we use a common radio resource management (CRRM) algorithm for RAT selection, based on the load, between these two technologies in a common coverage area. The scenario under study is presented in Fig. 14.1.

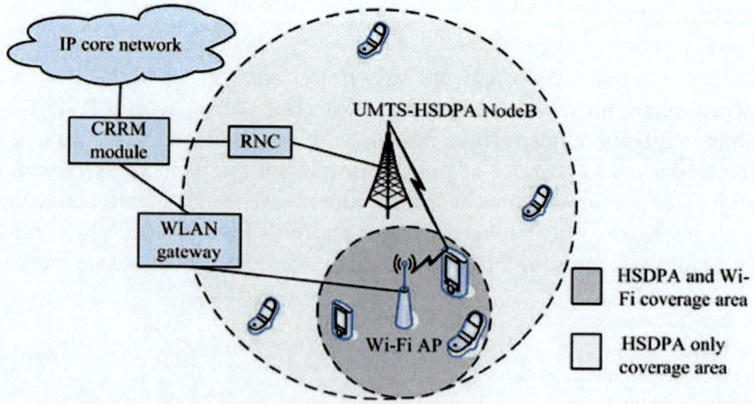

Fig. 14.1 Area covered by HSDPA and Wi-Fi in the interoperability scenario

An IP-based core network is assumed to act as the bridge between Wi-Fi and HSDPA. Within this bridge, a cooperative networking entity that logically communicates with HSDPA and Wi-Fi (referred as the CRRM entity) is responsible for:

- gathering system and user specific information;
- processing this information according to operator-specific criteria;
- triggering a new handover events according to the load balancing criteria.

Moreover, it is assumed that a common operator deploys both systems or those systems from different operators share a service level agreement.

The scenario addresses the delivery of near real-time video (NRTV) services that can be streamed either over the HSDPA or Wi-Fi systems. The end user is currently subscribing to an IPTV service, which is also currently being delivered over the Wi-Fi hotspot. This initial connection was chosen since it was deemed to be the most "fitting" network for the requested service. An example is shown in Fig. 14.1. The operator, which is monitoring both networking entities, observes a sudden surge in Wi-Fi subscribers overloading the Wi-Fi network, while UMTS is under-loaded and handling the usual voice services. The CRRM entity may decide that it would be more efficient to move some of the Wi-Fi users to the UMTS-HSDPA network, since this leads to better QoS provisioning and exploits the existing network capacity in a more efficient way. As a consequence, the CRRM triggers a series of handover events that ensure an even load distribution across both networks. When a user is triggered for handover, the multi-mode terminal initiates a new connection with UMTS-HSDPA, terminating the existing connection with the Wi-Fi system. The handover events occur in a seamless manner.

14.2.2 Simulation Results

The interoperability scenario is based on a test field covered by the HSDPA and Wi-Fi technologies, assuming high-priority NRTV video traffic characterised by the 3GPP model [4] at 64 kbps. The generation of NRTV calls is modelled by a Poisson distribution, while the call duration is exponentially distributed with an average of 180 s. The scenario is deployed in our custom-made simulator created by the IST-UNITE project [5]. Details on the HSDPA simulator structure and features are presented in [6], while details for the IEEE 802.11e package are given in [7]. The main simulation parameters are presented in Table 14.1. Since the Wi-Fi network supports

Table 14.1 Values for the simulations

Parameter	HSDPA	Wi-Fi
Mode	FDD (Tx mode)	EDCA (MAC Tx mode)
Scheduler	MaxCI	Round-robin
Link Adaptation	BLER max = 10%	Similar to [8]
Radio propagation model	3GPP indoor + FF	ITU 2 GHz propagation (Path Loss)
Cell type	Omni directional	Omni directional
Number HS- PDSCH (data codes)	15	–
Bandwidth	5 MHz	Variable with the user SNR
Initial number of users	20	11

a larger throughput than HSDPA, we started it with four FTP, four voice and three NRTV users. This enables to load the network from the beginning of the simulation and reach a stable point without having an excessively long simulation time.

Users are distributed uniformly in the area of HSDPA coverage. This area is larger than the Wi-Fi zone. The HSDPA cell radius (R) is variable so that, in the best case situation, Wi-Fi covers 50% of the HSDPA area, while in the worst case it covers just 13% of it. It is assumed that NRTV users prefer to use HSDPA. If the load surpasses a given threshold the value of the user suitability is calculated to help on RAT selection decisions. If a user is more suited to be within the Wi-Fi, then the CRRM entity may decide to move him to Wi-Fi, depending on the coverage. To analyse the benefits obtained by having CRRM procedures, two scenarios were considered:

- No vertical handover;
- The users position is fully known and only the HSDPA users that are within Wi-Fi zone may be switched (from HSDPA to Wi-Fi).

All the results obtained are presented with a 95% confidence interval. Figure 14.2 presents results for the throughput without considering the CRRM entity, as a function of the total number of users for a cell radius of 50 m (R = 50 m). The over the air (OTA) throughput represents the number of bits that have been transmitted in the cell, the service (Serv) throughput represents the number of bits that have been transmitted and correctly received (without packet errors), and the quality-of-service (QoS) throughput represents the number of bits correctly received (without packet errors) within the allowed delay. Details on the formulation of these evaluation metrics can be found in [9].

After 37 users (11 within Wi-Fi and 26 within HSDPA), the QoS throughput starts to decrease. The number of unsatisfied users for the NRTV is given by:

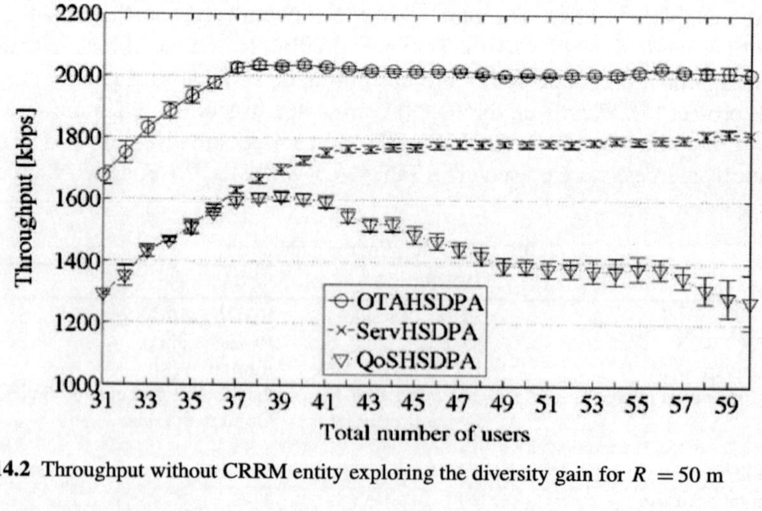

Fig. 14.2 Throughput without CRRM entity exploring the diversity gain for R = 50 m

$$\frac{64 \times number\ of\ users - QoS_{\mathrm{throughput}}}{64} \tag{14.1}$$

When 60 users are active (11 in Wi-Fi and 49 in HSDPA), 28 users (approximately 57%) are unsatisfied, since they are being served with poor quality, e.g., long queueing time.

The objective is to reach the optimal load balance between the two RATs. By analysing the values of the load when the QoS starts to decrease in each system, we get to the conclusion that the most appropriate load thresholds are $LTh_0 = 0.6$ for HSDPA (RAT 0) and $LTh_1 = 0.53$ for Wi-Fi (RAT 1). Figure 14.3 presents the results for the throughput in HSDPA for $LTh_0 = 0.6$.

Other values were considered for the load threshold. However, using $LTh_0 = 0.7$ (with results presented in Fig. 14.4) and $LTh_0 = 0.8$ (with results presented in Fig. 14.5) resulted in users exceeding the QoS delay threshold, i.e., 300 ms. For example, considering $LTh_0 = 0.8$ and 60 users, around 350 kbps are delivered above the delay threshold, corresponding to 10% of unsatisfied users.

In Wi-Fi, the service class that suffers the most degradation by adding NRTV users is the background one, since it is the one with less priority. The delays suffered either by the voice or by the video service classes are always lower than the respective thresholds specified in the literature, i.e., 30 ms for voice and 300 ms for video. For the background application, we considered a delay threshold of 10 s. Our results show that this delay threshold is overcome when there are more than 13 users in the Wi-Fi system, as shown in Fig. 14.6. This corresponds to a Wi-Fi load threshold of $LTh_1 = 0.53$.

Figure 14.7 compares the overall QoS throughput with the offered load in the presence and absence of the CRRM. It increases with the offered load until the HSDPA system capacity is reached. By taking the characterisation of the available

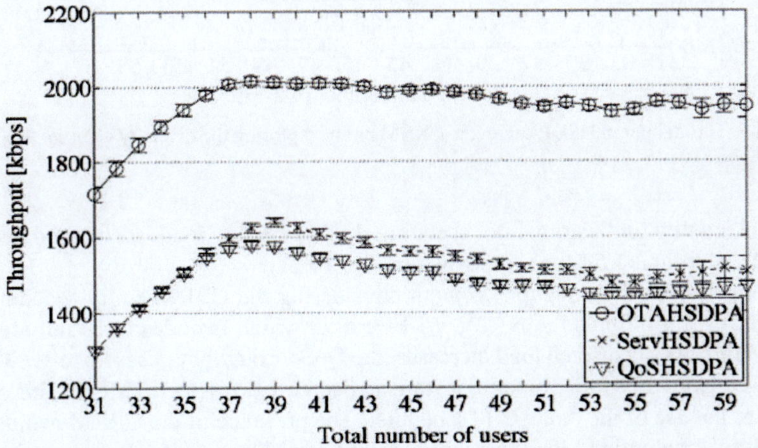

Fig. 14.3 Throughput in HSDPA with the CRRM entity exploring the diversity gain, for $R = 50$ m

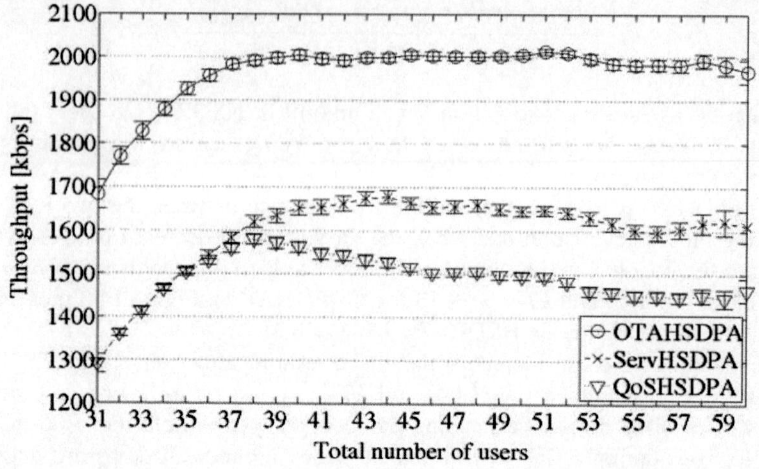

Fig. 14.4 Throughput in HSDPA with the CRRM entity exploring the diversity gain, for $R = 50\,\mathrm{m}$ for $LTh_0 = 0.7$

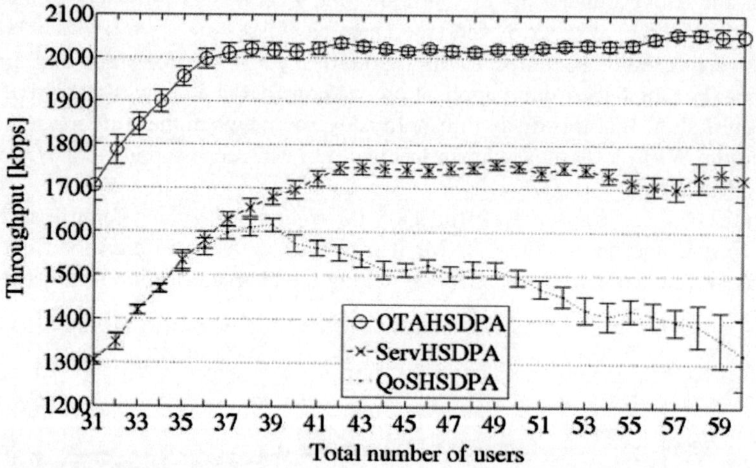

Fig. 14.5 Throughput in HSDPA with the CRRM entity exploring the diversity gain, for $R = 50\,\mathrm{m}$ for $LTh_0 = 0.8$

channel quality indicators into account, the maximum load accommodated by HSDPA is around 1.6 Mbps (as shown in Fig. 14.2).

In Fig. 14.7, the throughput without considering the CRRM entity accounts for the HSDPA traffic plus the NRTV Wi-Fi traffic, which includes three initial users from Wi-Fi. As the offered load increases, the QoS throughput raises up to 1.7 Mbps. However, from this point forward, it starts to decrease down to 1.5 Mbps. This effect is due to the use of the "max C/I" scheduler. The presence of the CRRM avoids this decrease. If the optimal load threshold is reached ($LTh_0 = 0.6$), by comparing the throughput between the presence and the absence of CRRM, the observable gain

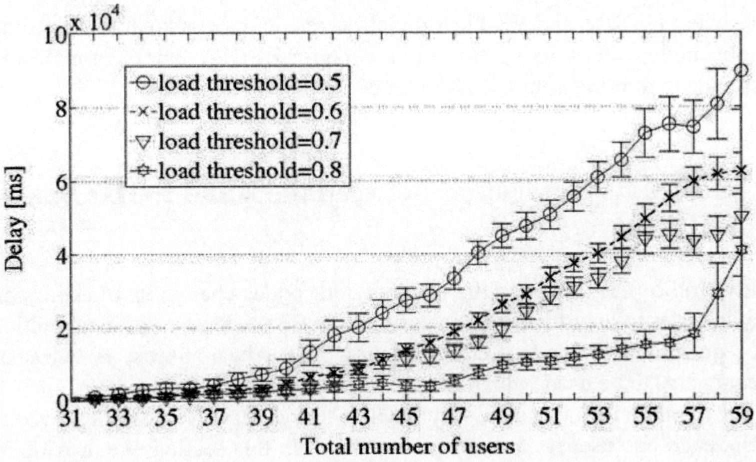

Fig. 14.6 Delay for background traffic as a function of the total number of users, for several values of load threshold

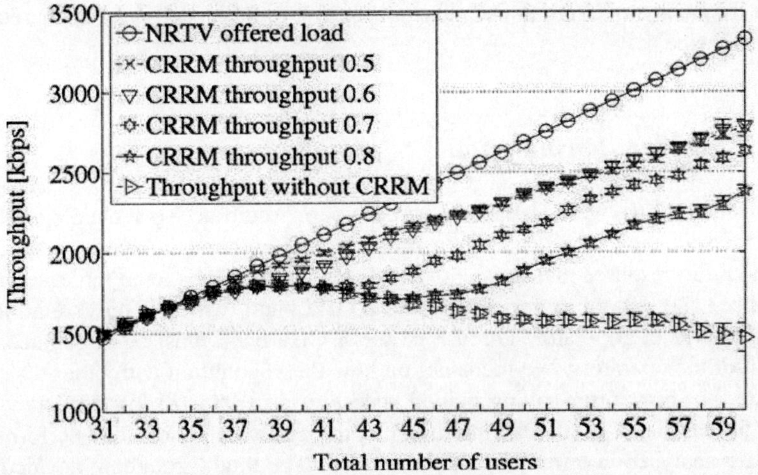

Fig. 14.7 Comparison of the total throughput with and without the CRRM entity exploring the diversity gain.

is 2700/1500 = 80% (at 60 users). The use of lower load thresholds is not advised, since it causes the Wi-Fi system to overload "faster," causing problems to the background traffic.

14.2.3 Lessons Learned from RAT Selection

In this section we found the optimal load threshold that maximises the total QoS throughput for the cooperation between two wireless networks of different technologies: HSDPA and Wi-Fi. Simulation results have shown that optima load thresholds in this interworking scenario are 0.6 for HSDPA and 0.53 for Wi-Fi. This corresponds to a throughput CRRM gain of 80%, with 60 users.

14.3 Spectrum Aggregation Between the 2 and 5 GHz Bands in HSDPA

Cognitive radio (CR) refers to the way network nodes change its transmission frequency in order to avoid interference or to increase the amount of bandwidth it can use. It requires that the node is equipped with the hardware necessary to transmit in the frequencies to be used, which is paid back in terms of QoS.

In our hypothetical scenario, we consider that several NBs are deployed in the scenario, each one uses its own frequency band. In this section, we provide results for testing a topology just considering two NBs. In this scenario, an operator not only has an exclusive license at 2 GHz but also has access to the 5 GHz band. The latter is shared with other operators. The operator's management entity defines which frequency to assign to each user, according to its capacity and users requesting the service.

14.3.1 Problem Formulation

In our simulations, we consider the same type of radio access technology (RAT) for both frequency bands, trying to prove that integration of dynamic spectrum use and radio resource management (RRM) techniques leads to an increase of performance. The system prefers to use the 2 GHz band, which is the one allocated exclusively to the operator. The use of the 5 GHz band must be negotiated with the rest of the operators. We focus not on how the negotiation with other operators is made, but how, after having gained access to a certain (fixed) frequency pool in the 5 GHz band, to allocate the users in the available bands. The performance gains are analysed in terms of data throughput. The total throughput is calculated as a function of the radio channel qualities for each user in the considered bands. The channel quality indicator (CQI) depends on the path loss, which depends not only on the distance from the NB but also on the carrier frequency adopted. The operator applying Multi-Band Scheduling (MBS) will reach considerable improvements when the users have heterogeneous spatial distribution in the cell (variable distances from the NB) and different channel qualities in the considered spectrum bands.

The problem of scheduling the users into two bands (2 GHz and part or all of the frequency pool at 5 GHz) can be formulated as an integer programming (IP) optimisation problem [10]. The profit function (PF) to be maximised via a single objective maximisation problem is the total throughput of the operator (fairness and QoS requirements of the service classes are not considered). However, multiple objectives can be easily introduced and implemented in the problem, such as maximising the total throughput while minimising the QoS satisfaction indexes for each service class [11]. Solving multiple objectives general assignment problem (MO-GAP) can be very difficult and usually the objectives are combined together via a linear combination, called "scalarisation" [12]. The GMBS problem can be solved considering users with the added capability of transmitting and receiving in multiple frequency simultaneously (the user equipment has multiple transceivers) or when it can just chose one band among all the bands in the network and choosing one of the transceiver configurations available at the radio. In both cases, we have a formulation of an IP problem with different constrains, as described in [13].

14.3.2 System Modelling

The resource allocation (RA) component is responsible for allocating the available radio resources to the user traffic in a effective manner and includes a scheduling mechanism, link adaptation, code allocation policy, and an hybrid automatic repeat request (H-ARQ) scheme to improve service throughput for users at the cell edge. The HSDPA network was simulated using the simulator described in Section 14.2.2, with the following functionalities and characteristics:

- Multi omni-directional cell deployment model, hexagonal cells, consisting in three tiers, for interference purposes (results are presented only for the centre cell);
- NRTV streaming traffic model from [4] at 64 kbps;
- RRM schemes, including of adaptive modulation and coding (AMC), n-parallel channel H-ARQ using chase combining and round-robin scheduling algorithm;
- ITU-based radio propagation models: the channel loss between the user and the NB is modelled by path loss, shadowing loss by log-normal distribution and fast fading using approximated Jakes model [14];
- The interference in the user is calculated considering the signal strength received from the neighbour NBs and the thermal noise;
- The simulator as an input uses a BLER table provided by the link layer simulations of [15].

Each time transmission interval (TTI) is associated with a sub-frame duration that corresponds to an HSDPA frame duration of 2 ms, with three time slots of 0.67 ms each. The HSDPA physical layer [16] provides 15 orthogonal codes available for data transmission within a sub-frame. The available data rates are summarised in Table 14.2. For each CQI identifier, the modulation scheme, the block sizes, the number of transport channels, and the transport rate are given. The CQI is a mapping of the averages of the signal-to-interference ratio (SIR) recorded over time. The direct mapping between CQI_{bu} and the $R(CQI_{bu})$ can be expressed as:

$$R(\mathrm{CQI}_{bu}) = \begin{cases} 188.5 \text{ } if \text{ } CQI_{bu} = 5 \\ 198.0 \text{ } if \text{ } CQI_{bu} = 8 \\ 331.9 \text{ } if \text{ } CQI_{bu} = 15 \\ 716.8 \text{ } if \text{ } CQI_{bu} = 22 \end{cases} \tag{14.2}$$

Table 14.2 Transport block size and bit rate associated to CQI

CQI	Modulation	Transport block size [bits]	Number of HS-PDSCH	R(CQI) [kbps]
CQI 5	QPSK	377	1	188.5
CQI 8	QPSK	792	2	396.0
CQI 15	QPSK	3319	5	1659.5
CQI 22	16-QAM	7168	5	3584.0

14.3.3 Resource Allocation

The RA algorithm allocates the user packets to the available radio resources in order to satisfy the user requirements and to ensure efficient packet transport to maximise spectral efficiency. The RA is an entity within the set of RRM algorithms that should have inherent tuning flexibility to maximise the spectral efficiency of the system for any type of traffic QoS requirements. The adopted RA algorithm maps packets of variable size into variable length radio blocks for transmission over the PHY layer, and the length is dependent on the channel quality. The following events are performed:

1. User packets awaiting transmission are prioritised according to the scheduling algorithm criteria;
2. A CQI identifier is selected according to the link adaptation algorithm, using the available CQI options from the PHY layer;
3. The scheduler calculates the number of MAC transport blocks required to transmit the scheduled packet. The number of channels is calculated according to:

$$\text{Number of blocks} = \left\lceil \frac{\text{PacketSize}}{\text{BlockSize(CQI)}} \right\rceil \tag{14.3}$$

   where $\lceil x \rceil$ is the lowest integer higher or equal to x;
4. An idle ARQ channel j is selected to hold and manage the ARQ transmission;
5. The packet is transmitted and received at the user equipment. Soft retransmissions are combined with previous packet transmissions (chase combining) and the ARQ messages are generated accordingly. These are then signalled to the NB, and the ARQ processes are released if the messages are positive acknowledgements (ACKs).

14.3.4 2 and 5 GHz Usage Under a CRRM Approach

The 2 and 5 GHz frequency bands are characterised over the same HSDPA architecture by assuming the model summarised in Table 14.3, which is according to [17].

By adopting a CRRM [18] to manage the extra resources, we propose to use HSDPA at 2 and at 5 GHz. The CRRM entity keeps track of the CQI in both frequencies. This scenario is presented in Fig. 14.8, where the darker areas represent better coverage.

14.3.5 Results

For the results we focus on the throughput of the network, which is the total number of bits that have been transmitted and correctly received by all users in the cell. Users are displaced in the cell with an uniform distribution within a distance of 900 m so that both frequencies can cover the whole cell. The NRTV calls are modelled by a Poisson distribution, the call duration is exponentially distributed with an average of 180 s. The bars on the curves represent a 95% confidence interval, which is achieved by running at least 50 simulations for each case. In some cases, additional runs were made to achieve the required accuracy.

Figure 14.9 shows the results for the throughput without MBS. In this case, the two bands are managed separately, the call requests are divided into the two bands. It is not possible to switch a service from one band to the other. The "Overall Serv"

Table 14.3 Parameters and models used for 2 and 5 GHz bands

Carrier frequency	2 GHz	5 GHz
Bandwidth	5 MHz	5 MHz
Path loss model:	$128.1 + 37.6\log(d_{[\mathrm{km}]})$	$141.52 + 28\log(d_{[\mathrm{km}]})$
Shadowing decorrelation length	5 m	20 m

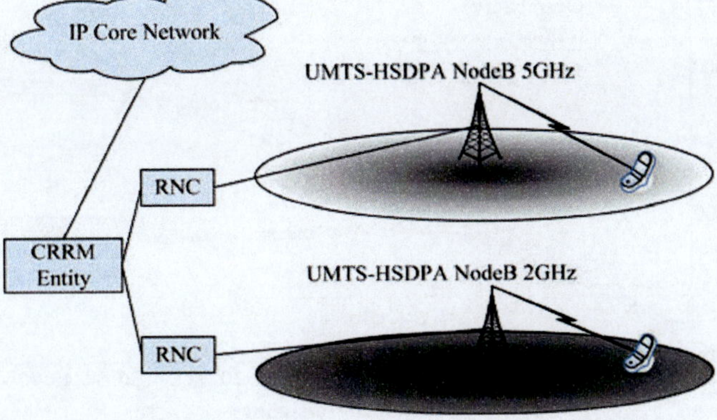

Fig. 14.8 CRRM in the context of two separated frequencies

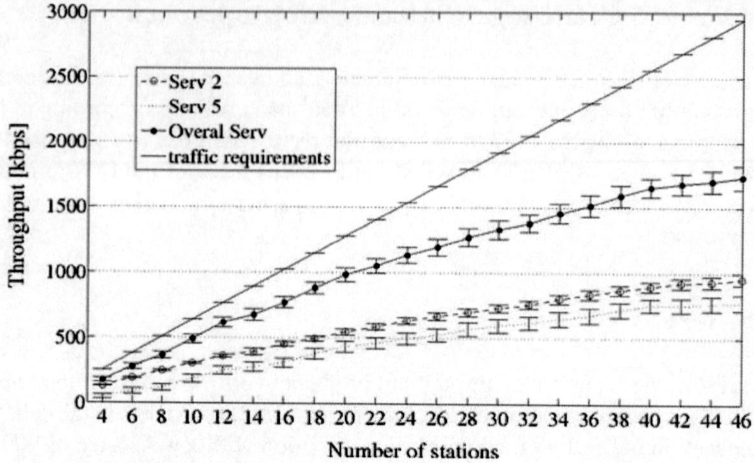

Fig. 14.9 Average throughput without MBS

throughput is the sum of the service throughput in both frequency bands. The traffic requirement is the traffic required to satisfy all the users (i.e., the NRTV required rate – 64 kbps – multiplied by the number of users in the system). From the number of users presented in the abscissas, half of them are operating in the 2 GHz and the others in the 5 GHz band.

Figure 14.10 shows the results when the MBS algorithm is applied. The CRRM entity can decide on which band the user can be allocated in order to maintain control over the network resources. After 18 users, as the 2 GHz band has reached a threshold for the load, the MBS algorithm handles the user allocation.

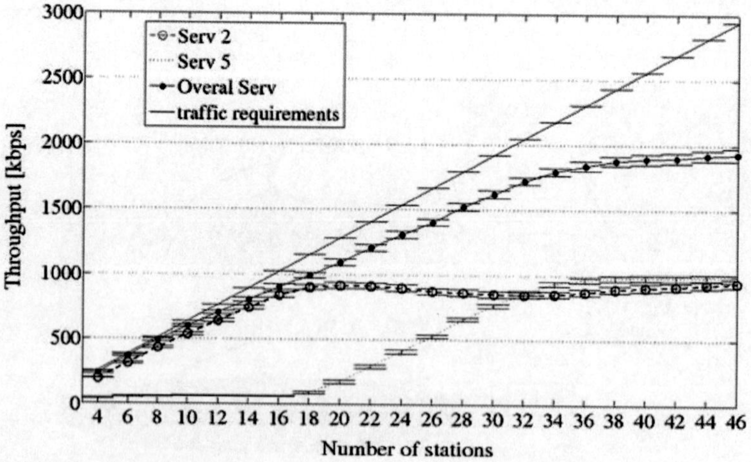

Fig. 14.10 Average throughput with MBS

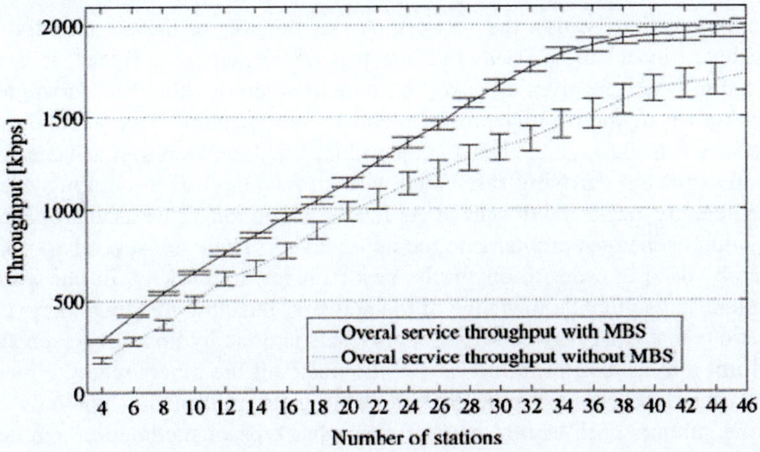

Fig. 14.11 Service throughput in the presence and absence of MBS

When the MBS is applied over the two bands, a higher throughput is achieved due to the switching of the users based on their respective channel qualities. The curves from Fig. 14.11 enable a comparison between the results with and without the use of MBS. It is shown that the MBS algorithm increases the performance of the system. This is more evident when more than 24 stations are in the scenario topology, corresponding to a gain up to 400 kbps. This happens because, up to a given load, both bands can deal with the required traffic. However, after this value for the load, retransmissions start to make the difference.

14.3.6 Summary and Conclusions

In this section we analysed how CR can be used in the context of our work. The use of the same RAT with widely separated spectrum resources can increase the system performance without generating interference and obtaining a gain of 22%. This gain was obtained by optimally managing radio resources based on a integer optimisation procedure, where inputs come from active sensing on all the probable useful spectrum and past receptions. This, however, occurs at the cost of some additional resources, a common radio resource management entity that manages resource, and compatible radios at the user side.

14.4 IEEE 802.11e Ad Hoc Networking

In this section, we present several approaches to determine the best path for packets flowing in an IEEE 802.11e ad hoc network. Unlike the infrastructure mode, the IEEE 802.11e ad hoc mode does not use an entity responsible for managing the communications within the network. As so, the role of the access point (AP) is distributed by all the stations that are part of the network. Hence, it is up to each station which receives a packet to know to where it should be forwarded to. This is known as the routing problem, and its optimisation is very important so that packets flow through the best path possible, enabling them to reach destination efficiently (meaning arriving safely and with a low delay). If the routing does not work efficiently, packets will start to get lost or taking longer paths and/or entering loops, which generates problems to the upper layers. There are several approaches that can be used in order to obtain the best path for the packets. In our work, all the stations have global knowledge of the network, meaning that they know all the connections that exist between the stations. This is done by impelling each station to perform a transmission in full power and make all the others read the received signal-to-interference-noise-ratio (SINR). This gives information not only about which neighbours each station has but also what type of modulation and coding schemes can be used in each link. We use that information to compute the routing table for all stations, whose entries indicate the next hop for each destination.

In our study, we focus not on how the information about the links quality is spread across the network, but on the best way to, given a table of the connections between all the stations, determine which path the packets should take to arrive promptly at the destination. For this purpose, we address the optimisation problem in two different ways: the empirical approach and the Genetic Algorithm (GA) one. In the empirical approach, we manually define the cost functions for each link. In turn, GAs are an optimisation approach that enables to reach the optimal solution without having to study the entire space of possible solutions.

All the results presented in this section were obtained by running the simulations in the custom-made IEEE 802.11e simulator from [19]. This simulator was developed by the Instituto de Telecomunicações – Covilhã Delegation team and tightly follows the amendment "e" of IEEE 802.11 standard, which embraces the support for quality-of-service in these networks. QoS support is achieved by mapping all packets into one of the four existing access categories (VI, VO, BK, and BE) and unequally treating them according to this classification. This means that the MAC layer scheduling will give, for example, priority to a packet marked as video over another one marked as background.

14.4.1 Empirical Approach

In our first approach, we manually assign a cost to each link. Then, Dijkstra's Algorithm (DA) [20] is run in a table containing the cost of all links, which allows for getting the least-cost path from each station to all the others. This information is inserted in the routing table which, for each pair source/destination, indicates the ID of the "next hop." For the first tests we wanted to verify how the signal strength in each link could be used to compute the best path. If a link has a strong signal, it is more robust to the outside interference, and it can use a modulation (and coding rate) that allows for transmitting at a higher data rate, allowing for faster transmissions,

Table 14.4 The relation between the SINR, modulation, and data rate

Mode	Modulation	Code rate	Min. SINR	Link throughput [Mbit/s]
1	BPSK	1/2	4.1	6
2	BPSK	3/4	–	9
3	QPSK	1/2	7.9	12
4	QPSK	3/4	11.0	18
5	16-QAM	1/2	14.8	24
6	16-QAM	3/4	17.8	36
7	64-QAM	2/3	22.8	48
8	64-QAM	3/4	24.2	54

as shown in Table 14.4. However, strong signal can only be achieved if the stations are close to each other, which means that the overall progress of the packet towards the destination will be slower than if a longer link was used. Of course, the use of a longer link implies a lower data rate, and since the signal is weaker, more interference will exist.

In order to define if the best approach was to use longer, robuster, or intermediary links, we proposed the following cost function:

$$cost = abs\,(data\_rate \ - set\,point) \tag{14.4}$$

The setpoints used are presented in Table 14.5.

Three types of traffic sources were chosen. The traffic source parameters are presented in Table 14.6, as well as the access categories (AC) of each type of traffic. Seven traffic streams were put in the scenario, one VI, three VO, and three BK streams. VI and BK traffic are unidirectional, while VO is bidirectional.

Simulations were run for a random deployment of 30 stations on a $150 \times 150\ \text{m}^2$ field. The obtained results for a simulation time of 15 s are summarised in Table 14.7. These results are only for the video traffic that is being generated in station 3 and has station 10 as the destination.

From the results, one can conclude that the approach which privileges links with higher SINR allows to deliver more packets and with a lower latency.

Table 14.5 The setpoints in study

Function	Setpoint [Mbit/s]	Privileges
Alfa	$(data\_rate + 1)$	Paths with less hops/longer links
Beta	21	Intermediary links
Gama	30	Intermediary links
Delta	56	Robuster links

Table 14.6 Traffic parameters

AC	Voice (VO)	Video (VI)	Background (BK)
Packet size	1280 bit	10240 bit	18430 bit
Packet interval	20 ms	10 ms	12.5 ms

Table 14.7 Results for each cost function

Function	Latency [ms]	Packets delivered	Packets delivered [%]
Alfa	6547	74	74/1500 = 0.049(3)
Beta	5675	197	197/1500 = 0.131(3)
Gama	5704	98	98/1500 = 0.065(3)
Delta	5361	282	282/1500 = 0.188

14.4.2 Genetic Algorithms Approach

In the second approach, we replaced the determination of the best path for the GAs approach. The challenges presented by GAs are the following ones:

Codification: the chromosomes are the paths. A chromosome is a vector of integers that represents the set of node IDs through which a packet has to go from the origin to the destination. The first locus of the chromosome is the origin while the last one is the destination. The ones in the middle have an order that the packet being transmitted needs to follow. For example, the chromosome [A N1 N2 B] establishes that a packet going from A to B has to go first to node N1, then to N2, and finally to B. The maximum size of the chromosomes is the maximum number of stations in the scenario.

Fitness metric: the fitness represents chromosome quality. It must reflect as precisely as possible the quality of a chromosome. In our case, the fitness value has some noise provided by the randomisation features of the simulator. To reduce this noise and not discard a good chromosome (or use a bad chromosome as a parent more often), we evaluate a chromosome with a given degree of freedom. In this work, the fitness value is provided by the number of packets delivered in a given period. However, we could have considered the delay or any other metric for the fitness parameter, and the algorithm would work the same way.

Population initialisation: an heuristic method in which the costs of the links in the DA follow an exponential distribution was used.

Selection procedure: the selection procedure chooses the chromosomes to be parents of an offspring. The parents with best fitness will give the best offspring [21]. Therefore, if we want to increase the fitness while getting closer to the optimal solution, the chromosomes with higher fitness should have higher probability of being selected, i.e., the selection procedures focus on the exploitation of promising areas of the solution space. From the several tested the rank selection was the one that presented better results. It involves the following steps:

1. The chromosomes are ranked according to the fitness;
2. The top 1/4 of the chromosomes are selected;
3. To generate offspring, two chromosomes are selected randomly from the set found in step 2.

Crossover between genes: two chromosomes are two solutions for the problem. The crossover brings a new solution derived from them. In our case, the crossover is done by exchanging partial route of two chosen chromosomes (paths).

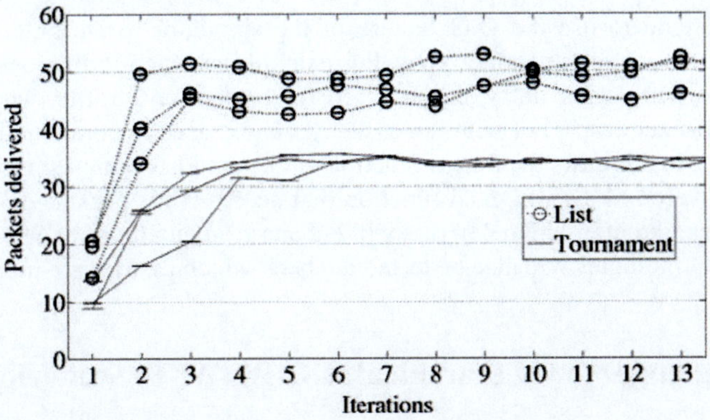

Fig. 14.12 Packets delivered in topology 1 with the tournament selection (*solid lines*) and list selection (*dashed lines*) algorithms

Mutations: a mutation is performed by randomly changing the genes in the chromosome. The mutation in our case is not a standard case since we cannot change a given locus randomly without any constraint; otherwise we could get a path without connection. Mutation is only possible if, when requested, there is an alternative path to the one being used from the locus over which the mutation was requested.

Results for the rank/list and tournament selection algorithms were obtained from 3 s simulations and are presented in Fig. 14.12 for the same topology as in the previous test. Several simulation runs were considered for the list selection (dashed lines) and for the tournament selection (solid lines) algorithms.

It can be observed that the list selection algorithm presents better results than the tournament selection algorithm, as it delivers more packets. The list selection algorithm also converges faster than the tournament selection one. The list selection algorithm delivers approximately 16% of the packets, but its performance is not as high as the delta function from the empirical approach, which delivers 18.8% of the packets (but it can be improved).

14.4.3 Conclusions and Future Work

The goal of the routing algorithms proposed in this work was to find a path that delivers the highest number of packets in an IEEE 802.11e ad hoc network with the lowest delay. This optimisation was tested for a single video stream while considering background and voice traffic streams to increase the network load. A novel hybrid method to initialise the GA chromosomes was proposed. The simulations with the best initial population size (40) and the best selection algorithm (List/Rank) always converge to a single solution. In this case, the GA approach did not manage to reach better results than the best empirical one, but from our simulations in more random generated topologies in most of the cases GA outperformed the other approach. However, we only present the initial results in this chapter. As a future work, we intend to use the QoS feature of the simulator to attribute a different cost function to different traffic types. For example, a function that delivers more packets, but with larger delay, can be used for background traffic, while another with more packet drops, but with lower delay, can be used to compute the path for video traffic. In this case, the fitness function of GA could account for more than the packets delivered like a weighted function that accounts for packets delivered and other metrics like jitter and delay that will enhance the quality of experience of the users. More topologies will also be tested to check which metrics are more suitable.

14.5 Challenges for Hierarchical HSDPA/Wi-Fi Scenario

In our envisaged scenario we are looking for a way to provide the best service to the end user. One possibility for our scenario is depicted in Fig. 14.13.

The user is requesting NRTV while moving on a public transportation service. This can be a bus moving inside the town or a train in the suburbs. On the one hand, the mobility (which is higher in the train scenario) is presented as a great challenge, since it requires frequents handover as the user travels outside the area that was covered by the antenna he was connected to. On the other hand, the natural and man-made (e.g., buildings) obstacles interfere with the signal, also requiring a well-planned handover strategy since a massive building can temporary block the connection between the user and his serving antenna. The service must be provided with QoS assurance, meaning that the video displayed on the user terminal must have enough quality (which include acceptable throughput and no service interrupts). For that purpose, it can be served by an HSDPA NB, or by a nearby Wi-Fi hotspot, or even another user that is closer, depending on which can provide the (best) service. The train has an efficient outdoor receiver that can better receive the

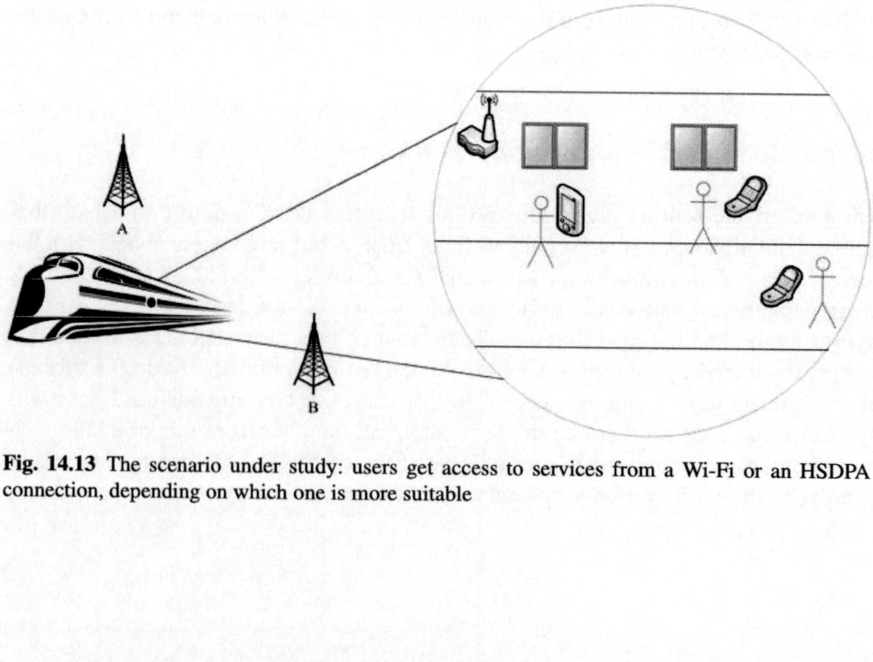

Fig. 14.13 The scenario under study: users get access to services from a Wi-Fi or an HSDPA connection, depending on which one is more suitable

signal than any of the user terminals inside and is connected to the Wi-Fi hotspot in the coach. If the train is passing some problematic spots, e.g., very tall buildings or mountain area, the user terminal may start getting a lousy signal from the NB and to ensure the QoS it can be switched to the Wi-Fi network. In this process, spectrum aggregation is applied to decide if the HSDPA backbone communications is performed either in the 2 GHz band or in the 5 GHz band while giving access to the users trough the IEEE 802.11e ad hoc network. For users that are far from the wireless router, for example if they are in another coach, the connection can be made by using another terminal as a relay.

Nowadays, the technology allows mobile phones to have 3G, Wi-Fi, and even WiMAX communications capabilities, all integrated into one equipment. As so, the availability of different technologies in the end-user equipment is not an issue for our scenario. The true issue is how to decide to which technology should the terminal be connected to and how to make the different RATs cooperate and not interfere. This is integrated in the larger challenge which is the routing one: the packets have to be routed through several hops, each one using its own technology, before arriving into the core network. We designate these routes as cognitive paths (CPs). The entity that controls the flow of information and define the CPs must have accurate and up-to-date information regarding all users in the network, their connections quality and user requirements. This involves a good amount of information flowing to the routing coordination entity, and enough processing capacity required to it, in order to efficiently serve all the network. The algorithm that generates the CPs may involve complex calculations if the network has to serve a large number of users and several connection options. However, nowadays both software and hardware solutions are enoughly developed to materialise this scenario.

14.6 Conclusions

In this chapter we aimed to provide the best service possible to a user requesting high-quality and time-dependent services, such as video or voice. The core of the service is provided by standard HSDPA NBs, whose coverage is improved indoor by using Wi-Fi ad hoc connections. The Wi-Fi uses the standard IEEE 802.11 with its amendment "e" to prioritise the delay sensitive traffic. The solution presented is not yet implemented nor simulated, but we provide isolated tests for each component of the system.

First we studied the inter interoperability between these two wireless technologies. Since the operation is performed in different frequency bands, these technologies can coexist without interfering with each other. From the tests run in our simulator, we demonstrated that the Wi-Fi can improve the coverage of the HSDPA without the need of a complex management system, providing a better service for the users.

Then we studied the HSDPA system in the context of cognitive radio, by using two NBs operating at different frequencies. The operator has access to the 2 GHz and the 5 GHz bands. The system automatically chooses the frequency to serve the user. It was shown that having a secondary band can increase the throughput of the system and serve more users.

In the Wi-Fi ad hoc network, the goal was to find the path that delivered the highest number of packets and with the lowest delay. The optimisation was tested for a video stream, while considering background and voice traffic streams to increase the network load. We first defined our cost functions manually, but ended-up by using an automatic optimisation procedure in the form of GAs. Choosing the path that minimises the delay while maintaining the delivery ratio is very important to provide QoS to the user, and our optimal approach manages to reach these two objectives.

The main challenge for our cognitive radio scenario for public transportation is to discover and maintain the paths between the service provider and the end user, which we call the cognitive paths. We have shown that, isolated, each piece of the network can be optimised to provide the best performance possible, and we expect to use the optimisation performed in the global management entity. The simulator to be used is being conceived and adapted for the tests and we expect to obtain results soon.

Acknowledgements The authors would like to acknowledge the following projects who provided financial support: IST-UNITE (a Specific Targeted Research Project supported by the European 6th Framework Programme, Contract number IST-FP6-STREP-026906), Marie Curie European Reintegration Grant PLANOPTI (Planing and Optimization for the Coexistence of Mobile and Wireless Networks Towards Long Term Evolution, FP7-PEOPLE-2009-RG), UBIQUIMESH (Cross-Layer Optimization in Multiple Mesh Ubiquitous Networks ref PTDC/EEA-TEL/105472/2008), CROSSNET (a Portuguese Foundation for Science and Technology, FCT, POSC project with FEDER funding), Marie Curie Intra-European Fellowship OPTIMOBILE (Cross-layer Optimization for the Coexistence of Mobile and Wireless Networks Beyond 3G, FP7-PEOPLE-2007-2-1-IEF), and Projecto de Re-equipamento Científico REEQ/1201/EEI/2005 (an FCT project). João Ferro and Orlando Cabral acknowledge the Ph.D. grants from FCT ref. SFRH/BD/36742/2007 and SFRH/BD/28517/2006, respectively. Authors also acknowledge the COST Action 2100 - Pervasive Mobile & Ambient Wireless Communications, the Portuguese project Smart-Clothing, Valdemar Monteiro, Jonathan Rodrigues, Filippo Meucci, and Albena Mihovska.

References

1. G. Hiertz, D. Denteneer, S. Max, R. Taori, J. Cardona, L. Berlemann, and B. Walke. Ieee 802.11s: The WLAN mesh standard. *IEEE Wireless Communications*, 17(1):104–111, 2010.
2. Caution – capacity utilization in cellular networks of present and future generation. IST project Website, 2010. http://www.telecom.ntua.gr/caution/start.html.
3. Aroma – advanced resource management solutions for future all IP heterogeneous mobile radio environments. IST project Website, 2010. http://www.aroma-ist.upc.edu/.
4. 3rd Generation Partnership Project, Technical Specification Group Radio Access Network. *3GPP TR 25.892 v6.0.0, Feasibility Study for Orthogonal Frequency Division Multiplexing (OFDM) for UTRAN enhancement*, June 2004.
5. Unite: Virtual distributed testbed for optimisation and coexistence of heterogeneous systems. IST project, 2008. http://cordis.europa.eu/fetch?CALLER=PROJ_ICT\&ACTION=D\&CAT=PROJ\&RCN=80685.

6. CEA-LETI and al. *D1.1.1 Definition of Cross-Layer and Cross-System Framework Scenarios*. IST-4-026906 UNITE, Deliverable D1.1.1, IST Central Office, Brussels, Belgium, September 2006.
7. O. Cabral, A. Segarra, and F.J. Velez. Event-driven simulation for IEEE 802.11e optimization. *IAENG International Journal of Computer Science*, 35(1):161–173, 2008.
8. D. Qiao and S. Choi. Goodput enhancement of IEEE 802.11a wireless lan via link adaptation. In *Proc. of IEEE ICC'2001*, pages 1995–2000, Helsinki, Finland, 2001.
9. R. Prasad, O. Cabral, F.J. Felez, J. Rodriguez, V. Monteiro, and A. Gameiro. Optimal load suitability based RAT selection for HSDPA and IEEE 802.11e. In *Proc. of 1st International Conference on Wireless Communication, Vehicular Technology, Information Theory and Aerospace and Electronic Systems Technology, Wireless VITAE 2009*, pages 722–726, Aalborg, Denmark, 2009.
10. J. K. Karlof. *Integer Programming: Theory and Practice*. 1st edition, CRC Press, Boca Raton, FL 2005.
11. F. Meucci, A. Mihovska, B. Anggorojati, and N. R. Prasad. Joint Resource Allocation and Admission Control Mechanism for an OFDMA-Based System. In *Proc. of 11th International Symposium on Wireless Personal Multimedia Communications (WPMC 2008)*, Lapland, Finland, 2008.
12. H. Kellerer, U. Pferschy, and D. Pisinger. *Knapsack Problems*. Springer, Heidelberg, Germany, 2005.
13. F. Meucci, O. Cabral, F. J. Velez, A. Mihovska, and N. R. Prasad. Spectrum aggregation with multi-band user allocation over two frequency bands. In *Proc. of 2009 IEEE conference on Mobile WiMAX*, MWS'09, pages 81–86, Piscataway, NJ, 2009.
14. Nortel Networks, 3GPP TSG-RAN-1 Meeting #31. *R1-03-0249, Validation of System-Level HSDPA Results for CDMA and OFDM in a Flat Fading Channel*, 18–21 February 2003.
15. Ericsson. *3GPP2-C30-20030429-010, Effective SNR mapping for modeling frame error rates in multiple-state channels*.
16. 3GPP. *TR25.211: Physical channels and mapping of transport channels onto physical channels (FDD)*, 5.7.0 edition, June 2005.
17. *IST MATRICE-2001-32620, D4.5 Layer 2& 3 reference simulation results dynamic resource allocation algorithms and IP transport*, September 2004. http://www.ist-matrice.org/.
18. R. Skehill, M. Barry, W. Kent, M. O'Callaghan, N. Gawley, and S. Mcgrath. The common rrm approach to admission control for converged heterogeneous wireless networks. *IEEE Wireless Communications Magazine*, 14(2):48–56, 2007.
19. O. Cabral, A. Segarra, and F.J. Velez. Event-driven simulation for ieee 802.11e optimization. *IAENG International Journal of Computer Science (IJCS)*, 35(1):161–173, 2008.
20. E.W Dijkstra. A note on two problems in connexion with graphs. *Numerische Mathematik*, 1:269–271, 1959.
21. M. Mitchell. *An Introduction to Genetic Algorithms*. The MIT Press, Cambridge, MA, 1999.

Chapter 15
An Autonomous Access Point for Cognitive Wireless Networks

Bheemarjuna Reddy Tamma, B.S. Manoj, and Ramesh Rao

Abstract In this chapter, we present an application of the Cognitive Networking paradigm to the design and development of autonomous Cognitive Access Point (CogAP) for Wi-Fi hotspots and home wireless networks. In these environments, we typically use only one AP per service provider/residence for providing wireless connectivity to the users. Here we can reduce the cost of autonomic network control by equipping the same AP with a cognitive functionality. We first present the architecture of autonomous CogAP which consists of two main modules: Traffic sensing module and cognitive controller module. The traffic sensing module uses an efficient packet sampling scheme to characterize traffic from all Wi-Fi channels with single wireless interface. The cognitive controller module consists of two submodules: traffic predictor and cognitive decision engine. The Neural Network-based traffic predictor module makes use of the historical traffic traces for traffic prediction on all channels. The cognitive decision engine makes use of traffic forecasts to dynamically decide which channel is best for CogAP to operate on. We have built a prototype CogAP device using off-the-shelf hardware components and obtained better performance with respect to state-of-the-art channel selection strategies.

15.1 Introduction

Access points (APs) are specially configured wireless devices that are connected to a wired LAN and act as central transceiver of radio signals in wireless local area networks (WLANs). Wireless devices, such as laptops and PDAs, having IEEE 802.11-based Wi-Fi network interface cards (NICs) connect to the wired LAN and/or Internet via an AP. WLAN deployment is a hard and laborious task even with a moderate number of APs; today, this is commonly carried out by a manual operation with several reconfiguration cycles. The deployment involves selecting locations for placing APs and (for each AP) selecting operating radio channel (802.11*b/g* and *a* based APs have 3 and 13 orthogonal channels, respectively, to choose from), transmit power (which decides the coverage area of AP), data rate, and medium access control (MAC) layer-related parameters such as beacon interval,

---

B.R. Tamma (✉)
Indian Institute of Technology Hyderabad, 502205, India
e-mail: tbr@iith.ac.in

F.R. Yu (ed.), *Cognitive Radio Mobile Ad Hoc Networks*,
DOI 10.1007/978-1-4419-6172-3_15, © Springer Science+Business Media, LLC 2011

RTS/CTS threshold, contention window (CW) parameters, and retry limits. However, due to the dynamic and shared nature of wireless medium (shared with APs in the same WLAN, with APs in other WLANs, and with devices that are not APs at all including Bluetooth devices and microwave ovens), parameters controlling access to the wireless medium on each AP must be monitored frequently and modified in a coordinated fashion to maximize WLAN performance. Manually monitoring the network traffic and determining an optimum configuration for the parameters related to the wireless medium is a task that takes significant time and effort; for this reason, autonomic network control has attracted a lot of attention from the research community.

Autonomic network control can be seen as the automation of the network reconfiguration process carried out by an intelligent centralized network controller without manual interactions. In order to achieve performance optimization through automated control, the recently proposed cognitive networking concept can be utilized. Cognitive networking involves developing wireless systems that will have much deeper awareness about their own operations and the network environment, learn relationships among network parameters, network protocols, and the network environment, plan and make decisions in order to achieve local, end-to-end, and network-wide performance, as well as resource management, goals. In cognitive network paradigm, all network elements track the spatial, temporal, and spectral dynamics of their own behavior and the behavior associated with the environment and report that information to a cognitive controller. The information so gathered is used by cognitive controller to learn, plan, and act in a way that meets network or application requirements [1, 2]. The cognitive network paradigm is originally conceived for wireless networks, thus the aim of the cognition process is to provide the network with enhanced adaptability and re-configurability to cope with the challenges of radio communications. Cognitive networking differs from cognitive radios or cognitive radio networking [3–5] in that the latter two apply cognition only to the physical layer to dynamically detect and use spectrum holes in licensed bands, focusing strictly on dynamic spectrum access, whereas the objective of cognitive networks is to apply cognition to all layers of the network protocol stack and network components for achieving network-wide performance goals. While cognitive radio networks need to avoid harmful interference to primary users of licensed bands, cognitive networking paradigm does not take into such heterogenous usage of network resources by primary and secondary users, instead focuses on equipping networks with intelligence for autonomous operation.

The small-scale WLAN systems such as Wi-Fi hotspots and home networks typically contain only one AP per service provider/residence for providing wireless connectivity to the users. In these wireless network environments, we can reduce the cost of autonomic network control by equipping the same AP with cognitive controller functionality, hence called as Cognitive AP (CogAP). For the purpose of characterizing traffic from multi-channel wireless environment, we equip CogAP with a *traffic sensing module* designed using an efficient packet sampling scheme. Cognitive functionality is realized with a *cognitive controller module* which is made up of two sub-modules: a neural network-based traffic prediction and decision engine.

In this chapter, we demonstrate how the proposed CogAP system predicts future traffic loads on each of the Wi-Fi channels and dynamically decides which channel is best for efficiently serving wireless clients. We have prototyped CogAP system using off-the-shelf hardware components and implementing sensing and cognitive controller modules in software. Finally, we show the performance of CogAP system in a real-world wireless environment.

Rest of the chapter is organized as follows: Section 15.2 presents related work in this area. In Section 15.3, we outline architecture of CogAP system. The CogAP's traffic sensing and cognitive controller modules design are given in Sections 15.4 and 15.5, respectively. The CogAP prototype development is presented in Section 15.6. Section 15.7 presents performance evaluation of CogAP's channel selection strategy. Finally, Section 15.8 contains concluding remarks.

15.2 Related Work

In literature, several studies addressed the issue of characterizing wireless LANs (see [6–8] and the references in them). Most of them, however, relied upon wired monitoring and/or the use of SNMP statistics, and hence failed to characterize the wireless spectrum in a spatio-temporal fashion. Some work which employed wireless traffic sensors was focused on the merging of traces and the characterization of global network activity [6–8]. However, the cognitive networking paradigm requires the characterization of the traffic on all channels in a spatio-temporal fashion. Hence the wireless monitoring system should have the capability to sample a fraction of the traffic from all wireless channels. On the characterization of wide area network traffic, Claffy et al. in [9] presented a detailed study of the performance of various count-based packet sampling methods. However, implementing a count-based sampling method is very expensive in a multi-channel wireless network environment as it requires one dedicated wireless interface for monitoring each channel of interest. In [10], the authors proposed two methods for time-based sampling in IEEE 802.11b/g networks; however, the effect of sampling parameters on the accuracy of sampling, which is one of the most important aspects, is not studied in their paper.

Time-based sampling methods are not studied thoroughly in the literature. Hence, in this work, we study the accuracy of time-based sampling schemes for characterizing traffic in multi-channel wireless networks. Then we present some results on the characterization of a campus 802.11 network environment in a spatio-temporal-spectral fashion.

Cognitive networking is a relatively new networking paradigm [1]. Cognitive networking follows the evolution of the cognitive radio systems in [5]. However, though cognitive radios and cognitive networking have a lot of differences, they are similar in one aspect: cognition from their respective operating environments. In [11], authors formulated wireless network protocol stack configuration as an optimization problem and then proposed the overall architecture of cognitive engine, which uses simulated annealing as optimization algorithm, for solving it. In order to improve the convergence rate of the basic simulated annealing algorithm, they applied machine learning techniques to construct graphical models on the perceived relations between network stack parameters and application-specific network utilities. In [12], Bayesian network models are used to create a representation of the dependency relationships between significant parameters spanning transport and medium access control (MAC) layers in multi-hop wireless network environments. The authors then proposed a cognitive network node architecture with the goal of optimizing the network performance by carefully tuning the value of controllable parameters within the network protocol stack, through sensing of the observable parameters and the exploitation of knowledge acquired from historical network behavior. Examples of observable parameters are the number of packets transmitted at the MAC layer and throughput seen at the transport layer. Controllable parameters are, for example, the TCP congestion window and the MAC contention window. Work on an early centralized cognitive networking system for enterprise WLANs is suggested in [13]. In that work, focus was on providing a cognitive network solution for a centralized system where a collection of enterprise WLAN APs gather network environmental information and transfer that to a centralized cognitive controller system. The centralized cognitive controller then uses neural network approaches to determine the best channel for each of the APs in the network. In this work, our focus is on designing autonomous cognitive APs for small-scale wireless networks such as Wi-Fi hotspots and home WLANs. In order to reduce cost in these small-scale networks, we propose a cognitive network architecture which performs cognitive controller functionality locally in CogAP.

15.3 CogAP Architecture

Figure 15.1 shows the schematic diagram of autonomous cognitive AP. It has two main modules: a sensing & serving module and a cognitive controller module.

15.3.1 Sensing and Serving Module

This module helps CogAP to obtain sensorial information about the surrounding communication environment. This sensorial information is expected to be available in the form of measurements of different types, e.g., traffic information, signal and noise power measurements, as well as time and location coordinates. This module uses two wireless NICs, one for sensing or monitoring the traffic-related information on each of the 802.11 channels in the 2.4 GHz spectrum and another for communication and serving wireless clients associated with the AP. The main components of sensing module are the time-based sensing control module, traffic sensor module, network error sensor module, and the synchronization and control module. The time-based sensing control module controls the time for which the traffic samples

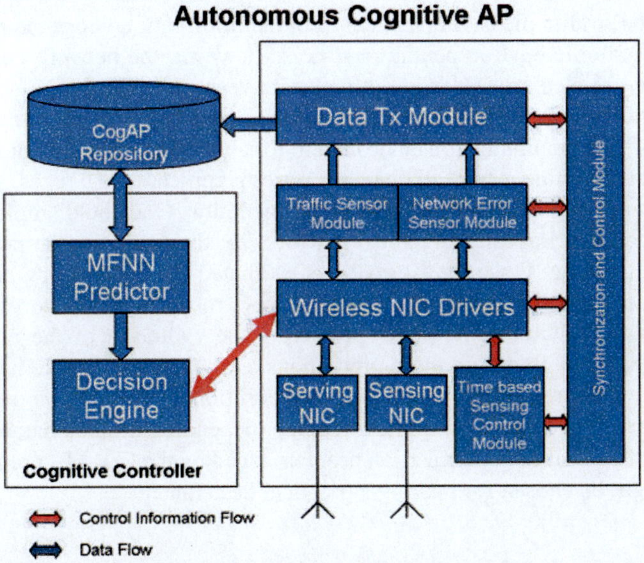

Fig. 15.1 Architecture of the autonomous Cognitive AP

from each of the channels are to be collected. This module enables the traffic sensor module to sample traffic in all the channels in a predefined static or dynamic fashion.

The synchronization and control module performs synchronization of the CogAP to a public NTP server to achieve time synchronization. The network error sensor module is used to gather additional measurements which can be used to better characterize the conditions of the wireless medium. In particular, we are interested in collecting additional statistics like the number of CRC errors per unit time (i.e., failed packet receptions due to a bad CRC in the MAC trailer) and PHY errors per unit time (failed packet receptions due to the CRC failure on the PLCP header). The data transfer module dumps all these traffic-related measurements in a local database (CogAP repository) and thereby provides historical as well as instantaneous traffic characteristics for the cognitive controller module.

15.3.2 Cognitive Controller Module

This module is designed to identify the status of the network and the impact of different configuration settings on the network performance. It is, of course, possible to provide directly the controller with this knowledge, e.g., by hard coding the actions to be taken in response to different network conditions. However, a cognitive controller is expected to be able to learn these dependencies, thus relieving the engineering effort in providing the needed knowledge to the controller. The cognitive controller module has two sub-modules: prediction and decision engine.

Prediction: traffic measurements can provide information about the current and past wireless traffic environmental conditions; however, the network configuration needs to be selected so that it is optimal with respect to the future status of the environment. To cope with this issue, CogAP is expected to be able to predict the evolution of the communication environment for all the channels. Learning the communication environment over all channels poses significant challenges, since many external factors play an important role in this evolution. One could employ artificial neural networks or Bayesian graphical networks for the prediction purpose.

Decision making: Once the CogAP has estimated the dependency between the network status and the performance with respect to different network settings, it is provided with suitable means of predicting the evolution of the environment across all channels. Based on these predictions, CogAP needs to take its decisions, i.e., to select the most desirable network configuration. Depending on the complexity of the reconfiguration to be performed, i.e., depending on the characteristics of the solution space of the optimization problem faced by the CogAP, a proper search strategy might be chosen for making a practical decision.

15.4 Traffic Sensing Module Design

The traffic in WLANs is spread across a number of channels; however, a single wireless NIC that we have in APs for traffic sensing can typically monitor only one channel at a time. For example, the 802.11b/g-based wireless networks operate on ISM band and have 11 channels. Even though orthogonal channels are typically used for configuring APs, in some cases (e.g., non-802.11 sources such as Bluetooth, microwave ovens, and noise render an orthogonal channel useless) other channels are also being used in the configuration of APs. In small-scale WLAN systems such as home WLANs, APs belong to different WLANs co-exist on the same channel and compete for radio resources in the same geographic region. To characterize traffic in such network environments, the wireless monitoring system should have the capability of monitoring all wireless channels in a spatio-temporal fashion. In order to characterize wireless traffic on all channels, the traffic sensing module has either to have a dedicated wireless NIC per channel in order to measure all the traffic in that channel or to switch a single NIC across all the channels in a round-robin fashion, thus measuring a fraction of traffic on each channel. In the first case, the traffic sensing module would be very complex due to the presence of large number of channels in the wireless environment, and moreover storage and analysis of the captured packets from a large number of simultaneous channels could be problematic for single board computer-based CogAP device. Therefore, such a multi-interface complete capture solution would be very expensive and would scale poorly to multi-channel wireless network environments. The multi-channel packet sampling scheme is therefore essential [14]. In the following sections, we review existing works in literature on traffic sensing and characterization in wireless networks, discuss the importance of packet-level traffic sampling in characterizing the traffic in multi-channel wireless networks, and compare the accuracy of various sampling schemes using traffic traces collected in real WLAN environment.

15.4.1 Strategies for Accurate Sampling of Wireless Traffic

As a scalable means of observing network traffic, packet sampling has attracted much attention from the industrial and research communities. Sampling is a form of passive traffic measurement, in which not all packets are measured, but only a fraction of them, which is selected based on the sampling method and parameters associated with the sampling process. Count-based systematic sampling methods such as "1 out of N packets" represent a popular sampling design employed in Cisco and Juniper routers.

Sampling methods can be characterized by the sampling algorithm, which describes the basic process for the selection of packets from each sampling interval, and the type of trigger used to start the packet capture. Based on the sampling algorithm, there are three main classes of sampling methods: simple random sampling, systematic sampling, and stratified random sampling [15]. As shown in Table 15.1, for each class, one can use either packet counts or timers to trigger the selection of packets for inclusion in a sample. In the case of simple random sampling, packets are selected randomly from the parent population by treating the whole trace as single sampling interval. In other two methods, the whole population trace is divided into several small sampling intervals and some packets are chosen from each such interval. In systematic sampling, packets are selected deterministically from each sampling interval, whereas Stratified random sampling involves selecting packets randomly from each sampling interval.

Based on the type of trigger used to start the packet capture, sampling methods can be broadly classified into count-based and time-based sampling schemes. In count-based sampling, the packet count triggers the start of a sampling interval.

Table 15.1 Sampling schemes

Sampling scheme	Sampling algorithm	Trigger type (Sampling duration, sampling period)	Example
SR	Simple random	Count-driven count-based (only one interval)	k pkts randomly out of *ALL* pkts
SCC	Systematic	Count-driven count-based	k pkts out of every N pkts
SRCC	Stratified random	Count-driven count-based	k pkts randomly out of every N pkts
SCT	Systematic	Count-driven time-based	k pkts in every 11 s
SRCT	Stratified random	Count-driven time-based	k pkts randomly in every 11 s
STT	Systematic	Timer-driven time-based	t secs in every 11 s
SRTT	Stratified random	Timer-driven time-based	t secs randomly in every 11 s

Here, the sampling period (SP) is defined in terms of the number of packets. The sampling duration (SD) or length is defined as the number of packets selected for inclusion in the sample from each sampling interval. In time-based sampling, a timer triggers the start of a sampling interval or sampling period. Hence, SP is defined as a time interval. However, SD can be either timer-driven or count-driven for time-based sampling schemes. Hence sampling schemes can be further classified into timer-driven time-based sampling and count-driven time-based sampling schemes.

Figure 15.2 shows an example of simple random sampling with SD = k pkts and SP = N pkts, where N is the total number of packets in the complete packet trace. Each pattern filled circle represents one packet selected by sampling scheme from the population trace. Since this kind of sampling requires availability of complete packet trace in advance, it is useful for only off-line studies and hence not a candidate for implementing in real-time traffic sensor systems. An example of count-driven count-based sampling is to select two packets out of every five packets in the packet stream. Figures 15.3 and 15.4 show systematic and stratified random versions of this sampling (namely SCC and SRCC), respectively. Unlike in SCC, two packets are selected randomly from the pool of five packets in each instance of sampling period in case of SRCC scheme. An example of count-driven time-based sampling is to sample one packet every 1 s. Figures 15.5 and 15.6 show systematic and stratified random versions of this kind of sampling (namely SCT and SRCT), respectively. It is to be noted that different SPs may contain different number of packets available due to burstyness in network traffic. Sometimes it may even become impractical to realize count-driven time-based sampling schemes for certain combinations of sampling parameters. For example, when the number of packets available in SP is less than the number of packets to be sampled as per given value of SD. This is a problem because one parameter is defined in terms of packets (SD) while other one is defined as time interval (SP). But this kind of problem will not arise in case of timer-driven time-based sampling. An example of it is to capture whatever packets that arriving in the first 100 ms of every 1 s. Figures 15.7 and 15.8 show systematic and stratified random versions of this sampling (namely STT and SRTT), respectively.

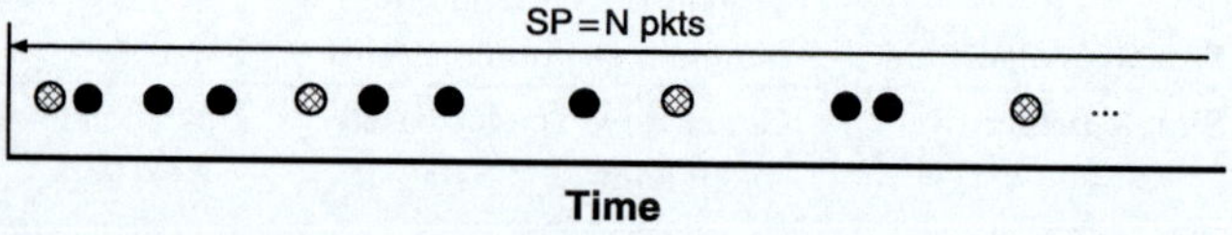

Fig. 15.2 Simple random: SD = k pkts and SP = N pkts

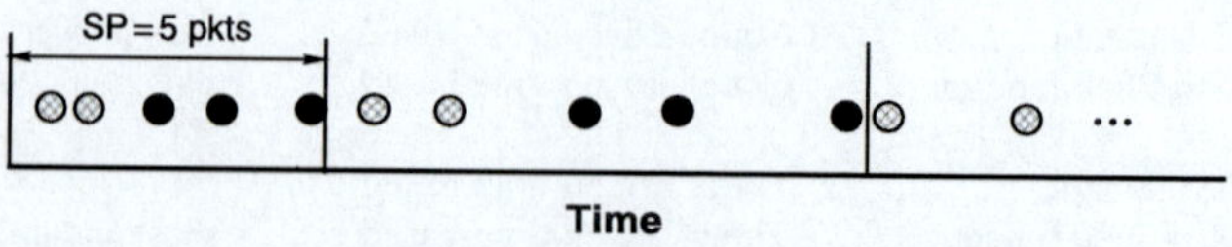

Fig. 15.3 Systematic count-driven count-based: SD = 2 pkts and SP = 5 pkts

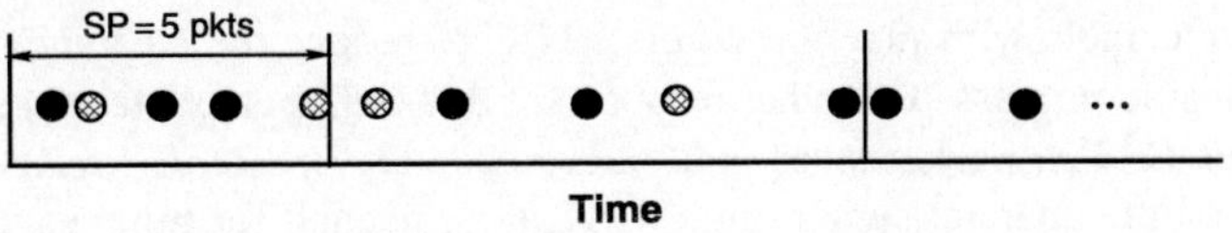

Fig. 15.4 Stratified random count-driven count-based: SD = 2 pkts and SP = 5 pkts

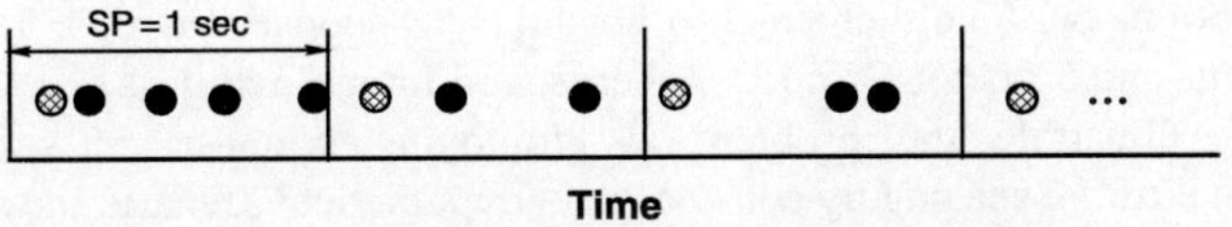

Fig. 15.5 Systematic count-driven time-based: SD = 1 pkt and SP = 1 s

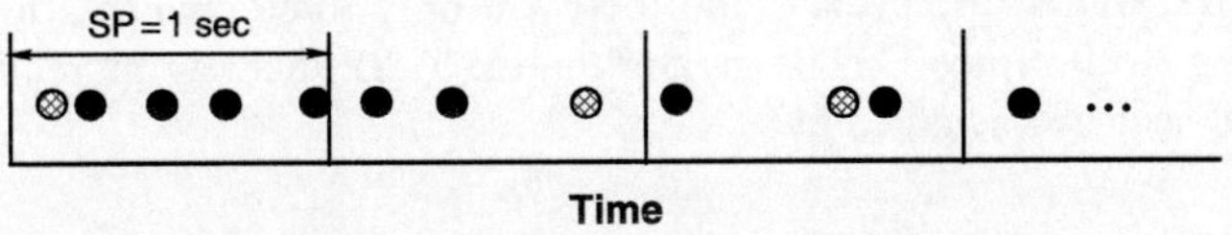

Fig. 15.6 Stratified random count-driven time-based: SD = 1 pkt and SP = 1 s

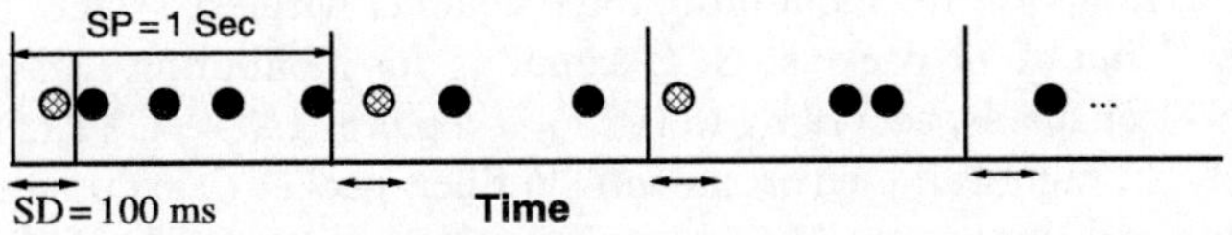

Fig. 15.7 Systematic timer-driven time-based: SD = 100 ms and SP = 1 s

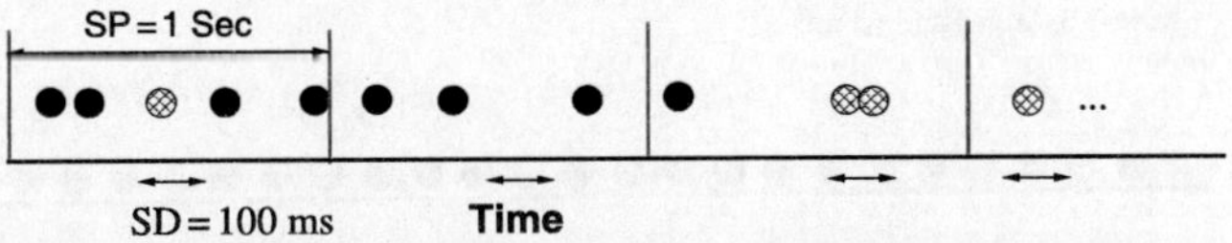

Fig. 15.8 Stratified random timer-driven time-based: SD = 100 ms and SP = 1 s

15.4.2 Why Not Count-Driven Count-Based Sampling for Multi-channel Wireless Traffic Sensing?

We now answer the question on why we cannot employ count-based sampling schemes like SCC and SRCC, very popular in Internet domain for traffic sensing and characterization, for multi-channel wireless network domain. One of the main reasons is that there are multiple channels to be monitored and we want to sample traffic from these channels with just one wireless NIC to reduce the cost and complexity of traffic sensing process. The other reasons are that traffic characteristics like inter-arrival times (IATs), packet sizes, and data rates vary *spectrally*, *temporally*, and *spatially*, and the channel switch time (CST, time needed for tuning wireless NIC from one channel to other) is not negligible. Although hardware specifications of several manufacturers indicate CST in the order of 200 μs, actual times may be higher as a result of driver code paths and delays in waiting for any pending receive DMA operations on the old channel to finish [16]. A conservative CST estimate is between 1 ms and 5 ms for Orinoco, Atheros, and Intersil wireless chipsets [17].

We now illustrate the problem in employing count-based sampling for multi-channel traffic sensing by considering a hypothetical scenario in which traffic characteristics *do not vary* spectrally and temporally. More specifically, we assume that IAT is constant and it is also same for all channels that we want to monitor. Let us also assume that packet transmission times are negligible, CST = 5 ms, IAT = 625 μs, SD is one packet, and there are only three channels to sample as shown in Fig. 15.9. Since CST is greater than IAT, SP that can be realized in this hypothetical scenario, is given by

$$SP \geq CST \times \text{Number of Channels}/IAT \tag{15.1}$$

Hence SP should be at least 24 packets and we can implement "1 out of 24 packets" SCC sampling for monitoring three channel wireless system. For implementing the "1 out of 11 packets" SCC sampling for monitoring IEEE 802.11b/g system with 11 channels, according to (15.1), it requires IAT $\geq$ CST (5 ms) which is not realistic in high traffic environments. Further, packet transmission times are not negligible and they affect IATs in real wireless systems. The IATs also vary widely across channels, and even in any given channel they are not constant and

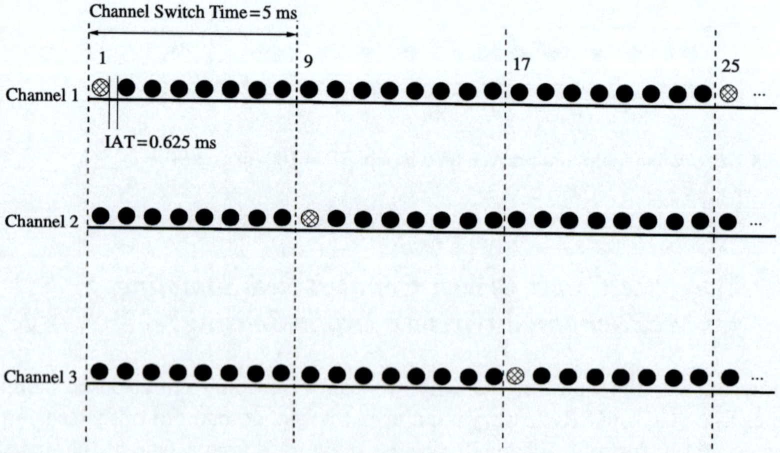

Fig. 15.9 Systematic count-driven count-based: SD = 1 pkt and SP = 24 pkts

vary widely on per packet basis from a few microseconds to several seconds due to burstyness of network traffic. Lack of constant IAT on a channel means we do not know in advance when to switch back to that channel for collecting next sample. In such chaotic real wireless systems, (15.1) does not hold any more and it is impractical to realize any count-based sampling scheme even when CST is zero. Hence, in practical multi-channel wireless network scenarios, count-based sampling schemes are not realizable with single wireless NIC.

We conclude that implementing a count-based traffic sampling method is very expensive in a multi-channel wireless network environment, as we do not know in advance at what times packets will appear on the channel and packet arrival rates vary across the channels. Therefore, it requires one dedicated wireless NIC to sense each channel. The time-based sampling methods seem to offer a cost-effective and scalable solution by reducing the cost of resources necessary to accurately characterize the wireless traffic. For example, the use of timer-driven time-based sampling enable us to make use of a single wireless interface to sample multiple channels in a round-robin fashion. However, in order to achieve a high sampling accuracy, we need to identify the right set of parameters to be used for time-based sampling schemes.

15.4.3 Evaluating the Accuracy of Sampling

In order to determine and quantify the accuracy of different traffic sampling strategies, we need a metric which tells us how close the distributions of the data obtained by sampling are to the actual distributions of the parent populations. We employ a popular statistical metric, the Normalized Kullback–Leibler Divergence (NKLD), to measure the accuracy of sampling schemes. NKLD [18] quantifies the distance, or relative entropy, between two probability distributions. Let $p(x)$ denote the probability mass function for the distribution of the parent population. The probability mass function defines the probability that the traffic metric being considered (e.g., packet size, RSSI) lies in certain intervals. Let $q(x)$ denote probability mass function for the distribution of the sampled packet trace obtained by sampling. The NKLD is defined as

$$\mathrm{NKLD}(p(x)||q(x)) = \frac{\mathrm{KLD}(p(x)||q(x))}{H(p(X))} \tag{15.2}$$

where

$$\mathrm{KLD}(p(x)||q(x)) = \sum_{x\in\chi} p(x)\log\frac{p(x)}{q(x)}$$

$$\text{and}\quad H(p(X)) = \sum_{x\in\chi} p(x)\log\frac{1}{p(x)}$$

$H(p(X))$ is the entropy of the random variable X with distribution $p(X)$. The KLD is zero when the distributions are identical and strictly positive otherwise.

The calculation of the NKLD requires the selection of a set of intervals (χ) for the traffic metric distribution under investigation. In our experiments, for each traffic metric distribution we choose χ in such a way that each interval contains at least 1% of the data from the population data set. For example, in case of packet size distribution we start from the lowest packet size value found in the population data set and increase the value until we get at least 1% of data to obtain the boundary of the first interval. Then, beginning with the boundary of the first interval, we repeat above procedure to obtain the boundary of the second interval, and so on. Once intervals are known, it is straightforward to obtain the probability mass functions for the population packet trace and sampled packet traces.

15.4.4 Performance Results on the Accuracy of Traffic Sampling

To get samples at different granularities for various time-based sampling schemes and measure their sampling accuracy using the NKLD measure, we need population (i.e., complete) packet traces. Therefore, we configured a few CogAPs (refer Section 15.6 for details on CogAP prototype implementation) to do continuous packet capture on one particular orthogonal channel for 4 weeks. We treat these traces as our parent population data sets and generate different sample traces by varying sampling duration and sampling period for various time-based sampling methods. In this study our target per-packet traffic distributions are packet sizes, inter-arrival times (IATs), packet PHY data rates, and received signal strength indicators (RSSIs). In the graphs all performance results are shown with 95% confidence intervals.

Effects of the Sampling Duration: Figure 15.10 shows the NKLD measures for the distribution of the packet size metric calculated from various samples obtained

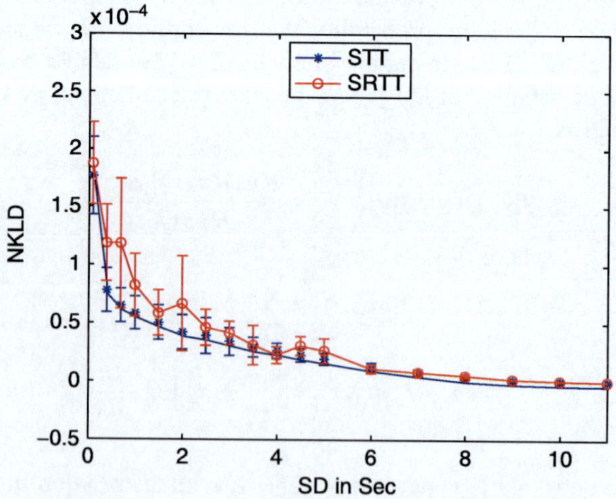

Fig. 15.10 NKLD of STT and SRTT schemes vs SD for packet size distribution

from STT and SRTT sampling schemes. In this experiment, we kept SP constant at 11 s and varied SD from 100 ms to 11 s to obtain various samples from the parent population packet size trace. As SD increases (i.e., the sample size increases) the values of NKLD gradually decrease. That is, we see a small divergence between the population packet size trace and sampled packet size traces obtained from STT and SRTT schemes. In addition, when the SD goes beyond 1 s, the variation in the NKLD decreases. In other words, even samples generated with a SD of 1 s are very closely matching the distribution of parent population. From these figures we can also observe that the performance of the STT and the SRTT sampling schemes is almost identical. However, as SRTT requires the generation of pseudo-random numbers, it is more expensive to implement in real sampling systems.

In Fig. 15.11, we show the NKLD measures of count-driven time-based schemes, SCT and SRCT. For these sampling schemes SD is defined as the number of packets to capture for inclusion in the sample during SP. As shown in the plot, the performance of SCT and SRCT schemes is almost identical. However, when we compare these schemes against the timer-driven time-based schemes (STT and SRTT), the NKLD measures are significantly lower for the timer-driven time-based schemes than for the count-driven time-based sampling schemes (SCT and SRCT). This is because in count-driven time-based schemes the sample size is fixed (as SD is defined in terms of packet count) and does not grow linearly with the traffic load in the network, while in the case of the STT and SRTT schemes sample size grows proportionally to the traffic load, as SD is defined as a time interval. From the above results, we conclude that the STT sampling scheme is the best sampling strategy in terms of sampling accuracy and ease of implementation in real systems. We now employ STT and compute NKLD measures for samples obtained from the population of packet PHY data rates and IATs. Figures 15.12 and 15.13 show their corresponding NKLD scores. Comparing these figures with Fig. 15.10,

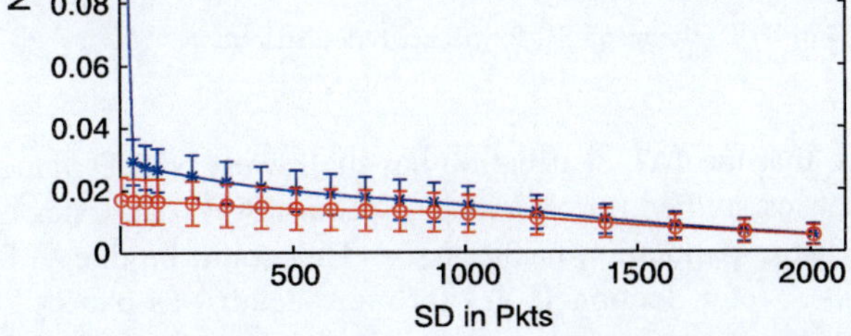

Fig. 15.11 NKLD of SCT and SRCT schemes vs SD for packet size distribution

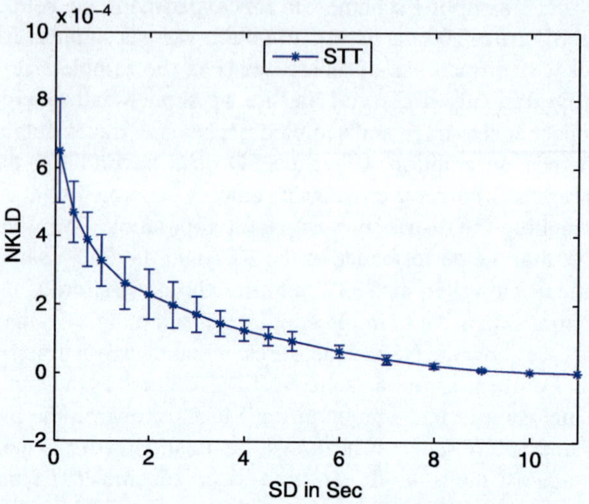

Fig. 15.12 NKLD of STT scheme vs SD for packet PHY data rate distribution

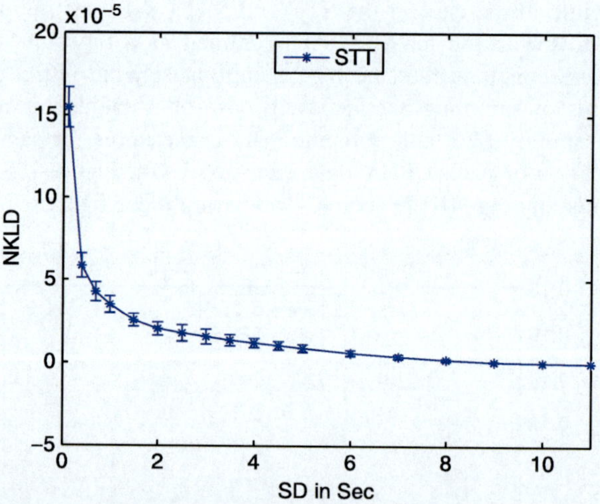

Fig. 15.13 NKLD of STT scheme vs SD for packet IAT distribution

we can observe that the IAT distribution has the lowest NKLD values. Out of the three traffic metrics studied, the samples of packet PHY data rates exhibit higher divergence from the population packet trace. This could be due to fewer number of intervals ($|\chi$—, refer Section 15.4.3 for more details) as packet PHY data rate can take only a fixed number of pre-determined values for 802.11 b/g networks. Since the maximum divergence is still very small and negligible, we can configure the traffic sensor module with STT sampling to accurately sense and characterize all the traffic metrics in multi-channel wireless network environments such as WLANs.

Effects of the Sampling Period: In this experiment we study the effect of sampling period on the sampling accuracy for various traffic metrics. We kept SD constant at 1 s and compute the NKLD for the samples obtained using the STT sampling scheme by varying SP from 1 s to 32 s. Figures 15.14, 15.15, 15.16, and 15.17 show the corresponding NKLD values for the traffic metrics under study. As expected, for all the traffic metrics the NKLD increases when SP increases due to decrease in sample size. We also conducted similar experiments by varying channel number for 4 weeks in our campus; the results are very similar to what we presented here and have, therefore, been omitted due to space limitations.

In order to reduce the cost of wireless traffic characterization, we would like to reduce the value of SD and increase the value of SP to maximum extents without sacrificing sampling accuracy. Based on these sampling accuracy results, we can conclude that even though there are 11 channels to be monitored for traffic in IEEE 802.11b/g spectrum, when we employ timer-driven time-based sampling with sampling parameters SD = 1 s and SP = 11 s, one wireless interface would be sufficient for sensing all the channels with very good sampling accuracy.

15.4.5 Traffic Characterization

We implemented the STT sampling scheme in several CogAPs with SD = 1 s and SP = 11 s and collected samples for several months in a large six-story academic building. We make use of the sampled traffic to estimate population traffic statistics like mean packet rate, mean traffic load, mean data rate, and traffic intensity. We now

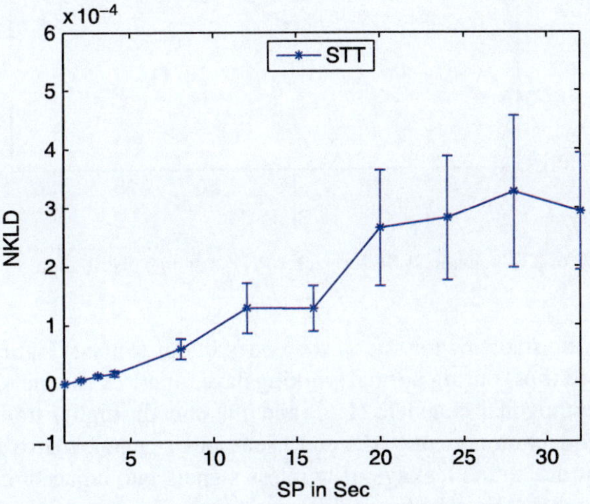

Fig. 15.14 NKLD of STT scheme vs SP for packet size distribution

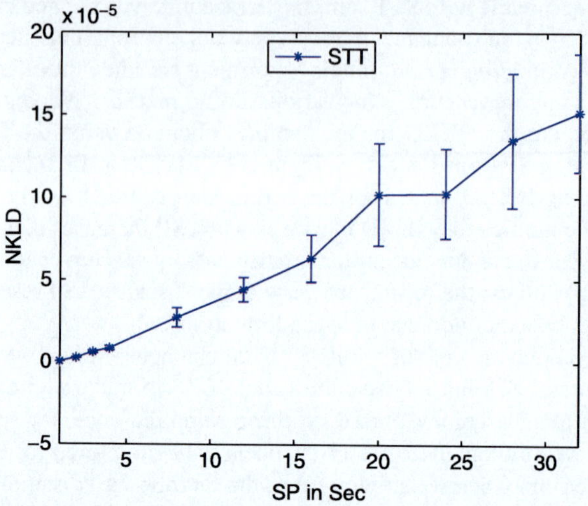

Fig. 15.15 NKLD of STT scheme vs SP for packet IAT distribution

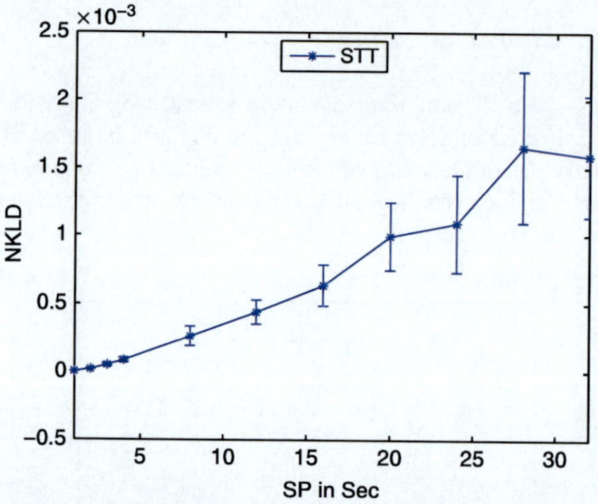

Fig. 15.16 NKLD of STT scheme vs SP for packet PHY data rate distribution

show a few traffic statistics for one of the nodes in our testbed. Figure 15.18 show mean traffic load (bps) during normal working days for all 11 channels, respectively. Even though orthogonal channels (1, 6, and 11) contain higher traffic due to the presence of production network APs and users, other channels also face traffic to a certain extent due to the leakage of wireless signals into adjacent channels. Furthermore, we observed a significant difference in traffic patterns of working days and holidays. While the mean traffic increases during business hours of working

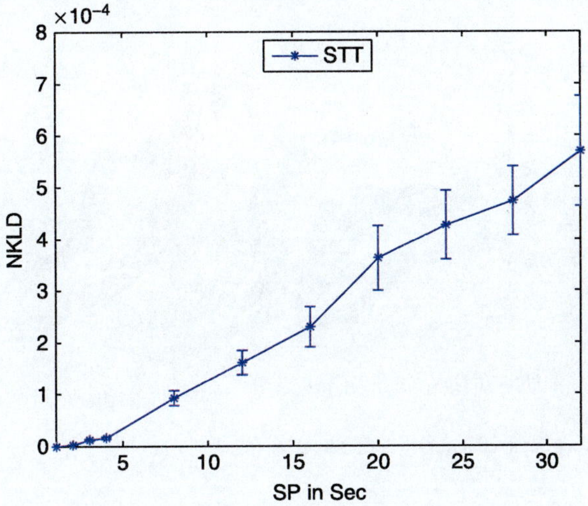

Fig. 15.17 NKLD of STT scheme vs SP for packet RSSI distribution

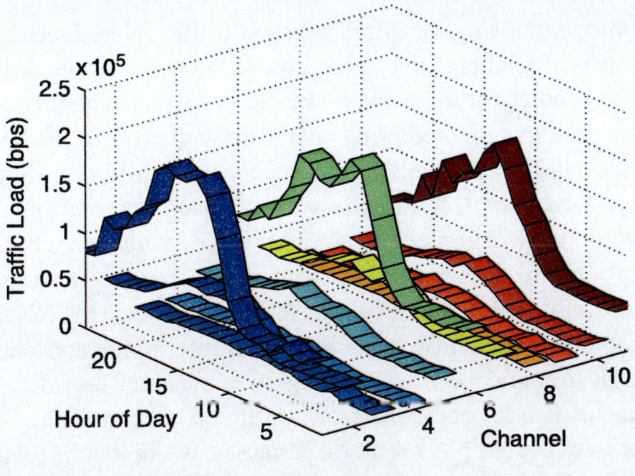

Fig. 15.18 Working days: Traffic load vs channel vs time

days, it stays pretty much flat for the whole day during non-working days. It is to be noted that traffic is non-zero for non-working hours as the APs of the campus production network always send periodic beacons. The small variation in traffic is due to presence of some experimental wireless networks in the vicinity of our testbed infrastructure. We now show mean traffic load variation in a spatio-temporal fashion by fixing the operating channel constant. CogAPs 1–7 and 8–11 are deployed on the 6th and 4th floors of academic building, respectively. Figure 15.19 shows mean traffic loads of all nodes in channel 6. From this plot we can observe that the traffic is also varying in spatial dimension and the traffic is low in some parts of the building due to low user density.

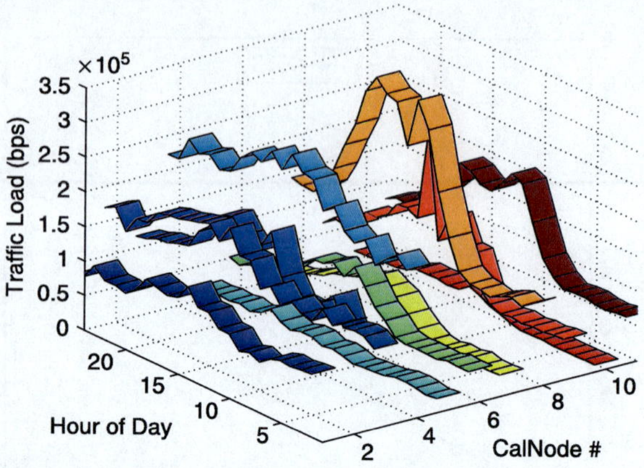

Fig. 15.19 Working days, channel 6: Traffic load vs space vs time

15.5 Cognitive Controller Module Design

Traffic sensing module of CogAP stores wireless traffic-related information in local database repository which helps the traffic prediction sub-module of Cognitive controller module to predict future network traffic across all channels. In this work, we consider the problem of predicting mean network traffic load[1] for next time interval on each of the 11 channels. We consider three different time intervals for traffic prediction: 1-h interval, 5-min interval, and 1-min interval [19]. With the predicted information, the decision engine could then see which channel would be less crowded and reconfigure the serving NIC's channel accordingly. In order to accomplish this, we used Multilayer Feedforward Neural Networks (MFNNs) because they can model the complex relationship between multiple inputs and outputs in a way similar to biological neural networks. Using such a neural network, we could use our historical traffic data to predict network traffic on each channel.

Before we present the MFNN predictor schemes, we first define various parameters that are used as inputs/outputs of predictors.

- *Channel*: Operating channel which ranges from 1 to 11.
- *Weekday*: Day of week ranging from 1 (Monday) to 7 (Sunday).
- *Hour*: Hour of day ranging from 1 to 24.
- *Traffic(t-i)*: Average traffic observed in time interval $(t-i-1, t-i]$, $i \geq 0$, measured in *Kbps*.
- *Traffic(t+j)*: Average traffic estimated over future time interval $(t, t+j]$, $j \geq 1$, measured in *Kbps*.

---

[1] Traffic load is defined as the ratio of the sum of sizes (in Kilo bits) of packets exchanged in the network over a time interval to the number of time units (in seconds) in that time interval.

Table 15.2 Different MFNN predictors

Predictor name	*Channel*	*Weekday*	*Hour*	Input traffic #1	Input traffic #2	Input traffic #3	Output traffic
MLP(3,1)	√	√	√	*Traffic (M-2)*	*Traffic (M-1)*	*Traffic(M)*	*Traffic (M+1)*
HLP(3,1)	√	√	√	*Traffic (H-2)*	*Traffic (H-1)*	*Traffic(H)*	*Traffic (H+1)*
MILP(3,1)	√	√	√	*Traffic (MI-2)*	*Traffic (MI-1)*	*Traffic(MI)*	*Traffic (M+1)*

We designed three kinds of WLAN traffic prediction schemes, namely MLP (Minute Level Prediction), MILP (Minute Interval Level Prediction), and HLP (Hourly Level Prediction) schemes (refer Table 15.2). All schemes use same number of input parameters (six) and same number of output parameters (one which is given in last column in Table 15.2). All schemes use environmental parameters *Channel, Weekday, and Hour* as their inputs in addition to three traffic-related parameters which are based on time scale of prediction. In MLP scheme, we use input parameters to predict mean value of traffic load over next 1 min interval. We denote this scheme as MLP(3,1) scheme because here we are using traffic from previous 3 min traffic as the three 1-min input traffic parameters (i.e., *Traffic(M-2)*, *Traffic(M-1)*, and *Traffic(M)*) to the MFNN model for predicting the next one minute's traffic as the output parameter (i.e., *Traffic(M+1)*). Similarly, in HLP(3,1) scheme we aggregate traffic on hourly time scale and use previous 3 h traffic as the three 1-h input traffic parameters for predicting next 1-h traffic. However, in the case of MILP, we aggregate input traffic in every 5 min to be a 5-min interval and predict traffic load over next 5-min interval. For MILP(3,1) scheme, we use previous 15 min traffic as the three 5-min traffic input parameters (i.e., *Traffic(MI-2)*, *Traffic(MI-1)*, and *Traffic(MI)*) to the MFNN model.

We made use of MATLAB neural network toolbox to implement MFNN-based traffic predictors and study their prediction accuracy. We constructed MFNN predictors as two-layer feedforward backpropagation networks with one hidden layer and one output layer. Hidden layer has 20 neurons with tan-sigmoid transfer function and output layer has one neuron with linear transfer function. Multiple layers of neurons with linear and non-linear transfer functions allow the MFNN model to learn linear and non-linear relationships between input environmental parameters, traffic load parameters, and the output traffic load parameter. In our experiments, the backpropagation training is done by employing Levenberg–Marquardt algorithm [20] which uses adaptive learning rate and the number of epochs is set to 100. Since MFNN predictors require historical traffic traces for the training phase of neural network model, we configured traffic sensor modules of CogAP devices to collect traffic samples for 3 months (January–March 2009) in the CALIT2 building of University of California San Diego campus. As we would like to predict traffic generated by other WLANs operating in the wireless environment, we process the collected packet traces to exclude CogAP's self-traffic (i.e., management, control, and data traffic generated by CogAP and its associated wireless clients). Such processed traces are grouped together based on *Channel, Weekday, and Hour* parameters. This process generates input/output tuples which are randomly divided into three sets. 70% of the tuples are first used to train the neural network during the training phase. 15% of the tuples are used to validate how well the network generalized. Finally, the remaining 15% of the tuples provide an independent test of network generalization to data that the network has never seen during the testing phase.

In order to compare performance of MFNN-based traffic prediction models with traditional autoregressive (AR) models, we also designed Auto Regressive Integrated Moving Average (ARIMA) and Fractional ARIMA (FARIMA)-based models to predict network traffic from historical traffic traces. We employ Box-Jenkins methodology [21] in order to find an appropriate ARIMA model for a given network traffic trace. For model identification, the first step is to make the traffic trace stationary, which implies the window mean and standard deviation are same as the overall traffic trace mean and standard deviation. In order to achieve stationarity, we may need to calculate the difference of the traffic trace several times and typically we use d to denote the number of differencing required. At the second step, with a stationary data set, the autocorrelation function (ACF) and partial autocorrelation function (PCF) are helpful to identify the order of the model. Typically, for an MA process of order q the ACF decays to zero after the lag q, and for an AR process of order p the PCF goes to zero after the lag p. We can model the traffic data set as an ARIMA(p, d, q) model by observing the ACF and PCF functions. Finally, we calculate the estimate of the traffic value $X(t)$ using the following equation:

$$A(Z) * (1 - Z)^d * X(t) = B(Z) * W(t) \tag{15.3}$$

$$A(Z) = 1 - a(1) * Z - a(2) * Z^2 - a(3) * Z^3 -, \ldots, -a(p) * Z^p \tag{15.4}$$

$$B(Z) = 1 + b(1) * Z + b(2) * Z^2 + b(3) * Z^3 +, \ldots, +b(q) * Z^q \tag{15.5}$$

Note that $A(Z)$ and $B(Z)$ are polynomials of order p and q, respectively. The coefficients $a(1), a(2), \cdots, a(p)$ and $b(1), b(2), \cdots, b(q)$ can be calculated using Yule–Walker equation. The Z is the backward-shift operator, i.e., $X(t) * Z^p = X(t - p)$. The $(1 - Z)^d$ means taking the difference of the data set d times and $W(t)$ is a white noise process. We predict network traffic value $X(t)$ as a linear combination of previous traffic values and white noise.

FARIMA prediction is very similar to ARIMA prediction. It also obeys (15.3), but the order of differencing, d, is fractional now. Typically, this value d is between $(-0.5, 0.5)$. In this case, we get:

$$(1 - Z)^d = 1 + C(1) * (-Z) + C(2) * (-Z)^2 + C(3) * (-Z)^3 + \cdots \tag{15.6}$$

where the coefficients $C(k)$ can be calculated in a recursive manner: $C(0) = 1$ and $C(k + 1) = (k + d) * C(k)/(k + 1)$.

15.5.1 Performance Results

The following metrics are used to evaluate prediction accuracy of MFNN-based and AR-based traffic predictor schemes.

1. *Regression Coefficient*, also called as R-value, is a measure of how well the variation in the predicted values is explained by the actual traffic values. If this value is equal to 1, then there is perfect correlation between predicted and actual values.
2. *Mean Squared Error (MSE)* measures error as the average of the squared difference between the predicted (output) traffic values and actual (target) traffic values, with ideal performance yielding an MSE value of zero. The RMSE (Root MSE) is the square root of MSE and it has the same units as the quantity being predicted.
3. *Relative Error (RE)* is a measure of the proportion that the predicted traffic drifts from the actual traffic.
4. *Channel Selection Accuracy (CSA)* is the percentage of right channel selection decision that our predictor makes. We calculate the percentage in the following manner: the cognitive controller chooses the channel with the lowest predicted traffic value among all channels and compares that with the channel with the lowest actual traffic a posteriori. If the selected channel is the same as that channel, then we say the traffic predictor scheme makes a correct channel selection. We divide the number of correct channel selections by the total number of selections in order to get the CSA ratio.

Figure 15.20a shows traffic prediction performance of MLP(3,1) and corresponding ARIMA and FARIMA schemes in terms of RE, CSA, and R-values. All the predictors are fed with the same traffic data set for fair comparison. We can see from figure that ANN-based MLP scheme outperforms the other two schemes. We also compared MILP(3,1) and HLP(3,1) schemes with corresponding ARIMA and FARIMA schemes in Fig. 15.20b, c, respectively. Again, ANN models outperform ARIMA-based models. Hence, we can conclude that ANN-based prediction schemes have better prediction accuracy than traditional ARIMA and FARIMA

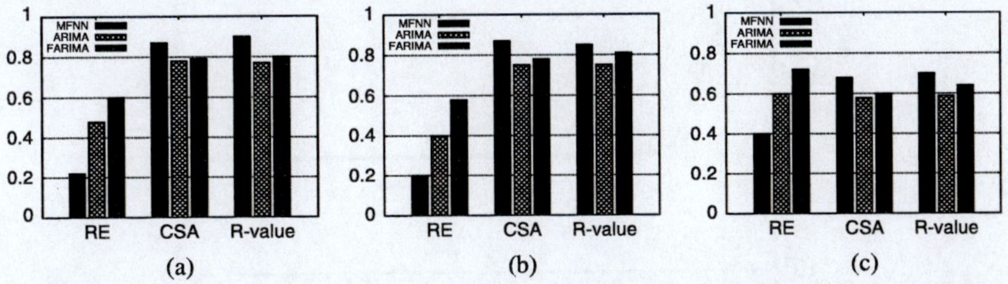

Fig. 15.20 Comparison of prediction schemes: **(a)** MLP(3,1) vs ARIMA(2,1,0) vs FARIMA (2,0.3,0); **(b)** MILP(3,1) vs ARIMA(1,1,1) vs FARIMA(2,0.4,0); **(c)** HLP(3,1) vs ARIMA(1,1,1) vs FARIMA(2,0.3,0)

prediction schemes. The reason for that is the ability of ANNs to model complex nonlinear relationship between inputs and outputs. We can also see that FARIMA model performs better than ARIMA model, this indicates that FARIMA model is more suitable than ARIMA model to predict wireless network traffic which has self-similarity. The performance of MILP(3,1) is about the same as that of MLP(3,1), while HLP(3,1) has poor performance than MLP(3,1) and MILP(3,1) schemes. This is because for any given trace data set the total number of training samples is much smaller compared to other schemes. Further, traffic-related input and output parameters are separated by 1-h intervals for HLP(3,1) and, therefore, these samples do not have strong correlation.

We now take MLP(3,1) scheme and show its performance as a function of epochs in Fig. 15.21. The plot shows the MSE of the network starting at a large value and decreasing to a smaller value. In other words, it shows that the network is learning and converging at a faster rate. This plot is also useful for seeing how quickly the predictor learns. Best validation performance (minimum MSE of 164.25) is obtained at epoch 24, which demonstrates the predictor's quick learning capability.

Figure 15.22 shows regression coefficient between predicted traffic load and actual (target) traffic load in the testing phase of MLP(3,1) scheme. The network outputs are plotted, versus the targets, as open circles. The best linear fit is indicated by a solid line. The perfect fit (output equal to targets) is indicated by the dashed line. In this case, it is difficult to distinguish the best linear fit line from the perfect fit line because the fit is so good. Figure 15.23 shows time-series comparison between real traffic values and predicted values given by MLP(3,1) scheme. Here also we can see that there is very good matching between predicted and actual traffic values.

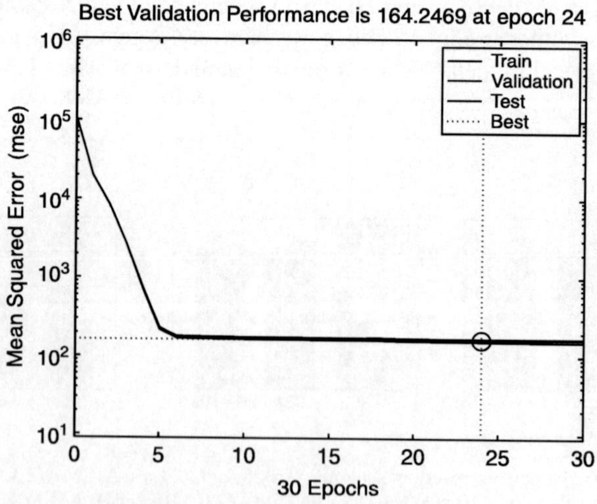

Fig. 15.21 MSE vs Epochs for MLP(3,1) scheme

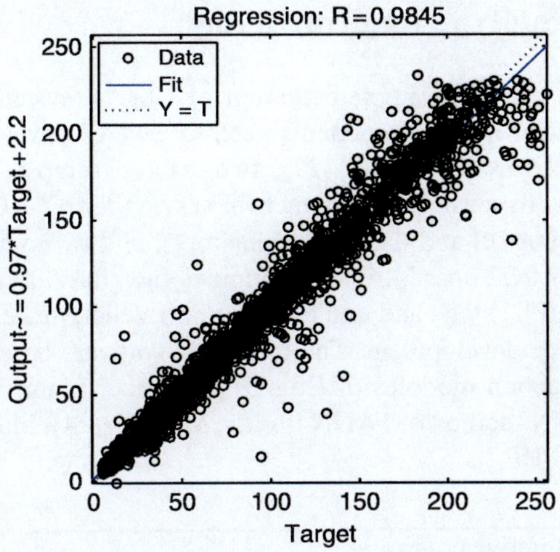

Fig. 15.22 Regression performance in testing phase for MLP(3,1) scheme

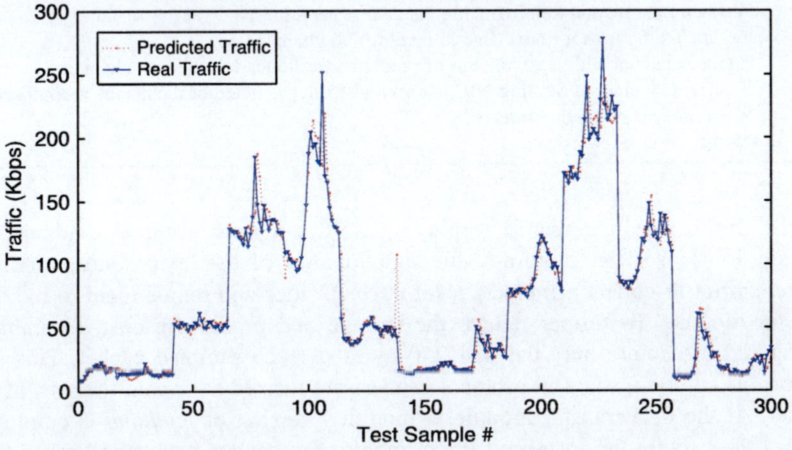

Fig. 15.23 A sample trace of predicted and real traffic for MLP(3,1) scheme

15.5.2 Decision Making

Decision engine part of Cognitive controller module makes use of traffic forecasts made by MFNN predictor scheme to decide which channel is best for serving NIC to operate on. Decision engine periodically (hourly/minute wise/five-minute wise) instructs prediction module to forecast traffic for the next time interval for each of the channels. It then chooses the channel with least amount of predicted traffic load as the best channel. Pseudo code of Cognitive controller module is given in Algorithm 1.

15.6 CogAP Prototype Implementation

We now present a prototype implementation of CogAP system described in the previous sections. Hardware components used in CogAP development include an ALIX2C2 embedded system board [22], two Atheros chipset-based 802.11 b/g miniPCI cards (one for sensing and another for serving users), 8 GB Compact Flash (CF) card (for storing OS and database repository), and two omni-directional pig-tail antennas. ALIX 2C2 board, from PC Engines Inc., has 500 MHz AMD Geode processor, two miniPCI slots, and one Ethernet port which make it a perfect choice for CogAP prototype development. The following software base is used for developing cognitive function modules of CogAP prototype: Linux Voyage 0.5.2 [23], MATLAB with ANN toolbox or FANN library, MySQL [24], MadWiFi driver [25], *tcpdump* [26], and PHP.

Algorithm 1 : Cognitive controller

```
Fetch historical input/output tuples from CogAP database repository
Train traffic predictor with historical tuples
loop
  Perform Incremental training using instantaneous tuples.
  Predict Traffic load for next time interval for all channels.
  Choose channel with least amount of predicted traffic load as the best channel
  If current channel of Serving NIC is different from predicted best channel, reconfigure it.
  Sleep till next prediction interval.
end loop
```

The traffic sensing module employing STT sampling scheme (explained in Section 15.4) uses the capture-to-file functionality of the open source *tcpdump* packet sniffer to collect sampled packet traces in files and remits them to the data transfer module. To further reduce the storage and processing cost, *tcpdump* is configured to capture only the first 250 bytes of each sampled packet. This is a reasonable solution, since all protocol headers are located at or near the start of the packet. At the data transfer module, a modified version of *tcpdump* is employed to read the capture file to extract Prism monitoring header fields and header field values from the MAC through transport layers of the TCP/IP protocol stack. These values are stored in local CogAP database repository as instantaneous traffic records, implemented using MySQL server, from which they can be queried to extract the training and testing datasets, as well as for generic data analysis purposes. In our measurement period of 3 months, we found a lot of variation in traffic load across working days, holidays, and days during which academic examinations were conducted. Sampled packet traces contain minimum of 2 million packets and maximum value goes up to 20 million packets per day. Such large size packet traces result in huge size tables (up to 4 GB per day) in CogAP database repository. Since CogAP's storage space is very limited, it creates historical traffic records from instantaneous traffic records, by averaging data referring to the same channel over 1-min intervals for the entire day. Historical records help in achieving scalability by reducing storage cost and speeding up subsequent reprocessing of information by not needing to process the complete traffic records every time (which is a very time-consuming operation).

Implementation of the time-based sensing module is done using a combination of shell scripts running *wireless-tools* and MadWiFi tools. These are used to periodically switch the sensing NIC's channel setting in order to gather traffic samples from all 802.11 b/g channels; channel switching is done in parallel to the traffic sensing activity.

Cognitive controller outlined in Algorithm 1 is implemented in MATLAB using neural network toolbox[2]. MATLAB-MySQL interface is used to query MySQL server from MATLAB environment. We installed MATLAB in the embedded CogAP device in order to implement MFNN predictors as part of CogAP. We faced a number of issues in the implementation of MATLAB-based Cognitive controller because it consumes a lot of computing power of CogAP device during the initial training phase. As given in Algorithm 1, historical tuples are used for initial training of MFNN predictor. However, the initial training needs to performed only once and it took approximately 15 min for CogAP device to complete this job using historical records that span a duration of 3 months. Incremental training and other controller tasks are performed hourly basis and CogAP took only a few seconds for completing these tasks. If historical records are not available[3], CogAP skips initial training phase and does only incremental training. In such scenarios, due to the lack of enough traffic samples for training the MFNN predictor, channel selection based on such predictions may not be optimal. However, over a period of time the prediction accuracy improves and Cognitive controller will make optimal reconfigurations.

15.7 Performance Results

In this section, we study the performance of CogAP in terms of uplink and downlink throughput obtained by wireless clients. The testbed consists of one CogAP device implementing MLP(3,1) scheme and one wireless LAN client, which are separated by a physical distance of 10 m (non-line-of-sight). This setup was placed in an academic laboratory building where many production network APs operating on fixed channels contend for spectrum resources. Wireless test client was built using the same hardware as that of CogAP, but having a single wireless NIC, in managed mode, used to connect to the CogAP. The test WLAN client runs a modified version of the `iperf` software to carry out active measurements by performing TCP data transfers both in the uplink and in the downlink direction. We ran these active measurements for 1 week. In this active measurement setup CogAP and test WLAN client pair measures performance of all channels by switching serving NIC's channel to one of the 11 channels after every measurement in a round-robin fashion.

---

[2] As a lighter and open-source alternative to MATLAB neural network toolbox, we also used Fast ANN (FANN) library [27], using C programming language, for implementing the Cognitive controller.

[3] This happens when CogAP is deployed in a new geographical area for the first time.

In this way application-layer throughput can be measured on each of the channels which is useful to determine the best performing channel during a posteriori *analysis*. We compared throughput performance of CogAP with three other schemes which are outlined below.

- *Random Scheme*: In typical Wi-Fi hotspot environments, AP's serving NIC's channel is fixed and it is configured on one of the available orthogonal 802.11 channels. Hence to simulate such a scenario, in Random Scheme we periodically (on hourly basis) switch the NIC to one of channels randomly (1, 6, and 11) which is then considered as the best channel.
- *Weighted Average Scheme (WAS)*: It uses lags and shifts in the historical traffic load to uncover patterns and predict the future network traffic load. In our study, we use it to predict future traffic on all channels, *Traffic(t+1)* from past values *Traffic(t-2), Traffic(t-1), Traffic(t)* with weights 0.3, 0.3, and 0.4, respectively.

$$\text{Traffic}(t+1) = 0.3 \times \text{Traffic}(t-2) + 0.3 \times \text{Traffic}(t-1) + 0.4 \times \text{Traffic}(t)$$

  Channel that is associated with the least predicted value for traffic load is considered as the best channel.
- *Best*: Since we measure throughput on all channels, channel that is having highest a posteriori throughput is considered as the best channel.

The uplink and downlink throughput performance for the channel selection schemes just described is reported in Fig. 15.24. Throughput measurements are averaged over the whole measurement period of 1 week. Random and WAS schemes perform very poorly compared to CogAP channel selection scheme. Random scheme does not have any knowledge on current network conditions and randomly changes its channel, which may cause it operating on non-optimal channel. In case of WAS scheme, as it depends only on the moving average of past traffic to estimate future traffic it could not able to take into account the effect of environment parameters like Day and Hour on future traffic. It is also noted that CogAP's performance is

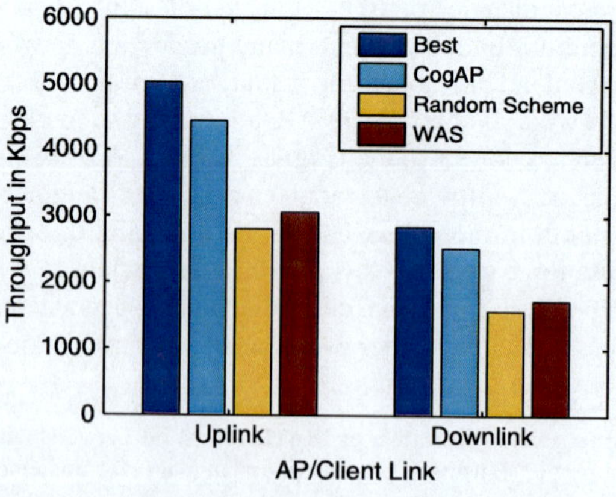

Fig. 15.24 Throughput performance of different channel selection schemes

slightly less compared to best a posteriori performance. This can be attributed to prediction error of MLP(3,1) scheme. The “best” performance can be better than the one achievable by any possible predictor, especially when there is a high short-term variance in the network conditions, in the performance measurements, or in both. However, the “best" strategy cannot be realizable in real-world multi-channel networks because the channel selection decision can only be made a posteriori. It is also interesting to note that downlink performance is much less compared to the uplink performance in our experiments. This is mainly due to spatial variation in network traffic conditions at AP and test WLAN client locations. Therefore, the comparison between the relative performance of different schemes for either uplink or downlink is the focus of our result in Fig. 15.24. These results as well as other similar measurements that have been omitted here due to space constraints show that CogAP with its MFNN-based channel selection scheme outperforms legacy channel assignment strategies used in Wi-Fi hotspots and home WLANs, thus proving that the MFNN-based approach to autonomic cognitive network control is practical and effective.

15.8 Conclusions

Wireless LAN optimization is too complex to carry out manually. Automation of the network optimization demands incorporation of cognitive capability within the network elements. We presented the architecture of CogAP, studied various time-based sampling schemes for multi-channel traffic characterization, designed the cognitive controller module using neural network-based traffic prediction scheme, and developed a prototype CogAP device using commercial off-the-shelf hardware components. We studied the sampling accuracy of time-based sampling schemes such as STT, SRTT, SCT, and SRCT schemes by using packet traces collected in our campus 802.11 network. From our experiments, we found that the Systematic Timer-driven Time-based (STT) sampling strategy is the best sampling strategy for traffic sensing and characterization in multi-channel wireless networks. We also showed that neural network-based traffic predictors exhibit higher prediction accuracy compared with traditional ARIMA- and FARIMA-based predictors. Results from performance evaluation carried out in our testbed on a prototype CogAP device showed that Neural network-based predictors can indeed help the CogAP to find the best channel which gives higher and more sustained bandwidth. Further, our solution achieves performance close to the best possible performance.

References

1. R. W. Thomas, D. H. Friend, L. A. DaSilva, and A. B. MacKenzie, “Cognitive networks: Adaptation and learning to achieve end-to-end performance objectives,” *IEEE Communications Magazine*, vol. 44, no. 12, pp. 51–57, 2006.
2. B. S. Manoj, R. Rao, and M. Zorzi, “Architectures and protocols for next generation cognitive networking,” in *Cognitive Wireless Networks: Concepts, Methodologies and Visions*, M. Katz and F. Fitzek, Eds. Springer, Netherlands, 2007.

3. I. F. Akyildiz, W.-Y. Lee, M. C. Vuran, and S. Mohanty, "Next generation/dynamic spectrum access/cognitive radio wireless networks: A survey," *Computer Networks*, vol. 50, no. 9, pp. 2127–2159, 2006.
4. T. Weiss and F. Jondral, "Spectrum pooling: An innovative strategy for the enhancement of spectrum efficiency," *IEEE Communications Magazine*, vol. 42, no. 3, pp. 8–14, 2004.
5. J. Mitola, "Cognitive radio: An integrated agent architecture for software defined radio," Ph.D. dissertation, Royal Institute of Technology (KTH), 2000.
6. M. Balazinska and P. Castro, "Characterizing mobility and network usage in a corporate wireless local-area network," in *Proc. of ACM Mobisys*, 2003, pp. 303–316.
7. J. Yeo, M. Youssef, T. Henderson, and A. Agrawala, "An accurate technique for measuring the wireless side of wireless networks," in *Proc. of Workshop on Wireless Traffic Measurements and Modeling*, 2005.
8. Y.-C. Cheng, J. Bellardo, P. Benkö, A. C. Snoeren, G. M. Voelker, and S. Savage, "Jigsaw: Solving the puzzle of enterprise 802.11 analysis," *SIGCOMM Computer Communication Review*, vol. 36, no. 4, pp. 39–50, October 2006.
9. K. C. Claffy, G. C. Polyzos, and H. W. Braun, "Application of sampling methodologies to network traffic characterization," *Computer Communication Review*, vol. 23, no. 4, pp. 194–203, October 1993.
10. U. Deshpande, T. Henderson, and D. Kotz, "Channel sampling strategies for monitoring wireless networks," in *Proc. of the International Symposium on Modeling and Optimization in Mobile, Ad Hoc and Wireless Networks*, April 2006.
11. E. Meshkova, J. Riihijärvi, A. Achtzehn, and P. Mähönen, "Exploring simulated annealing and graphical models for optimization in cognitive wireless networks," in *Proc. of IEEE GLOBECOM*, 2009, pp. 4939–4946.
12. G. Quer, H. Meenakshisundaram, B. R. Tamma, B. S. Manoj, R. Rao, and M. Zorzi, "Cognitive network adaptation using bayesian networks," in *Proc. of IEEE MILCOM*, 2010.
13. N. Baldo, B. R. Tamma, B. S. Manoj, R. Rao, and M. Zorzi, "A neural network based cognitive controller for dynamic channel selection," in *Proc. IEEE ICC 2009*, June 2009.
14. B. R. Tamma, N. Baldo, B. S. Manoj, and R. Rao, "Multi-channel wireless traffic sensing and characterization for cognitive networking," in *Proc. of IEEE ICC*, June 2009.
15. N. Duffield, "Sampling for passive internet measurement: A review," *Statistical Science*, vol. 19, no. 3, pp. 472–498, 2008.
16. A. Sharma and E. M. Belding, "Freemac: Framework for multi-channel mac development on 802.11 hardware," in *Proc. of ACM workshop on Programmable Routers for Extensible Services of Tomorrow*, 2008, pp. 69–74.
17. D. Murray, M. Dixon, and T. Koziniec, "Scanning delays in 802.11 networks," in *Proc. of International Conference on Next Generation Mobile Applications, Services and Technologies*, 2007, pp. 255–260.
18. S. Kullback, in *Information Theory and Statistics*. Wiley, New York, NY 1959.
19. Y. Liu, B. R. Tamma, B. S. Manoj, and R. R. Rao, "On cognitive network channel selection and the impact on transport layer performance," in *Proc. of IEEE Globecom*, December 2010.
20. D. W. Marquardt, "An algorithm for least-squares estimation of nonlinear parameters," *Journal of the Society for Industrial and Applied Mathematics*, vol. 11, no. 2, pp. 431–441, 1963.
21. G. Box, G. Jenkins, and G. Reinsel, *Time Series Analysis: Forecasting and Control*, 3rd ed. Prentice Hall, Englewood Cliffs, NJ, 1994.
22. http://pcengines.ch/.
23. http://linux.voyage.hk/.
24. http://www.mysql.com/.
25. http://madwifi.org/.
26. http://www.tcpdump.org/.
27. http://leenissen.dk/fann/.

Part VI
Game Theoretic Approach for Modeling and Optimization

Chapter 16
Economic Approaches in Cognitive Radio Networks

Sabita Maharjan, Yan Zhang, and Stein Gjessing

Abstract Efficient resource allocation is one of the key concerns of implementing cognitive radio networks. Game theory has been extensively used to study the strategic interactions between primary and secondary users for effective resource allocation. The concept of spectrum trading has introduced a new direction for the coexistence of primary and secondary users through economic benefits to primary users. The use of price theory and market theory from economics has played a vital role to facilitate economic models for spectrum trading. So, it is important to understand the feasibility of using economic approaches as well as to realize the technical challenges associated with them for implementation of cognitive radio networks.

With this motivation, we present an extensive summary of the related work that uses economic approaches such as game theory and/or price theory/market theory to model the behavior of primary and secondary users for spectrum sharing and discuss the associated issues. We also propose some open directions for future research on economic aspects of spectrum sharing in cognitive radio networks.

16.1 Introduction

Cognitive radio networks [1, 9, 21] have been proposed to facilitate effective use of the electromagnetic spectrum through dynamic spectrum access and sharing. Cognitive radio networks should coexist with licensed users without providing them significant interference. The licensed users are called primary users and the users of the cognitive radio network are called secondary users. To peacefully coexist with primary users, secondary users should have timely and accurate information about the usage of primary user spectrum. There are two different approaches for secondary users to get this information:

- Through Spectrum Sensing
  In this case, secondary users perform sensing of primary user spectrum in order to detect the vacant spectra called spectrum holes. Spectrum sensing is a crucial function for such opportunistic spectrum access.

S. Maharjan (✉)
Simula Research Laboratory, 1325 Lysaker, Norway
e-mail: sabita@simula.no

F.R. Yu (ed.), *Cognitive Radio Mobile Ad Hoc Networks*,
DOI 10.1007/978-1-4419-6172-3_16, © Springer Science+Business Media, LLC 2011

- Exclusive Information from Primary Users
  In this case, the primary users explicitly provide information about the available spectrum to secondary users. In this model, the primary users get monetary or some other kinds of benefit by allowing the secondary users to use the spectrum.

In an opportunistic spectrum access scenario, there is no motivation for primary users to participate in the spectrum sharing process because they do not get any benefit by letting secondary users use their spectrum. In this approach, the primary users are inflexible and the overall responsibility of maintaining peaceful coexistence with primary users is on the secondary users, thus making the implementation aspect more complex and guaranteeing the performance harder. On the other hand, a resource trading-based approach of spectrum sharing is that primary users can lease the spectrum to secondary users whenever and wherever they are not using the particular bands which in turn gives the primary users monetary or other benefits from secondary users.

Figure 16.1 shows spectrum trading and resource allocation as two different issues in cognitive radio networks. Effective resource allocation is the key to efficient spectrum sharing. Resource allocation can be in terms of frequency band, channel access time, transmission power, etc. and can be between primary users and secondary users and among secondary users. Spectrum trading is the economic aspect of spectrum sharing in an incentive-driven framework of coexistence of primary and secondary users. In a spectrum trading scenario, while primary users compete to sell the spectrum in order to maximize their revenue, secondary users compete to get the spectrum according to their needs at better price to maximize their satisfaction. Spectrum trading can be between primary and secondary users or can be among secondary users only.

There is a crucial need to study the competitive and cooperative strategies of users for multiplayer optimization of the resource allocation problem. Meanwhile, understanding the pricing issues and market structures for spectrum trading is not less important either for practical implementation of cognitive radio networks. Thus, our motivation for this work stems from the need to establish a framework to understand the possibilities and challenges of using economic approaches for deploying cognitive radio networks.

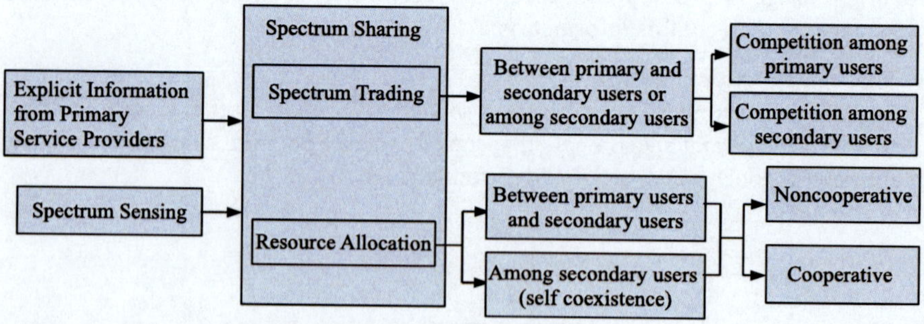

Fig. 16.1 Spectrum sharing issues in cognitive radio networks

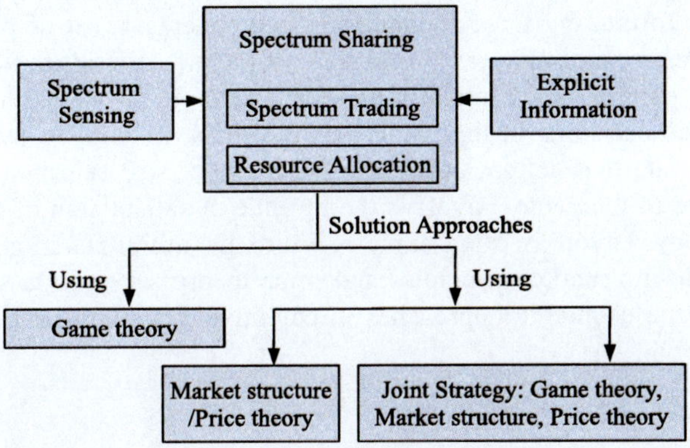

Fig. 16.2 Solutions for spectrum sharing in cognitive radio networks

Different approaches that have been used to model the strategic interactions for spectrum sharing are shown in Fig. 16.2. Game theory serves as a powerful tool in order to model the strategic behavior of primary and secondary users for their coexistence. The economic models include principles such as setting the price of the spectrum available in order to maximize revenue of the primary users, choosing the best seller for the spectrum in order to maximize the satisfaction from the usage of the spectrum for secondary users, modeling market competition, etc. Therefore, many of the existing literature on modeling the economic interactions in wireless networks use a combination of game theory, price theory, and market theory.

The rest of the chapter is organized as following. Different kinds of games that can be applicable to model economic interactions in cognitive radio networks, the concept of equilibrium, inefficiency of Nash equilibrium, and equilibrium selection are described in Section 16.2. An overview of the related work along with the research challenges of using game theory for spectrum sharing are presented in the same section. Price theory and market principles to model the trading activities in cognitive radio networks, related work, and the corresponding research challenges are discussed in Section 16.3. An overview of the work using a combination of game theory, price theory, and market theory is presented in Section 16.4. Section 16.5 presents a classification of related work based on the issues addressed and models used from economic approaches. Open problems for future research are introduced in Section 16.6. Section 16.7 concludes the paper.

16.2 Game Theory

Game theory is the study of conflict and cooperation among individuals, groups, or firms. It provides an analytical framework with a set of mathematical tools for the analysis of interactive decision-making processes. It is a multi-player optimization approach and the concept applies whenever the actions of several players are interdependent.

A game is formed by three fundamental components: a set of players (N), a set of strategies for each player A_i, and a set of payoffs or utilities u_i for player i. Thus, a game is represented in a strategic form as $< N, A_i, u_i >$. A player is the one that makes decisions in the game. A strategy is a complete contingent plan or a decision rule that defines an action that a player will select in every distinguishable state of the game. Payoff is the revenue or satisfaction of the player for a given strategy. Payoff is often expressed through utility functions. Game theory combined with market principles and price theory serves as a strong ground for modeling the economic approaches in cognitive radio networks for spectrum sharing.

16.2.1 Cooperative and Non-cooperative Games

Games can be classified into different categories based on different criteria. Broadly, games can be classified as cooperative and non-cooperative games.

16.2.1.1 Cooperative game

In a cooperative game, there is no competition between players in a group and they act as a single entity to maximize the total group utility. An example is a bargaining game, which is often used to formulate the interaction among cooperative players provided that a player can influence the action of other players. In a bargaining game, the players can negotiate and bargain with each other. A general solution of the bargaining game is the Nash bargaining solution, which can ensure efficiency as well as fairness among the players.

16.2.1.2 Non-cooperative game

A non-cooperative game is the one in which players are selfish and each individual player makes decisions independently. In a non-cooperative environment, players have different (often conflicting) interests. Non-cooperative game theoretical framework is used to obtain an equilibrium solution that optimizes the payoff of all players. One of the most widely used solutions for non-cooperative games is the Nash equilibrium.

16.2.2 Equilibrium: Existence, Refinement and Selection

16.2.2.1 Nash equilibrium

Nash equilibrium is the strategy set for each player at which any player in the game cannot achieve a better solution by deviating unilaterally, given the actions of the other players. In other words, Nash equilibrium is the solution of a non-cooperative game where each player plays its best strategy while the strategy of other players is given. For a two player game, a Nash equilibrium is a pair of strategies $(a_1{}^*, a_2{}^*)$ such that

$$\begin{cases} u_1\,(a_1{}^*, a_2{}^*) \geq u_1\,(a_1, a_2{}^*)\ \forall a_1 \in A_1 \\ u_2\,(a_1{}^*, a_2{}^*) \geq u_2\,(a_1{}^*, a_2)\ \forall a_2 \in A_2, \end{cases} \tag{16.1}$$

where $u_1\,(a_1{}^*, a_2{}^*)$ $(u_2\,(a_1{}^*, a_2{}^*))$ is the payoff of user 1 (2) for strategy $a_1{}^*$ $(a_2{}^*)$ when the best response of user 2 (1) is given.

In many games however, players' strategies may not be deterministic. In such cases, players take pure strategies with certain probabilities (their behavior is randomized) and the equilibrium strategy is called mixed strategy Nash equilibrium or randomized Nash equilibrium.

If the action set of each player is a_1, a_2, a general non-deterministic behavior specifies using a_1 with probability p and a_2 with probability $(1 - p)$. A mixed strategy σ is the probability $p(s)$ of using each of the pure strategies $s \in S$. Let the set of strategies be $S = \{s_a, s_b, s_c, \ldots\}$, then a mixed strategy can be represented as a vector of probabilities: $\sigma = (p(s_a), p(s_b), p(s_c), \ldots)$.

Although, Nash equilibrium is a widely employed solution concept for non-cooperative games, how to find Nash equilibrium and how players converge to the strategy set given by Nash equilibrium is a big issue, especially when the system is implemented in a distributed manner. Apart from that, while some games may not have pure Nash equilibrium, some games may have multiple Nash equilibria. When players do not have information about the moves of other players, the lack of knowledge affects the choice of the equilibrium strategy and this effect is significant especially for one-stage games. Moreover, Nash equilibrium is often inefficient, yielding poor payoffs for the players. In order to address these issues, concepts, such as correlated equilibrium, evolutionary equilibrium, potential games, etc., have emerged. These are explained next.

16.2.2.2 Correlated equilibrium

When a game has multiple Nash equilibria, the choice of the equilibrium that yields maximum payoff for each player is an issue in such cases. A correlated equilibrium is a solution concept widely used to improve the efficiency of Nash equilibrium. It is more general than Nash equilibrium and is based on the assumption of the presence of a trusted authority, which instructs the players which strategy to choose based on an unbiased probabilistic rule. If no individual player has an incentive to deviate from the recommended strategy assuming that the others do not deviate either, such a distribution is called correlated equilibrium.

Consider a simple example: the game *Battle of Sexes*. This game has two players, both of them either want to go to a game G_1 or to a game G_2. The payoff matrix is shown in Table 16.1. The first entry in a cell denotes the payoff of player 1 and the second entry denotes the payoff of player 2. It is clear that (G_1, G_1) and (G_2, G_2) are two pure Nash equilibria, with payoffs $(2, 1)$ and $(1, 2)$, respectively. A mixed strategy Nash equilibrium consists of player 1 choosing G_1 with probability $\frac{2}{3}$ and

Table 16.1 *Battle of Sexes* game: Payoff table

Strategies	G_1	G_2
G_1	(2, 1)	(0, 0)
G_2	(0, 0)	(1, 2)

G_2 with probability $\frac{1}{3}$ and player 2 choosing G_1 with probability $\frac{1}{3}$ and G_2 with probability $\frac{2}{3}$. The expected payoff in this case is $\left(\frac{2}{3}, \frac{2}{3}\right)$, which is fair but lower than even the worst reward in the pure strategy equilibrium.

Now, let us assume that a trusted authority tosses a fair coin and based on the outcome of the coin toss, tells the players what strategies they should take. For example, if the coin shows heads, both players are asked to choose G_1 and both players are recommended to choose G_2 when the outcome is tails. It is important to note that the expected rewards are now higher $\left(\frac{3}{2}, \frac{3}{2}\right)$ compared to that of the mixed Nash equilibrium. Thus, correlated equilibrium can significantly improve the efficiency of the game for each player provided that an unbiased authority is present.

For a game, there can exist multiple correlated equilibria. The set of mixed strategy Nash equilibrium is a subset of the set of correlated equilibria too. A more formal definition of correlated equilibrium is as following:

A probability distribution P_d over $A_1 \times, \ldots, \times A_{|N|}$ is a correlated equilibrium if for every strategy $a_i^{'} \in A_i$ such that $p(a_i) > 0$ and for every alternative strategy $a_i^{'} \in A_i$,

$$\sum_{a_{-i} \in A_{-i}} p(a_i, a_{-i}) u(a_i, a_{-i}) \geq p(a_i, a_{-i}) u\left(a_i^{'}, a_{-i}\right), \tag{16.2}$$

where a_{-i} denotes the strategy chosen by for player $j : j \neq i$.

16.2.2.3 Evolutionary Equilibrium

In games of incomplete information, when multiple equilibria exist and when there is no authority to recommend the strategies to choose, the players are not certain about the strategy to take. Evolutionary game theory (EGT) [43] provides a good framework to address the strategic uncertainty of players. Evolutionary equilibrium is the concept used as the solution of evolutionary games. In a distributed scenario, despite facing the uncertainty, the players can gradually approach a robust equilibrium strategy by taking out-of-equilibrium behavior and learning through the strategic interactions with other players in repeated stages of the game. This equilibrium is called evolutionarily stable strategy (ESS) [38], which is a strategy such that, if all members of the population adopt it, then no mutant strategy can invade the population under the influence of natural selection.

Let $u(p, p^{'})$ denote the payoff of an individual using the strategy p against another individual using the strategy p'. Then, a strategy p^* is an ESS if and only if,

$$\begin{cases} u(p, p^*) \le u(p^*, p^*) \ \forall p \neq p^* \text{ and} \\ \text{If } u(p, p^*) = u(p^*, p^*), u(p, p) \le u(p^*, p) \ \ \forall p \neq p^* \end{cases} \tag{16.3}$$

16.2.2.4 Potential Games

The existence and uniqueness of equilibrium can be guaranteed if the game has certain special features. A game with those special features is called a potential game [22]. A game $< N, A_i, u_i >$ is an exact potential game if there is a potential function $P : A \to \mathbf{R}$ such that

$$P(a_i, a_{-i}) - P\left(a'_i, a_{-i}\right) = u(a_i, a_{-i}) - u\left(a'_i, a_{-i}\right) \ \ \forall i \in N,\ a \in A, \text{ and } a'_i \in A_i \tag{16.4}$$

where $u(a_i, a_{-i})$ denotes the utility of player i for taking action a_i, given the strategy of other players. The game is said to be an ordinal potential game if

$$sgn\left(P(a_i, a_{-i}) - P\left(a'_i, a_{-i}\right)\right) = sgn\left(u(a_i, a_{-i}) - u\left(a'_i, a_{-i}\right)\right) \ \ \forall i \in N, \\ a \in A, \text{ and } a'_i \in A_i \tag{16.5}$$

where $sgn(.)$ is the sign function.

16.2.3 Different Game Models

Some game models that have been used extensively for analyzing the strategic interactions among users for spectrum sharing are as following:

16.2.3.1 Stackelberg Game

The Stackelberg leadership model [5] is a strategic game in which there is at least one player defined as the leader who can make the decision and commit the strategy on the price before other players who are defined as followers. The players engage in Stackelberg competition if one has some kind of incentive to move first. The strategy chosen by the leader can be observed by the followers, and the followers can adapt their decisions accordingly. The leader can choose a strategy such that its profit is maximized, given that the followers will choose their best responses. This solution is called the Stackelberg equilibrium and can be obtained by backward induction technique. With backward induction, the best response of the follower is obtained first given the price set by the leader, i.e., $p_f^* = \mathcal{B}_f\left(p_l^*\right)$, where p_f^*, p_l^* are the price set by the follower based on its best response and the price set by the leader to maximize its profit, respectively, and $\mathcal{B}_f()$ is the best response of the follower. Then, this best response is used to compute the profit of the leader, and the leader chooses a strategy for which the profit is maximized, i.e., $p_l^* = argmax_{p_l} u(p_l, \mathcal{B}_f(p_l))$.

16.2.3.2 Bertrand Game

In a Bertrand game [5], there are a finite number of firms that decide on the service prices simultaneously. Given the price offered by a service provider, based on a demand function, the amount of commodity requested from the users can be determined. Then, the profit is computed and used in a profit maximization problem for a service provider to obtain the best response in terms of setting the service price. For a spectrum trading scenario, the service providers are the primary users, the consumers are the secondary users and the size of the spectrum will change according to the price set by the primary users. When the service providers offer their prices simultaneously, Nash equilibrium is the solution. At Nash equilibrium, the best response of service provider i in terms of the price it sets is given by $p_i^* = \mathcal{B}_i(p_{-i}{}^*)$, $\forall i$, where $p_{-i}{}^*$ denotes the set of best responses for player $j : j \neq i$. The best responses of the players/firms in Stackelberg model and Bertrand model are obtained in similar way. However, the first move gives the leader in Stackelberg game a crucial advantage. The interaction in Stackelberg game is more dynamic due to the timing in strategy adaptation compared to the Bertrand model. But there is an important assumption of the perfect information in Stackelberg game that the follower observes the strategy chosen by the leader. If this assumption is released and all firms decide their service prices simultaneously, Stackelberg model reduces to Bertrand model.

16.2.3.3 Cournot Game

In Cournot game [5], the competition is in terms of the quantity of the commodity. In a spectrum trading scenario, the secondary users are the consumers of the spectrum offered by the primary users. Each secondary user decides its strategy simultaneously based on the price set by the primary users, and the decision of each user is affected by the strategies of other users. At Nash equilibrium, the best response of secondary user i in terms of the size of spectrum requested is given by $b_i^* = \mathcal{B}_i\left(b_{-i}{}^*\right)$, $\forall i$, where $b_{-i}{}^*$ denotes the set of best responses for secondary user $j : j \neq i$.

16.2.3.4 Coalition Games

Cooperative game theory provides analytical tools to study the behavior of rational players when they collaborate. The group of cooperating players that can strengthen the players' position in the game is called a coalition, and all players forming a coalition act as a single entity. A coalition S is defined to be a subset of the total set of users N. The coalition form of a game is given by the pair (N, v), where v is a real-valued function, called the characteristic function. $v(S)$ is the value of the cooperation for coalition S. A coalition is stable if and only if no other coalition has the incentive and/or power to spoil it.

Based on application oriented approach, coalitional games can be classified into three categories [34]: canonical (coalitional) games, coalition formation games, and coalitional graph games. In canonical games, no group of players can do worse by joining a coalition than by acting non-cooperatively. In coalition formation games, forming a coalition brings advantage to its members but the gains are limited by a cost for forming the coalition. In coalitional graph games, the coalitional game is in graph form and the interconnection between the players strongly affects the characteristics as well as the outcome of the game.

16.2.3.5 Games with Learning

In games with incomplete information players do not have the knowledge about the strategies of other players. In fact, even if the information is available, exchange of this information can be a significant overhead. As a consequence, distributed implementations are preferred. For a distributed scenario, each player needs to learn about the strategy of other players through repeated interactions with them. Such games where players adapt their strategies based on the interactions with players are often called games with learning. These games are closely related to the concept of "repetition" in games. If the games are repeated, players can learn and adapt their behaviors and strategies in subsequent rounds of the game. Any of the specific game models described above can be a game with learning if it incorporates a learning algorithm so that the players can gradually approach the desired equilibrium based on the history of their won payoffs.

16.2.3.6 Repeated Games

Repeated games [5] are an important tool in order to understand the concepts of reputation and punishment in game theory. A repeated game allows a strategy to be contingent on the past moves, thus allowing threats and promises about future behavior to influence current behavior, which create possibilities for cooperation among greedy users. If a greedy user behaves selfishly and chooses the strategy to optimize his/her individual payoff, it can enjoy the benefit in one round. However, if this user has to depend on others as well for future rounds of the game, it will be punished by them. Players must therefore consider the effects that their chosen strategy in any round of the game will have on opponents' strategies in subsequent rounds. For an infinitely repeated game $< N, A_i, u_i >$, the payoff function for player i is the discounted average of the immediate payoffs from each round of the repeated game: $u_i(a^1, a^2, \ldots, a^t) = (1-\delta)\sum_{j=0}^{\infty} \delta^{(t-1)} u_i(a^t)$, where δ is the discount factor ($0 < \delta < 1$) and for every value of time t, the chosen action at t depends on the history $(a^1, a^2, \ldots, a^{t-1})$.

16.2.3.7 Stochastic games

Another dimension of interest in competition and cooperation games for spectrum resources is the state of the network. In a dynamic scenario where the spectrum opportunities and the surrounding radio opportunities keep changing with time, the interaction between the players can be better modeled by using the concept of stochastic games.

In addition to the players and their action sets $A_1, A_2, \ldots, A_{|N|}$, a stochastic game [37] comprises of a set of states, X with a stage game defined for each state. In each state x, player i can choose actions from a set $A_i(x)$. When players interact by playing a similar stage game numerous times, the game is called a repeated game. A stochastic game is a repeated game with stochastic transitions. In a stochastic game, every time a player executes its chosen action, the game moves to a new state with a transition probability which depends on the current state of the network and the action chosen by the player. For example, for a two player stochastic game, if the players are in state x and they choose actions $a_1 \in A_1(x)$ and $a_2 \in A_2(x)$ respectively, the players receive immediate payoffs $u_1(x, a_1, a_2)$ and $u_2(x, a_1, a_2)$ and the probability that they find themselves in state x' for the next decision is $p(x'|x, a_1, a_2)$.

16.2.4 Applications of Game Theory in Spectrum Sharing

Game formulations can be used for multiplayer optimization to achieve individual optimal solution for resource allocation. The use of game theoretic models for resource allocation has mainly focused on issues such as admission control, throughput optimization, power control, channel allocation. Table 16.2 summarizes the related work on the use of game theory for resource allocation, in terms of the specific issue addressed, approach/model(s) used, and the solution proposed. These works are explained next.

In [6], a game theoretical approach is proposed for distributed resource allocation in wireless networks. Power control at the user level and throughput control at the system level are linked through non-cooperative games.

In [7], a distributed non-cooperative game is proposed for joint subchannel assignment, adaptive modulation, and power control for multi-cell multi-user OFDMA networks. In order to improve the performance of Nash equilibrium points, a virtual referee is introduced in the system that can modify the rule of the resource competition game for efficient resource sharing.

In [10], the authors modeled the channel/power allocation for cognitive radios considering IEEE 802.22 framework. The strategic behavior of the system was studied considering the limit on the total interference from all opportunistic transmissions for each primary user as well as the minimum SINR requirement of the cognitive radios and a cooperative scheme based on Nash bargaining solution was proposed for optimal channel/power allocations. In [36], the self-coexistence of multiple overlapping IEEE 802.22 networks operated by multiple wireless service providers that compete for resources and try to seek a spectrum band without any interference from other coexisting IEEE 802.22 networks was investigated from a game theoretic perspective. The dynamic channel switching was modeled as a distributed modified minority game (MMG), in which each user has to decide whether

Table 16.2 Summary of related work on spectrum sharing using game theory

Paper	Issue(s) addressed	Approach/specific model(s) used	Solution
[6]	Power control, throughput control	Non-cooperative, distributed game	Nash equilibrium (Optimal for power, Optimal or suboptimal for network throughput)
[7]	Power control, rate adaptation, subchannel assignment	Non-cooperative, distributed game	Nash equilibrium (Transmitted power)
[10]	Channel/power allocation	Non-cooperative/ cooperative, distributed game	Nash equilibrium (Radio range)/Nash bargaining solution
[36]	Spectrum allocation (Channel switching)	Non-cooperative, distributed game (Modified minority game)	Nash equilibrium (Channel switching probability)
[45]	Channel access time	Cooperative Stackelberg game (between primary and secondary users), non-cooperative payment selection game (among secondary users), distributed game	Nash equilibrium (Payment vector)
[3]	Spectrum assignment	Cooperative, distributed, bargaining game	Bargaining-based solution (Spectrum usage)
[20]	Channel allocation	Cooperative, distributed game	Correlated equilibrium (Spectrum access)
[8]	Packet forwarding	Cooperative, distributed game (repeated coalitional game)	Min-max fairness, average fairness and market fairness investigated
[35]	Spectrum sensing	Cooperative, distributed game (non-transferable utility coalitional game)	Performance compared with non-cooperative and centralized scheme
[40]	Spectrum sensing	Non-cooperative, decentralized game (evolutionary game)	Evolutionarily stable strategy (probability of contributing in sensing)
[41]	Security/defence mechanism	Non-cooperative (zero-sum), distributed (stochastic game)	Minimax equilibrium
[23]	Spectrum allocation	Non-cooperative cooperative, distributed (potential game)	Nash equilibrium (pure strategy and mixed strategy)

to leave a particular band or to continue using it when another user also appears in the same band.

In [45], a cooperative cognitive radio framework is formulated as a Stackelberg game where primary user acting as the leader, selects some of the secondary users to be the cooperative relay, and in return, leases portion of the channel access time to them for their own data transmission. Selected secondary transmitters, acting as the followers, can use the wireless channel only if they cooperate with the primary link and meanwhile make a certain amount of payment to the primary system.

In [3], the authors propose a local bargaining approach to achieve distributed conflict-free spectrum assignment adapted to network topology changes assuming that there is a collaboration between network nodes to improve system utility. In this chapter, the authors propose fairness bargaining with feed poverty to improve fairness in spectrum assignment and have derived a lower bound on the spectrum assignment (poverty line) that each node can get from bargaining.

In [20], a decentralized dynamic spectrum access scheme is proposed for cognitive radios considering the application domain as a set of collision channels from game theoretical perspective. The authors proposed the use of an adaptive procedure called regret tracking, which converges even when multiple users are adapting their behavior simultaneously, for which correlated equilibrium (in terms of channel allocation) is investigated.

In [8], an approach based on coalition games is proposed for symbiotic cooperation between boundary nodes and backbone nodes in selfish packet forwarding wireless networks. Different fairness criteria are investigated including market fairness. In addition, a joint protocol is designed using both repeated games and coalition games for packet forwarding, and it has been shown that the network connectivity can be significantly improved using the proposed protocol compared to using pure repeated game approach.

In [35], distributed collaborative spectrum sensing algorithm is developed based on a dynamic coalition formation game among secondary users to improve the overall probability of miss detection with increase in the probability of false alarm as the cost for coalition formation.

In [40], the authors use evolutionary game framework to analyze the behavior dynamics of non-cooperative secondary users and to enforce cooperation among them for the spectrum sensing game. The authors derive the evolutionarily stable strategy for the secondary users in terms of the probability of contributing in the sensing game. In addition, the authors develop a distributed learning algorithm for the secondary users so that their strategy converges to the ESS based solely on the observation of their own payoffs.

When malicious attackers exist in the network whose objective is to cause harm to the legitimate users by reducing their benefit through the utilization of spectrum, securing effective spectrum sharing becomes an issue of critical importance. A recent work [41] proposes an anti-jamming defence mechanism for secondary users against cognitive attackers that can launch jamming attack to prevent efficient utilization of spectrum opportunities, using stochastic game modeling. Since the secondary users and attackers have opposite objectives, the anti-jamming game is modeled as a zero-sum game and the optimal policy of the secondary users, which maximizes the spectrum-efficient throughput, is obtained by the minimax-Q learning algorithm.

Nie and Comanici [23] is an example of the application of potential game for channel allocation in cognitive radio. In [23], the authors consider two different type of users: selfish users and cooperative users and define different objective functions for them for the spectrum sharing game. The utility functions are defined to satisfy the conditions of a potential game, and thus the channel allocation problem converges to a deterministic channel allocation Nash equilibrium point. The authors also propose a no-regret learning implementation and it is shown to have similar performance with the potential game when cooperation is enforced.

16.2.5 Research Challenges in Using Game Theory

Although game theory is a very powerful tool to model and analyze the interactions among primary and/or secondary users for spectrum sharing, its implementation in a distributed scenario faces numerous challenges, especially because of the need of dynamic access of the spectrum and heterogeneous requirements on the quality/quantity of the offered spectrum for different users in the network. Some of the challenges for specific game models are explained next.

The availability of perfect knowledge in Stackelberg model may be quite costly for distributed implementation of cognitive radio networks. How to efficiently manage the information flow among leaders and followers in a dynamic scenario is a big challenge. Even if the information can be made available to all players, the overhead due to this information exchange (which increases with the number of players in the game) cannot be ignored. Although, the prices and profits of some service providers may be higher at Stackelberg equilibrium than at Nash equilibrium [29], Bertrand/Cournot game models (thus Nash equilibrium solution) may incur relatively less overhead for the distributed approach because of the simultaneity of moves.

Cooperative framework is often used to model the spectrum sharing scenario in cognitive radio networks. However, finding users with common interest to form coalitions and get them to act cooperatively, itself is a big issue. Even if there exist users that can be symbiotic, the change in network topology, change in channel conditions, motivation toward cooperating with other users, etc., may cause the coalition not to be stable. Introducing incentive or monetary gain-based schemes can be quite expensive and inefficient due to divergence of user interests.

The concept of repeated games provides a good direction toward counteracting the possibility of collusive behavior in a network by certain selfish players and to adapt one's strategies accordingly. However, algorithms are necessary to effectively estimate long-term profits. In a practical network, the overhead to maintain the database of the strategies of each player for a number of stages of the game and to update it is a huge challenge in a distributed scenario. In addition, when the game is being played the second last time or the last time, the future profits are not meaningful. So, in such a case, the concept of repeated games may not be effective enough. Besides, if an individual deviates from the optimal strategy for the system for its own benefit, to find out which user it was, is another issue. It may be possible to locate the group from which the deviating action occurred but locating the exact player from the group may still be difficult considering the time limitations, especially when users have the right not to disclose their strategies.

16.3 Price Theory and Market Theory

16.3.1 Price Theory

Price theory explains how relative prices are determined and how prices function to coordinate the economic activities. For incentive/monetary gain-based spectrum sharing, appropriate pricing schemes are necessary for setting up the price of the spectrum, formulating economic models and maximizing the payoffs of both primary and secondary users. Pricing is an important issue not only to maximize the revenue of the service providers but also to prevent unnecessary competition (to reserve the resources) and to allocate the radio resource efficiently.

Auction and bargaining are the pricing strategies mostly used for resource trading. These are explained next.

16.3.1.1 Auction Theory

An auction [17] is a decentralized form of trading, widely known for providing efficient allocation of scarce resources. Sellers use auctions to improve revenue by dynamically pricing based on buyer demands. Buyers benefit since auctions assign resources to buyers who value them the most. In a game-theoretic auction model, the action set of each player is a set of bid functions or reservation prices. Each bid function maps the player's value (in case of a buyer) or cost (in case of a seller) to a bid price.

There are different kinds of auctions such as English auction, Dutch auction, sealed-bid first price auction, Vickrey auction, and double auction. English auction, which is the most common form of auction, is the ascending price auction. Participants bid openly against one another, with each subsequent bid higher than the previous bid. The auction ends when no participant is willing to bid further and the highest bidder gets the commodity at its bid. Dutch auction is an open descending price auction. In the traditional Dutch auction the auctioneer begins with a high asking price which is lowered until some participant is willing to accept the auctioneer's price. The winning participant pays the last announced price. In sealed-bid first-price auction, all bidders simultaneously submit sealed bids so that no bidder knows the bid of any other participant. The highest bidder pays the price they submitted. Vickrey auction is the sealed-bid second-price auction, in which the players submit sealed bids and the highest bidder wins, but pays only as much as the second-highest bid. Vickrey auction is faster than English auction and can be applied for spectrum management when secondary users are all bidding for the available spectrum offered by the primary user. When there are multiple items and multiple buyers, double auction is often used to model the double competition. Vickrey–Clarke–Groves (VCG) auction [17] is a generalization of a Vickrey auction for multiple items. A VCG auction is a type of sealed-bid auction where multiple items are up for bid, and each bidder submits a different value for each item. The auction system assigns the items in a socially optimal manner, while ensuring that each bidder receives at most one item. This mechanism charges each individual "the harm they cause to others" and ensures that the optimal strategy for a bidder is to bid the true valuations of the objects.

16.3.1.2 Bargain Theory

Bargaining is a type of negotiation in which the buyer and seller of a commodity or service dispute the price that will be paid and the exact nature of the transaction that will take place, and eventually come to an agreement. Bargaining is an alternative pricing strategy to fixed prices. In a bargaining scenario, the buyer's willingness to pay is dominant over the actual price of the commodity. It allows for capturing more consumer surplus as it allows price discrimination, a process whereby a seller can charge a higher price to the buyer who is more eager.

16.3.2 Market Theory

A market is the most efficient known mechanism for the allocation of goods and services. A market consists of sellers and buyers of a commodity or service. Appropriate pricing schemes are necessary to maintain the stability of the market. So, pricing is very closely related to market theory. Market theory can be used to investigate the market structure for resource trading in cognitive radio networks as well. For spectrum trading, primary users are the sellers and secondary users are the buyers. The stability of the trading and the revenues obtained by the primary and secondary users depends on the market structure, pricing policies as well as the strategies of all users. Some of the useful concepts of market theory used for spectrum trading are as following:

16.3.2.1 Monopoly

Monopoly is the simplest market structure where there is only one seller in the market. Since there is a single seller in this market structure, the seller can optimize the trading to achieve the highest profit based on the demand from buyers. This market structure can be either seller driven or buyer driven. In the former case, the seller sets the price and broadcasts the information on available spectrum. The buyer determines the spectrum demand and proceeds to buy the spectrum. Alternatively, the market can be buyer driven where the buyer proposes the price and specifies the requested spectrum. The seller chooses the buyer with the best incentive and then allocates the spectrum accordingly. This buyer-driven approach is auction.

16.3.2.2 Oligopoly

In an oligopoly market structure, a small number of firms dominate the market. Few firms compete with each other independently to achieve the highest profit by controlling the quantity or the price of the supplied commodity. Oligopoly differs from monopoly in the sense that there are multiple firms, but it is also different from perfect competition scenario because there are few firms only and each firm is large enough such that the effect of each firm in setting the market price is significant, thus making it necessary for all firms to take into account the strategies of all other firms as well.

16.3.2.3 Competitive Equilibrium

In a monopolistic context, the pricing is a single level of game between buyers. However, when there are multiple sellers in the market for the same commodity, the competition between them can highly affect the results of price determination. This competition introduces an additional level of game among the service providers. In double auction scenario, competitive equilibrium concept is often used to predict the possible outcomes of the game theoretically. Competitive market equilibrium is the economic equilibrium, appropriate for the analysis of commodity markets with flexible prices and many traders. Buyers demand less as the price of a commodity increases. On the other hand, sellers tend to produce more as the price increases. The corresponding demand and supply curves are shown in Fig. 16.3. The intersection of the two curves, i.e., the price at which the quantity supplied of a product/service and the quantity of it demanded are equal is called the market equilibrium or competitive equilibrium.

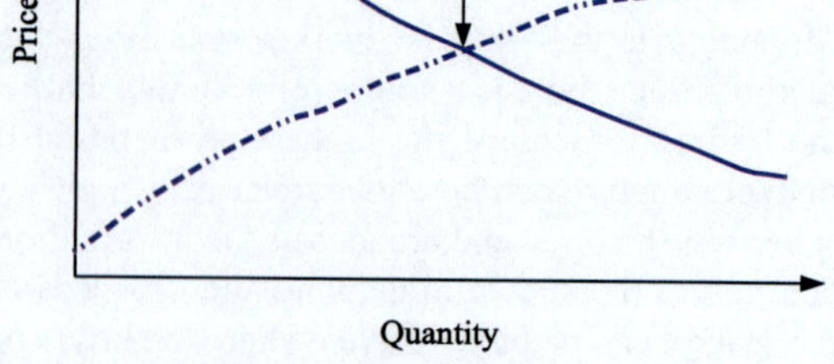

Fig. 16.3 Supply and demand curves: Competitive equilibrium

16.3.3 *Applications of Price Theory and Market Theory in Spectrum Sharing*

In this section, an overview of the existing work using price theory and market principles to model economic interactions for spectrum sharing is presented (summarized in Table 16.3 in terms of the specific issue addressed, structure, approach/model(s) used and the solution proposed). Although price theory and market theory provide us models to address pricing issues and market stability, to model the interdependency of the sellers and buyers in the market and their strategic interactions, game theory is used. So, many literature use a combination of price theory, market theory, and game theory for analyzing spectrum sharing. An overview of these works is presented in Section 16.4.

In [33], the problem of a CDMA operator participating in a dynamic spectrum allocation scheme is addressed in a cooperative framework based on multi-unit Vickrey auction. A spectrum manager implements DSA by periodically auctioning short-term spectrum licenses and a pricing-driven solution based on the willingness to pay of each user is introduced.

In [4], a framework based on an auction mechanism was presented for dynamic spectrum access using classical optimization approach. In the system model considered in [4], multiple spectrum buyers submit spectrum demand function, which is based on piecewise linear price demand (PLPD) to the spectrum owner that formulates an optimization problem to maximize revenue under an interference constraint. The authors propose to restrict the interference constraints and reduce them into a number that grows linearly with the number of buyers.

In [25], a spectrum trading model based on multiple markets for different frequency bands is proposed between the primary and the secondary services. The authors have investigated two different cases: the first one with equilibrium pricing where spectrum supply is equal to spectrum demand and the second one is the case where the sellers do not offer the equilibrium price and have proposed models for

Table 16.3 Summary of related work on spectrum sharing using price theory/market theory

Paper	Issue(s) addressed	Structure	Approach/specific model(s) used	Solution
[33]	Spectrum allocation	Cooperative, centralized	Optimization (multi-unit Vickrey auction)	Optimal price
[4]	Spectrum access	Cooperative, centralized	Optimization	Optimal price
[25]	Spectrum trading	Competitive (among secondary users), distributed	Joint pricing and marketing theory	Pricing solution obtained through market equilibrium and disequilibrium
[15]	Spectrum trading	Competitive (among secondary users), distributed	Optimization VCG auction	Optimal/ sub-optimal revenue

both cases using linear feedback time-invariant control systems. Classical control system stability techniques are used to analyze the dynamics of market behavior under both cases.

In [15], the authors have designed auction mechanisms for spectrum sharing between primary and secondary users that maximize the expected revenue of the primary users. The authors propose an optimal auction mechanism based on the concept of virtual valuation, which uses VCG mechanism to maintain the truthfulness of bidding and they have considered the interference constraints such that no two neighboring nodes are allocated the same channel. VCG auction, although, provides optimal revenue for primary user, it has high complexity. So, the authors propose sub-optimal auction scheme where the revenue is sub-optimal but it is computationally more efficient and maintains the truthful bidding.

16.3.4 Research Challenges of Using Price Theory and Market Theory

Though price theory/market theory can be applicable to model the spectrum trading for cognitive radio networks, the information exchange required for pricing and negotiation poses a big challenge. For example, Vickrey–Clarke–Groves auction is one of the most used auctions for resource trading. It can be used to achieve a socially optimal allocation. However, it requires gathering global information from the users. The communication overhead and computational complexity to gather and manage global information in a distributed scenario may be quite costly.

For bargaining, users form groups and bargain with other groups. However, the larger the groups are, the more is the complexity of bargaining due to high costs of synchronization and communication overhead. Efficient formation of bargaining groups and effective communication between them in a distributed spectrum sharing scenario is also another issue. In addition, the stability of the bargaining groups formed in a network with rapidly changing topology and other underlying conditions is hard to maintain. To dynamically split and merge to firm optimal coalitions with only local information available, is again a big problem.

The pricing theory with per unit price of the resource same for all users for any amount of resource demanded may not produce highest revenue for the sellers. So, a discriminatory pricing scheme as proposed in [4] may be better in terms of maximizing the revenue. However, the computational complexity of this kind of scheme is yet a big issue.

16.4 Joint Strategy: Game Theory, Market Theory, and Price Theory

While market principles and price theory are needed to model economic activities, game theory is necessary to analyze the interdependency and strategies of the users for spectrum sharing. So, many literature use a combination of these to investigate spectrum trading and resource allocation in cognitive radio networks. Table 16.4

Table 16.4 Summary of related work on spectrum sharing using joint strategy

Paper	Issue(s) addressed	Structure	Specific model(s) used	Solution
[2]	Power control	Non-cooperative, distributed	Non-cooperative game, pricing mechanism	Nash equilibrium (For uniformly strictly convex pricing function)
[11]	Channel access	Non-cooperative, distributed	Non-cooperative game, SINR auction, power auction	Nash equilibrium (Bidding profile, power profile)
[42]	Joint power/ channel allocation	Non-cooperative/ cooperative (among CR pairs), distributed	Non-cooperative game pricing-based cooperation	Nash equilibrium/ Pareto optimum boundary (Power vector)
[26]	Spectrum trading	Competitive (among primary services), centralized/distributed	Non-cooperative game, Bertrand game, Repeated game, Oligopoly market	Nash equilibrium, optimal spectrum price
[24, 27]	Spectrum trading	Competitive (among secondary users), centralized/distributed	Non-cooperative game, Cournot game, Oligopoly market	Nash equilibrium (spectrum size)
[29]	Spectrum pricing	Competitive (among service providers), centralized	Non-cooperative game, Stackelberg game, Bertrand game, Oligopoly market	Stackelberg equilibrium, Nash equilibrium (spectrum price)
[28]	Spectrum trading	Market equilibrium/ competitive/cooperative (among primary service providers), distributed	Non-cooperative game, Optimization, Oligopoly market	Market equilibrium, Nash equilibrium, optimal price
[32]	Spectrum trading	Competitive (among primary users, among secondary users), distributed	Non-cooperative game, (among primary users), Evolutionary game (among secondary users), Oligopoly market	Nash equilibrium (spectrum price), Evolutionary equilibrium (spectrum size)
[14]	Spectrum allocation	Non-cooperative, distributed	Non-cooperative game, Double auction, Pricing based allocation	Nash bargaining solution (lower bound on payoff), Competitive equilibrium (without user collusion)

Table 16.4 (continued)

Paper	Issue(s) addressed	Structure	Specific model(s) used	Solution
[18]	Slotted resource allocation	Competitive (among providers), distributed	Non-cooperative two stage pricing game	Nash equilibrium of the pricing game (under mild conditions)
[44]	Spectrum trading/access	Competitive (among primary users), centralized/distributed	Non-cooperative game	Nash equilibrium (spectrum price)
[13]	Spectrum leasing	Non-cooperative (between primary and secondary users)	Non-cooperative power control game	Nash equilibrium (transmission power)
[31]	Joint spectrum bidding and pricing	Non-cooperative, centralized	Non-cooperative game, Sealed-bid double auction	Nash equilibrium (spectrum price)
[30]	Spectrum trading	Non-cooperative (among TV broadcasters), (among WRAN users), distributed	Non-cooperative game, Generalized fading memory scheme, Microeconomic approach	Market equilibrium (spectrum price)

summarizes the related work using joint strategy in terms of the specific issue addressed, structure, approach/model(s) used, and the solution proposed, each of which is briefly explained next.

In [2], the CDMA uplink power control in a multicell CDMA wireless network model is addressed as a non-cooperative game. The game incorporates a pricing mechanism that limits the overall interference and preserves battery energy of mobiles. The concept of outage probability was introduced as a performance metric for the quality of the channel. Distributed iterative power algorithms are analyzed using an outage probability-based utility function for a generalized fading channel model.

In [11], a non-cooperative game was formulated to address the problem of spectrum sharing among users using spread spectrum signaling, in a distributed scenario to access the channel subject to a constraint on the interference temperature at a measurement point and two auction mechanisms, SINR auction and power auction, are proposed for allocating received power.

In [42], a joint power/channel allocation scheme is proposed using a distributed pricing approach for cognitive radio networks and a frequency-dependent power mask constraint is introduced for secondary users in addition to maximum transmission power constraint and minimum SINR constraint.

In [26], the issue of spectrum pricing in cognitive radio network is addressed for multiple primary services and 1 secondary service. The trading of spectrum between primary and secondary services was modeled as an oligopoly market. A Bertrand game model was applied for price competition among primary services to obtain the Nash equilibrium pricing. Distributed algorithms were presented to obtain the solution of the dynamic game, when primary services have to make decisions based only on the spectrum demand from the secondary service. A repeated game was formulated to analyze the behavior of selfish primary operators that try to deviate from the equilibrium point to increase their profit at the cost of lower profit to other primary operators.

In [24, 27], the competition among multiple secondary users for spectrum offered by 1 primary user was modeled using Cournot game. In [27], the problem is formulated as an oligopoly market competition and the spectrum allocation for secondary users is obtained through non-cooperative game. The competition is in terms of the size of the spectrum they request. A dynamic game is formulated in which the selection of strategy by the secondary users is solely based on the pricing information obtained from the primary user.

In [29], the authors modeled service competition and pricing in a WiMAX and WiFi based heterogeneous wireless access network using non-cooperative Stackelberg and Bertrand game models, respectively, and the performance was compared for the two models.

In [28], spectrum trading for cognitive radio networks was investigated considering multiple primary services that are willing to sell the available spectrum to the secondary service. Distributed algorithms were presented for three different pricing schemes: market equilibrium, competitive, and cooperative pricing models and the performance of all three schemes as compared.

In [32], the problem of spectrum trading with multiple primary users selling spectrum opportunities to multiple secondary users is considered. The competition among primary users is formulated as a non-cooperative game where each primary user sets the size of spectrum to be shared and the price of the spectrum such that its own payoff is maximized. It is assumed that the secondary users can evolve over time to buy the spectrum opportunities that provide the best payoff in terms of performance and price.

In [14], spectrum allocation among primary and secondary users is modeled as a bilateral pricing process to maximize the utilities of both primary and secondary users and a distributed collusion-resistant dynamic pricing approach with optimal reserve prices was proposed to achieve efficient spectrum allocation while combating user collusion. Double auction scenario was considered for the pricing game. A belief function was introduced that builds up certain belief of other players' future possible strategies for each user to assist its decision making.

In [18], the competition among providers was studied on a non-cooperative game theoretic framework. The authors introduced a pricing model dealing with how for fixed prices, total demand is split among providers following Wardrop's principle and determined the existence and uniqueness of Nash equilibrium under mild conditions.

In [44], the economic interactions are modeled considering both price of the offered spectrum and its quality. The analysis scenario consists of multiple self-interested spectrum providers operating with different technologies and offering the spectrum at different costs that compete with each other to get potential customers that are grouped into two categories: quality sensitive group and price sensitive group. In [44]. The authors have proposed a practical price updating strategy using structured stochastic learning for which the price is shown to converge to the optimal equilibrium.

In [13], a game theoretic framework was developed to facilitate dynamic spectrum leasing (DSL) in cognitive radio networks in which primary users are also included as active decision makers in a non-cooperative game with secondary users by selecting an interference cap on the total interference they are willing to tolerate.

In [31], the joint spectrum bidding and pricing scheme was proposed for dynamic spectrum access in the exclusive usage model for IEEE 802.22 based cognitive radio network. Multiple TV broadcasters offer the available TV bands and WRAN service providers bid for these TV bands. A sealed-bid double-auction scenario was considered for the procurement of TV bands from TV broadcasters in terms of the number of TV bands and the trading price. After buying TV bands, multiple WRAN service providers compete with each other to sell the spectrum to WRAN users. A non-cooperative game was formulated to model the competitive environment for bidding and pricing strategies.

In [30] the authors have proposed a market-equilibrium based model for spectrum trading between primary and secondary services using supply and demand functions. A non-cooperative game is formulated between primary and secondary users where a distributed generalized fading memory algorithm is used by the secondary service to estimate spectrum price and adjust spectrum demand accordingly so that the market equilibrium can be reached for the price and size of the spectrum allocated for the secondary service by the primary service.

16.5 Classification of Related Work Based on Issues and Solutions

In this section, we categorize the related work on modeling the economic interactions in cognitive radio networks on three different bases as following:

Figure 16.4 shows the classification based on different aspects of spectrum trading and resource allocation as described in Section 16.1. The related work on spectrum trading between primary and secondary users address competition among primary users and competition among secondary users. While [18, 26, 28, 44] deal with price competition among primary users/service providers for profit maximization, [24] and [27] address the competition among secondary users in terms of the size or quality of spectrum demanded. On the other hand, [14, 30, 32] address the competition among primary services as well as among secondary users for spectrum trading.

In terms of resource allocation, while [13] and [45] consider the coexistence between primary service providers and secondary users, [20, 23, 36, 42] explore

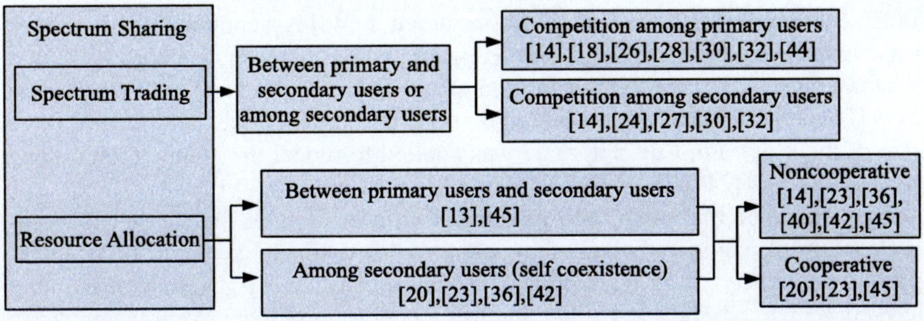

Fig. 16.4 Classification of related work on spectrum sharing based on resource allocation and spectrum trading

self-coexistence among secondary users. Although both [13] and [45] consider the coexistence between primary and secondary systems, the framework considered in [13] is non-cooperative, while [45] is based on the cooperation between primary and secondary users. The payment selection game among secondary users in [45] however, is a non-cooperative game. For self-coexistence among secondary users, while [20] considers a cooperative approach for channel allocation, [36] proposes a non-cooperative dynamic channel switching game. On the other hand, [42] considers a non-cooperative framework and a pricing-based cooperative approach for self-coexistence of secondary users. In [23], both cooperative and selfish spectrum sharing etiquettes are considered for the adaptive channel allocation problem based on the level of cooperation of a particular user. Similarly, in [40], collaboration is enforced among selfish secondary users for cooperative spectrum sensing.

Figure 16.5 shows the classification based on the particular game models described in Section 16.2. Both [29] and [45] use Stackelberg game model. However, while [29] uses Stackelberg game for price competition among WiMAX and

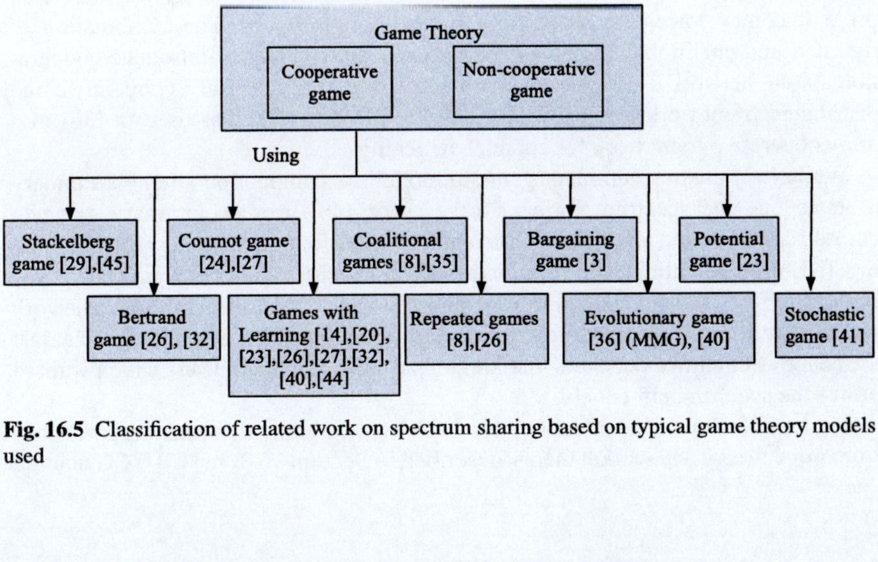

Fig. 16.5 Classification of related work on spectrum sharing based on typical game theory models used

WiFi service providers, the approach considered in [45] is a cooperative framework between primary and secondary users. However, when all players in the game should act simultaneously, the price competition game can be modeled using Bertrand game as in [26] and [29]. When the competition is not in terms of price, but in terms of the size of the spectrum, Cournot game was applied to model the competition among secondary users [24, 27].

In a distributed scenario with incomplete information, the players have access to local information only and they have to learn about the strategy of other players adaptively. Such interactions can be modeled using games with learning [14, 20, 23, 26, 27, 32, 40, 40, 44]. In [14], a belief-assisted approach (using belief function) was employed for selfish users to reduce pricing overhead. In [20], the authors have used a learning-based game called Regret Tracking for channel allocation among cognitive radio users. In [23], a distributed no-regret learning implementation is proposed that performs similar to the case when cooperation is enforced. In [26], when a primary service provider has the information about the demand from secondary service but no information about the current pricing strategy of other primary services, a non-cooperative game with learning was used to decide its strategy based on the past strategy of other primary services. Similar approach is applied but in case of secondary users in [27]. Niyato et al. [32] addresses the issue for both primary and secondary users when there are multiple primary services and multiple secondary users. In [40], a distributed learning algorithm based on replicator dynamics was proposed such that the secondary users can gradually converge to the ESS of the game based on the observation of their own payoffs only. In [44], a price updating strategy was proposed using structured stochastic learning when the sellers have no knowledge about each others' strategy and also about the consumer population.

Coalitional game was used in [35] for collaborative spectrum sensing. In [8], coalition game combined with repeated game was used to make the boundary nodes and backbone nodes collaborate for packet forwarding. In [26], repeated games were used to prevent selfish primary services from deviating unilaterally for individual profit that may lower the profit for other primary services. The combination of repeated and coalitional games can be a very effective way to introduce cooperation among network users. A bargaining game was used in [3] in a cooperative and distributed framework for spectrum assignment. An MMG was used in [36] in a non-cooperative framework for channel switching.

While [32] uses evolutionary game to model the competition among secondary users to buy the spectrum from primary users, [40] uses evolutionary game to enforce collaboration among selfish secondary users for cooperative spectrum sensing. [23] used potential game formulation to ensure the existence of the Nash equilibrium for the channel allocation problem, a stochastic game theoretic framework was deployed in [41] in order to develop anti-jamming defence mechanism against the adaptive cognitive attackers that aim to harm the secondary users which want to utilize the spectrum effectively.

Figure 16.6 depicts the classification based on the particular models/approaches from price theory and market theory described in Section 16.3. In [15] VCG auction

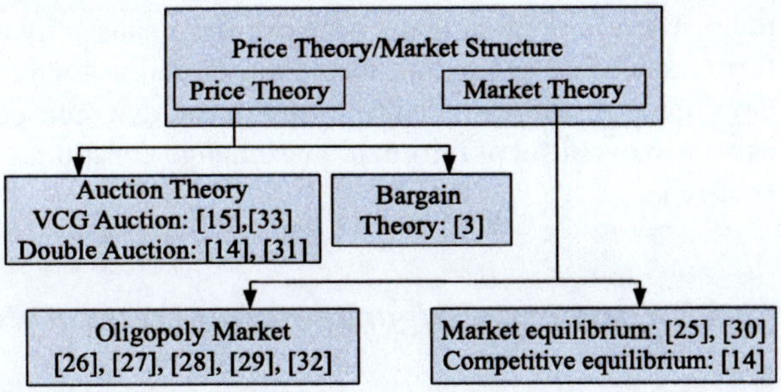

Fig. 16.6 Classification of related work on spectrum sharing based on typical price theory/market models used

was used to maximize the revenue of the primary user maintaining truthfulness of the bidders, while in [33] VCG auction was used for dynamic spectrum allocation in a cooperative and centralized framework. Double-auction scenario was considered in [14] and [31] in a non-cooperative framework. Bargaining game was used in [3] in a cooperative and distributed framework for spectrum assignment. Oligopoly market structure was considered for analyzing pricing schemes in [26–29, 32]. While [14] uses competitive equilibrium as the solution for the case without user collusion in the network, [25] and [30] considered market equilibrium as the pricing solution.

The applicability of particular game models or pricing schemes/market structure depends on many factors. The selfish behavior of network users and the limited amount of information available usually requires distributed schemes for spectrum sharing. On the other hand, the synchronization issues and the information exchange overhead in a distributed system is a huge research challenge. The implementation of pricing schemes can also cause significant communication overhead to the system. In many cases, Nash equilibriums are not efficient and Pareto optimal solutions provide higher revenues/payoffs. However, the cost to make the users act cooperatively in a distributed environment may be quite high, which leaves users no other choice but Nash equilibrium. The trade off is in terms of information exchange and computation complexity versus distributedness.

16.6 Open Research Problems

Some possible directions for future research in investigating the economic interactions for spectrum sharing in cognitive radio networks are as following:

16.6.1 Coalition Formation and Communication Overhead

Many of the literature that propose cooperative/coalitional strategies for primary/secondary users have not considered the overhead/cost for cooperation. The cost can be the power required for negotiation, delay because of the information exchange, etc. In a practical scenario, it is not reasonable to neglect this cost for spectrum management especially in a resource constrained network like cognitive radio. Therefore, investigating the overhead caused by the information exchange to form a coalition can be an interesting direction for future research. Alternatively, development of distributed algorithms for dynamic coalition formations without excessive overhead of information exchange constitutes a promising area for further research.

16.6.2 Distributed Algorithms for Truthful Bidding

Market-driven dynamic spectrum auctions can drastically improve spectrum utilization while providing substantial revenue to spectrum owners. However, they face significant challenges due to the fear of market manipulation. Bidder collusion [19] is very likely among secondary users to lower the price of the spectrum offered by the primary services and it is probably the most serious practical threat to the revenue of the primary users. Therefore, in auction-based markets, ensuring truthful bidding is of critical importance. A truthful or strategy-proof spectrum auction enforces players to bid their true valuations of the spectrum. Hence bidders can avoid the expensive overhead of strategizing over others and the auctioneers can maximize their revenue by assigning spectrum to bidders who value it the most. Conventional auction mechanisms like VCG auction can ensure the truthfulness of bidders. But, the amount of information exchange in VCG is tremendous and it is not computationally efficient. Because of the peculiar properties of spectrum trading (spectrum is reusable unlike conventional commodities), the optimal spectrum allocation problem is NP-complete. Therefore, real-time multi-user spectrum auctions must use greedy algorithms. However, under greedy allocations, VCG auction looses its strategy-proof nature. Some recent works [15, 46] have proposed distributed schemes that can maintain truthful bidding with less computational complexity. However, the work in this direction is still in its infancy. Distributed algorithms that can detect the presence of bidder collusion or that can enforce truthful bidding by designing a mechanism such that the payoff achievable for truthful bidding is always higher than the payoff otherwise are required.

16.6.3 Incentive-Driven Spectrum Sensing

Cooperative spectrum sensing has been recognized as a powerful approach to improve reliability and the detection performance. While sensing information is important for secondary users for opportunistic spectrum access, sensing takes some time energy which they can otherwise utilize for their data transmission. In most of the existing literature on cooperative spectrum sensing, there is an inherent assumption that all secondary users cooperate with each other to achieve greater good of the system [39]. However if the users are selfish, they do not collaborate with each other for cooperative sensing without any incentive or enforcement. The time that a user contributes in sensing is proportional to the reduction in throughput for that user. The energy spent on sensing may also be of critical importance for mobile users. As a consequence, if the secondary users are selfish and rational, each secondary user tries not to contribute in sensing as long as they can have a free ride. This kind of selfish behavior comes into the picture for different situations, e.g., if the secondary users belong to different systems, if their spectrum requirements are heterogeneous, if there is no central authority to assign the sensing responsibility, etc. Collaboration can be maintained in different ways. One possible approach for this can be the reputation-based method using repeated games where, if users do not contribute in sensing and try to enjoy the free ride in many stages of the game, they will be punished. If users have heterogeneous traffic, some users may take advantage of the heterogeneity over others. In such cases, distributed algorithms ensuring fairness to all users that can guarantee the convergence are necessary. Incentive/monetary gain-driven schemes for collaborative spectrum sensing can be another possible direction for further research.

16.6.4 Trust and Security

Security is one of the critical issues for the implementation of cognitive radio networks. In [16], the authors have shown that when at least as many sensors are Byzantine as are honest, the Byzantine sensors can completely defeat the sensor fusion so that no information can be transmitted reliably. The operation of cognitive radio users depends completely on the result of spectrum sensing. If the sensing information cannot be reliably transmitted and combined, the operation of secondary users can become a threat for primary transmission. Investigating different issues such as how can malicious users attack the sensors, how can sensors be adaptive, and how to counteract such attacks, the criteria that the security can be maintained despite the presence of malicious users, etc., using game theory can be a very interesting research area.

Similarly, attacks from malicious users such as pretending to be the primary users so that the secondary users detect the channels as occupied even when primary users are not active, with an objective of not allowing the secondary users to use the channels effectively, malicious secondary users acting like primary users and trying to compete for the resources with the same priority as the primary users, etc. also need to be looked further into. Some recent studies have proposed different approaches for security issues for dynamic spectrum access [12, 41]. However, there is a lot more to explore in this area. Game theory is a natural tool for defence mechanisms in a distributed non-cooperative environment. Stochastic games can be used to derive the optimal defence strategies against the attackers, without having the complete information about each others' strategies. Similarly, repeated games may be used to detect malicious users based on their reputation.

16.6.5 Assumption of Rationality and Complete Information

Although game theory is a powerful tool for analyzing resource allocation/optimization and security problems, it inherently assumes that the players are rational and often the payoff functions of all players are assumed to be common knowledge, which in many scenarios may not be realistic. Game theory merely suggests what strategies should be taken to maximize individual/total utility. It does not provide much insight into the analysis of what the players are most likely to do and how much rational the players are likely to be when the information available is incomplete or faulty. Therefore, further study is needed to investigate the factors that may make the players irrational and to predict the natural behavior of players in addition to the strategies that are toward maximizing the individual/total utility.

16.7 Conclusion

We have described different types/models of games and price theory/market principles that have been used to model the economic activities of primary and secondary users for resource allocation and spectrum trading. An extensive summary of the related work on economic approaches has been presented with the classification based on the spectrum sharing issues and solutions. We also discussed the research challenges of using game theory and price theory/market theory for their application in cognitive radio research. We discussed the open research problems and proposed some interesting directions for future research to model economic approaches in cognitive radio networks.

References

1. Akyildiz IF, Lee WY, Vuran MC, Mohanty S (2006) NeXt generation/dynamic spectrum access/cognitive radio wireless networks: a survey, computer networks, Vol. 50, no. 13, pp. 2127–2159
2. Alpca T, Basar T, Dey S (2006) A power control game based on outage probabilities for multicell wireless data networks. IEEE Transactions on Wireless Communications, Vol. 5, no. 4, pp. 890–899
3. Cao L, Zheng H (2005) Distributed spectrum allocation via local bargaining. Proceedings, IEEE Sensor and Ad Hoc Communications and Networks (SECON) 2005, Santa Clara, CA, pp. 475–486
4. Gandhi S, Buragohain C, Cao L, Zheng H, Suri S (2007) A general framework for wireless spectrum auctions. IEEE International Symposium on New Frontiers in Dynamic Spectrum Access Networks (DySPAN), Dublin, Ireland, 2007
5. Gibbons R (1992) A Primer in Game Theory. Harvester Wheatsheaf, Hemel Hempstead, UK
6. Han Z, Liu KJR (2005) Non-cooperative power-control game and throughput game over wireless networks. IEEE Transactions on Communications, Vol. 53, no. 10, pp. 1625–1629
7. Han Z, Ji Z, Liu KJR (2007) Non-cooperative resource competition game by virtual referee in multi-cell OFDMA networks. IEEE Journal on Selected Areas in Communications, Vol. 25, no. 6, pp. 1079–1090
8. Han Z, Poor HV (2009) Coalition games with cooperative transmission: a cure for the curse of boundary nodes in selfish packet-forwarding wireless networks. IEEE Transactions on Communications, Vol. 57, no. 1, pp. 203–213
9. Haykins S (2005) Cognitive radio: brain-empowered wireless communications. IEEE Journal on Selected Areas in Communications, Vol. 23, no. 2, pp. 201–220
10. Hosseinabadi G, Manshaei H, Hubaux JP (2008) Spectrum sharing games of infrastructure-based cognitive radio networks, Technical Report LCA-REPORT-2008-027

11. Huang J, Berry RA, Honig ML (2006) Auction-based spectrum sharing. ACM Mobile Networks and Applications J., Vol. 11, no. 3, pp. 405–418
12. Husheng L, Zhu H (2010) Catching attacker(s) for collaborative spectrum sensing in cognitive radio systems:an abnormality detection approach. IEEE Symposium on New Frontiers in Dynamic Spectrum Access Networks (DySPAN), Singapore, 2010
13. Jayaweera SK, Tianming L (2009) Dynamic spectrum leasing in cognitive radio networks via primary-secondary user power control games. IEEE Transactions on Wireless Communications, Vol. 8, no. 6, pp. 3300–3310
14. Ji Z, Liu KJR (2008) Multi-stage pricing game for collusion-resistant dynamic spectrum allocation. IEEE Journal on Selected Areas in Communications, Vol. 26, no. 1, pp. 182–191
15. Jia J, Zhang Q, Zhang Q, Liu M (2009) Revenue generation for truthful spectrum auction in dynamic spectrum access. Proceedings of the tenth ACM International Symposium on Mobile Ad Hoc Networking and Computing, New Orleans, LA, pp. 3–12
16. Kosut O, Tong L (2006) Capacity of cooperative fusion in the presence of byzantine sensors. Proceedings of the 44th Annual Allerton Conference on Communication, Control and Computation, Ithaca, NY
17. Krishna V (2002) Auction Theory. Academic, London
18. Maille P, Tuffin B (2008) Analysis of price competition in a slotted resource allocation game. Proceedings, IEEE Infocom 2008, Phoenix, AZ, pp. 1561–1569
19. Marshalla RC, Marx LM (2006) Bidder Collusion. Journal of Economic Theory, Elsevier, Vol. 133 (available online: 3rd Feb. 2006, published: 2007) pp. 374–402
20. Maskery M, Krishnamurthy V, Zhao Q (2009) Decentralized dynamic spectrum access for cognitive radios: cooperative design of a non-cooperative game. IEEE Transactions on Communications, Vol. 57, no. 2, pp. 459–469
21. Mitola J (1999) Cognitive radio for flexible mobile multimedia communications. Proceedings, IEEE Workshop on Mobile Multimedia Communication, San Diego, CA, pp. 3–10
22. Monderer D, Shapley L (1996) Potential games. Games and economic behavior, Vol. 14, no. 1, pp. 124–143
23. Nie N, Comaniciu C (2005) Adaptive channel allocation spectrum etiquette for cognitive radio networks. IEEE International Symposium on New Frontiers in Dynamic Spectrum Access Networks (Dyspan), Baltimore, MD, pp. 269–278
24. Niyato D, Hossain E (2007) A game-theoretic approach to competitive spectrum sharing in cognitive radio networks. Proceedings of IEEE WCNC, Hongkong, 2007
25. Niyato D, Hossain E (2007) Equilibrium and disequilibrium pricing for spectrum trading in cognitive radio: a control theoretic approach. Proceedings of IEEE GLOBECOM, Washington DC, 2007
26. Niyato D, Hossain E (2008) Competitive pricing for spectrum sharing in cognitive radio networks: dynamic game, inefficiency of Nash equilibrium and collusion. IEEE Journal on Selected Areas in Communications, Vol. 26, no. 1, pp. 192–202
27. Niyato D, Hossain E (2008) Competitive spectrum sharing in cognitive radio networks: a dynamic game approach. IEEE Transactions on Wireless Communications, Vol. 7, no. 7, pp. 2651–2660
28. Niyato D, Hossain E (2008) Market-equilibrium, competitive and cooperative pricing for spectrum sharing in cognitive radio networks: analysis and comparison. IEEE Transactions on Wireless Communications, Vol. 7, no. 11, pp. 4273–4283
29. Niyato D, Hossain E (2008) Competitive pricing in heterogeneous wireless access networks: issues and approaches. IEEE Network, Vol. 22, no. 6, pp. 4–11
30. Niyato D, Hossain E (2008) Spectrum trading in cognitive radio networks: a market-equilibrium based approach. IEEE Wireless Communications Magazine, Vol. 15, no. 6, pp. 71–80
31. Niyato D, Hossain E, Han Z (2009) Dynamic spectrum access in IEEE 802.22-based cognitive wireless networks: a game theoretic model for competitive spectrum bidding and pricing. IEEE Wireless Communications, Vol. 16, no. 2, pp. 16–23

32. Niyato D, Hossain E, Han Z (2009) Dynamics of multiple-seller and multiple-buyer spectrum trading in cognitive radio networks: a game theoretic modeling approach. IEEE Transactions on Mobile Computing, Vol. 8, no. 8, pp. 1009–1022
33. Rodriguez V, Moessner K, Tafazolli R (2005) Auction driven dynamic spectrum allocation: optimal bidding, pricing and service priorities for multi-rate, multi-class CDMA. IEEE PIMRC 2005, Vol. 3, pp. 1850–1854
34. Saad W, Han Z, Debbah M, Hjrungnes A, Basar, T (2009) Coalitional game theory for communication networks: a tutorial. IEEE Signal Processing Magazine, Special Issue on Game Theory, Vol. 26, pp. 77–99
35. Saad W, Han Z, Debbah M, Hjrungnes A, Basar T (2009) Coalitional games for distributed collaborative spectrum sensing in cognitive radio networks. Proceedings, Annual IEEE Conference on Computer Communications, INFOCOM, Rio de Janeiro, Brazil, 2009
36. Sengupta S, Chandramouli R, Brahma S, Chatterjie M (2008) A game theoretic framework for distributed self-coexistence among IEEE 802.22 networks. IEEE Globecom, New Orleans, LA, 2008
37. Shapley L (1953) Stochastic games. Proceedings of the National Academy of Sciences of the United States of America, Vol. 39, no. 10, pp. 1095–1100
38. Smith JM (1982) Evolution and the Theory of Games. Cambridge University Press, UK
39. Unnikrishnan J, Veeravalli VV (2007) Cooperative spectrum sensing and detection for cognitive radio. IEEE Globecom, Washington, DC, USA, 2007
40. Wang B, Liu KJR, Clancy TC (2010) Evolutionary cooperative spectrum sensing game: how to collaborate? IEEE Transactions on Communications Vol. 58, no. 3, pp. 890–900
41. Wang B, Wu Y, Liu KJR (2010) An anti-jamming stochastic game for cognitive radio networks. http://www.rsc.org/dose/http://www.ece.umd.edu/~bebewang/jsac09_bw.pdf
42. Wang F, Krunz M, Cui S (2008) Price based spectrum management in cognitive radio networks. IEEE Journal of Selected Topics in Signal Processing, Vol. 2, no. 1, pp. 74–87
43. Weibull, JW (1995) Evolutionary Game Theory. MIT Press, Cambridge
44. Xing Y, Chandramouli R, Cordeiro, CM (2007) Price dynamics in competitive agile spectrum access markets. IEEE Journal on Selected Areas in Communication, Vol. 25, no. 3, pp. 613–621
45. Zhang J, Zhang Q (2009) Stackelberg game for utility-based cooperative cognitive radio networks, Proceedings of the tenth ACM International Symposium on Mobile Ad hoc Networking and computing (MobiHoc) 2009, New Orleans, LA, pp. 23–32
46. Zhou X, Gandhi S, Suri S, Zheng H (2008) eBay in the sky: Strategy-proof wireless spectrum auctions. Proceedings of the 14th ACM International Conference on Mobile Computing and Networking, San Francisco, CA, pp. 2–13

Chapter 17
Game Based Self-Coexistence Schemes in Cognitive Radio Networks

Sajal K. Das, Vanessa Gardellin, and Luciano Lenzini

Abstract Cognitive radio networks are seen as the key enabling technology to address the spectrum shortage problem in wireless applications and services. One of the major challenges for implementing cognitive radio networks is to guarantee self-coexistence among devices, which means address interference issues among devices operating under the same set of rules and sharing the same resources. Among the several mathematical tools used to address the self-coexistence problem, we recognize the game theoretic approach as the most powerful. In this chapter, first we present an overview of cognitive radio technology focusing on the importance of guaranteed self-coexistence among cognitive devices. Then, we analyze the pros and cons of several game theoretic approaches proposed in the literature in order to model the self-coexistence problem. We conclude by describing non-cooperative and cooperative game paradigms to model the self-coexistence problem in cognitive radio networks.

17.1 Challenges and Terminology

Several studies on the frequency spectrum have recently been conducted. The motivation behind these studies is the proliferation of wireless applications and services operating in unlicensed bands, which is causing what is known as *spectrum overcrowding*. To tackle the problem of spectrum overcrowding, the Federal Communications Commission (FCC) in the United States conducted a careful analysis on the overall spectrum. This analysis highlighted the fact that, in contrast to unlicensed bands, most licensed bands are under-utilized [16].

Figure 17.1 shows how the frequency spectrum from 3 kHz to 300 GHz is assigned by the FCC to different services [26][1]. The figure highlights how every chunk of the frequency spectrum has been assigned to a specific use and that the

[1] A detailed description of these services can be found at http://www.ntia.doc.gov/osmhome/spectrumreform/.

V. Gardellin (✉)
Italian National Research Council, Pisa, Italy
e-mail: v.gardellin@iit.cnr.it

F.R. Yu (ed.), *Cognitive Radio Mobile Ad Hoc Networks*,
DOI 10.1007/978-1-4419-6172-3_17, © Springer Science+Business Media, LLC 2011

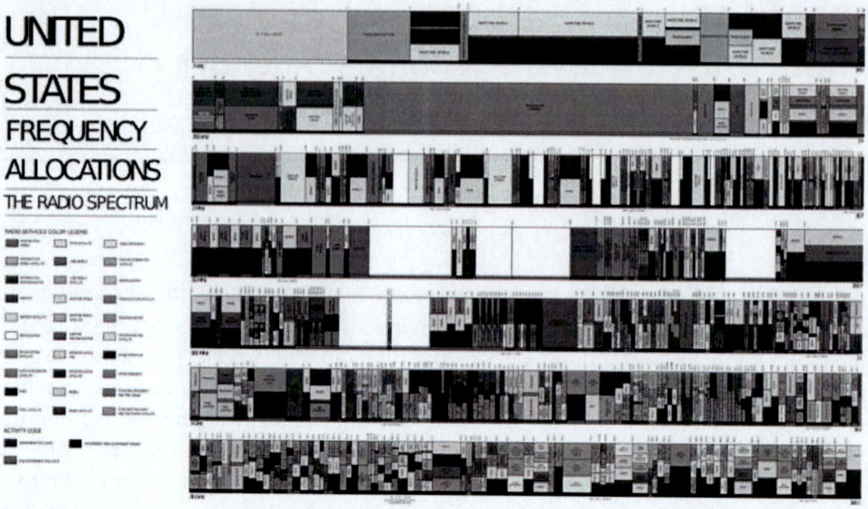

Fig. 17.1 Spectrum utilization in the United States [26]. The *white chunks* (broadcasting) are TV channels

same frequency can be shared between different services (vertical division of frequencies). However, in order to deal with the increasing demands for spectrum, the FCC proposed the adoption of a new regulatory spectrum policy, which allows the use of licensed spectra by unlicensed users. This new regulatory spectrum policy was set up to the develop a new family of unlicensed users, called *cognitive radios* (CR) [34].

Cognitive radio technology focuses on ranges of the frequency spectrum allocated to broadcasting services, i.e., TV channels. In Fig. 17.1, we recognize these ranges as [54, 88] MHz, [174, 216] MHz, and [470, 806] MHz, where each TV channel has a width of 6 MHz. When CRs identify one or more unused TV channels, they can decide to use either the entire TV channel width, part of it or the 6 MHz can even be shared among different CRs.

Studies focusing on the characterization of unused licensed frequency spectrum have been conducted [7, 46]. These studies identified the possibility of occupying *television white spaces* (TVWS) for data transmission among CRs. In fact, CRs can analyze channels and TVWS that are being used in any location and time instant as shown in [48], where the authors characterized the frequency spectrum in Chicago's business district. In [48], TV channels in the range [470, 806] MHz were occupied for only half of their capacity. Hence, the possibility of using TVWS is an important resource for unlicensed users in order to transmit data.

Using TVWS is not without its challenges. In fact, the major problem is the variability in time and space of the unused TV channels. Hence, CRs are increasingly attracting the research community who want to provide CRs with the capability of exploiting unused channels and dealing with variability issues.

17.1.1 Cognitive Radios and Cognitive Radio Networks

Cognitive radios (CRs) have the ability to sense the external environment, learn from history, and make intelligent decisions adjusting their transmission parameters dealing with situations that had not been planned for during the network design phase. A CR senses the wireless medium and, among other features, identifies spectrum opportunities, i.e., chunks of the spectrum unused by licensed users, which are usually called *channels*. This identification is based on beacons and overheard messages sent by licensed users.

The CR terminology species two types of users: *primary users* (PUs) and *secondary users* (SUs) or *cognitive devices*. PUs are licensed paying users, and example PUs are digital TV, analog TV, and licensed low-power auxiliary devices such as wireless microphones. On the contrary, SUs are unlicensed users and hence have no rights on the frequency spectrum. In fact, SUs do not pay any fee and thus are not allowed to cause any harmful interference to PUs, that is, they have to operate on a *non-interfering base*. Example SUs are base stations and end-users, which provide services in rural and remote areas, cognitive devices used in order to create emergency networks in disaster areas, and in general any device that wants to operate into the licensed spectrum but does not have any right. The FCC rules establish that SUs have to be invisible to PUs, i.e., no complexity has to be added to PUs which, indeed, do not require any change in their spectrum management. Therefore, SUs have to deploy rules to opportunistically operate in the licensed spectrum without causing any degradation on PU's transmissions.

Although design CRs are challenging from an electronic point of view, they are becoming a reality due to recent "Moore's law" advances in programmable integrated circuits that have created the opportunity to develop radios that can adapt to a wide variety of interference conditions and multiple protocol standards. Resulting in a collaboration between otherwise incompatible systems.

From cognitive radios to cognitive radio networks is a short step. Cognitive radio networks (CRNs) [31] are easily maintainable networks which are continuously improved and upgraded without human intervention. As opposite of CRs, in CRNs the overall network works following a cognitive process where it can sense current environment, plan for the future, make a decision, and act accordingly. The basic requirement of CRNs is the need to evolve over time which imposes a flexible and modular network that lead in a highly scalable infrastructure. Moreover, CRNs have a *proactive* behavior thanks to the abilities of learning and planning to reach end-to-end goals.

17.1.2 Self-Coexistence and Channel Assignment

A challenging problem in implementing CRNs is the coexistence among SUs that use the same resources under the same set of rules, known as the *self-coexistence problem*. Self-coexistence is the ability to access chunks of spectrum on a non-interfering basis with respect to PUs and SUs. Hence, it can be seen as the ability to guarantee that there is no degradation in PU services and no interference between SUs.

Self-coexistence can be addressed using centralized or distributed approaches depending on the nature of the network. If a centralized approach is used, common signaling among SUs or coordination entities is required. In contrast, if a distributed approach is used, SUs require self-organization and self-management abilities.

Centralized approaches offer easy tractability because they allow coordination among devices and hence agreements on the chunks of the spectrum used by each device. However, coordination mechanisms are difficult to guarantee in CRNs because these networks could house cognitive devices that respond to different standards (IEEE 802.11, IEEE 802.22, etc.) and/or belong to different Internet service providers (ISPs). Hence, devices may not have a common communication protocol. Moreover, devices are built by different manufactures, who may have an incentive to develop products with a selfish behavior, so that they perform better than products developed by other manufacturers.

Distributed approaches are characterized by SUs that access chunks of the spectrum in a *distributed* manner by acting selfishly. A selfish behavior implies that devices are worried only about their own outcome instead of the overall network performance. To obtain self-coexistence using distributed approaches, devices need protocols and algorithms under which they are able to self-organize and self-manage their own resources taking into account the existence of other devices using the same resources.

In wireless environments and therefore in CRNs, the most important resource is the frequency spectrum. As previously mentioned, the frequency spectrum is divided into chunks, called *channels*, and distinct channels do not cause interfere with each other if they are orthogonal. Hence, devices operating in orthogonal channels do not interfere. Therefore, self-coexistence can be seen as the problem of assigning channels to devices for communication purposes.

In the literature, the ability to assign channels to devices avoiding co-channel interference is referred as the *channel assignment problem*. The key concept behind an efficient channel assignment is to find appropriate channels in such a manner that devices can coexist without causing any harmful interference and network objectives are met. Usually the objectives include QoS satisfaction, high spectrum utilization, and traffic throughput.

In conclusion, the self-coexistence problem in CRNs can be regarded as a channel assignment problem where channels are spectrum opportunities identified by cognitive devices and used for communication purposes on a non-interfering basis. In fact in both self-coexistence and channel assignment problems, a channel must communicate without causing interference to any other device, licensed or unlicensed.

17.1.2.1 Interference Models

In order to model channel assignment algorithms among cognitive devices, we need an interference model which describe how devices influence each other. We identify two interference models: protocol and physical interference models.

The *protocol interference model* is a binary paradigm where interferences are considered on a pair basis and hence it provides a simple and easily tractable approach. However, the protocol interference model fails in properly capture the interference generated by the entire network. In fact, using a binary model it is possible to associate at each device the exact number of overlapping competitors, but doing so, some solutions are ruled out. For example the protocol interference model do not consider solutions where two cognitive devices do not interfere; however, if a third one is transmitting on the same channel, then the interference destroy the communication.

The *physical interference model* [55], instead, is a cumulative paradigm and hence capture how the entire network influence a single devices guaranteeing a more realistic model of the interactions among devices. Usually the physical interference model use the *signal to interference and noise ratio* (SINR) as in (17.1).

$$\mathrm{SINR}_{i,j}(t_1, t_2) = \frac{P_{i,j}}{\sum\limits_{h \in \mathcal{T}(t_1,t_2) \setminus i} P_{h,j} + \mathcal{W}} \tag{17.1}$$

Here $\mathcal{T}(t_1, t_2)$ is the larger set of devices transmitting in the time interval $[t_1, t_2]$; $\mathcal{W}$ is the background noise power; and, $P_{i,j} = \frac{P^{\mathrm{T}}(\lambda/4\pi)^2}{\Delta_{i,j}^{\kappa}}$ is the received power at device j when device i is transmitting the packet. $P_{i,j}$ depend on: $\Delta_{i,j}$, the Euclidean distance between the devices i and j; P^{T}, the transmitted power of device i; κ, the path loss exponent; and, λ, the signal wavelength. To effectively capture transmissions the SINR on the receiver device (j) has to be greater than or equal to its *capture threshold*, indicated by γ_j.

17.1.3 Outline of the Chapter

CRNs are obtaining a significant success thank to their adaptability to various scenarios and their large range of applicability. However, the self-coexistence problem has to be addressed because it afflicts each type and size of CRN. Even small-scale emergency networks are afflicted by self-coexistence problems. An example was the fireworks depot explosion in the 2000 in Netherlands [45] where fire brigades, police, and relief workers of the medical teams experienced a great deal of communication breakdown, both internally and with each another due to lack of common standards for each disaster relief group and overloaded emergency frequency bands.

In Section 17.2 and Section 17.3, we address the self-coexistence problem in more details starting from how IEEE standard communities address the problem and how, together with the functionalities provided by standards, the self-coexistence problem can be modeled as a channel assignment problem and hence addressed at a higher level using game theoretic approaches.

In Section 17.4 we highlight the advantages and disadvantage of several game theoretic approaches used in the literature. We distinguish between two large families of games: non-cooperative, described in Section 17.4.1, and cooperative, described in Section 17.4.2. Of the non-cooperative games, we identify minority games, auction games, and potential games, to highlight how the most important game approaches can be used to address the self-coexistence problem. For the cooperative games, we identify bargain games and coalitional games to describe the self-coexistence problem.

Finally, we conclude the chapter in Section 17.5.

17.2 Self-Coexistence: From the MAC Layer up to the Network Layer

The self-coexistence problem involves network design and architecture from the physical layer up to the networking layer. Hence, in the literature several approaches have been used. Section 17.2.1 describes efforts of the networking communities in addressing the self-coexistence problem, while Section 17.2.2 focuses on network architectures used for CRNs. We conclude with a description of the IEEE 802.22 standard which is the most mature standard for self-coexistence in Section 17.2.3.

17.2.1 IEEE 802 Standards for Self-Coexistence

The importance of cognitive devices and the criticality of self-coexistence have been recognized also by IEEE standard communities that are mainly working on two standards to guarantee self-coexistence: IEEE 802.19 and IEEE 802.22.

IEEE 802.19 [25] specifies radio technology methods to enable the family of IEEE 802 wireless standards to use the most effectively TVWS by providing standard self-coexistence methods. They define recommended wireless coexistence metrics and methods for computing them.

IEEE 802.22 [27] targets wireless broadband access in rural and remote areas using TVWS in very high-frequency (VHF) and ultra high-frequency (UHF) bands. Most of the work in IEEE 802.22 focuses on develop and improve spectrum sensing techniques to detect and avoid PUs. Moreover, this standard is characterized by a *pro-active* approach that includes self-coexistence protocols and algorithms as conception and definition of the initial standard. In contrast, traditional IEEE standards take *re-active* approaches where the self-coexistence problem is addressed when the specification is already finalized.

Additionally, the IEEE 802.11 group is working on the new IEEE 802.11*af* standard focusing on TVWS and its initial draft is planned by the end of 2010.

17.2.2 Network Architectures for Cognitive Radio Networks

In CRNs, we distinguish infrastructure-based [18] and infrastructure-less [10] network architectures as shown in Fig. 17.2.

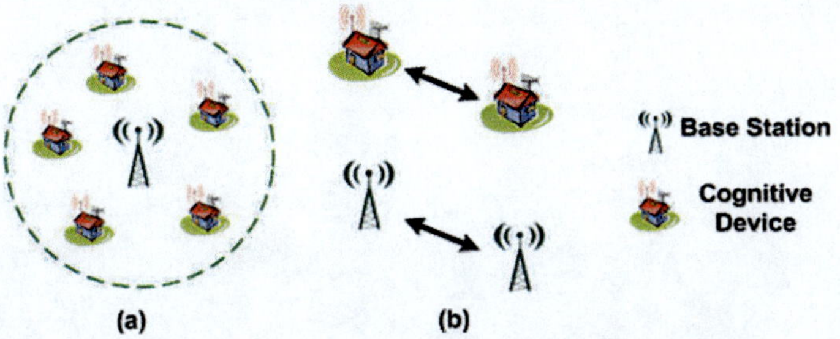

Fig. 17.2 Network architectures. (**a**) Infrastructure-based; (**b**) Infrastructure-less

Infrastructure-based algorithms consider point-to-multipoint communication paradigms where a base station supports multiple cognitive devices. Base station and cognitive devices are assumed to be a single entity, called *cell*, in the self-coexistence problem and hence the communication channel is assignment to the entire cell. Communication between cells may or may not exist, hence an infrastructure-based network does not imply a centralized entity. Infrastructure-based network architectures are commonly used for last-mile broadband accesses.

Infrastructure-less algorithms consider source-destination pairs and hence point-to-point communication paradigms. Hence, self-coexistence is seen how the ability of assign channels to source-destination pairs. In infrastructure-less algorithms, channels are assigned to pairs of devices, hence channels are effectively assigned to links. Infrastructure-less network architectures are used to establish direct connections between network devices.

17.2.3 IEEE 802.22 Standard

In Section 17.2.1, we introduced several standards on which the IEEE networking community is working on (IEEE 802.19, IEEE 802.22, and IEEE 802.11af).

Hereafter we focus on the IEEE 802.22 standard because it is the most mature standard to guarantee self-coexistence among SUs. The reason behind its wide develop in self-coexistence techniques is its much larger transmission range compared with other IEEE 802 standards. In fact, larger transmission ranges turn into larger interference ranges and hence heavier mutual interference among devices. Therefore, robust interference avoidance mechanisms, like self-coexistence techniques, are necessary.

IEEE 802.22 defines a standard to develop cognitive devices operating under a point-to-multipoint communication paradigm. The network architecture is characterized by cells, called *wireless regional area networks* (WRANs), which contain two types of SUs: *base stations* (BSs) and *consumer premises equipments* (CPEs). Figure 17.3 shows an example of network architecture defined by IEEE 802.22 standard. Each WRAN is managed by a BS which offers connectivity to its end-users,

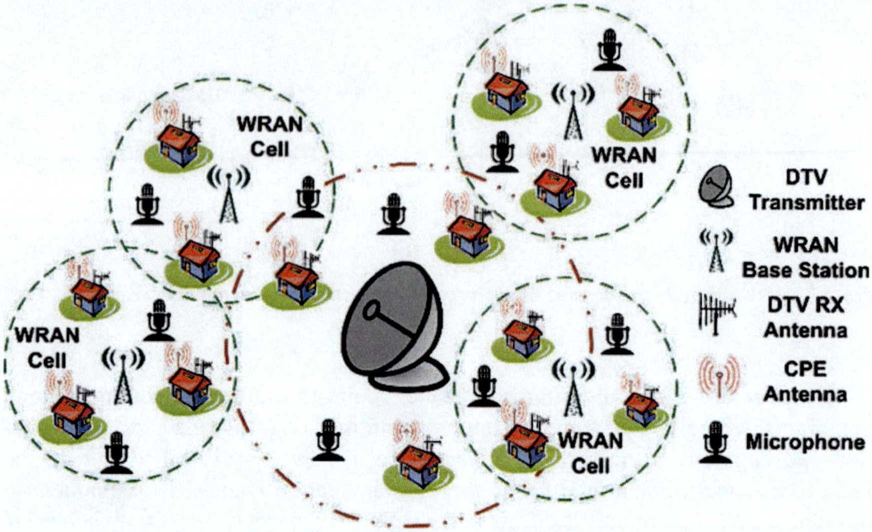

Fig. 17.3 IEEE 802.22 cognitive radio network architecture

i.e., CPEs, which are cognitive devices as well. In fact, each BS can also count on its CPEs to perform channel sensing and report their availability.

BSs have to perform two steps in order to address the self-coexistence problem: network discovery and channel assignment. We briefly describe network discovery mechanisms offered by IEEE 802.22 standard in Section 17.2.3.1, while in Section 17.2.3.2 we explain how self-coexistence is addressed using channel assignment techniques.

17.2.3.1 Network Discovery

Network discovery indicates the set of techniques defined by the IEEE 802.22 standard to pro-actively address the self-coexistence problem, thus improving our knowledge of the surrounding analysis and providing mechanisms at the medium access control (MAC) layer. Based on these, we can identify two techniques provided by the standard: surrounding analysis and communication protocols.

Surrounding analysis includes the discovery of TVWS and channel occupancies of neighboring WRANs. In order to improve our knowledge of the environment, each BS is equipped with a GPS and is connected to a centralized server to obtain information about TVWS in each location at a given time.

Several works have been carried out to analyze the environment and understand the potential and challenges of CRNs. The authors in [22] see the relationship between SU and PU as *protection* or *pollution* viewpoints. Protection means that a CR can only operate in locations where it cannot generate "harmful interference" to PU receivers [18]. In contrast, pollution allows interference caused by SUs and that afflict PUs. However, interference has to be under a given threshold to be non-disruptive. Hence the limits on the amount of acceptable interference that a PU can tolerate have been studied [24]. To establish these thresholds, SUs need to have knowledge regarding PUs, which may or may not be possible depending on the type and location of SUs.

On the other hand, *communication protocols* are addressed at the MAC layer, where self-coexistence is obtained using coexistence beacon protocol (CBP) and self-coexistence window (SCW).

The CBP is a communication protocol that uses beacon transmissions between WRANs for PU protection, as well as for self-coexistence. The SCW is a time window in each MAC frame introduced as an innovation by the standard with the purpose of reducing the collision probability between BSs and CPEs. The SCW can be scheduled whenever necessary by a BS, which defines what should be done during this time. Moreover, the SCW is employed by CBP packets to signal key informations and to carry out geolocation informations among CPEs of the same WRAN cell.

There are several studies in the literature that address the self-coexistence problem at the MAC layer where a new coexistence mechanism is defined to replace the existing one [5] or where an integration is proposed [2]. However, these mechanisms require some modifications in the IEEE 802.22 MAC layer, which are out of the scope of this chapter.

17.2.3.2 Channel Assignment

Channel assignment is based on algorithms act to efficiently use channels in order to minimize interference among devices.

There have been extensive researches on channel assignment algorithms in multi-channel ad hoc and mesh networks based on IEEE 802.11 standard [1, 11, 30, 38, 47, 54, 60].

Commonly used approaches are packet-based, link-based [8, 63], and flow-based [58]. In *packet-based* channel assignments, channels are assigned on a per-packet basis between sender and receiver using a common control channel. Frequent channel switching, i.e., the time required by a radio to change communication channel, synchronization requirements, and scheduling overhead make this approach not suitable with the current technology. In *link-based* channel assignments, channels are assigned at the granularity of a link between sender and receiver, that is, all packets between the same sender and receiver are transmitted on the same channel. As the packet-based approach, the link-based channel assignment is afflicted by a significant channel switching delay incurred when a sender serves two receivers on different channels. In *flow-based* approaches, instead, channels are assigned at the granularity of a flow, that is packets of a flow are scheduled on the same channel. The major concern is that different flows between the same sender and receiver may operate on different channels and hence require channel switching on both devices introducing switching delay and synchronization issues.

Figure 17.4 is taken as example to explain the different channel assignment approaches. We consider a CRN with 6 nodes, 4 flows, 2 primary users, and

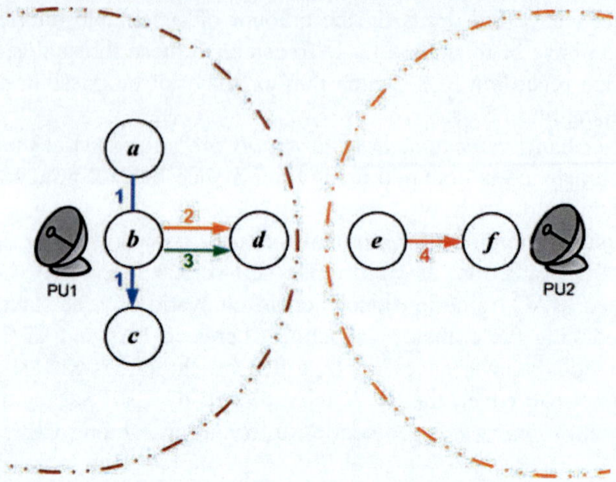

Fig. 17.4 Cognitive radio network with 6 nodes, 4 flows, 2 primary users, and 4 channels

4 channels. Channel 1 is occupied by the PU on the left and channel 2 is occupied by the PU on the right in Fig. 17.4. Hence, nodes a, b, c, and d can use channels 2, 3, and 4, while nodes e and f can use channels 1, 3, and 4. In the packet-based approach, for each packet a channel needs to be assigned, hence the channel assignment does not depend on flows. In the link-based approach, a different channel could be assigned to serve flow 1 and flows 2 and 3. Hence, node b switches channel to serve different neighbors. In the flow-based approach node b switches channel to satisfy flows 1, 2, and 3.

A different channel assignment approach, respect to the earlier presented, assumes that each device has a radio tuned on each channel [1, 14, 29, 38, 39]. However in CRNs this would result in prohibitively high cost due to the expected not negligible cost of a single cognitive radio.

In summary, channel assignment algorithms for multi-radio multi-channel ad hoc and mesh networks assume that all devices are "well-behaved" or "cooperative" and that the available resources do not vary in time and space, i.e., they require a static set of channels accessible at every device. Although a static environment offers a great convenience for coordination in channel negotiation and switching, it is not realistic for CRNs where PU transmissions can occur in every moment and hence the set of available channels vary depending on location and time instant. Unfortunately, this variability adds another dimension of complexity to the channel assignment problem. Consequently, CRNs require channel assignment algorithms with the ability to dynamically use available channels as the radio channel utilization changes.

Recalling packet/link/flow-based channel assignments, the authors in [6] proposed a channel assignment strategy for CRN that uses the granularity of segments. A segment is defined as the group of devices having common set of available channels and housing source and destination of each flow. For example, applying the segment-base approach to the network in Fig. 17.4 we can identify two segments. Segment 1 is the set $seg(1) = \{a, b, c, d\}$ and segment 2 is the set $seg(2) = \{e, f\}$. Hence, no channel switching is required because nodes a, b, c, d are set to transmit and receive on the same channel. In fact, devices are grouped in order to use the same channel and hence avoid channel switching issues. However, synchronization issues are not addressed and the channel assignment could be not efficient.

Several methodologies have been proposed to analyze self-coexistence as channel assignment, among them we recall graph coloring [8, 21], linear integer programming techniques [56], and game theory [18, 61]. In this chapter we focus on game theory as methodology to solve the channel assignment problem thanks to its capabilities to well describe the iteration among devices that share common resources. In Section 17.3, we describe the general ideas behind game theory and in Section 17.4 we take an overview of several studies that use game theoretic approaches as instrument to solve self-coexistence.

17.3 Game Theory

Game theory (GT) is a powerful mathematical tool developed for the purpose of analyzing the interactions in decision processes. GT has been extensively applied in microeconomics but recently has received attention as a useful tool to design and analyze distributed resource allocation algorithms [19, 33].

In this section we present the general components that characterize a game and we introduce some game definitions. In Section 17.4, instead, we compare the most popular games used to describe distributed resource sharing among cognitive devices in order to address the self-coexistence problem.

We can mathematically define a game in its normal form as $G = \{\mathcal{N}, \mathcal{S}, \mathcal{U}\}$ where G is a game, $\mathcal{N}$ is the finite set of *players*; $\mathcal{S}$ is the non-empty set of *strategies*; and $\mathcal{U}$ is the set of *utility functions* or payoffs. To understand how these components describe a game let us singularly illustrate them.

17.3.1 Set of Players

Players are entities participating in the game. We consider a set $\mathcal{N}$ representing players that compete for shared resources and we indicate with N the cardinality of this set.

Several classifications are given to characterize a player. All the players are *rational*, that is each player always selects the strategy that yields it the greatest payoff. In terms of their impact on other players they can be *myopic* or *foresighted.* Myopic players always act to maximize their immediate achievable reward. They ignore the impact of their competitors' reactions over their own performance and determine their responses to gain the maximal immediate rewards. Foresighted players, instead, behave by taking into account the long-term impacts of their actions on their rewards. They anticipate how the other users will react and maximize their performance by considering the responses of the other players. As consequence, foresighted players require additional knowledge about the other players to assist their decision process.

We classify players distinguishing in Section 17.3.1.1 between cooperative and non-cooperative behavior, while in Section 17.3.1.2 we describe single stage games and repeated games, which are two way how players can engage into a game.

17.3.1.1 Cooperative vs. Non-cooperative

The type of players reflects in the coordination mechanism that can be *cooperative* [37] or *non-cooperative* [3]. They both are defined as centralized or distributed. If a centralized paradigm is used, the scalability is an issue because players require a centralized coordination entity and/or a common signaling paradigm.

Cooperative and non-cooperative terminologies can be deceptive because they may suggest that there is not space for cooperation in the former and no conflict or competition, in the latter. Instead, the difference is in the way how behaviors are imposed. In non-cooperative game, players self-enforce, while in cooperative game there is an "external" entity that impose the way to act. From a purely game theoretic prospective, a non-cooperative game specifies strategies that are available to players while cooperative games describe the resulting outcomes when players engage the game together in different combinations. However, cooperative players do not have the same interests.

A cooperative game is characterized by players that behave cooperatively to obtain the most out of the game. The major concern about cooperative games is the necessity of a centralized coordination entity or a common signaling, which, however, are afflicted by scalability issues. To address these issues *coalition games* have been defined. Coalitional games are an emerging class of games that address the scalability defining cooperative subgroups, i.e., *coalitions*, of the original set of players. Thus reducing the original centralized problem in distributed sub-problems where players belonging to the same coalition cooperate but there is not cooperation between coalitions. This means that players cooperate in a coalition do not have a centralized coordination entity but they use distributed transmissions.

Coalitional games can be divided based on the payoff distributions into *transferable utility* (TU) [41] and *non-transferable utility* (NTU) [37]. In TU coalitional games, player belonging to the same coalition divide the total payoff among them. In NTU coalitional games, instead, each player has a different payoff based on its advantage of belonging to a coalition and cannot be shared with other players. From this the concept of *self-interested* or *selfish* players, which are characterized by making their own decision independently in order to maximize their own payoff without necessarily respecting system objectives.

Usually, cooperative games are used where there is the need of fairly share a scarce resource among competing players. Concepts such as *bargaining games* embody specific notions of fairness and take into account the strategic interests of competing users. However, due to scalability issues non-cooperative games are preferred.

A non-cooperative game is characterized by players that make decisions independently and can easily deviate from the network protocol to seek for more benefit for themselves. This can lead to a solution that is not social efficient, because players can increase their performance degrading the others players' payoff. One of the most used technique to provide incentives for selfish players to behave cooperatively is the payment method [28]. The payment method introduces a way to influence players' behaviors. Players assume that there is some kind of virtual currency in the system and that each player has to pay some virtual money to the central entity based on payoffs. However, the payment method is hardly scalable due to the necessity of a centralized entity. For this reason a distributed approach is desirable also in non-cooperative games.

17.3.1.2 Single Stage Game vs. Repeated Game

Another important factor that a player has to decide is how to engage into the game. The way how a game can be played distinguish single stage games and repeated games [32]. Given the base game G, called *stage game*, a single stage game is a game where each player engages only one time into the game. On contrary, a repeated game is characterized by finite or infinite repetitions of the same stage game. Both, single and repeated games, can be simultaneous or sequential which means that players can play all in the same time instant or one after another, respectively.

Repeated games are a simplification of a bigger family of games called *stochastic games* (SGs) [53]. SGs are repeated games with probabilistic/stochastic transitions, i.e., the game moves to a new state with a certain probability. The new state depends on the previous state and the actions chosen by players. In SGs each player knows its own state and strategies, but it does not know states and strategies taken by other players. The set of strategies distinguish repeated and stochastic games. In fact, in the latter the set of strategies depends on the current state, while in the former is equal at each stage.

The work flow in repeated and stochastic games is the same. At the begining of each stage, the game is in a certain state. Players select their strategies and receives a reward that depends on both current state and selected strategies. We distinguish internal and external strategies. Given a player $i \in \mathcal{N}$, external strategies are strategies of the other $\mathcal{N} - i$ players, while the internal strategy is the strategy chosen by the player i. Therefore, the state transition of each player is directly impacted by its own internal actions and indirectly impacted by the external actions of all players through the resource competition.

In [57], the authors proposed a stochastic game where the decisions that need to be taken are based on the players' incomplete and asymmetric information about the environment and other players' strategies. Based on their information, each player can develop *beliefs* about the current state of the network and its evolution over time. Based on these beliefs players can pro-actively select the optimal policy for interacting with other devices such that they maximize their utilities.

In Section 17.4 we describe several families of games more in details and we explain how different types of players can be used to model cognitive devices.

17.3.2 Set of Strategies

Strategies are the choices that a player can make. We assume the existence of a strategy set S_i for each player $i \in \mathcal{N}$ thus we have $\mathcal{S} = \{S_1, S_2, \ldots, S_N\}$ for the set of players $\mathcal{N} = \{1, 2, \ldots, N\}$. The *strategy profile* of the game [17], instead, is given by $\mathbf{s} = \{s_1, s_2, \ldots, s_N\}$ where player 1 chooses strategy $s_1 \in S_1$, player 2 chooses strategy $s_2 \in S_2$ and so on. For every different combination of individual strategies, we have a different strategy profile $\mathbf{s}$ and the set of all such strategy profiles is $\mathbf{S} = S_1 \times S_2 \times \cdots \times S_N$.

17.3.3 Set of Utility Functions

An utility function and the resulting payoff decides how good a strategy profile is. At each player $i \in \mathcal{N}$ is associated a strategy set $S_i \in \mathcal{S}$ and a payoff set $U_i \in \mathcal{U}$. Results that to a strategy $s_i \in S_i$ correspond a payoff $u_i \in U_i, \forall i \in \mathcal{N}$.

To properly choose the utility function several factors have to be taken into consideration. The utility function has to reflect system characteristics as well as physical properties of the environment. In addiction, the utility function has to satisfy mathematical properties with the objective to guarantee equilibrium convergence. To determine if a convergence point exist, the *Nash equilibrium* point (NE) has been defined [43, 44].

A NE point correspond to a strategy profile where no player has interest to deviate. Given a strategy profile $\mathbf{s}^* = \left(s_i^*, s_{-i}^*\right) \in \mathbf{S}$, a NE point is defined as in (17.2) for each player $i \in \mathcal{N}$.

$$u_i\left(s_i^*, s_{-i}^*\right) \geq u_i\left(s_i, s_{-i}^*\right), \quad \forall s_i \in S_i \tag{17.2}$$

where $u_i(s_i, s_{-i})$ is the utility function of player i when it uses the strategy s_i and the other players use strategies s_{-i}. By carefully designing utility function and strategies, the game can be balanced at a unique socially optimal NE, where the summation of all payoffs is maximized.

Stronger notions of equilibrium also exist. A commonly used concept is the *strongly dominant strategy equilibrium* (SDSE). In SDSE, the convergence is to a overall system optimum. Given a strategy profile $\mathbf{s}^* \in \mathbf{S}$, a SDSE is defined as in (17.3).

$$\begin{cases} \forall s_{-i} \in S_i, \forall s_i \neq s_i^*, \; u_i\left(s_i^*, s_{-i}\right) \geq u_i(s_i, s_{-i}) \\ \exists s_{-i} \in S_i, \forall s_i \neq s_i^*, \; u_i\left(s_i^*, s_{-i}\right) > u_i(s_i, s_{-i}) \end{cases} \tag{17.3}$$

This means that at least for one player the strategy has to be strictly dominant.

17.4 Families of Games for Cognitive Radio Networks

Many researchers are currently engaged in designing efficient protocols for cognitive radio networks. These studies cover a wide range of issues including channel assignment, power control, call admission control, and interference avoidance. A tutorial survey describing how game theory has been applied to CRNs is proposed in [59] where the authors do not focus on a specific problem, but instead describe game concepts and paradigms in detail. However in this chapter, we only focus on game theoretic studies and methodologies related to the self-coexistence problem in CRNs, and we describe network architectures and characteristics of cognitive devices.

In order to model interactions among players, several families of games with different objectives have been used. We focus on the channel assignment problem as a technique to guarantee self-coexistence among cognitive devices, by dividing approaches based on the family of games used. We distinguish between non-cooperative and cooperative games from a game theoretic prospective, and of these games we highlight the network architecture by differentiating between infrastructure-based and infrastructure-less networks.

Figure 17.5 shows how games are used to treat the self-coexistence problem.

17.4.1 Non-cooperative Games

Non-cooperative games are characterized by players that self-enforce a behavior, i.e., the game does not explicitly say the payoff for a group of players if they play a specific strategy profile but, instead, each player knows only its own payoff when its strategy changes. This means that the game is under *incomplete information* because each player has not information about the other players strategies.

In this section we present three families of non-cooperative games extensively used in the literature to model the cognitive device ability of reconfigure transmission parameters. We present *minority games* [51], *auction games* [10, 20, 52], and *potential games* [18, 42] giving a general introduction of the game along with examples from the literature.

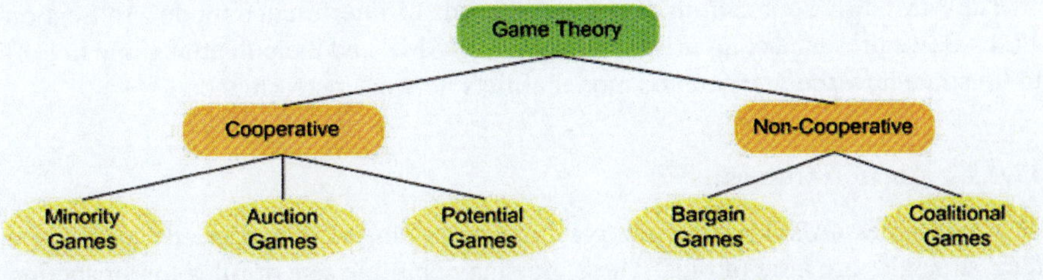

Fig. 17.5 Game theoretic approaches used to address self-coexistence in cognitive radio networks

The reason behind the application of different families of games is that, depending on characteristics and objectives of players, a type of game can better model a system respect to another.

17.4.1.1 Minority Games

Minority games (MGs) are characterized by a group of $\mathcal{N}$ players that have to independently decide between a binary set of strategies. Originally MGs have been proposed by Challet and Zhang [9], is a branch of GT for studying competition and self-imposed cooperation in a non-cooperative environment with limited resources. Players in a MG do not interact or negotiate with each other directly, i.e., the game is under incomplete information. The goal of MGs is help all the players to make better decisions even without direct knowledge of other players' strategies.

The most famous MG is the El Farol game. In this game players have to decide if to go or not to go to the El Farol bar on Friday night. Going to the bar is enjoyable only if the bar is not too crowded and at the same time an empty bar is not a desirable condition. For this reason, if all the $\mathcal{N}$ players decide to go, thinking that the bar will be empty, then the bar will be overcrowded. In contrast, if they all decide not to go, the bar will be empty. In the same way, if all the devices in a wireless network decide to transmit on the same channel, the situation is not enjoyable because no transmission is successful due to interference and if no device transmits on a channel, the resource is waste.

In [51] the authors proposed a *modified minority game with mixed strategies* (MMGMS) to model self-coexistence among non-cooperative players. They identify players with WRANs, which act under a non-cooperative paradigm in a distributed manner, i.e., without centralized authority or common signaling. The authors modeled the spectrum band switching game as an infinitely repeated game where WRANs try to minimize their cost in finding a clear channel, i.e., minimize the number of repetitions where transmissions fail. They propose a mixed strategy where the competing WRANs must adhere to in order to achieve a NE point. As interference model the authors used the protocol interference model, which associate to each WRAN the exact number of overlapping competitors. Their choice of interference model is given by a better tractability of the problem in spite of a more realistic model that can be obtained using the physical interference model as explained in Section 17.1.

The MMGMS approach in [51] fails in terms of interference model. In Section 17.4.1.3, we present a comparison between MMGMS and the potential game in [18] to illustrate how the interference model afflicts network performance.

17.4.1.2 Auction Games

Auction games (AGs) study how players interact in auction markets, i.e., where their strategies are a set of bids. There are many possible sets of rules for an auction and usually the objectives are efficiency and equilibrium of the bidding strategies.

Several are also the type of auctions which depend on how players engage the game and the amount of "money" payed by winning players. The term *money* in AGs is used to refer actual money transaction, virtual credits, or virtual currency to indicate a quantity of a good that players have to pay in order to use a resource.

Among AGs we recall first-price sealed-bid auctions, Vickrey auctions, English auctions, and Dutch auctions. Unfortunately these single unit auctions do not properly model situations where multiple winners emerge as in resource sharing problems. Therefore, to model a channel assignment different types of auction games have to be used. One of these is the knapsack auction mechanism.

The *knapsack auction mechanism* is characterized by players that want to place objects in a knapsack. Here, the knapsack represents available channels and the objects are the amount of data to transmit. Each player evaluates the placement of an object in the knapsack and its bid is related to the amount of money that players want to pay to use a resource.

In [52], the authors modeled the coexistence of licensed and unlicensed users as a sealed-bid knapsack auction, which dynamically allocates channels to devices based on their bids. Their objective is to maximize the spectrum utilization for every player. The authors took into consideration service providers and end-users. However, the similarity with BSs and CPEs is straightforward. The disadvantage of this approach is the need of a centralized entity, which knows bids and amount of data that any device wants to send, and a common signaling to communicate with the centralized entity. Due to the necessity of a centralized entity, the sealed-bid knapsack auction in [52] is hardly applicable to CRNs, where in fact the existence of a centralized entity is not guaranteed and often not desirable. A centralized entity is not guaranteed because cognitive devices could belong to different IEEE standards, i.e., they do not follow the same paradigms but want to self-coexist. The need of a common signaling, instead, is not desirable due to the overhead that brings into the system.

Another AG used to model the spectrum allocation is the Anglo-Dutch auction game. In [20], the authors proposed a two rounds mechanism as follows. In the first round, N players compete for K resources using an English auction increasing their bids until $K+1$ players remain. In the second round, each remaining player submits a sealed bid at or above the bid at which the first round had stopped. Players with the K highest bids win the auction and pay either their respective bids or the highest bid. However, the Anglo-Dutch auction in [20] is inefficient to model the channel assignment problem because does not consider that more than one player can be on the same resource (channel).

AGs can be used also to model source-destination cognitive pairs, as in [10] where the authors proposed a non-cooperative multiple-PU multiple-SU auction game. In this game, SUs share the available spectrum of licensed PUs subject to the interference temperature constraint at each PU, i.e., they use a pollution model as described in Section 17.2.3.1. The authors proposed a distributed algorithm in which each SU updates its strategy based on local informations to converge to an equilibrium point. Their algorithm required the exchange of a small amount of information among nearby SUs.

In [10], the *ping-pong effect* is also studied. This means that the authors took into consideration configurations where free channels exist and players could infinitely jump from a free channel to another, without realizing that both the channel assignments yield the same outcome, or they could jump all together on the same channel that hence become overcrowded. In this work, the authors addressed the ping-pong effect using the no-regret learning [23], which converge to the same equilibrium given by the auction game. The basic idea under the no-regret learning is that the probability of choosing a strategy is proportional to the "regret" for not having chosen other strategies. This approach requires that SUs have knowledges regarding PUs and other SUs, hence is not suitable in every type of CRN.

Another interesting game is the more competitive Stackelberg game [62] in which PUs choose their prices to maximize their revenue, i.e., PUs sell excessive spectrum to SUs for monetary return. However, this is in contrast with the assumption for which SUs are transparent to PUs as explained in Section 17.1.1 and hence can be applied only to network where PUs and SUs are coordinated.

In summary, AGs can be used to model channel assignment problems but they also require the existence of a centralized entity, which, however, is not guaranteed and desirable in CRNs.

17.4.1.3 Potential Games

Potential games (PGs) are characterized by a finite set of players that can engage in a finite set of strategies and the incentives of all players are mapped into one function, called *potential function*. A potential function is used to analyze NE points. In [35] has been proved that PGs guarantee the convergence to an NE point. To obtain a potential game the following requirements have to be satisfied.

(i) Players have to choose from a finite set of strategies.
(ii) Players' payoffs have to depend on the number of other players choosing the same strategy.
(iii) A potential function $Pot : \mathcal{S} \rightarrow \mathbb{R}$ mapping strategies into real numbers has to be defined. A potential function is an increasing function that exactly reflects any unilateral change in players' payoffs. To guarantee the convergence to an NE point, the potential function has to increase at each stage game. Hence, players have to engage into the game one at time. From this the last requirement.
(iv) Players have to play in a sequential order.

PGs can be divided based on how changes in players' payoffs reflect in the potential function. We identify exact, weighted, and ordinal potential games where the variation in the potential function is respectively equal, proportional, or has the same sign respect to the player's payoff improvement. An overview on potential games and their application to interference, power, and waveforms game formulations for cognitive devices is given in [40].

PGs have been used to model the channel assignment problem in CRNs where the set of players ($\mathcal{N}$) is the set of cognitive devices competing for a finite set of TVWS. Hence, the set of strategies of each player (S_i) is the TVWS chosen to transmit, i.e., the *channel*. We indicate with $s_i = \{s_i^1, \ldots, s_i^k, \ldots, s_i^K\}$ the strategy profile of a player $i \in \mathcal{N}$ where $k \in \mathcal{K}$, $\mathcal{K}$ is the set of TVWS identified in the network and K is its cardinality. Given the strategy profile described above, the interference produced by a strategy profile in the wireless environment seems to be the proper utility function because it represents the quality of a channel.

We now briefly analyze three potential games that consider both, point-to-point and point-to-multipoint, communication paradigms.

In [42], the point-to-point interaction between source–destination cognitive pairs is described. The authors proposed a potential game which describes a distributed and dynamic channel assignment scheme for selfish players under cooperative and non-cooperative scenarios. However, their main results are for cooperative devices only. Hence, no further explanations and proprieties are brought out for non-cooperative potential games.

The point-to-multipoint communication paradigm was modeled in [24] as a N-player game, but the authors had as objective the maximization of the number of CPEs for each WRAN, which is not comparable with the self-coexistence problem in this chapter.

In [18], the point-to-multipoint communication paradigm, as described in the IEEE 802.22 standard, is taken into consideration. The authors proposed a non-cooperative repeated game, called *NoRa*, which models the utility function for each WRAN depending on the interference experienced by each cognitive devices and on the number of other devices that are using the same channel. Moreover, to obtain a sequential repeated game, they defined a novel backoff mechanism to ensure that only one player changes its strategy at each stage and hence the potential function increases from a stage to the following. The interference model used in [18] is the physical interference model, which helps the authors to have a realistic viewpoint of the surrounding environment.

In order to understand how the interference model used afflicts the network performance we show in Fig. 17.6 the throughput obtained by MMGMS [51] in comparison with the throughput obtained by *NoRa* [18] when 28 BSs, 280 CPEs, 20 PUs, and 8 channels are considered. We can notice how the throughput is greater when *NoRa* is used. This result is due to the following two factors.

(i) Physical interference model in opposition to protocol interference model, i.e., a more realistic description of the environment is conducted using *NoRa*.
(ii) Reduced number of stages where transmissions fail. In fact, *NoRa* converges faster to an NE point compared to the MMGMS approach.

In conclusion, the algorithm presented in [18] is a fully distributed approach and hence does not require any centralized entity so reducing the computational complexity of the channel assignment scheme. Moreover, *NoRa* guarantees the convergence to an NE point thanks to the properties given by potential games.

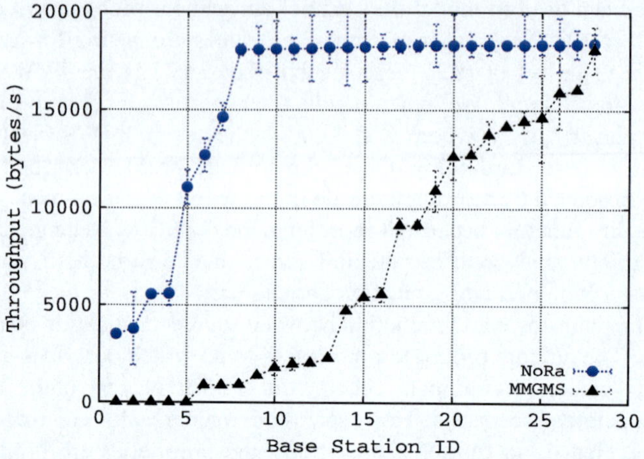

Fig. 17.6 Comparison in terms of throughput between *NoRa* [18] and MMGMS [51]. 28 BSs, 280 CPEs, 20 PUs, and 8 channels are considered

17.4.1.4 Conclusions on Non-cooperative Games

Non-cooperative game theoretic approaches are used to efficiently model distributed environments where players have only local knowledge because centralized entities or global signal paradigms do not exist.

The absence of centralized or common coordination mechanisms means that non-cooperative games adapt quickly to the surrounding environment. However, they can lead to a bad equilibrium and hence poor performance due to the lack of mechanisms that soften the selfish behavior of cognitive devices. In order to address the bad equilibrium problem in the following section we analyze cooperative games.

17.4.2 Cooperative Games

Cooperative games define the outcome for each player if a group of players follows a specific strategy profile. These games analyze situations where the players' objectives are partially cooperative and partially conflicting, i.e., players have an interest in cooperating, in order to achieve the greatest possible total payoff but at the same time they have conflicting goals in sharing the resources obtained.

Generally the game theoretic approaches used to model cooperative games are *bargaining games* [8, 20, 24] and *coalitional games* [4]. The former uses a centralized approach, meanwhile the latter can be centralized or distributed.

In Section 17.4.2.1 we describe bargain games and their application to solve spectrum management problems, while in Section 17.4.2.2 we describe and propose solutions obtained using coalitional games.

17.4.2.1 Bargain Games

Bargain games (BGs) are situations where players want to reach an agreement regarding how to distribute resources. Each player prefers the agreement that maximize its own payoff. In BGs, players analyze their own rewards if the agreement is reached or if players fail to do so. Hence, the only analysis of the single player is not sufficient to model the entire bargain process.

A BG is defined using a set of possible agreements (set of strategies) and a disagreement point. Players ask for portions of the resource under consideration. If the sum of the requested portions of the resource is less than the total resource, then all players obtain the resource otherwise they obtain nothing. The amount of requested resource is the strategy of each player, while results of the bargaining process and disagreement point are payoffs. Therefore, a BG forces cooperation among players regarding on how to divide a resource. In self-coexistence problems players are cognitive devices, while the resource to share is the frequency spectrum.

A common concept used to model bargaining interactions is the *Nash bargain solution* (NBS) [36]. A NBS provides a way to divide a resource and defines four axioms that a solution has to satisfy to guarantee the convergence to an optimal equilibrium point.

(i) Linearity.
(ii) Independence of irrelevant alternatives.
(iii) Symmetry.
(iv) Pareto optimality.

Let us consider a two players game to explain how a bargain game works. Player a makes an offer, say $x(a)$, to player b, which can either accept or reject the offer. If player b accepts the offer, this offer takes place. Otherwise, player b applies a discount on its utility and makes an offer, say $x(b)$, to player a. Note that $x(a)$ and $x(b)$ are in the range $[d, z]$, where z is the total resource and d is the disagreement point, i.e. the minimum obtainable utility by both players. Moreover, we have that the offers of the two players have to satisfy $x(a) + x(b) \leq z$. The bargain game continues until player a or player b accepts the offer, that is, maximize $x(b)$ and $x(a)$.

In [24], the authors proposed a cooperative scheme using the NBS. Given a set of channels and a set of CPEs, they assign channels to WRANs so that the number of CPEs managed by each WRAN is maximized. They show that using the NBS, the number of CPEs into each WRAN increases compared to a game where players do not cooperate. This means that each WRAN uses resources more efficiently. Unfortunately the game presented in [24] does not scale because the complexity of this game grows when the number of channels and/or the number of CPEs increases. In fact, the authors used a centralized approach where each WRAN has to find all the possible channel assignments.

To reduce the complexity of a centralized approach, the authors in [8] proposed local bargaining groups. They addressed the spectrum management problem in cognitive radios considering that players self-organize the network into bargaining groups. They shown how their approach significantly reduces the algorithm complexity compared to graph-coloring solutions. The problem with this approach is the implicit willingness of collaboration among devices which is not realistic in practical systems.

A different variation of BGs used to model a distributed implementation of dynamic spectrum allocation is the Rubinstein–Stahl bargaining model [49]. This bargaining approach consists of a game where players want to reach an agreement on how to share a resource and at each negotiation step where they fail in reach an agreement, part of the resource is waste for all players.

In [20], the authors proposed a bargaining Rubinstein–Stahl based algorithm. They considered a decentralized game to reflect the decentralized nature of the network, and hence they modeled a bargain game under incomplete information. This game theoretic model can lead to a waste of resources and hence is not a desirable approach.

In conclusion, BGs are mainly centralized approaches which require communication among cognitive devices, but how previously explained common communication paradigms are not guaranteed in CRNs. Moreover, BGs could produce situations where a non-negligible amount of resources is wasted. In fact, until players do not reach an agreement point, the resources are not utilized. Hence, properly modeled BGs can be effective in configurations where the time is not important; however, the channel assignment problem is not among these.

In the next section we analyze coalitional games aimed at reducing the computational complexity of bargaining centralized approaches. Moreover, the problem of reaching an agreement point in coalitional games does not take place.

17.4.2.2 Coalitional Games

Coalitional games (CGs) have been extensively used to model distributed cooperation among players sharing resources.

CGs are characterized by the formation of cooperative *sub-groups* of devices, referred to as *coalitions* [37]. The formation of coalitions includes two activities: *structure generation* and *optimization.*

(i) Structure generation is the formation of coalitions such that players, within each coalition, coordinate their activities, but players do not coordinate between coalitions. Therefore, the number of coalitions and the cardinality of each subgroup need to be defined.
(ii) The optimization problem that has to be solved is the maximization of the difference of gain given by the cooperation minus the cooperation cost.

In a coalitional game, we call: $\mathcal{C}$ the coalition structure; C the number of coalitions; $\mathcal{C}_p$ one of the coalitions with $p \in \{1, \ldots, C\}$; and C_p the size of coalition $\mathcal{C}_p$.

A coalition structure can be characterized by overlapped or disjoint coalitions, and partial or exhaustive involvement of players in the coalition structure. When coalitions are disjoint and exhaustive, they are called *partitions*. This means that given two partitions, $\mathcal{C}_p$ and $\mathcal{C}_q$, where $p, q \in \{1, \ldots, C\}$ and $p \neq q$, we have $\mathcal{C}_p \cap \mathcal{C}_q = \emptyset$ and $\bigcup_{p=1}^{C} \mathcal{C}_p = \mathcal{N}$. Where $\mathcal{N}$ is the set of all the players.

To clarify the difference between coalitions and coalition structure [50], let us consider $\mathcal{N} = \{1, 2, 3\}$, disjoint coalitions and exhaustive involvement of players. In this case we have $2^N - 1 = 7$ possible coalitions: $\{1\}, \{2\}, \{3\}, \{1, 2\}, \{1, 3\}, \{2, 3\}, \{1, 2, 3\}$ and 5 coalition structures: $\mathcal{C} = \{\ \{\{1\}, \{2\}, \{3\}\},\ \{\{3\}, \{1, 2\}\},\ \{\{2\}, \{1, 3\}\},\ \{\{1\}, \{2, 3\}\},\ \{\{1, 2, 3\}\}\ \}$.

Moreover, every coalition $\mathcal{C}_p$ is characterized by a *coalition value*, denoted by $v(\mathcal{C}_p)$, which quantifies the coalition's payoff in a game. The coalition value is particularly important because it determines form and type of the game.

Let us use the previous example to illustrate what a coalition value is. We consider, for sake of simplicity a without loss of generality, a game where coalition values are an a priori knowledge and are given as follows:

(i) Coalitions of single players: $v(\{1\}) = 0$; $v(\{2\}) = 0$; $v(\{3\}) = 0$.
(ii) Coalition of players' pairs: $v(\{1, 2\}) = 150$; $v(\{1, 3\}) = 150$; $v(\{2, 3\}) = 150$.
(iii) Grand coalition: $v(\{1, 2, 3\}) = 120$.

This means that players on their own have a null payoff, any pair of players has a payoff equal to 150 and the coalition containing all the players, called *grand coalition*, has an overall payoff equal to 120. Note that the coalition value has to be divided among the members of the coalition. Assuming that players divide the coalition value in equal parts among them, the individual payoff for the members of the grand coalition is lesser than the payoffs of the members of coalitions with two players. This condition is due to cooperation costs, therefore players have the interest to cooperate but they have to consider also their costs in cooperation.

The *grand coalition* concept is strictly related to CGs and has a considerable importance. Often the grand coalition is seen as the optimal solution because no coalition has a value greater than the sum of all the players' payoffs. However, in games where a coalition brings gains to its members, but gains are limited by the cost in forming the coalition, the grand coalition is seldom the optimal structure. In fact, large coalitions increase the complexity of the game due to high synchronization and communication costs.

In [4], the authors proposed a repeated coalitional game to model the competitive behavior between independent wireless networks in allocating a common shared radio channel. Players are wireless networks, which play repeatedly in a radio resource sharing games without direct coordination or information exchange. Strategies determine whether competing radio networks cooperate or ignore the presence of other radio networks. Players have to make decisions about when and how often to attempt to access the wireless medium in order to maximize their observed utility. The approach used in [4] is not comparable with channel assignment approaches. In fact, the authors do not consider channels, but they proposed a distributed coordination algorithm to schedule transmissions on the wireless medium.

In [13] the authors proposed a CG where players are selfish and the architecture is infrastructure based. The proposed game is classified as *Hedonic Coalitional Formation game* [15], which is characterized by disjoint coalitions. The cooperation is enforced among WRANs belonging to the same coalition and the cooperation cost of a coalition increases with the increment in number of WRANs in that coalition. The authors highlighted the difficulty in obtaining cooperation among selfish devices in a distributed environment. In fact, devices can easily deviate from seeking more benefits for themselves. They shown that cooperation among devices achieves a better performance but at the cost of higher computational complexity, which leads to a loss of advantages in cases where rapid changes occur in the channels occupancy by PUs.

Figure 17.7 shows the throughput in bytes/s for each of the 28 BSs when 280 CPEs and 20 TVs exist. We compare two cases: no TV changes its transmission channel and all the TVs change their transmission channels during a simulation time of 50 s. The channel assignment algorithms considered are the non-cooperative game *NoRA* [18] and the coalitional game *HeCtor* [13], described above.

We observe that the cooperation among players in *HeCtor* attains higher throughput when no changes occur. However, when TVs change channels, the throughput is comparable to the non-cooperative game. This is due to the fact that *HeCtor* is computationally more intensive than *NoRa*, i.e., *HeCtor* requires more time to reallocate channels to BSs. During this time BSs cannot transmit, thus implying that the throughput decreases. On the other hand, *NoRa* has the same throughput whether or not the TVs change transmission channels. This means that in the considered games,

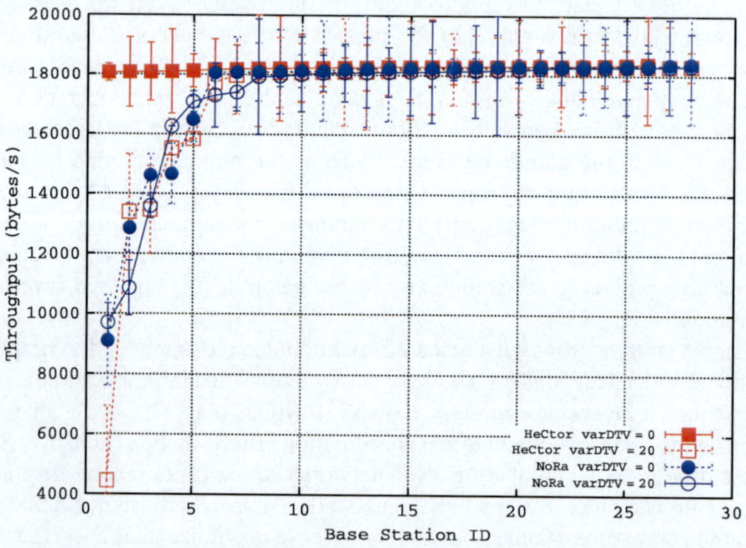

Fig. 17.7 Throughput for each one of the 28 BS in the CRN, when 280 CPEs and 9 channels and 20 TVs are considered. We compare the case where no TV changes its channels, with the case where all the DTVs change their channels

the non-cooperative behavior adapts faster when changes occur in the surrounding environment.

CGs are also used to address the self-coexistence problem from a power allocation prospective [12] or using a joint power/rate control and channel assignment solution [56]. These approaches are beyond the scope of this chapter and consequently we will postpone their study for future work.

In conclusion, CGs are a valuable instrument to model channel assignment algorithms and hence to guarantee self-coexistence among cognitive devices. However, CG complexity increases when changes occur in the surrounding environment due to a slower adaptability compared to non-cooperative games. They should therefore be applied in scenarios where PUs do not rapidly change transmission channels and where slower adaptability is not an issue.

17.4.2.3 Conclusions on Cooperative Games

Cooperative games perform better than non-cooperative games and avoid bad equilibrium conditions which non-cooperative games suffer from. However, they have a higher computational complexity. For this reason distributed cooperative games have been proposed in the literature [13].

Distributed cooperative approaches soften the selfish behavior of cognitive devices and at the same time keep the complexity low. However, they suffer when rapid changes occur in the network surrounding environment.

17.5 Conclusions

In this chapter we have analyzed the self-coexistence problem in cognitive radio networks focusing on how game theoretic approaches can be applied to address the problem. We have identified two big families of games, non-cooperative and cooperative games, which can be used to model several network environments and player characteristics.

In order to solve the self-coexistence problem, we have identified minority games, auction games, and potential games among non-cooperative games and bargaining games and coalitional games among cooperative games. For all these families we have proposed and analyzed several studies presented in the literature.

Non-cooperative game theoretic approaches efficiently model distributed environments where centralized entities or global signal paradigms do not exist. In fact, distributed non-cooperative games adapt quickly to the surrounding environment. However, they can lead to a bad equilibrium and hence poor performance. For this reason cooperative games have been proposed into the literature.

Cooperative games soften the selfish behavior of cognitive devices. However, they are characterized by a higher computational complexity. Bargain games are mainly centralized approaches, which require communication coordination and produce situations where a non-negligible amount of resources is wasted. Instead, coalition games are a valuable instrument to model channel assignment algorithms and hence to guarantee self-coexistence among cognitive devices. However, they are slower to adapt than non-cooperative games when changes occur in the network surrounding environment.

References

1. A. Adya, P. Bahl, J. Padhye, A. Wolman, and L. Zhou. A multi-radio unification protocol for IEEE 802.11 wireless networks. In *Proc. of ICST BroadNets 2004*, pages 344–354, San José, CA, 25–29 Oct. 2004.
2. R. Al-Zubi, M. Z. Siam, and M. Krunz. Coexistence Problem in IEEE 802.22 Wireless Regional Area Networks. In *Proc. of IEEE GLOBECOM 2009*, pages 5801–5806, Honolulu, HI, 30 Nov.–4 Dec. 2009.
3. T. Başar and G. J. Olsder. *Dynamic Non-cooperative Game Theory*. Society for Industrial and Applied Mathematics, Philadelphia, PA, 1998.
4. L. Berlemann, G. R. Hiertz, B. H. Walke, and S. Mangold. Radio resource sharing games: enabling QoS support in unlicensed bands. *IEEE Network*, 19(4):59–65, 2005.
5. K. Bian and J.-M. Park. A coexistence-aware spectrum sharing protocol for 802.22 WRANs. In *Proc. of IEEE ICCCN 2009*, pages 1–6, San Francisco, CA, 2–6 Aug. 2009.
6. K. Bian and J. M. Park. Segment-based channel assignment in cognitive radio ad-hoc networks. In *Proc. of CROWNCOM*, pages 327–335, Orlando, FL, 31 Jul.–3 Aug. 2007.
7. T. X. Brown. An analysis of unlicensed device operation in licensed broadcast service bands. In *Proc. of IEEE DySPAN 2005*, pages 11–29, Baltimore, MD, 8–11 Nov. 2005.
8. L. Cao and H. Zheng. Distributed spectrum allocation via local bargaining. In *Proc. of IEEE SECON 2005*, volume 5, pages 475–486, Santa Clara, CA, 26–29 Sept. 2005.
9. D. Challet and Y. C. Zhang. Emergence of cooperation and organization in an evolutionary game. *Physica A: Statistical and Theoretical Physics*, 246(3–4):407–418, 1997.
10. M. Chen, S. Chang Liew, Z. Shao, and C. Kai. Markov approximation for combinatorial network optimization. In *Proc. of IEEE INFOCOM 2010*, pages 1–9, San Diego, CA, 14–19 Mar. 2010.
11. C. Cicconetti, V. Gardellin, L. Lenzini, and E. Mingozzi. PaMeLA: A joint channel assignment and routing algorithm for multi-radio multi-channel wireless mesh networks with grid topology. In *Proc. of IEEE MASS 2009*, Macau, 12–15 Oct. 2009.
12. N. Clemens and C. Rose. Intelligent power allocation strategies in an unlicensed spectrum. In *Proc. of DySPAN 2005*, pages 37–42, Baltimore, MD, 8–11 Nov. 2005.
13. V. Gardellin, S. K. Das, L. Lenzini, Cooperative vs. Non-Cooperative: Self-Coexistence among Selfish Cognitive Devices, Proceedings of the IEEE WoWMoM Conference, Work in Progress (WiP) paper, June 20–23, 2011, Lucca, Italy, 2011.
14. R. Draves, J. Padhye, and B. Zill. Routing in multi-radio, multi-hop wireless mesh networks. In *Proc. of ACM MobiCom 2004*, pages 114–128, 26 Sept.–1 Oct. 2004.
15. J. H. Dreze and J. Greenberg. Hedonic coalitions: Optimality and stability. *Econometrica*, 48(4):987–1003, 1980.
16. Facilitating Opportunities for Flexible, Efficient, and Reliable Spectrum Use Employing Cognitive Radio Technologies FCC Report and Order, Mar. 2005. FCC-05-57A1.
17. D. Fudenberg and J. Tirole. *Game Theory*. MIT Press, Cambridge, MA, 1991.
18. V. Gardellin, S. K. Das, and L. Lenzini. A fully distributed game theoretic approach to guarantee self-coexistence among WRANs. In *IEEE INFOCOM-2010 Workshop on Cognitive Wireless Communications and Networking*, San Diego, CA, 19 Mar. 2010.
19. D. Goodman and N. Mandayam. Network assisted power control for wireless data. In *Proc. of IEEE Vehicular Technology Conference*, volume 6, pages 1022–1026, 2001.
20. D. Grandblaise, C. Kloeck, T. Renk, P. Bag, P. Levine, K. Moessner, J. Yang, M. Pan, and K. Zhang. Microeconomics inspired mechanisms to manage dynamic spectrum allocation. In *Proc. of IEEE DySPAN 2007*, pages 452–461, Dublin, Ireland, 17–20 Apr. 2007.

21. M. M. Halldórsson, J. Y. Halpern, L. Li, and V. S. Mirrokni. On spectrum sharing games. In *Proc. of ACM Symposium on Principles of Distributed Computing*, 25–28 Jul. 2004.
22. K. Harrison, S. M. Mishra, and A. Sahai. How much white-space capacity is there? In *Proc. of IEEE DySPAN 2010*, Singapore, 6–9 Apr. 2010.
23. S. Hart and A. Mas-Colell. A simple adaptive procedure leading to correlated equilibrium. *Econometrica*, 68(5):1127–1150, Sept. 2000.
24. G. Hosseinabadi, M. H. Manshaei, and J.-P. Hubeaux. Spectrum sharing games of infrastructure-based cognitive radio networks. Technical report, Sept.
25. http://www.ieee802.org/19/.
26. http://www.ntia.doc.gov/osmhome/osmhome.html.
27. IEEE P802.22/DRAFTv2.0: Draft Standard for Wireless Regional Area Networks Part 22: Cognitive Wireless RAN Medium Access Control (MAC) and Physical Layer (PHY) Specifications: Policies and Procedures for Operation in the TV Bands, May 2009.
28. H. Ishibuchi, T. Nakashima, H. Miyamoto, and C. H. Oh. Fuzzy Q-learning for a multi-player non-cooperative repeated game. In *Proc. of the Sixth IEEE International Conference on Fuzzy Systems*, volume 3, pages 1573–1579, Barcelona, Spain, 1–5 Jul. 1997.
29. N. Jain, S. R. Das, and A Nasipuri. A multichannel CSMA MAC protocol with receiver-based channel selection for multihop wireless networks. In *Proc. of IEEE IC3N 2001*, volume 1, pages 432–439, Scottsdale, AZ, 15–17 Oct. 2001.
30. P. Kyasanur and N. H. Vaidya. Routing and interface assignment in multi-channel multi-interface wireless networks. In *Proc. of IEEE WCNC 2005*, volume 4, pages 2051–2056, New Orleans, LA, 13–17 Mar. 2005.
31. Q. H. Mahmoud. *Cognitive Networks: Towards Self-Aware Networks*. Wiley-Interscience, New York, NY, 2007.
32. G. J. Mailath and L. Samuelson. *Repeated Games and Reputations: Long-Run Relationships*. Oxford University Press, New York, NY.
33. R. Menon, A. B Mackenzie, R. M. Buehrer, and J. H Reed. Game theory and interference avoidance in decentralized networks. In *Proc. of the Software Defined Radio Technical Conference and Product Exposition*, Phoenix, AZ, 15–18 Nov. 2004.
34. J. Mitola. Cognitive radio an integrated agent architecture for software defined radio dissertation. *Ph.D. dissertation, Computer Communication System Laboratory, Department of Teleinformatics, Royal Institute of Technology (KTH)*, May 2000.
35. D. Monderer and L. S. Shapley. Potential games. *Games and Economic Behavior*, 14(1): 124–143, 1996.
36. A. Muthoo. *Bargaining theory with applications*. Cambridge University Press, New York, NY, 1999.
37. R. B. Myerson. *Game Theory: Analysis of Conflict*. Harvard University Press, Cambridge, MA, 1991.
38. A. Nasipuri and SR. Das. Multichannel CSMA with signal power-based channel selection for multihop wireless networks. In *Proc. of IEEE Vehicular Technology Conference*, volume 1, pages 211–218, 24–28 Sept. 2000.
39. A. Nasipuri, J. Zhuang, and S. R. Das. A multichannel CSMA MAC protocol for multihop wireless networks. In *Proc. of IEEE WCNC 1999*, volume 2, pages 1402–1406, New Orleans, LA, 21–24 Sept. 1999.
40. J. O. Neel, J. H. Reed, and R. P. Gilles. The role of game theory in the analysis of software radio networks. In *Symposium A Quarterly Journal In Modern Foreign Literatures*, 2002.
41. J. Von Neumann and O. Morgenstern. *Theory of Games and Economic Behavior*. Princeton University Press, Princeton, NJ, 1944.
42. N. Nie and C. Comaniciu. Adaptive channel allocation spectrum etiquette for cognitive radio networks. *Mobile Networks and Applications*, 11(6):779–797, 2006.
43. M. J. Osborne and A. Rubinstein. *A Course in Game Theory*. MIT Press, Cambridge, MA, 1994.
44. G. Owen. *"Game Theory"*, London, UK: Academic Press, 3rd edition, Oct. 1995.

45. P. Pawelczak, R. Venkatesha Prasad, L. Xia, and I.G.M.M. Niemegeers. Cognitive radio emergency networks – Requirements and design. In *Proc. of IEEE DySPAN 2005*, pages 601–606, Baltimore, MD, 8–11 Nov. 2005.
46. J. D. Poston and W. D. Horne. Discontiguous OFDM considerations for dynamic spectrum access in idle TV channels. In *Proc of IEEE DySPAN 2005*, pages 607–610, Baltimore, MD, 8–11 Nov. 2005.
47. A. Raniwala, K. Gopalan, and T. C. Chiueh. Centralized channel assignment and routing algorithms for multi-channel wireless mesh networks. *SIGMOBILE Mobile Computer Communication Review*, 8(2):50–65, 2004.
48. D. A. Roberson, C. S. Hood, J. L. LoCicero, and J. T. MacDonald. Spectral occupancy and interference studies in support of cognitive radio technology deployment. In *Proc. of IEEE Workshop on Networking Technologies for Software Defined Radio Networks*, pages 26–35, Reston, VA, 25–25 Sept. 2006.
49. A. Rubinstein. Perfect equilibrium in a bargaining model. *Econometrica: Journal of the Econometric Society*, 50:97–109, 1982.
50. T. Sandholm, K. Larson, M. Andersson, O. Shehory, and F. Tohmé. Coalition structure generation with worst case guarantees. *Artificial Intelligence*, 111(1–2):209–238, 1999.
51. S. Sengupta, R. Chandramouli, S. Brahma, and M. Chatterjee. A game theoretic framework for distributed self-coexistence among IEEE 802.22 networks. In *Proc. of IEEE GLOBECOM 2008*, pages 1–6, New Orleans, LA, 30 Nov.–4 Dec. 2008.
52. S. Sengupta, M. Chatterjee, and S. Ganguly. An economic framework for spectrum allocation and service pricing with competitive wireless service providers. In *Proc. of IEEE DySPAN 2007*, pages 89–98, Dublin, Ireland, 17–20 Apr. 2007.
53. L. S. Shapley. Stochastic games. *Proceedings of the National Academy of Sciences*, 39(10):1095–1100, 1953.
54. J. So and N.-H. Vaidya. A routing protocol for utilizing multiple channels in multi-hop wireless networks with a single transceiver. Technical report, Oct. 2004.
55. G. L. Stüber. *Principles of Mobile Communication*. 2nd ed. Kluwer, Norwell, MA, 2001.
56. S. Tao and M. Krunz. Coordinated channel access in cognitive radio networks: A multi-level spectrum opportunity perspective. In *Proc. of IEEE INFOCOM 2009*, pages 2976–2980, Rio de Janeiro, Brazil, 19–25 Apr. 2009.
57. M. Van Der Schaar and F. Fu. Spectrum access games and strategic learning in cognitive radio networks for delay-critical applications. *Proceedings of the IEEE*, 97(4):720–740, Apr. 2009.
58. R. Vedantham, S. Kakumanu, S. Lakshmanan, and R. Sivakumar. Component based channel assignment in single radio, multi-channel ad hoc networks. In *Proc. of ACM MobiCom 2006*, pages 378–389, New York, NY, 24–29 Sept. 2006.
59. B. Wang, Y. Wu, and K. J. R. Liu. Game theory for cognitive radio networks: An overview. *Computer Networks*, 54(14):2537–2561, 2010.
60. S. L. Wu, C. Y Lin, Y. C. Tseng, and J. P. Sheu. A new multi-channel MAC protocol with on-demand channel assignment for multi-hop mobile ad-hoc networks. In *Proc. of IEEE Symposium on Parallel Architectures, Algorithms and Networks*, pages 232–237, Dallas/Richardson, TX, 7–9 Dec. 2000.
61. S. L. Wu, Y. C. Tseng, C. Y. Lin, and J. P. Sheu. A multi-channel MAC protocol with power control for multi-hop mobile ad hoc networks. *The Computer Journal*, 45(1):101–110, 2002.
62. Y. Xu, J. C. S. Lui, and D. M. Chiu. On oligopoly spectrum allocation game in cognitive radio networks with capacity constraints. *Computer Networks*, 54(6):925 – 943, 2009.
63. J. Zhao, H. Zheng, and G. H. Yang. Distributed coordination in dynamic spectrum allocation networks. In *Proc. of IEEE DySPAN 2005*, pages 259–268, Baltimore, MD, 8–11 Nov. 2005.

Index

F.R. Yu (ed.), *Cognitive Radio Mobile Ad Hoc Networks*,
DOI 10.1007/978-1-4419-6172-3, © Springer Science+Business Media, LLC 2011

D